YEN LIN CHIA

EE508

Stanford University

M000282217

06/02

Springer Series in Statistics

Advisors:
P. Bickel, P. Diggle, S. Fienberg, K. Krickeberg,
I. Olkin, N. Wermuth, S. Zeger

Springer
New York
Berlin
Heidelberg
Barcelona
Hong Kong
London
Milan
Paris
Singapore
Tokyo

Springer Series in Statistics

Andersen/Borgan/Gill/Keiding: Statistical Models Based on Counting Processes.
Atkinson/Riani: Robust Diagnotstic Regression Analysis.
Berger: Statistical Decision Theory and Bayesian Analysis, 2nd edition.
Bolfarine/Zacks: Prediction Theory for Finite Populations.
Borg/Groenen: Modern Multidimensional Scaling: Theory and Applications
Brockwell/Davis: Time Series: Theory and Methods, 2nd edition.
Chen/Shao/Ibrahim: Monte Carlo Methods in Bayesian Computation.
David/Edwards: Annotated Readings in the History of Statistics.
Devroye/Lugosi: Combinatorial Methods in Density Estimation.
Efromovich: Nonparametric Curve Estimation: Methods, Theory, and Applications.
Fahrmeir/Tutz: Multivariate Statistical Modelling Based on Generalized Linear Models, 2nd edition.
Farebrother: Fitting Linear Relationships: A History of the Calculus of Observations 1750-1900.
Federer: Statistical Design and Analysis for Intercropping Experiments, Volume I: Two Crops.
Federer: Statistical Design and Analysis for Intercropping Experiments, Volume II: Three or More Crops.
Fienberg/Hoaglin/Kruskal/Tanur (Eds.): A Statistical Model: Frederick Mosteller's Contributions to Statistics, Science and Public Policy.
Fisher/Sen: The Collected Works of Wassily Hoeffding.
Friedman et al: The Elements of Statistical Learning: Data Mining, Inference and Prediction
Glaz/Naus/Wallenstein: Scan Statistics.
Good: Permutation Tests: A Practical Guide to Resampling Methods for Testing Hypotheses, 2nd edition.
Gouriéroux: ARCH Models and Financial Applications.
Grandell: Aspects of Risk Theory.
Haberman: Advanced Statistics, Volume I: Description of Populations.
Hall: The Bootstrap and Edgeworth Expansion.
Härdle: Smoothing Techniques: With Implementation in S.
Harrell: Regression Modeling Strategies: With Applications to Linear Models, Logistic Regression, and Survival Analysis
Hart: Nonparametric Smoothing and Lack-of-Fit Tests.
Hartigan: Bayes Theory.
Hedayat/Sloane/Stufken: Orthogonal Arrays: Theory and Applications.
Heyde: Quasi-Likelihood and its Application: A General Approach to Optimal Parameter Estimation.
Huet/Bouvier/Gruet/Jolivet: Statistical Tools for Nonlinear Regression: A Practical Guide with S-PLUS Examples.
Ibrahim/Chen/Sinha.: Bayesian Survival Analysis.
Kolen/Brennan: Test Equating: Methods and Practices.
Kotz/Johnson (Eds.): Breakthroughs in Statistics Volume I.
Kotz/Johnson (Eds.): Breakthroughs in Statistics Volume II.
Kotz/Johnson (Eds.): Breakthroughs in Statistics Volume III.

(continued after index)

Joseph G. Ibrahim
Ming-Hui Chen
Debajyoti Sinha

Bayesian Survival Analysis

With 51 Illustrations

Springer

Joseph G. Ibrahim
Department of Biostatistics
Harvard School of Public
Health and Dana–Farber
Cancer Institute
44 Binney Street
Boston, MA 02115
USA
ibrahim@jimmy.harvard.edu

Ming-Hui Chen
Department of Mathematical Sciences
Worcester Polytechnic Institute
100 Institute Road
Worcester, MA 01609-2280
USA
mhchen@wpi.edu

Debajyoti Sinha
Department of Biometry
and Epidemiology
Medical Universtiy of South Carolina
135 Rutledge Ave
PO Box 250551
Charleston, SC 29425
USA
sinhad@musc.edu

Library of Congress Cataloging-in-Publication Data
Ibrahim, Joseph George.
Bayesian survival analysis / Joseph G. Ibrahim, Ming-Hui Chen, Debajyoti Sinha.
p. cm. — (Springer series in statistics)
Includes bibliographical references and indexes.
ISBN 0-387-95277-2 (alk. paper)
1. Failure time data analysis. 2. Bayesian statistical decision theory. I. Chen Ming-Hui, 1961– II. Sinha, Debajyoti. III. Title. IV. Series.
QA276 .I27 2001
519.5′42—dc21 2001020443

Printed on acid-free paper.

© 2001 Springer-Verlag New York, Inc.
All rights reserved. This work may not be translated or copied in whole or in part without the written permission of the publisher (Springer-Verlag New York, Inc., 175 Fifth Avenue, New York, NY 10010, USA), except for brief excerpts in connection with reviews or scholarly analysis. Use in connection with any form of information storage and retrieval, electronic adaptation, computer software, or by similar or dissimilar methodology now known or hereafter developed is forbidden.
The use of general descriptive names, trade names, trademarks, etc., in this publication, even if the former are not especially identified, is not to be taken as a sign that such names, as understood by the Trade Marks and Merchandise Marks Act, may accordingly be used freely by anyone.

Production managed by MaryAnn Brickner; manufacturing supervised by Erica Bresler.
Photocomposed pages prepared from the authors' LaTeX2ε files.
Printed and bound by Edwards Brothers, Inc., Ann Arbor, MI.
Printed in the United States of America.

9 8 7 6 5 4 3 2 1

ISBN 0-387-95277-2 SPIN 10833390

Springer-Verlag New York Berlin Heidelberg
A member of BertelsmannSpringer Science+Business Media GmbH

To Joseph G. Ibrahim's parents, Guirguis and Assinat Ibrahim

Ming-Hui Chen's parents, his wife, Lan, and his daughters,
Victoria and Paula

Debajyoti Sinha's parents, Nemai Chand and Purabi Sinha,
and his wife, Sebanti

Preface

Survival analysis arises in many fields of study including medicine, biology, engineering, public health, epidemiology, and economics. Recent advances in computing, software development such as BUGS, and practical methods for prior elicitation have made Bayesian survival analysis of complex models feasible for both practitioners and researchers. This book provides a comprehensive treatment of Bayesian survival analysis. Several topics are addressed, including parametric and semiparametric models, proportional and nonproportional hazards models, frailty models, cure rate models, model selection and comparison, joint models for longitudinal and survival data, models with time-varying covariates, missing covariate data, design and monitoring of clinical trials, accelerated failure time models, models for multivariate survival data, and special types of hierarchical survival models. We also consider various censoring schemes, including right and interval censored data. Several additional topics related to the Bayesian paradigm are discussed, including noninformative and informative prior specifications, computing posterior quantities of interest, Bayesian hypothesis testing, variable selection, model checking techniques using Bayesian diagnostic methods, and Markov chain Monte Carlo (MCMC) algorithms for sampling from the posterior and predictive distributions.

The book will present a balance between theory and applications, and for each of the models and topics mentioned above, we present detailed examples and analyses from case studies whenever possible. Moreover, we demonstrate the use of the statistical package BUGS for several of the models and methodologies discussed in this book. Theoretical and applied problems are given in the exercises at the end of each chapter. The book is structured so that the methodology and applications are presented in the main body of each chapter and all rigorous proofs and derivations are placed in Appendices. This should enable a wide audience of readers to use the book without having to go through the technical details. Without compromising our main goal of presenting Bayesian methods for survival analysis, we have tried to acknowledge and briefly review the relevant frequentist methods. We compare the frequentist and Bayesian techniques whenever possible and discuss the advantages and disadvantages of Bayesian methods for each topic.

Several types of parametric and semiparametric models are examined. For the parametric models, we discuss the exponential, gamma, Weibull, log-normal, and extreme value regression models. For the semiparametric models, we discuss a wide variety models based on prior processes for the cumulative baseline hazard, the baseline hazard, or the cumulative baseline distribution function. Specifically, we discuss the gamma process, beta process, Dirichlet process, and correlated gamma process. We also discuss frailty survival models that allow the survival times to be correlated between subjects, as well as multiple event time models where each subject has a vector of time-to-event variables. In addition, we examine parametric and semiparametric models for univariate survival data with a cure fraction (cure rate models) as well as multivariate cure rate models. Also, we discuss accelerated failure time models and flexible classes of hierarchical survival models based on neural networks. The applications are all essentially from the health sciences, including cancer, AIDS, and the environment.

The book is intended as a graduate textbook or a reference book for a one- or two-semester course at the advanced masters or Ph.D. level. The prerequisites include one course in statistical inference and Bayesian theory at the level of Casella and Berger (1990) and Box and Tiao (1992). The book can also be used after a course in Bayesian statistics using the books by Carlin and Louis (1996) or Gelman, Carlin, Stern, and Rubin (1995). This book focuses on an important subfield of application. It would be most suitable for second- or third-year graduate students in statistics or biostatistics. It would also serve as a useful reference book for applied or theoretical researchers as well as practitioners. Moreover, the book presents several open research problems that could serve as useful thesis topics.

We would like to acknowledge the following people, who gave us permission to use some of the contents from their work, including tables and figures: Elja Arjas, Brad Carlin, Paul Damien, Dipak Dey, Dario Gasbarra, Robert J. Gray, Paul Gustafson, Lynn Kuo, Sandra Lee, Bani Mallick, Nalini Ravishanker, Sujit Sahu, Daniel Sargent, Dongchu Sun, Jeremy Taylor, Bruce Turnbull, Helen Vlachos, Chris Volinsky, Steve Walker, and Marvin Zelen. Joseph Ibrahim would like to give deep and special thanks to Marvin Zelen for being his wonderful mentor and friend at Harvard, and to whom he feels greatly indebted. Ming-Hui Chen would like to give special thanks to his advisors James Berger and Bruce Schmeiser, who have served as his wonderful mentors for the last ten years. Finally, we owe deep thanks to our parents and our families for their constant love, patience, understanding, and support. It is to them that we dedicate this book.

Joseph G. Ibrahim, Ming-Hui Chen, and Debajyoti Sinha
March 2001

Contents

1
Introduction

1.1 Aims

The analysis of time-to-event data, generally called survival analysis, arises in many fields of study, including medicine, biology, engineering, public health, epidemiology, and economics. Although the methods we present in this book can be used in all of these disciplines, our applications will focus exclusively on medicine and public health. There have been several textbooks written that address survival analysis from a frequentist perspective. These include Lawless, Cox and Oakes (1984), Fleming and Harrington (1991), Lee (1992), Andersen, Borgan, Gill, and Keiding (1993), and Klein and Moeschberger (1997). Although these books are quite thorough and examine several topics, they do not address Bayesian analysis of survival data in depth. Klein and Moeschberger (1997), however, do present one section on Bayesian nonparametric methods. Bayesian analysis of survival data has received much recent attention due to advances in computational and modeling techniques. Bayesian methods are now becoming quite common for survival data and have made their way into the medical and public health arena.

The focus of this book is to examine Bayesian approaches to survival analysis. We discuss several types of models, including parametric and semiparametric models, proportional and nonproportional hazards models, frailty models, cure rate models, joint models for longitudinal and survival data, models with time-varying covariates, missing covariate data, accelerated failure time models, models for multivariate survival data, and flexible classes of hierarchical models. We will also consider various censoring schemes, including right and interval censored data. Several additional topics will be discussed, including noninformative and informative prior specifications, computing posterior quantities of interest, Bayesian hypothesis testing, model selection, model checking techniques, Markov chain Monte Carlo (MCMC) algorithms for sampling from the posterior and predictive distributions, and Bayesian design and monitoring of clinical trials. The book will present a balance between theory and applications, and for each class of models mentioned above, we present detailed examples and analyses from case studies whenever possible. Moreover, we demonstrate the use of the statistical package BUGS for several of the models discussed in this book.

1.2 Outline

Chapter 2 examines the analysis of parametric Bayesian models for univariate censored survival data, including the exponential, Weibull, extreme value, log-normal, and gamma regression models. We illustrate Bayesian analyses and computations for each model. Also, we explore various types of prior distributions for the model parameters and discuss MCMC techniques for carrying out the computations. Chapter 3 examines semiparametric models for survival data, and serves as one of the central chapters of this book. We first discuss piecewise exponential models based on grouped survival data, and then examine several types of prior processes for the cumulative baseline hazard and the baseline hazard in the Cox model, including gamma processes, beta processes, correlated gamma processes, and the Dirichlet process. For each type of prior process, we discuss the construction of the likelihood function, properties of the model, prior elicitation strategies, computational methods, and applications. In Chapter 4, we examine Bayesian methods for frailty models and other multivariate survival models. Here, we study various survival models based on different frailty distributions, including the gamma frailty and the positive stable frailty. In addition, multiple event time and panel count data models are also examined in this chapter and several applications are presented. In Chapter 5, we examine the cure rate model. Here we derive the cure rate model, review parametric and semiparametric cure rate models, examine the properties of the models, and discuss the construction of the likelihood function, prior elicitation strategies, and computational methods. We also present multivariate extensions of the cure rate model. Chapter 6 presents methods for model comparison, including variable subset selection, Bayes factors for hypothesis testing involving nested and non-nested models, and criterion based methods for model assessment and adequacy, including the L measure and the Conditional Predictive Ordinate (CPO).

Chapter 7 is devoted to joint modeling of longitudinal and survival data. In this chapter, we present several methods for joint modeling of longitudinal and survival data, and present applications to AIDS and cancer clinical trials. Specifically, we examine several types of models, the construction of the likelihoods, prior distributions, and computational implementation for these models. Applications to AIDS and cancer vaccine clinical trials are discussed. In Chapter 8, we discuss Bayesian methods for missing data. Specifically, we examine missing covariate and/or response data for various types of survival models that arise in applications. We discuss methods for modeling the covariate distributions, methods for Bayesian inference, model checking for the covariate distributions, Gibbs sampling, and computational implementation. In Chapter 9, we examine Bayesian methods for the design and monitoring of clinical trials. We discuss methods for sample size determination in phase III clinical trials, conditional and predictive power, interim analyses, prior distributions, Bayesian monitoring, and prediction. Finally, Chapter 10 is devoted to special topics. These include proportional hazards models built from monotone functions, models with time varying covariates, accelerated failure time models, change-point models for survival data, special types of hierarchical models, Bayesian model diagnostic techniques (including Bayesian latent residuals and prequential methods), and future research topics.

1.3 Motivating Examples

The following examples of real data will help motivate the various survival models which are examined later in this book.

Example 1.1. Multiple myeloma data. Multiple myeloma is a hematologic cancer characterized by an overproduction of antibodies. A study involving a chemotherapy involving alkylating agents was undertaken by the Eastern Cooperative Oncology Group (ECOG). This study, denoted E2479, had $n = 479$ patients. The results of E2479 are available in Kalish (1992). Several covariates were measured for this study at diagnosis. These were blood urea nitrogen, hemoglobin, platelet count, age, white blood cell count, bone fractures, percentage of the plasma cells in bone marrow, proteinuria, and serum calcium. These are typical covariates measured in multiple myeloma studies. An important issue for this disease is identifying which covariates are important predictors of survival, that is, *variable subset selection*. The inherent difficulties in Bayesian variable selection are:

(i) the specification of a prior distribution for all of the parameters arising from all possible models in the model space;

(ii) a prior on the model space; and

(iii) computations.

In Chapter 6, we consider these issues in detail. For these data, we consider a proportional hazards model, i.e., a Cox model (Cox, 1972), and specify a gamma process prior for the baseline hazard. Also, we specify a class of informative prior distributions for the regression coefficients (see Section 1.7). Using this semiparametric model, we present methods for Bayesian variable subset selection. The prior distributions for the parameters arising from all of the possible models are specified in a semi-automatic fashion using historical data from a similar previous study, and the prior distribution for the model space is also constructed from historical data. Using the methods of Ibrahim and Chen (1998), and Ibrahim, Chen, and MacEachern (1999), we demonstrate how to compute posterior model probabilities efficiently by using samples only from the full model. This method results in an efficient Gibbs sampling scheme for computing the posterior model probabilities for all possible subset models.

Example 1.2. Melanoma data. Melanoma incidence is increasing at a rate that exceeds all solid tumors. Although education efforts have resulted in earlier detection of melanoma, patients who have deep primary melanoma ($> 4\ mm$) or melanoma metastatic to regional draining lymph nodes, classified as *high-risk melanoma* patients, continue to have high relapse and mortality rates of 60% to 75% (see Kirkwood et al., 2000). Recently, several post-operative (adjuvant) chemotherapies have been proposed for this class of melanoma patients, and the one which seems to provide the most significant impact on relapse-free survival and survival is interferon alpha-2b (IFN). This chemotherapy was used in two recent Eastern Cooperative Oncology Group (ECOG) phase III clinical trials, E1684 and E1690. The first trial, E1684, was a two-arm clinical trial comparing high-dose interferon (IFN) to observation (OBS). There were a total of $n_0 = 286$ patients enrolled in the study, accrued from 1984 to 1990, and the study was unblinded in 1993. The results of this study suggested that IFN has a significant impact on relapse-free survival (RFS) and survival (OS), which led to U.S. Food and Drug Administration (FDA) approval of this regimen as an adjuvant therapy for high-risk melanoma patients. The results of the E1684 trial have been published in Kirkwood et al. (1996).

Figure 1.1 shows the Kaplan-Meier survival curve of the E1684 data. We see that the right tail of the survival curve appears to "plateau" after sufficient follow-up. Such a phenomenon has become quite common in melanoma as well as other cancers, and the models used to analyze such data are called *cure rate models*. The cure rate model has been used for modeling time-to-event data for various types of cancers, including breast cancer, non-Hodgkins lymphoma, leukemia, prostate cancer, melanoma, and head and neck cancer, where for these diseases, a significant propor-

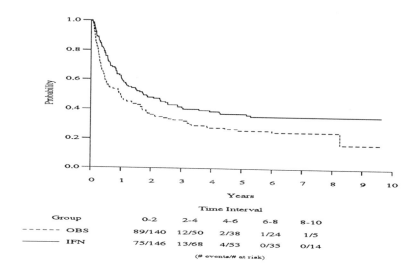

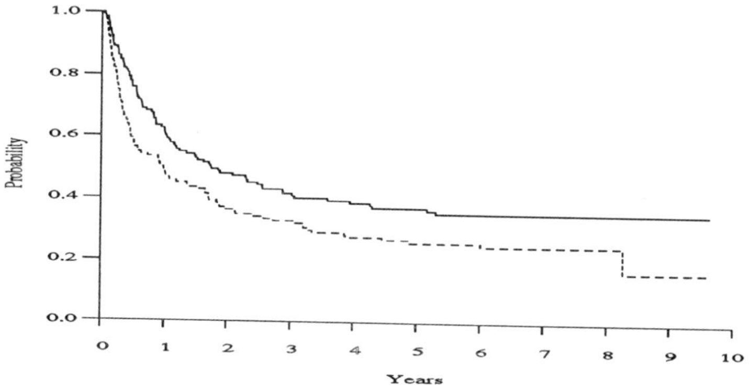

Group	Time Interval				
	0-2	2-4	4-6	6-8	8-10
------ OBS	89/140	12/50	2/38	1/24	1/5
——— IFN	75/146	13/68	4/53	0/35	0/14

(# events/# at risk)

FIGURE 1.1. Kaplan-Meier RFS plots for E1684.

tion of patients are "cured" after sufficient follow-up. In Chapter 5, we give a complete Bayesian treatment of the cure rate model and examine several topics, including informative prior elicitation, semiparametric and parametric models, inclusion of covariates, and Gibbs sampling techniques.

The treatment effect favoring IFN seen in E1684 was larger than expected and was accompanied by substantial side effects due to the high-dose regimen. As a result, ECOG began a second trial (E1690) in 1991 to attempt to confirm the results of E1684 and to study the benefit of IFN given at a lower dose. The ECOG trial E1690 was a three-arm phase III clinical trial, and had treatment arms consisting of high-dose interferon, low-dose interferon, and observation. This study had $n = 427$ patients on the high-dose interferon arm and observation arm combined. E1690 was initiated right after the completion of E1684. The E1690 trial accrued patients from 1991 until 1995, and was unblinded in 1998. The E1690 trial was designed for exactly the same patient population as E1684, and the high-dose interferon arm in E1690 was identical to that of E1684. This example nicely demonstrates the use of historical data. Here, the E1684 data can be used as historical data for analyzing the E1690 data. We discuss informative prior elicitation in cure rate models using historical data in Chapter 5. An additional important issue with these models is the estimation of the cure rate and the assessment of the goodness of fit of the cure rate model compared to the usual Cox model. We also address these issues in Chapter 6, where we examine model assessment techniques such as the Conditional Predictive Ordinate (CPO) and the L measure for assessing the fit of a model.

TABLE 1.1. Summary of E1684 Data.

Completely Observed Variables			Missing Covariates		
x_1 (frequency)	OBS	140	x_4 (frequency)	mean	0.92
	IFN	145		std dev	0.88
x_2 (years)	mean	47.1		missing	30
	std dev	13.1	x_5 (frequency)	mean	2.73
x_3 (frequency)	male	171		std dev	0.82
	female	114		missing	45
y (frequency)	censored	89	x_6 (frequency)	nodular	129
	relapsed	196		other	122
				missing	34

Source: Chen, Ibrahim, and Lipsitz (1999).

TABLE 1.2. Summary of E1690 Data.

Completely Observed Variables			Missing Covariates		
x_1 (frequency)	OBS	212	x_4 (frequency)	mean	1.04
	IFN	215		std dev	0.83
x_2 (years)	mean	47.9		missing	10
	std dev	13.2	x_5 (frequency)	mean	2.59
x_3 (frequency)	male	268		std dev	0.83
	female	159		missing	80
y (frequency)	censored	186	x_6 (frequency)	nodular	209
	relapsed	241		other	171
				missing	47

Source: Chen, Ibrahim, and Lipsitz (1999).

Moreover, in Chapter 8, we discuss missing data methods in survival analysis. Missing covariate data is quite common in clinical trials, and appropriate methods need to be developed for handling these situations. Tables 1.1–1.3 summarize the response variable (y) and several covariates for the E1690 and E1684 studies, respectively. The response variable is relapse-free survival (RFS), which may be right censored. The covariates are treatment (x_1: IFN, OBS), age (x_2), sex (x_3), logarithm of Breslow depth (x_4), logarithm of size of primary (x_5), and type of primary (x_6). Covariates x_1, x_3, and x_6 are all binary covariates, whereas x_2, x_4, and x_5 are all continuous. Also, the covariates x_1, x_2, and x_3 are completely observed, and x_4, x_5, and x_6 have missing values. The total missing data

TABLE 1.3. Median RFS for E1684 and E1690.

Study	OBS	IFN	Overall
E1684	0.98	1.73	1.35
E1690	1.62	2.47	1.94

fractions for these three covariates in E1684 and E1690 are 27.4% and 28.6%, respectively.

TABLE 1.4. Intervals of Cosmetic Deterioration (Retraction) for Early Breast Cancer Patents.

Radiotherapy			Radiotherapy and Chemotherapy		
$(45,-]$	$(25,37]$	$(37,-]$	$(8,12]$	$(0,5]$	$(30,34]$
$(6,10]$	$(46,-]$	$(0,5]$	$(0,22]$	$(5,8]$	$(13,-]$
$(0,7]$	$(26,40]$	$(18,-]$	$(24,31]$	$(12,20]$	$(10,17]$
$(46,-]$	$(46,-]$	$(24,-]$	$(17,27]$	$(11,-]$	$(18,21]$
$(46,-]$	$(27,34]$	$(36,-]$	$(17,23]$	$(33,40]$	$(4,9]$
$(7,16]$	$(36,44]$	$(5,11]$	$(24,30]$	$(31,-]$	$(11,-]$
$(17,-]$	$(46,-]$	$(19,35]$	$(16,24]$	$(13,39]$	$(14,19]$
$(7,14]$	$(36,48]$	$(17,25]$	$(13,-]$	$(19,32]$	$(4,8]$
$(37,44]$	$(37,-]$	$(24,-]$	$(11,13]$	$(34,-]$	$(34,-]$
$(0,8]$	$(40,-]$	$(32,-]$	$(16,20]$	$(13,-]$	$(30,36]$
$(4,11]$	$(17,25]$	$(33,-]$	$(18,25]$	$(16,24]$	$(18,24]$
$(15,-]$	$(46,-]$	$(19,26]$	$(17,26]$	$(35,-]$	$(16,60]$
$(11,15]$	$(11,18]$	$(37,-]$	$(32,-]$	$(15,22]$	$(35,39]$
$(22,-]$	$(38,-]$	$(34,-]$	$(23,-]$	$(11,17]$	$(21,-]$
$(46,-]$	$(5,12]$	$(36,-]$	$(44,48]$	$(23,32]$	$(11,20]$
$(46,-]$			$(14,17]$	$(10,35]$	$(48,-]$

Source: Finkelstein and Wolf (1985).

Example 1.3. Breast cancer data. We consider breast cancer data from Finkelstein and Wolfe (1985), which consists of a data set of (case-2) interval censored data. In this data set, 46 early breast cancer patients receiving only radiotherapy (covariate value $x = 0$) and 48 patients receiving radio-chemotherapy ($x = 1$) were monitored for cosmetic change through weekly clinic visits. Because patients missed some of their weekly visits, the data on survival time are typically recorded as, for example, [7,18] (at the 7th week clinic-visit patient had shown no change and then in the next clinic visit at the 18th week the patient's tissue showed that the change had already occurred). Since the clinic visits of different patients occurred at different times, the censoring intervals in the data set are found to be often overlapping and nondisjoint. The data from this study are given in Table 1.4. We are interested in the effect of the covariate x on the survival time y.

In Chapter 3, we discuss Bayesian semiparametric analyses of interval censored survival data, and for these data, we develop nonproportional hazards related to a discretized version of the Cox model (Cox, 1972). In Chapter 6, we examine several models for these data and discuss prior elicitation, model selection and assessment, and computational implementation.

TABLE 1.5. Recurrence Times of Infections in 38 Kidney Patients Using Portable Dialysis Machine.

Recurrence Times	Event Types	Sex	Recurrence Times	Event Types	Sex
8, 16	1, 1	1	22, 159	0, 0	2
15, 108	1, 0	2	22, 28	1, 1	1
447, 318	1, 1	2	402, 24	1, 0	2
13, 66	1, 1	2	24, 245	1, 1	1
7, 9	1, 1	1	12, 40	1, 1	1
113, 201	0, 1	2	53, 196	1, 1	2
15, 154	1, 1	1	34, 30	1, 1	2
2, 25	1, 1	1	141, 8	1, 0	2
96, 38	1, 1	2	27, 58	1, 1	2
5, 43	0, 1	2	536, 25	1, 0	2
17, 4	1, 0	1	190, 5	1, 0	2
119, 8	1, 1	2	292, 114	1, 1	2
6, 78	0, 1	2	63, 8	1, 0	1
33, 13	1, 0	2	152, 562	1, 1	1
30, 12	1, 1	1	39, 46	1, 0	2
511, 30	1, 1	2	132, 156	1, 1	2
7, 333	1, 1	2	130, 26	1, 1	2
149, 70	0, 0	2	152, 30	1, 1	2
185, 117	1, 1	2	54, 16	0, 0	2

Source: McGilchrist and Aisbett (1991).

Example 1.4. Kidney infection data. Table 1.5, reproduced from McGilchrist and Aisbett (1991), shows the times of infection from the time of insertion of the catheter for 38 kidney patients using portable dialysis equipment. The first column gives the time of first and second infection. The second column gives the event indicators for the first and second infections, respectively. Occurrence of infection is indicated by 1, and 0 means that the infection time is right censored. The last column describes the sex (1 indicating male and 2 indicating female). Other covariates, such as age, and disease types, given in the original work of McGilchrist and Aisbett (1991), have been omitted here. Previous analyses by McGilchrist and Aisbett (1991) and McGilchrist (1993) found sex as the only covariate with a possible significant effect on the infection times. In this example, the infection times from each patient share the same values of the covariate. After the occurrence or censoring of the first infection, sufficient time was allowed for the infection to be cured before the second insertion was allowed. This dataset is one of the most widely analyzed datasets for demonstrating frailty models. We make extensive use of this dataset in Chapter 4 and other chapters for illustrating the various frailty models and their properties.

TABLE 1.6. Survival Times (weeks) of One Exposed (E) and Two Control Rats ($C1, C2$) (* right censored times).

Litter	E	$C1$	$C2$	Litter	E	$C1$	$C2$
1	101.0*	49.0	104.0*	26	89.0*	104.0*	104.0*
2	104.0*	102.0*	104.0*	27	78.0*	104.0*	104.0*
3	104.0*	104.0*	104.0*	28	104.0*	81.0	64.0
4	77.0*	97.0*	79.0*	29	86.0	55.0	94.0*
5	89.0*	104.0*	104.0*	30	34.0	104.0*	54.0
6	88.0	96.0	104.0*	31	76.0*	87.0*	74.0*
7	104.0	94.0*	77.0	32	102.8	73.0	83.9
8	95.9	104.0*	104.0*	33	101.9	104.0*	80.0*
9	82.0*	77.0*	104.0*	34	79.9	104.0*	73.0*
10	70.0	104.0*	77.0*	35	45.0	79.0*	104.0*
11	88.9	91.0*	90.0*	36	94.0	104.0*	104.0*
12	91.0*	70.0*	92.0*	37	104.0*	104.0*	104.0*
13	39.0	45.0*	50.0	38	104.0*	101.0	94.0*
14	102.9	69.0*	91.0*	39	76.0*	84.0	78.0
15	93.0*	104.0*	103.0*	40	80.0	80.9	76.0*
16	85.0*	72.0*	104.0*	41	72.0	95.0*	104.0*
17	104.0*	63.0*	104.0*	42	72.9	104.0*	66.0
18	104.0*	104.0*	74.0*	43	92.0	104.0*	102.0
19	81.0*	104.0*	69.0*	44	104.0*	98.0*	73.0*
20	67.0	104.0*	68.0	45	55.0*	104.0*	104.0*
21	104.0*	104.0*	104.0*	46	49.0*	83.0*	77.0*
22	104.0*	104.0*	104.0*	47	89.0	104.0*	104.0*
23	104.0*	83.0*	40.0	48	88.0*	79.0*	99.0*
24	87.0*	104.0*	104.0*	49	103.0	91.0*	104.0*
25	104.0*	104.0*	104.0*	50	104.0*	104.0*	79.0

Source: Clayton (1991).

Example 1.5. Animal carcinogenesis data. Survival data is often correlated between subjects or groups of subjects. To illustrate this, we consider a data set discussed in Clayton (1991) involving an experiment with litter-matched time-to-event data. The data are given in Table 1.6. In this study, three rats from each of 50 litters were matched, where one rat was treated with a putative carcinogen and the other serving as controls. The time to tumor occurrence or censoring in weeks is recorded. In these data, the three rats in each litter that are matched have correlated event times. This data structure leads to a *frailty* model. In Chapter 4, we discuss Bayesian methods for the frailty model, and examine various types of frailty models, such as the gamma and positive stable law frailty models. One of the interesting features in Table 1.6 is that the association between the control rats within the same litter is perhaps different than the association between the control and exposed rats within the same litter. We will study modeling strategies for these types of situations in Chapter 4. Other topics in Chapter 4 include multiple event and panel count data models, model synthesis, identifiability issues, prior elicitation, and Gibbs sampling from the posterior distribution.

TABLE 1.7. Mammary Tumor Data: Part I.

Rat	Group	Time to tumor (days)
1	1	182
2	1	182*
3	1	63, 68, 182*
4	1	152, 182*
5	1	130, 134, 145, 152, 182*
6	1	98, 152, 182
7	1	88, 95, 105, 130, 137, 167, 182*
8	1	152, 182*
9	1	81, 182*
10	1	71, 84, 126, 134, 152, 182*
11	1	116, 130, 182*
12	1	91, 182*
13	1	63, 68, 84, 95, 152, 182*
14	1	105, 152, 182*
15	1	63, 102, 152, 182*
16	1	63, 77, 112, 140, 182*
17	1	77, 119, 152, 161, 167, 182*
18	1	105, 112, 145, 161, 182
19	1	152, 182*
20	1	81, 95, 182*
21	1	84, 91, 102, 108, 130, 134, 182*
22	1	182*
23	1	91, 182*

Source: Gail, Santner, and Brown (1980).

Example 1.6. Mammary tumor data. In certain medical studies, a subject can experience multiple events, such as multiple times of occurrence of a tumor. An example of such a situation is given by Gail, Santner, and Brown (1980). For these data, the observations are the times of occurrence of mammary tumors in 48 rats injected with a carcinogen and subsequently randomized to receive either a treatment or a control; the events are occurrences of tumors, observed to develop during a certain observation period. The data are given in Tables 1.7 and 1.8. In Tables 1.7 and 1.8, a * indicates that this was a censorship time at which no new tumor was found, group 1 denotes the retinoid group, and group 2 denotes the control group. These data serve as an example of *multiple event time* data. In Chapter 4, we examine semiparametric Bayesian methods for multiple event time data. We derive the likelihood for such data, examine properties of the models, and discuss posterior inference using the Gibbs sampler.

TABLE 1.8. Mammary Tumor Data: Part II.

Rat	Group	Time to tumor (days)
24	2	63, 102, 119, 161, 161, 172, 179, 182*
25	2	88, 91, 95, 105, 112, 119, 119, 137, 145, 167, 172, 182*
26	2	91, 98, 108, 112, 134, 137, 161, 161, 179, 182*
27	2	71, 174, 182*
28	2	95, 105, 134, 137, 140, 145, 150, 150, 182*
29	2	66, 68, 130, 137, 182*
30	2	77, 85, 112, 137, 161, 174, 182*
31	2	81, 84, 126, 134, 161, 161, 174, 182*
32	2	68, 77, 98, 102, 102, 102, 182*
33	2	112, 182*
34	2	88, 88, 91, 98, 112, 134, 137, 137, 140, 140, 152, 152, 182*
35	2	77, 179, 182*
36	2	112, 182*
37	2	71, 71, 74, 77, 112, 116, 116, 140, 140, 167, 182*
38	2	77, 95, 126, 150, 182*
39	2	88, 126, 130, 130, 134, 182*
40	2	63, 74, 84, 84, 88, 91, 95, 108, 134, 137, 179, 182*
41	2	81, 88, 105, 116, 123, 140, 145, 152, 161, 161, 179, 182*
42	2	88, 95, 112, 119, 126, 126, 150, 157, 179, 182*
43	2	68, 68, 84, 102, 105, 119, 123, 123, 137, 161, 179, 182*
44	2	140, 182*
45	2	152, 182, 182
46	2	81, 182*
47	2	63, 88, 134, 182*
48	2	84, 134, 182

Source: Gail, Santner, and Brown (1980).

Example 1.7. Bivariate survival data. In survival analysis, it is often of interest to model jointly several types of failure time random variables, such as time to cancer relapse at two different organs, time to cancer relapse and time from relapse to death, times to first and second infection, and so forth. To illustrate this, we consider the E1684 melanoma data discussed in Example 1.2. We consider the two failure time random variables T_1 = time to relapse from randomization and T_2 = time from relapse to death. Table 1.9 summarizes (T_1, T_2). It is common in these settings that T_1 and T_2 have joint and marginal survival curves that plateau beyond a certain period of follow-up, and therefore it is important in these situations to develop a joint cure rate model. In Chapter 5, we examine bivariate cure rate models and discuss their properties. We develop the likelihood function, discuss prior elicitation, examine methods for inducing a correlation structure between T_1 and T_2, and present Gibbs sampling procedures for carrying out posterior inference.

TABLE 1.9. Summary of Survival and Relapse Times.

Variable	Years		Frequency	
Time to Relapse (T_1)	median	0.537	censored	88
	IQR	1.247	relapse	186
Relapse to Death (T_2)	median	0.660	censored	110
	IQR	1.014	death	164

Source: Chen, Ibrahim, and Sinha (2001).

Example 1.8. Cancer vaccine data. Vaccine therapies have received a great deal of attention recently as potential therapies in cancer clinical trials. One reason for this is that they are much less toxic than chemotherapies and potentially less expensive. However, currently little is known about the biologic activity of vaccines and that of finding good markers for assessing the efficacy of these vaccines and predicting clinical outcome. The antibody immune measures IgG and IgM have been proposed as potential markers in melanoma clinical trials because of their observed correlation with clinical outcome in pilot studies. To better understand the role of the IgG and IgM antibodies for a particular vaccine, we must better understand the relationship of clinical outcome to an individuals antibody (IgG and IgM titres) history over time. To illustrate this, we consider a phase II melanoma clinical trial conducted by the Eastern Cooperative Oncology Group (ECOG), labeled here as E2696. For adjuvant studies in melanoma, it has been conjectured that vaccines may be more efficacious than chemotherapies. This phase II trial was conducted to test this hypothesis. The treatment arms consist of a combination of interferon (IFN) and the ganglioside vaccine (GMK), which we label as A (IFN + GMK). The other treatment arm consists of GMK alone, labeled as B. There were 35 patients on each treatment arm, resulting in a total sample size of $n = 70$ patients. The survival endpoint is relapse-free survival (RFS), measured in months. The IgG and IgM antibody titre measurements were taken at the five time points—0, 4, 6, 12, and 52 weeks. Table 1.10 shows a statistical summary of the IgG

TABLE 1.10. Summary of IgG and IgM Measures.

Measure	Treatment	Week	0	4	6	12	52
IgG	A	Median	0.00	3.38	4.39	3.71	7.15
		Mean	0.00	2.84	4.17	3.05	6.74
		SD	0.00	2.53	2.18	2.40	2.15
		Number of Missing	3	1	5	5	13
	B	Median	0.00	3.04	3.71	3.04	6.46
		Mean	0.00	2.23	3.12	2.23	5.87
		SD	0.00	2.35	2.47	2.39	1.86
		Number of Missing	7	2	0	6	16
IgM	A	Median	0.00	5.08	5.08	3.71	4.73
		Mean	0.14	5.07	5.26	3.31	3.53
		SD	0.78	1.84	1.54	2.29	2.59
		Number of Missing	3	1	5	5	13
	B	Median	0.00	5.08	5.08	4.39	3.71
		Mean	0.58	5.02	4.99	3.85	2.99
		SD	2.04	1.94	1.40	2.04	2.46
		Number of Missing	7	2	0	6	16

Source: Ibrahim, Chen, and Sinha (2000).

and IgM measures for each treatment arm for each time point, and it also shows the fraction of missing antibody titres at each time point. In Chapter 7, we examine joint models for longitudinal and survival data. For these models, we discuss suitable models for the longitudinal and time to event components, prior elicitation strategies, interpretation of parameters, and computational methods. The E2696 dataset is used to demonstrate some of the methodologies.

1.4 Survival Analysis

Let T be a continuous nonnegative random variable representing the survival times of individuals in some population. All functions, unless stated otherwise, are defined over the interval $[0, \infty)$. Let $f(t)$ denote the probability density function (pdf) of T and let the distribution function be

$$F(t) = P(t \le t) = \int_0^t f(u)du. \tag{1.4.1}$$

The probability of an individual's surviving till time t is given by the survivor function:

$$S(t) = 1 - F(t) = P(T > t).$$

We note that $S(t)$ is a monotone decreasing function with $S(0) = 1$ and $S(\infty) = \lim_{t\to\infty} S(t) = 0$. The hazard function, $h(t)$, is the instantaneous rate of failure at time t, and is defined by

$$h(t) = \lim_{\Delta t\to 0} \frac{P(t < T \leq t + \Delta t | T > t)}{\Delta t} = \frac{f(t)}{S(t)}. \qquad (1.4.2)$$

In particular, $h(t)\Delta t$ is the approximate probability of failure in $(t, t + \Delta t]$, given survival up to time t. The functions $f(t)$, $F(t)$, $S(t)$, and $h(t)$ give mathematically equivalent specifications of the distribution of T. It is easy to derive expressions for $S(t)$ and $f(t)$ in terms of $h(t)$. Since $f(t) = -\frac{d}{dt}S(t)$, (1.4.2) implies that

$$h(t) = -\frac{d}{dt}\log(S(t)). \qquad (1.4.3)$$

Now integrating both sides of (1.4.3), and then exponentiating, we are led to

$$S(t) = \exp\left(-\int_0^t h(u)du\right). \qquad (1.4.4)$$

The cumulative hazard, $H(t)$, is defined as $H(t) = \int_0^t h(u)du$, which is related to the survivor function by $S(t) = \exp(-H(t))$. Since $S(\infty) = 0$, it follows that $H(\infty) = \lim_{t\to\infty} H(t) = \infty$. Thus, the hazard function, $h(t)$, has the properties

$$h(t) \geq 0 \quad \text{and} \quad \int_0^\infty h(t)dt = \infty.$$

Finally, in addition to (1.4.4), if follows from (1.4.3) that

$$f(t) = h(t)\exp\left(-\int_0^t h(u)du\right). \qquad (1.4.5)$$

Example 1.9. Weibull distribution. Suppose T has pdf

$$f(t) = \begin{cases} \alpha\gamma t^{\alpha-1}\exp(-\gamma t^\alpha) & \text{for } t > 0,\ \alpha > 0,\ \gamma > 0, \\ 0 & \text{otherwise.} \end{cases} \qquad (1.4.6)$$

This is a Weibull distribution with parameters (α, γ), denoted $\mathcal{W}(\alpha, \gamma)$. For the $\mathcal{W}(\alpha, \gamma)$ distribution, $h(t)$ is monotonically increasing when $\alpha > 1$ and monotonically decreasing when $0 < \alpha < 1$. It follows easily from (1.4.4) that the survivor function of T is $S(t) = \exp(-\gamma t^\alpha)$. From (1.4.3), the hazard function is $h(t) = \gamma\alpha t^{\alpha-1}$, and the cumulative hazard function is $H(t) = \gamma t^\alpha$.

1.4.1 Proportional Hazards Models

The hazard function depends in general on both time and a set of covariates, some of which may be time dependent. The proportional hazards model (Cox, 1972) separates these components by specifying that the hazard at time t for an individual whose covariate vector is $\boldsymbol{x}$ is given by

$$h(t|\boldsymbol{x}) = h_0(t) \exp \{G(\boldsymbol{x}, \boldsymbol{\beta})\},$$

where $h_0(t)$ is called the baseline line hazard function and $\boldsymbol{\beta}$ is a vector of regression coefficients. The second term is written in exponential form because it must be positive. This model implies that the ratio of the hazards for two individuals is constant over time provided that the covariates do not change over time. It is conventional to assume that the effect on the covariates is multiplicative, leading to the hazard function

$$h(t|\boldsymbol{x}) = h_0(t) \exp(\boldsymbol{x}'\boldsymbol{\beta}), \tag{1.4.7}$$

where $\eta = \boldsymbol{x}'\boldsymbol{\beta}$ is called the linear predictor. The model in (1.4.7) implies that the ratio of hazards for two individuals depends on the difference between their linear predictors at any time.

1.4.2 Censoring

Survival data are often right censored—that is, survival times are known for only a portion of the individuals under study, and the remainder of the survival times are known only to exceed certain values. Specifically, an observation is said to to be right censored at c if the exact value of the observation is not known but only that it is greater than or equal to c. Similarly, an observation is said to be left censored at c if it is known only that the observation is less than or equal to c. Also, an observation is said to be interval censored if it is known only that the observation is in the interval (c_1, c_2). Right censored and interval censored survival data are quite common in clinical trials and medical settings, whereas left censored data is quite rare. In this book, we shall only focus on right or interval censored survival data. Throughout the book, we will also assume that censoring is noninformative in the sense that inferences do not depend on the censoring process. Examples 1.1, 1.2, and 1.4–1.8 all involve right censored survival data, whereas Example 1.3 involves interval censored data.

We construct the likelihood function for a proportional hazards model with right censored data as follows. Suppose that there are n subjects under study, and that associated with the ith individual is a survival time t_i and a fixed censoring time c_i. The t_i's are assumed to be independent and identically distributed (i.i.d.) with density $f(t)$ and survival function $S(t)$. The exact survival time t_i of an individual will be observed only if $t_i \leq c_i$. The data in this framework can be represented by the n pairs of random variables (y_i, ν_i), where

$$y_i = \min(t_i, c_i) \tag{1.4.8}$$

and

$$\nu_i = \begin{cases} 1 & \text{if } t_i \leq c_i, \\ 0 & \text{if } t_i > c_i. \end{cases} \tag{1.4.9}$$

Then the likelihood function for $(\boldsymbol{\beta}, h_0(.))$ for a set of right censored data on n subjects is given by

$$\begin{aligned} L(\boldsymbol{\beta}, h_0(t)|D) &\propto \prod_{i=1}^{n} [h_0(y_i)\exp(\eta_i)]^{\nu_i} \left(S_0(y_i)^{\exp(\eta_i)}\right) \\ &= \prod_{i=1}^{n} [h_0(y_i)\exp(\eta_i)]^{\nu_i} \exp\left\{-\sum_{i=1}^{n} \exp(\eta_i) H_0(y_i)\right\}, \end{aligned} \tag{1.4.10}$$

where $D = (n, \boldsymbol{y}, X, \boldsymbol{\nu})$, $\boldsymbol{y} = (y_1, y_2, \ldots, y_n)'$, $\boldsymbol{\nu} = (\nu_1, \nu_2, \ldots, \nu_n)'$, $\eta_i = \boldsymbol{x}_i'\boldsymbol{\beta}$ is the linear predictor for subject i, $\boldsymbol{x}_i$ is the $p \times 1$ vector of covariates for subject i, X is the $n \times p$ matrix of covariates with i^{th} row $\boldsymbol{x}_i'$, and $S_0(t)$ is the baseline survivor function, which is related to $h_0(\cdot)$ by $S_0(t) = \exp\left(-\int_0^t h_0(u)\,du\right) = \exp(-H_0(t))$.

1.4.3 Partial Likelihood

Cox's (Cox, 1972) version of the proportional hazards model is semiparametric in the sense that the baseline hazard function $h_0(t)$ is not modeled as a parametric function of t. Due to Cox's development of partial likelihood (Cox, 1972, 1975), $h_0(t)$ is allowed to take on arbitrary values since it does not enter into the estimating equations for the parameters. Cox's partial likelihood proceeds as follows. Suppose that data are available on n individuals, and assume from these that we have d distinct event times and $n - d$ right censored survival times. For ease, we assume here that only one individual dies at each time, so that there are no ties in the data. Denote the ordered distinct survival times by $y_{(1)}, y_{(2)}, \ldots, y_{(d)}$, so that $y_{(j)}$ is the jth ordered survival time. The set of individuals who are at risk at time $y_{(j)}$ will be denoted by $\mathcal{R}_j$, so that $\mathcal{R}_j$ is the set of individuals who are event-free and uncensored at a time just prior to $y_{(j)}$. The quantity $\mathcal{R}_j$ is called the risk set. Cox's partial likelihood for $\boldsymbol{\beta}$ is thus given by

$$\text{PL}(\boldsymbol{\beta}|D) = \prod_{j=1}^{d} \frac{\exp(\boldsymbol{x}_{(j)}'\boldsymbol{\beta})}{\sum_{\ell \in \mathcal{R}_j} \exp(\boldsymbol{x}_\ell'\boldsymbol{\beta})}, \tag{1.4.11}$$

where $\boldsymbol{x}_{(j)}$ denotes the $p \times 1$ vector of covariates for the individual who has an event at the ordered survival time $y_{(j)}$. The summation in the denominator of (1.4.11) is the sum of the values of $\exp(\boldsymbol{x}_i'\boldsymbol{\beta})$ over all individuals who are at risk at time $y_{(j)}$. Note that the product is taken over the the individuals for whom event times have been recorded. Individuals for which the survival times are censored do not contribute to the numerator of (1.4.11), but they do enter into the summation over the risk sets at event times that occur before a censored observation. Moreover, this likelihood depends only on the ranking of the event times, since this determines the risk set at each event time. Consequently inferences about $\boldsymbol{\beta}$ depend only on the rank order of the survival times.

The likelihood function in (1.4.11) can be alternatively written as

$$\mathrm{PL}(\boldsymbol{\beta}|D) = \prod_{i=1}^{n} \left(\frac{\exp(\boldsymbol{x}'_{(i)}\boldsymbol{\beta})}{\sum_{\ell \in \mathcal{R}_i} \exp(\boldsymbol{x}'_{\ell}\boldsymbol{\beta})} \right)^{\nu_i}, \tag{1.4.12}$$

where ν_i is defined by (1.4.9). The partial maximum likelihood estimate of $\boldsymbol{\beta}$ can be obtained by maximizing (1.4.11) with respect to $\boldsymbol{\beta}$. This can be accomplished using numerical methods such as the Newton-Raphson procedure. The statistical software SAS fits (1.4.11) with the PHREG procedure.

1.5 The Bayesian Paradigm

The Bayesian paradigm is based on specifying a probability model for the observed data D, given a vector of unknown parameters $\boldsymbol{\theta}$, leading to the likelihood function $L(\boldsymbol{\theta}|D)$. Then we assume that $\boldsymbol{\theta}$ is random and has a *prior* distribution denoted by $\pi(\boldsymbol{\theta})$. Inference concerning $\boldsymbol{\theta}$ is then based on the *posterior* distribution, which is obtained by Bayes' theorem. The posterior distribution of $\boldsymbol{\theta}$ is given by

$$\pi(\boldsymbol{\theta}|D) = \frac{L(\boldsymbol{\theta}|D)\pi(\boldsymbol{\theta})}{\int_{\Theta} L(\boldsymbol{\theta}|D)\pi(\boldsymbol{\theta})\, d\boldsymbol{\theta}}, \tag{1.5.1}$$

where Θ denotes the parameter space of $\boldsymbol{\theta}$. From (1.5.1) it is clear that $\pi(\boldsymbol{\theta}|D)$ is *proportional* to the likelihood multiplied by the prior,

$$\pi(\boldsymbol{\theta}|D) \propto L(\boldsymbol{\theta}|D)\pi(\boldsymbol{\theta}),$$

and thus it involves a contribution from the observed data through $L(\boldsymbol{\theta}|D)$, and a contribution from prior information quantified through $\pi(\boldsymbol{\theta})$. The quantity $m(D) = \int_{\Theta} L(\boldsymbol{\theta}|D)\pi(\boldsymbol{\theta})\, d\boldsymbol{\theta}$ is the *normalizing constant* of $\pi(\boldsymbol{\theta}|D)$, and is often called the *marginal* distribution of the data or the prior predictive distribution.

In most models and applications, $m(D)$ does not have an analytic closed form, and therefore $\pi(\boldsymbol{\theta}|D)$ does not have a closed form. This difficulty leads us to the following question: How do we sample from the multivariate distribution $\pi(\boldsymbol{\theta}|D)$ when no closed form is available for it? This question has led to an enormous literature for computational methods for sampling from $\pi(\boldsymbol{\theta}|D)$ as well as methods for estimating $m(D)$. Perhaps one of the most popular computational methods for sampling from $\pi(\boldsymbol{\theta}|D)$ is called *the Gibbs sampler.* The Gibbs sampler is a very powerful simulation algorithm that allows us to sample from $\pi(\boldsymbol{\theta}|D)$ *without* knowing the normalizing constant $m(D)$. We discuss this method along with other Markov chain Monte Carlo (MCMC) sampling algorithms in detail in the next section.

The quantity $m(D)$ arises in model comparison problems, specifically in the computation of Bayes factors and posterior model probabilities. Thus, in addition to being able to simulate from $\pi(\boldsymbol{\theta}|D)$, one must also estimate the ratio $m_1(D)/m_2(D)$ for comparing two models, say, 1 and 2. To this end, there exists extensive literature for efficient estimation of ratios of normalizing constants using samples from the posterior distribution of $\boldsymbol{\theta}$, and therefore this is another major topic discussed in Chapter 6 as well as in other chapters of this book. For many of the applications discussed in this book, the prior $\pi(\boldsymbol{\theta})$ itself may not have a closed analytic form. This makes sampling from the posterior especially challenging, and novel computational methods are required to carry out the posterior computations. We discuss such methods throughout the book.

A major aspect of the Bayesian paradigm is prediction. Prediction is often an important goal in regression problems, and usually plays an important role in model selection problems. The *posterior predictive* distribution of a future observation vector $\boldsymbol{z}$ given the data D is defined as

$$\pi(\boldsymbol{z}|D) = \int_{\Theta} f(\boldsymbol{z}|\boldsymbol{\theta})\pi(\boldsymbol{\theta}|D)\, d\boldsymbol{\theta}, \tag{1.5.2}$$

where $f(\boldsymbol{z}|\boldsymbol{\theta})$ denotes the sampling density of $\boldsymbol{z}$, and $\pi(\boldsymbol{\theta}|D)$ is the posterior distribution of $\boldsymbol{\theta}$. We see that (1.5.2) is just the posterior expectation of $f(\boldsymbol{z}|\boldsymbol{\theta})$, and thus sampling from (1.5.2) is easily accomplished via the Gibbs sampler from $\pi(\boldsymbol{\theta}|D)$. This is a nice feature of the Bayesian paradigm since (1.5.2) shows that predictions and predictive distributions are easily computed once samples from $\pi(\boldsymbol{\theta}|D)$ are available.

1.6 Sampling from the Posterior Distribution

During the last decade, Monte Carlo (MC) based sampling methods for evaluating high-dimensional posterior integrals have been rapidly developing. Those sampling methods include MC importance sampling (Hammersley and Handscomb, 1964; Ripley, 1987; Geweke, 1989; Wolpert, 1991), Gibbs sampling (Geman and Geman, 1984; Gelfand and Smith, 1990), Metropolis–Hastings sampling (Metropolis et al., 1953; Hastings, 1970; Green, 1995), and many other hybrid algorithms.

The Gibbs sampler may be one of the best known MCMC sampling algorithms in the Bayesian computational literature. As discussed in Besag and Green (1993), the Gibbs sampler is founded on the ideas of Grenander

(1983), while the formal term is introduced by Geman and Geman (1984). The primary bibliographical landmark for Gibbs sampling in problems of Bayesian inference is Gelfand and Smith (1990). A similar idea termed *data augmentation* is introduced by Tanner and Wong (1987). Casella and George (1992) provide an excellent tutorial on the Gibbs sampler.

Let $\boldsymbol{\theta} = (\theta_1, \theta_2, \ldots, \theta_p)'$ be a p-dimensional vector of parameters and let $\pi(\boldsymbol{\theta}|D)$ be its posterior distribution given the data D. Then, the basic scheme of the Gibbs sampler is given as follows:

Step 0. Choose an arbitrary starting point $\boldsymbol{\theta}_0 = (\theta_{1,0}, \theta_{2,0}, \ldots, \theta_{p,0})'$, and set $i = 0$.

Step 1. Generate $\boldsymbol{\theta}_{i+1} = (\theta_{1,i+1}, \theta_{2,i+1}, \ldots, \theta_{p,i+1})'$ as follows:

- Generate $\theta_{1,i+1} \sim \pi(\theta_1|\theta_{2,i}, \ldots, \theta_{p,i}, D)$;
- Generate $\theta_{2,i+1} \sim \pi(\theta_2|\theta_{1,i+1}, \theta_{3,i}, \ldots, \theta_{p,i}, D)$;

  ... ...
- Generate $\theta_{p,i+1} \sim \pi(\theta_p|\theta_{1,i+1}, \theta_{2,i+1}, \ldots, \theta_{p-1,i+1}, D)$.

Step 2. Set $i = i + 1$, and go to Step 1.

Thus each component of $\boldsymbol{\theta}$ is visited in the natural order and a cycle in this scheme requires generation of p random variates. Gelfand and Smith (1990) show that under certain regularity conditions, the vector sequence $\{\boldsymbol{\theta}_i,\ i = 1, 2, \ldots\}$ has a stationary distribution $\pi(\boldsymbol{\theta}|D)$. Schervish and Carlin (1992) provide a sufficient condition that guarantees geometric convergence. Other properties regarding geometric convergence are discussed in Roberts and Polson (1994).

The Metropolis–Hastings algorithm was developed by Metropolis et al. (1953) and subsequently generalized by Hastings (1970). Tierney (1994) gives a comprehensive theoretical exposition of this algorithm, and Chib and Greenberg (1995) provide an excellent tutorial on this topic.

Let $q(\boldsymbol{\theta}, \boldsymbol{\vartheta})$ be a proposal density, which is also termed as a *candidate-generating density* by Chib and Greenberg (1995), such that

$$\int q(\boldsymbol{\theta}, \boldsymbol{\vartheta})\, d\boldsymbol{\vartheta} = 1.$$

Also let $\mathcal{U}(0,1)$ denote the uniform distribution over $(0,1)$. Then, a general version of the Metropolis–Hastings algorithm for sampling from the posterior distribution $\pi(\boldsymbol{\theta}|D)$ can be described as follows:

Step 0. Choose an arbitrary starting point $\boldsymbol{\theta}_0$ and set $i = 0$.

Step 1. Generate a candidate point $\boldsymbol{\theta}^*$ from $q(\boldsymbol{\theta}_i, \cdot)$ and u from $\mathcal{U}(0,1)$.

Step 2. Set $\boldsymbol{\theta}_{i+1} = \boldsymbol{\theta}^*$ if $u \le a(\boldsymbol{\theta}_i, \boldsymbol{\theta}^*)$ and $\boldsymbol{\theta}_{i+1} = \boldsymbol{\theta}_i$ otherwise, where the acceptance probability is given by

$$a(\boldsymbol{\theta}, \boldsymbol{\vartheta}) = \min\left\{\frac{\pi(\boldsymbol{\vartheta}|D)q(\boldsymbol{\vartheta}, \boldsymbol{\theta})}{\pi(\boldsymbol{\theta}|D)q(\boldsymbol{\theta}, \boldsymbol{\vartheta})}, 1\right\}. \tag{1.6.1}$$

Step 3. Set $i = i + 1$, and go to Step 1.

The performance of a Metropolis–Hastings algorithm depends on the choice of a proposal density q. In the context of the random walk proposal density, which is of the form $q(\boldsymbol{\theta}, \boldsymbol{\vartheta}) = q_1(\boldsymbol{\vartheta} - \boldsymbol{\theta})$, where $q_1(\cdot)$ is a multivariate density, Roberts, Gelman, and Gilks (1997) show that if the target and proposal densities are normal, then the scale of the latter should be tuned so that the acceptance rate is approximately 0.45 in one-dimensional problems and approximately 0.23 as the number of dimensions approaches infinity, with the optimal acceptance rate being around 0.25 in six dimensions. For the *independence chain*, in which we take $q(\boldsymbol{\theta}, \boldsymbol{\vartheta}) = q(\boldsymbol{\vartheta})$, it is important to ensure that the tails of the proposal density $q(\boldsymbol{\vartheta})$ dominate those of the target density $\pi(\boldsymbol{\theta}|D)$, which is similar to a requirement on the importance sampling function in Monte Carlo integration with importance sampling. The Metropolis algorithm can be used within the Gibbs sampler when direct sampling from the full conditional posterior is difficult. Also, if the conditional posterior is log-concave, one can readily use the adaptive rejection algorithm of Gilks and Wild (1992) within the Gibbs sampler to sample from the full conditional distributions. We also note that the ratio of uniforms method (see Ripley, 1987) is an alternative method for sampling from the full conditional distributions.

Another useful MCMC algorithm discussed by Gustafson (1997) is called the hybrid Monte Carlo method. The hybrid Monte Carlo method is an MCMC algorithm which originates in the statistical physics literature. The approach is due to Duane, Kennedy, Pendleton, and Roweth (1987), who build upon ideas of Andersen (1980) and Alder and Wainwright (1959). Neal (1993a,b, 1994a,b) examines the hybrid Monte Carlo method in detail, from a perspective which is more accessible to statisticians. The algorithm is not well known in the statistical community, however, so a cursory discussion follows here.

Let q denote the entire parameter vector, with $\dim(q) = m$. Notationally suppressing the conditioning on the observed data, let $E(q)$ be the negative log posterior density, up to an arbitrary additive constant. Thus the goal is to sample from the posterior density proportional to $\exp\{-E(q)\}$. In fact the hybrid algorithm operates on an augmented parameter vector (q, p), where $\dim(p) = m$. The additional parameter vector p has no statistical meaning; it is simply a computational device. The Markov chain is constructed so that its equilibrium distribution has the density

$$\pi(q, p) \propto \exp\left[-\left\{E(q) + K(p)\right\}\right], \tag{1.6.2}$$

where $K(p) = (1/2)\sum_{i=1}^{m} p_i^2$. That is, q and p are independent under (1.6.2), with q distributed as the posterior distribution and p distributed as independent standard normal variates. Therefore, the q marginal of a simulated chain is treated as a dependent sample from the posterior.

The specifics of the chain construction are best motivated via the physical analogy underlying the algorithm's origins. Under the analogy, q and p are respectively the position and momentum components of a physical system, while $E(q)$ and $K(p)$ are, respectively, the potential energy and kinetic energy of the system. The distribution (1.6.2) on (q, p) is the Boltzmann or steady-state distribution over the phase space. It is convenient to let $H(q,p) = E(q) + K(p)$ denote the total system energy.

The key to the hybrid algorithm is updating under which (q,p) evolves according to the Hamiltonian formulation of Newtonian dynamics. The updating is over time, denoted τ, which is an entirely fictitious quantity from a statistician's point of view. The dynamics are governed by the differential equations:

$$\frac{dq_i}{d\tau} = p_i \text{ and } \frac{dp_i}{d\tau} = -\nabla E(q),$$

where $\nabla E(q)$ is the gradient of E evaluated at q. In practice the dynamics must be followed along a discretized time grid. Treating q and p as functions of time, this is achieved via

$$q(\tau + \epsilon) = q(\tau) + \epsilon\{p(\tau) - (\epsilon/2)\nabla E(q(\tau))\}, \tag{1.6.3}$$

$$p(\tau + \epsilon) = p(\tau) - (\epsilon/2)\{\nabla E(q(\tau)) + \nabla E(q(\tau + \epsilon))\}. \tag{1.6.4}$$

This is the leapfrog discretization, which has important properties of time-reversibility and preservation of phase-space volume. When it is applied iteratively, there is a simpler form for the updating, where the updates for p "leapfrog" in time over those for q, and vice versa. See Neal (1993b) for a detailed discussion of the discretization.

The exact dynamics leave the total energy H unchanged, but the discretization causes H to fluctuate. This is compensated for by a Metropolis–Hastings accept-reject step (see Step 4 below). As ϵ is made smaller, the discretized dynamics better approximate the exact dynamics. So the acceptance rate can be made sufficiently high by keeping ϵ small. Additionally, a stochastic update of the momentum components p ensures that the chain visits states of varying total energy (see Step 5 below).

In the present implementation, one iteration of the Markov chain is carried out as follows:

Step 0. (q,p) is the current state.

Step 1. Simulate the step size, ϵ, from a symmetric distribution centered at zero.

Step 2. Simulate the number of steps, k, from a distribution on the positive integers.

Step 3. Generate a candidate state (q^*, p^*) via a discretized time evolution from $\tau = 0$ to $\tau = k\epsilon$. That is, (q^*, p^*) is obtained from (q,p) via k-fold iteration of the update given by (1.6.3) and (1.6.4).

Step 4. Calculate the acceptance probability $\omega = \min(\exp[-\{H(q^*, p^*) - H(q, p)\}], 1)$. With probability ω, replace (q, p) with (q^*, p^*). With probability $1 - \omega$, leave (q, p) unchanged.

Step 5. Replace p with simulated independent standard normal variates.

Step 6. (q, p) is the new state.

The symmetry of the distribution in step 1 is required for time-reversibility. This ensures that the acceptance probability has the simple form given in Step 4. For a detailed justification that (1.6.2) is the equilibrium distribution for a chain with transitions as described above, see Neal (1993b).

The random selection of k allows the algorithm to use a range of "trajectory lengths." Neal (1993b) advocates relatively large values of k, since long trajectories avoid the inefficient random walk behavior associated with some MCMC sampling schemes. The tradeoff is that larger values of k require more evaluations of ∇E per trajectory.

Recently, several Bayesian software packages have been developed. These include `BUGS` for analyzing general hierarchical models via MCMC (`http://www.mrc-bsu.cam.ac.uk/bugs/`), `BATS` for Bayesian time series analysis (`http://www.stat.duke.edu/~mw/bats.html`), `MATLAB`, and `MINITAB` contains Bayesian computational algorithms for introductory Bayesian analysis (`http://www-math.bgsu.edu/~albert/`), and many others. A more complete listing and description of pre-1990 Bayesian software can be found in Goel (1988). A listing of some of the Bayesian software developed since 1990 is given in Berger (2000).

1.7 Informative Prior Elicitation

Prior elicitation perhaps plays the most crucial role in Bayesian inference. In the context of survival analysis with covariates, the most popular choice of informative prior for $\boldsymbol{\beta}$ is the normal prior, and the most common choice of noninformative prior for $\boldsymbol{\beta}$ is the uniform prior. We discuss these choices of priors in detail for a wide variety of models in Chapters 2 and 3 as well as in several other chapters throughout the book. Although noninformative and improper priors may be useful and easier to specify for certain problems, they cannot be used in all applications, such as model selection or model comparison, as it is well known that proper priors are required to compute Bayes factors and posterior model probabilities. In addition, it is well known that Bayes factors are generally quite sensitive to the choices of hyperparameters of noninformative proper priors, and thus one cannot simply specify noninformative proper priors in model selection contexts to avoid informative prior elicitation. In addition, noninformative priors may cause instability in the posterior estimates and lead to convergence problems for the Gibbs sampler. Moreover, noninformative priors do not make use of real prior information that one may have for a specific application. Thus, informative priors are essential in these situations, and in general, they are useful in applied research settings where the investigator has access to previous studies measuring the same response and covariates as the current study. For example, in many cancer and AIDS clinical trials, current studies often use treatments that are very similar or slight modifications of treatments used in previous studies. We refer to data arising from previous similar studies as *historical data* throughout. In carcinogenicity studies, for example, large historical databases exist for the control animals from previous experiments. In all of these situations, it is natural to incorporate the historical data into the current study by quantifying it with a suitable prior distribution on the model parameters. The methodology discussed here can be applied to each of these situations as well as in other applications that involve historical data.

From a Bayesian perspective, historical data from past similar studies can be very helpful in interpreting the results of the current study. For example, historical control data can be very helpful in interpreting the results of a carcinogenicity study. According to Haseman, Huff and Boorman (1984), historical data can be useful when control tumor rates are low and when marginal significance levels are obtained in a test for dose effects. Suppose, for example, that 4 of 50 animals in an exposed group develop a specific tumor, compared with 0 of 50 in a control group. This difference is not statistically significant (p=0.12, based on Fisher's exact test). However, the difference may be biologically significant if the observed tumor type is known to be extremely rare in the particular animal strain being studied. By specifying a suitable prior distribution on the control response rates that reflect the observed rates of a particular defect over a large series of past studies, one can derive a modified test statistic that incorporates historical information. If the defect is rare enough in the historical series, then even the difference of 4/50 versus 0/50 will be statistically significant based on a method that appropriately incorporates historical information.

To fix ideas, suppose we have historical data from a similar previous study, denoted by $D_0 = (n_0, y_0, X_0)$ where n_0 is the sample size of the historical data, y_0 is the $n_0 \times 1$ response vector, and X_0 is the $n_0 \times p$ matrix of covariates based on the historical data. The power prior is defined to be the likelihood function based on the historical data D_0, raised to a power a_0. Here, $0 \le a_0 \le 1$ is a scalar parameter that controls the influence of the historical data on the current data. Further details of the power prior will follow shortly. One of the most useful applications of the power prior is for model selection problems, since these priors inherently automate the informative prior specification for all possible models in the model space. They are quite attractive in this context, since specifying meaningful informative prior distributions for the parameters in each model is a difficult task requiring contextual interpretations of a large number of parameters.

In variable subset selection, for example, the prior distributions for all possible subset models are automatically determined once the historical data D_0 and a_0 are specified. Berger and Mallows (1988) refer to such priors as "semi-automatic" in their discussion of Mitchell and Beauchamp (1988). Chen, Manatunga, and Williams (1998) use the power prior for heritability estimates from human twin data. Chen, Ibrahim, and Yiannoutsos (1999) demonstrate the use of the power prior in variable selection contexts for logistic regression. Ibrahim, Chen, and Ryan (2000) and Chen, Ibrahim, Shao, and Weiss (2001) develop the power prior for the class of generalized linear mixed models. Ibrahim and Chen (1998), Ibrahim, Chen, and MacEachern (1999), Chen, Ibrahim, and Sinha (1999), and Chen, Dey, and Sinha (2000) develop the power prior for various types of models for survival data.

We consider the power prior for an arbitrary regression model. Let the data from the *current* study be denoted by $D = (n, \boldsymbol{y}, X)$, where n denotes the sample size, $\boldsymbol{y}$ denotes the $n \times 1$ response vector, and X denotes the $n \times p$ matrix of covariates. Further, denote the likelihood for the current study by $L(\boldsymbol{\theta}|D)$, where $\boldsymbol{\theta}$ is a vector of indexing parameters. Thus, $L(\boldsymbol{\theta}|D)$ is a general likelihood function for an arbitrary regression model, such as a generalized linear model, random effects model, nonlinear model, or a survival model with censored data. Now suppose we have historical data from a similar previous study, denoted by $D_0 = (n_0, \boldsymbol{y}_0, X_0)$. Further, let $\pi_0(\boldsymbol{\theta}|.)$ denote the prior distribution for $\boldsymbol{\theta}$ before the historical data D_0 is observed. We shall call $\pi_0(\boldsymbol{\theta}|.)$ the *initial prior* distribution for $\boldsymbol{\theta}$. Given a_0, we define the *power prior* distribution of $\boldsymbol{\theta}$ for the current study as

$$\pi(\boldsymbol{\theta}|D_0, a_0) \propto L(\boldsymbol{\theta}|D_0)^{a_0} \pi_0(\boldsymbol{\theta}|c_0), \tag{1.7.1}$$

where c_0 is a specified hyperparameter for the initial prior, and a_0 is a scalar prior parameter that weights the historical data relative to the likelihood of the current study. The prior parameter c_0 controls the impact of $\pi_0(\boldsymbol{\theta}|c_0)$ on the entire prior, and the parameter a_0 controls the influence of the historical data on $\pi(\boldsymbol{\theta}|D_0, a_0)$. The parameter a_0 can be interpreted as a relative precision parameter for the historical data. It is reasonable to restrict the range of a_0 to be between 0 and 1, and thus we take $0 \le a_0 \le 1$. One of the main roles of a_0 is that it controls the heaviness of the tails of the prior for $\boldsymbol{\theta}$. As a_0 becomes smaller, the tails of (1.7.1) become heavier. Setting $a_0 = 1$, (1.7.1) corresponds to the update of $\pi_0(\boldsymbol{\theta}|c_0)$ using Bayes theorem. That is, with $a_0 = 1$, (1.7.1) corresponds to the posterior distribution of $\boldsymbol{\theta}$ from the previous study. When $a_0 = 0$, then the prior does not depend on the historical data, and in this case, $\pi(\boldsymbol{\theta}|D_0, a_0 = 0) \equiv \pi_0(\boldsymbol{\theta}|c_0)$. Thus, $a_0 = 0$ is equivalent to a prior specification with no incorporation of historical data. Therefore, (1.7.1) can be viewed as a generalization of the usual Bayesian update of $\pi_0(\boldsymbol{\theta}|c_0)$. The parameter a_0 allows the investigator to control the influence of the historical data on the current study. Such control is important in cases where there is heterogeneity between the previous and current study, or when the sample sizes of the two studies are quite different.

The hierarchical power prior specification is completed by specifying a (proper) prior distribution for a_0. This leads to a joint power prior distribution for $(\boldsymbol{\theta}, a_0)$ of the form

$$\pi(\boldsymbol{\theta}, a_0|D_0) \;\propto\; L(\boldsymbol{\theta}|D_0)^{a_0}\pi_0(\boldsymbol{\theta}|c_0)\pi(a_0|\gamma_0), \tag{1.7.2}$$

where γ_0 is a specified hyperparameter vector. A natural choice for $\pi(a_0|\gamma_0)$ is a beta prior. However, other choices, including a truncated gamma prior or a truncated normal prior, can be used. These three priors for a_0 have similar theoretical properties, and our experience shows that they have similar computational properties. In practice, they yield similar results when the hyperparameters are appropriately chosen. Thus, for a clear focus and exposition, we will use a *beta* distribution for $\pi(a_0|\gamma_0)$ throughout the book. The beta prior for a_0 appears to be the most natural prior to use and leads to the most natural elicitation scheme. The prior in (1.7.2) does not have a closed form in general, but it has several attractive theoretical and computational properties for the classes of models considered here. One attractive feature of (1.7.2) is that it creates heavier tails for the marginal prior of $\boldsymbol{\theta}$ than the prior in (1.7.1), which assumes that a_0 is a fixed value. This is a desirable feature since it gives the investigator more flexibility in weighting the historical data. In addition, our construction of (1.7.2) is quite general, with various possibilities for $\pi_0(\boldsymbol{\theta}|c_0)$. If $\pi_0(\boldsymbol{\theta}|c_0)$ is proper, then (1.7.2) is guaranteed to be proper. Further, (1.7.2) can be proper even if $\pi_0(\boldsymbol{\theta}|c_0)$ is an improper uniform prior. Specifically, Ibrahim, Ryan, and Chen (1998) and Chen, Ibrahim, and Yiannoutsos (1999) characterize the propriety of (1.7.2) for generalized linear models, and also show that for fixed a_0, the prior converges to a multivariate normal distribution as $n_0 \to \infty$. For the class of generalized linear mixed models, Ibrahim, Chen, and Ryan (2000), Chen, Ibrahim, Shao, and Weiss (2001), Chen, Ibrahim, and Shao (2000), and Chen, Dey, and Sinha (2000) characterize the propriety of (1.7.2) and derive various other theoretical properties of the power prior. Ibrahim, Chen, and MacEachern (1999), and Ibrahim and Chen (1998) characterize various properties of (1.7.2) for proportional hazards models, and Chen, Ibrahim, and Sinha (1999) examine various theoretical properties of (1.7.2) for a class of cure rate models. In this book, we will summarize the conditions for propriety as well as some of the properties for the above-mentioned models, but refer the reader to these articles for details and proofs.

The power prior defined in (1.7.2) can easily be generalized to multiple historical data sets. If there are N_0 historical studies, we define $D_{0k} = (n_{0k}, y_{0k}, X_{0k})$ to be the historical data based on the kth study, $j = 1, 2, \ldots, N_0$, and $D_0 = (D_{01}, D_{02}, \ldots, D_{0N_0})$. In this case, it may be desirable to define a weight parameter a_{0k} for each historical study, and take the a_{0k}'s to be i.i.d. beta random variables with hyperparameters $\gamma_0 = (\delta_0, \lambda_0)'$, $j = 1, 2, \ldots, N_0$. Letting $a_0 = (a_{01}, a_{02}, \ldots, a_{0N_0})$, the prior in (1.7.2) can be generalized as

$$\pi(\boldsymbol{\theta}, a_0|D_0) \propto \left(\prod_{k=1}^{N_0} [L(\boldsymbol{\theta}|D_{0k})]^{a_{0k}} \; \pi(a_{0k}|\gamma_0) \right) \pi_0(\boldsymbol{\theta}|c_0). \qquad (1.7.3)$$

It can be shown that (1.7.1) is proper for many types of models (see Ibrahim and Chen, 2000) and is best suited for situations in which a historical data set is available. However, the power prior can still be used if historical data is not available for the current study. In this case, D_0 can be elicited from expert opinion or by using case-specific information about each individual in the current study, or perhaps by a theoretical model giving forecasts for the event times. The elicitation scheme is less automated than the one in which a previous study exists. Thus, in this situation, there are a number of ways one could elicit D_0, and this elicitation depends on the context of the problem. We do not attempt to give a general elicitation scheme for D_0 in this setting, but rather mention some general possibilities. In any case, the most straightforward specification of D_0 is to use historical data from a similar previous study in which the response vector $\boldsymbol{y}_0$ would be taken to be the vector of survival times from the historical data and the covariate matrix X_0 is taken to be the matrix of covariates from the historical data.

1.8 Why Bayes?

A natural question to ask is what advantages does the Bayesian paradigm offer over the frequentist paradigm in survival analysis. We have identified several. First, it is well known that survival models are generally quite hard to fit, especially in the presence of complex censoring schemes. With the use of the Gibbs sampler and other MCMC techniques, fitting complex survival models is fairly straightforward, and the availability of software like BUGS eases the implementation greatly. In addition, MCMC sampling enables us to make exact inference for *any* sample size without resorting to asymptotic calculations. In the frequentist paradigm, variance estimates, for example, usually require asymptotic arguments which can be quite complicated to derive and in some models are simply not available. Then there is always the issue of whether the sample size is large enough for the asymptotic approximation to be valid. In contrast, in the Bayesian framework, variance estimates, as well as any other posterior summary come out as a by-product of the Gibbs sampler, and therefore are trivial to obtain once samples from the posterior distribution are available.

The Bayesian paradigm also enables us to incorporate prior information in a natural way, whereas the frequentist paradigm does not. For example, when we have historical data, as is the case with most clinical trials, we can use the power priors discussed in the previous section to formally incorporate this data into the current analysis. Thus the Bayesian paradigm is a powerful tool for quantifying prior data, especially historical data. Also, for many models, frequentist inference can be obtained as a special case of Bayesian inference with many types of noninformative priors such as the uniform prior. For example, with a uniform prior, the posterior mode corresponds to the maximum likelihood estimate. In this sense, frequentist inference can be viewed as a special case of Bayesian inference.

Bayesian inference also has several advantages over frequentist methods in the availability and flexibility of model building and data analysis tools. For example, in the Bayesian paradigm, model comparisons of nested or non-nested models are easily entertained via Bayes factors or model selection criteria. Exact computations of Bayes factors or model selection criteria can be obtained via the Gibbs sampler. In the frequentist paradigm, there is no unified methodology for comparing non-nested models and comparisons of nested models usually require asymptotic arguments. Other data analysis tools such as predictive distributions and residuals can be more easily calculated for survival models under the Bayesian paradigm. Missing covariate or response data is another area for which the Bayesian paradigm has clear advantages over the frequentist paradigm, since the missing values essentially get treated as parameters in the Bayesian framework and only add one extra layer in the Gibbs sampler. Thus, the computational algorithms change slightly in the presence of missing data under the Bayesian paradigm. In contrast, frequentist methods for missing data usually involve more computationally intensive methods than those needed in complete data settings, and often can be quite complicated. In missing data problems, variance estimates are generally quite tricky to compute and require new derivations, whereas in the Bayesian paradigm, they are a by-product of the Gibbs sampler. In this book, we demonstrate these advantages throughout the chapters, and show that the analysis of survival data can be carried out under the Bayesian paradigm in a straightforward and unified way.

Exercises

1.1 Prove (1.4.2).

1.2 Prove (1.4.5) using (1.4.3)

1.3 Derive the joint density of (y_i, ν_i) in terms of $f(y_i)$ and $S(c_i)$, where y_i and ν_i are given by (1.4.8) and (1.4.9).

1.4 Using Exercise 1.1, derive (1.4.10).

1.5 Consider the Weibull model in Example 1.9. Suppose we have a random sample of n subjects, some of which may be right censored.

(a) Suppose $\alpha = 2$ and γ is unknown. Find the likelihood maximum likelihood estimate of γ.
(b) Suppose $\alpha = 2$, γ is unknown, and $\pi(\gamma) \propto \gamma^{-1}$. Derive the posterior distribution of γ.
(c) Under what conditions is the posterior distribution in part (b) proper?
(d) Consider the conditions of part (b), and suppose y_{n+1} is a future observation (not censored). Find the predictive density of $(y_{n+1}|D)$ where $D = (\boldsymbol{y}, \boldsymbol{\nu})$, $\boldsymbol{y} = (y_1, y_2, \ldots, y_n)$, and $\boldsymbol{\nu} = (\nu_1, \nu_2, \ldots, \nu_n)'$
(e) Suppose (α, γ) are both unknown. Write out the likelihood equations for (α, γ) based on n subjects, some of which may be right censored.

1.6 Again consider the Weibull model in Example 1.9. Suppose we have a random sample of n subjects, some of which may be right censored.

(a) Suppose $\alpha = 1$ and $\lambda_i = \exp(\boldsymbol{x}_i'\boldsymbol{\beta})$, where $\boldsymbol{x}_i$ is a $p \times 1$ vector of covariates for the ith subject, $\boldsymbol{\beta} = (\beta_1, \beta_2, \ldots, \beta_p)'$, and $\pi(\boldsymbol{\beta}) \propto 1$. Derive the conditional posterior distribution of $(\beta_j|D, \beta_i, i \neq j)$.
(b) Suppose $\alpha = 1$, $p = 1$, and $\pi(\beta) \propto 1$. Derive a 95% Highest Posterior Density (HPD) interval for β.
(c) Under the conditions of part (b), derive the posterior median of β.

1.7 For the proportional hazards model given by (1.4.7), derive the probability density function, the cumulative hazard function, and the survivor function.

1.8 Suppose we use the proportional hazards model of (1.4.7) to fit the interval censored survival data given in Table 1.4, where x corresponds to the treatment covariate defined in Example 1.3. We further assume that the baseline hazard function $h_0(t) = \lambda$, where $\lambda > 0$, and let β denote the regression coefficient of x. Consider the joint prior

$$\pi(\beta, \lambda) \propto \lambda^{\eta_0 - 1} \exp(-\tau_0 \lambda),$$

where $\eta_0 > 0$ and $\tau_0 > 0$ are two specified hyperparameters.

(a) Write out the likelihood function and the posterior distribution of (β, λ) based on the observed interval censored data.
(b) Compute the maximum likelihood estimates of β and λ.
(c) Derive the full conditionals of β and λ. Are these conditionals easy to sample from? Why or why not?
(d) Let $(a_{l_i}, a_{r_i}]$ $(a_{l_i} < a_{r_i})$ denote the observed interval and y_i denotes the survival time $(a_{l_i} < y_i < a_{r_i})$ for the ith subject. Also let $x_i = 1$ if the ith subject received radio-chemotherapy

and $x_i = 0$ otherwise. Write out the likelihood function based on the complete data $\{(y_i, x_i), i = 1, 2, \ldots, n\}$.

(e) Based on the complete data, derive all of the full conditionals for β, λ, and the y_i's required for the Gibbs sampler.

(f) Conduct a comprehensive Bayesian data analysis for the data given in Table 1.4 using the Gibbs sampling algorithm.

1.9 For the power prior $\pi(\boldsymbol{\theta}, a_0|D_0)$ given by (1.7.2), show that (i) the prior mode of $\boldsymbol{\theta}$ is invariant to the prior $\pi(a_0|\gamma_0)$, and (ii) the prior mode of $\boldsymbol{\theta}$ is identical to the maximum likelihood estimate of $\boldsymbol{\theta}$ based on the historical data D_0.

2
Parametric Models

Parametric models play an important role in Bayesian survival analysis, since many Bayesian analyses in practice are carried out using parametric models. Parametric modeling offers straightforward modeling and analysis techniques. In this chapter, we discuss parametric models for univariate right censored survival data. We derive the posterior and predictive distributions and demonstrate how to carry out Bayesian analyses for several commonly used parametric models. The statistical literature in Bayesian parametric survival analysis and life-testing is too enormous to list here, but some references dealing with applications to medicine or public health include Grieve (1987), Achcar, Bolfarine, and Pericchi (1987), Achcar, Bookmeyer, and Hunter (1985), Chen, Hill, Greenhouse, and Fayos (1985), Dellaportas and Smith (1993), and Kim and Ibrahim (2001).

2.1 Exponential Model

The exponential model is the most fundamental parametric model in survival analysis. Suppose we have independent identically distributed (i.i.d.) survival times $\boldsymbol{y} = (y_1, y_2, \ldots, y_n)'$, each having an exponential distribution with parameter λ, denoted by $\mathcal{E}(\lambda)$. Denote the censoring indicators by $\boldsymbol{\nu} = (\nu_1, \nu_2, \ldots, \nu_n)'$, where $\nu_i = 0$ if y_i is right censored and $\nu_i = 1$ if y_i is a failure time. Let $f(y_i|\lambda) = \lambda \exp(-\lambda y_i)$ denote the density for y_i, $S(y_i|\lambda) = \exp(-\lambda y_i)$ denotes the survival function and $D = (n, \boldsymbol{y}, \boldsymbol{\nu})$

denotes the observed data. We can write the likelihood function of λ as

$$L(\lambda|D) = \prod_{i=1}^{n} f(y_i|\lambda)^{\nu_i} S(y_i|\lambda)^{(1-\nu_i)} = \lambda^d \exp\left(-\lambda \sum_{i=1}^{n} y_i\right), \tag{2.1.1}$$

where $d = \sum_{i=1}^n \nu_i$. The conjugate prior for λ is the gamma prior. Let $\mathcal{G}(\alpha_0, \lambda_0)$ denote the gamma distribution with parameters (α_0, λ_0), with density given by

$$\pi(\lambda|\alpha_0, \lambda_0) \propto \lambda^{\alpha_0 - 1} \exp(-\lambda_0 \lambda).$$

Then, taking a $\mathcal{G}(\alpha_0, \lambda_0)$ prior for λ, the posterior distribution of λ is given by

$$\begin{aligned}
\pi(\lambda|D) &\propto L(\lambda|D)\pi(\lambda|\alpha_0, \lambda_0) \\
&\propto \left(\lambda^{\sum_{i=1}^n \nu_i} \exp\left\{-\lambda \sum_{i=1}^{n} y_i\right\}\right) \left(\lambda^{\alpha_0 - 1} \exp(-\lambda_0 \lambda)\right) \\
&= \lambda^{\alpha_0 + d - 1} \exp\left\{-\lambda(\lambda_0 + \sum_{i=1}^{n} y_i)\right\}.
\end{aligned} \tag{2.1.2}$$

Thus we recognize the kernel of the posterior distribution in (2.1.2) as a $\mathcal{G}(\alpha_0 + d, \lambda_0 + \sum_{i=1}^n y_i)$ distribution. The posterior mean and variance of λ are thus given by

$$E(\lambda|D) = \frac{\alpha_0 + d}{\lambda_0 + \sum_{i=1}^n y_i}$$

and

$$\text{Var}(\lambda|D) = \frac{\alpha_0 + d}{(\lambda_0 + \sum_{i=1}^n y_i)^2}.$$

The posterior predictive distribution of a future failure time y_f is given by

$$\begin{aligned}
\pi(y_f|D) &= \int_0^\infty \pi(y_f|\lambda)\pi(\lambda|D)\, d\lambda \\
&\propto \int_0^\infty \lambda^{\alpha_0 + d + 1 - 1} \exp\left\{-\lambda(y_f + \lambda_0 + \sum_{i=1}^{n} y_i)\right\} d\lambda \\
&= \Gamma(\alpha_0 + d + 1) \left(\lambda_0 + \sum_{i=1}^{n} y_i + y_f\right)^{-(d+\alpha_0+1)} \\
&\propto \left(\lambda_0 + \sum_{i=1}^{n} y_i + y_f\right)^{-(d+\alpha_0+1)}
\end{aligned} \tag{2.1.3}$$

The normalized posterior predictive distribution is thus given by

$$\pi(y_f|D) = \begin{cases} \frac{(d+\alpha_0)(\lambda_0+\sum_{i=1}^n y_i)^{(\alpha_0+d)}}{(\lambda_0+\sum_{i=1}^n y_i+y_f)^{(\alpha_0+d+1)}} & \text{if } y_f > 0, \\ 0 & \text{otherwise.} \end{cases} \tag{2.1.4}$$

In the derivation of (2.1.3) above, we need to evaluate a gamma integral, which thus led to the posterior predictive distribution in (2.1.4). The predictive distribution in (2.1.4) is known as an *inverse beta* distribution and is discussed in detail in Aitchison and Dunsmore (1975).

To build a regression model, we introduce covariates through λ, and write $\lambda_i = \varphi(\boldsymbol{x}_i'\boldsymbol{\beta})$, where $\boldsymbol{x}_i$ is a $p \times 1$ vector of covariates, $\boldsymbol{\beta}$ is a $p \times 1$ vector of regression coefficients, and $\varphi(.)$ is a known function. A common form of φ is to take $\varphi(\boldsymbol{x}_i'\boldsymbol{\beta}) = \exp(\boldsymbol{x}_i'\boldsymbol{\beta})$. Another form of φ is $\varphi(\boldsymbol{x}_i'\boldsymbol{\beta}) = (\boldsymbol{x}_i'\boldsymbol{\beta})^{-1}$. Feigl and Zelen (1965) also discuss this regression model. Using $\varphi(\boldsymbol{x}_i'\boldsymbol{\beta}) = \exp(\boldsymbol{x}_i'\boldsymbol{\beta})$, we are led to the likelihood function

$$\begin{aligned} L(\boldsymbol{\beta}|D) &= \prod_{i=1}^n f(y_i|\lambda_i)^{\nu_i} S(y_i|\lambda_i)^{(1-\nu_i)} \\ &= \prod_{i=1}^n \left[\exp(\boldsymbol{x}_i'\boldsymbol{\beta})\exp(-y_i\exp(\boldsymbol{x}_i'\boldsymbol{\beta}))\right]^{\nu_i} \left[\exp(-y_i\exp(\boldsymbol{x}_i'\boldsymbol{\beta}))\right]^{(1-\nu_i)} \\ &= \exp\left\{\sum_{i=1}^n \nu_i \boldsymbol{x}_i'\boldsymbol{\beta}\right\} \exp\left\{-\sum_{i=1}^n y_i \exp(\boldsymbol{x}_i'\boldsymbol{\beta})\right\}. \end{aligned} \tag{2.1.5}$$

In (2.1.5), we define $D = (n, \boldsymbol{y}, X, \boldsymbol{\nu})$, where X is the $n \times p$ matrix of covariates with i^{th} row $\boldsymbol{x}_i'$. Throughout the remainder of this chapter, $D = (n, \boldsymbol{y}, \boldsymbol{\nu})$ denotes the observed data without covariates, while $D = (n, \boldsymbol{y}, X, \boldsymbol{\nu})$ denotes the observed data for regression models. Common prior distributions for $\boldsymbol{\beta}$ include the uniform improper prior, i.e., $\pi(\boldsymbol{\beta}) \propto 1$, and a normal prior. In the regression setting, closed forms for the posterior distribution of $\boldsymbol{\beta}$ are generally not available, and therefore one needs to use numerical integration or Markov chain Monte Carlo (MCMC) methods. Before the advent of MCMC, numerical integration techniques were employed by Grieve (1987). However, due to the availability of statistical packages such as BUGS, the regression model in (2.1.5) can easily be fitted using MCMC techniques. Suppose we specify a p-dimensional normal prior for $\boldsymbol{\beta}$, denoted by $N_p(\boldsymbol{\mu}_0, \Sigma_0)$, where $\boldsymbol{\mu}_0$ denotes the prior mean and Σ_0 denotes the prior covariance matrix. Then the posterior distribution of $\boldsymbol{\beta}$ is given by

$$\pi(\boldsymbol{\beta}|D) \propto L(\boldsymbol{\beta}|D)\pi(\boldsymbol{\beta}|\boldsymbol{\mu}_0, \Sigma_0), \tag{2.1.6}$$

where $\pi(\boldsymbol{\beta}|\boldsymbol{\mu}_0, \Sigma_0)$ is the multivariate normal density with mean $\boldsymbol{\mu}_0$ and covariance matrix Σ_0. The posterior in (2.1.6) does not have a closed form in general, and thus MCMC methods are needed to sample from the posterior distribution of $\boldsymbol{\beta}$. The statistical package BUGS can be readily used for this model to do the Gibbs sampling. To carry out Gibbs sampling for this model, we need to write out the full conditional distributions. Let β_j denote the jth component of $\boldsymbol{\beta}$, and let $\boldsymbol{\beta}^{(-j)}$ denote the $\boldsymbol{\beta}$ vector without the jth component. Then, the jth full conditional can be written as

$$\pi(\beta_j|D,\boldsymbol{\beta}^{(-j)}) \propto L(\beta_j,\boldsymbol{\beta}^{(-j)}|D)\pi(\beta_j,\boldsymbol{\beta}^{(-j)}|\boldsymbol{\mu}_0,\Sigma_0) \tag{2.1.7}$$

for $j = 1, 2, \dots, p$. Note that each of the p full conditionals is proportional to the likelihood times the prior, where in each case, the random variable in (2.1.7) is β_j and $\boldsymbol{\beta}^{(-j)}$ is held fixed. To sample from each of the full conditionals in (2.1.7), we can use a rejection algorithm. An efficient algorithm is available here since each of the full conditionals is log-concave, and thus we can use the adaptive rejection algorithm of Gilks and Wild (1992) to sample from each of the full conditionals. To show log-concavity of (2.1.7), we need to show that

$$\frac{\partial^2}{\partial\beta_j^2}\log(\pi(\beta_j|D,\boldsymbol{\beta}^{(-j)})) \le 0$$

for $j = 1, 2, \dots, p$. The kernel of the log-posterior is given by

$$\log(\pi(\boldsymbol{\beta}|D)) = \sum_{i=1}^n \nu_i \boldsymbol{x}_i'\boldsymbol{\beta} - \sum_{i=1}^n y_i \exp(\boldsymbol{x}_i'\boldsymbol{\beta}) + \log(\pi(\boldsymbol{\beta}|\boldsymbol{\mu}_0,\Sigma_0)). \tag{2.1.8}$$

By taking two derivatives with respect to β_j in (2.1.8), we get

$$\begin{aligned}\frac{\partial^2}{\partial\beta_j^2}\log(\pi(\boldsymbol{\beta}|D)) &= -\sum_{i=1}^n y_i x_{ij}^2 \exp(\boldsymbol{x}_i'\boldsymbol{\beta}) + \frac{\partial^2}{\partial\beta_j^2}\log(\pi(\boldsymbol{\beta}|\boldsymbol{\mu}_0,\Sigma_0)) \\ &= -\sum_{i=1}^n y_i x_{ij}^2 \exp(\boldsymbol{x}_i'\boldsymbol{\beta}) - \sigma_0^{(jj)} \le 0, \end{aligned} \tag{2.1.9}$$

where $\sigma_0^{(jj)}$ is the jth diagonal element of Σ_0^{-1}. Thus, $\pi(\boldsymbol{\beta}|D)$ is log-concave in each component of $\boldsymbol{\beta}$. We note here that for the normal prior in (2.1.6)

$$\frac{\partial}{\partial\boldsymbol{\beta}}\log(\pi(\boldsymbol{\beta}|\boldsymbol{\mu}_0,\Sigma_0)) = -\Sigma_0^{-1}(\boldsymbol{\beta}-\boldsymbol{\mu}_0)$$

and

$$\frac{\partial^2}{\partial\boldsymbol{\beta}\partial\boldsymbol{\beta}'}\log(\pi(\boldsymbol{\beta}|\boldsymbol{\mu}_0,\Sigma_0)) = -\Sigma_0^{-1},$$

so that $\frac{\partial^2}{\partial\beta_j^2}\log(\pi(\boldsymbol{\beta}|\boldsymbol{\mu}_0,\Sigma_0)) = \sigma_0^{(jj)}$. We also note that $\pi(\boldsymbol{\beta}|D)$ will always be log-concave as long as the prior $\pi(\boldsymbol{\beta}|\boldsymbol{\mu}_0,\Sigma_0)$ in (2.1.6) is log-concave, and thus priors other than the normal prior can be used and still yield log-concavity of the posterior distribution of $\boldsymbol{\beta}$. Dellaportas and Smith (1993) demonstrate the log-concavity property for the class of generalized linear models and some parametric survival models. We refer the reader to their article for more details.

A posterior summary often of interest is the posterior distribution of the survival function at a particular value t. For the exponential model, the survival function of an individual with covariate vector $\boldsymbol{x}$ at the point t is given by $S(t) = \exp(-\exp(\boldsymbol{x}'\boldsymbol{\beta})t)$, so that if one obtains samples from the posterior distribution of $\boldsymbol{\beta}$, we can readily calculate posterior summaries of $S(t)$, such as the posterior mean, variance, and credible intervals.

Example 2.1. Melanoma data. To demonstrate the exponential model, we consider the E1684 melanoma clinical trial discussed in Example 1.2. This model can be easily fit in BUGS. The BUGS code for this model is given in the website "`http://www.springer-ny.com`." We use the likelihood in (2.1.5) with a single covariate, treatment, and therefore $\boldsymbol{\beta} = (\beta_0, \beta_1)'$, where β_0 denotes the intercept term and β_1 is the coefficient for the treatment covariate. Further, we take $\boldsymbol{\beta} \sim N_2(0, 10^4 I)$ a priori. Using a burn-in of 1000 samples, and then an additional 3000 Gibbs samples, we get the following posterior summaries of $\boldsymbol{\beta}$ from BUGS.

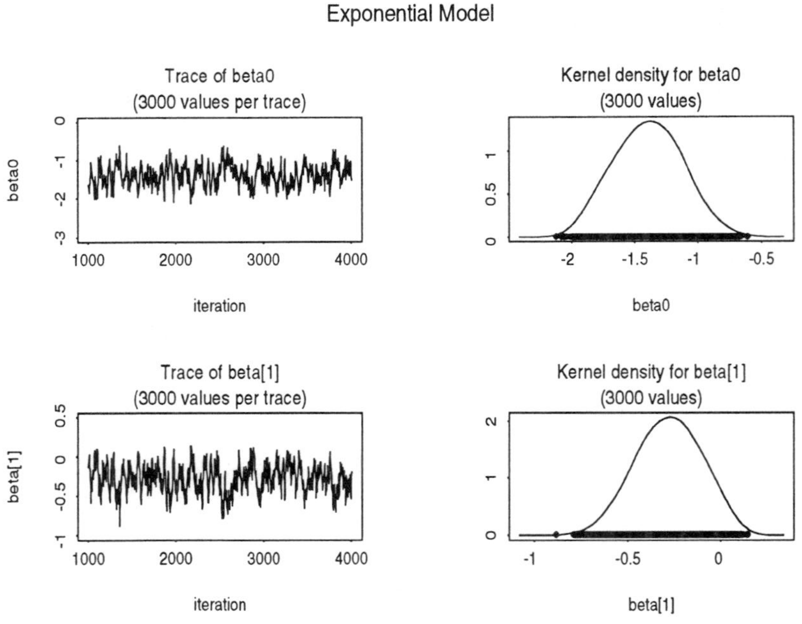

FIGURE 2.1. Trace plots and marginal posterior densities of β_0 and β_1 using the exponential model for E1684 data.

In Table 2.1, 2.5% and 95.5% correspond to the respective posterior percentiles of $\boldsymbol{\beta}$. The 95% credible interval for β_1 is thus $(-0.62, 0.04)$, and most of the mass for the posterior distribution of β_1 is to the left of 0, indicating a treatment effect in favor of IFN. This can be further illustrated in a plot of the marginal posterior density of β_1. This can easily be accomplished with the program CODA (Best, Cowles, and Vines, 1995). Figure 2.1 shows trace plots of the Gibbs samples and marginal posterior densities of β_0 and β_1 using the exponential model. The trace plots in Figure 2.1 also indicate that the Gibbs sampler is mixing well.

TABLE 2.1. Posterior Summaries of $\boldsymbol{\beta}$ Using the Exponential Model for E1684 Data.

Parameter	Mean	SD	2.5%	97.5%	Median
β_0	−1.40	0.26	−1.89	−0.88	−1.39
β_1	−0.28	0.17	−0.62	0.04	−0.27

2.2 Weibull Model

The Weibull model is perhaps the most widely used parametric survival model. Suppose we have independent identically distributed survival times $y = (y_1, y_2, \ldots, y_n)'$, each having a Weibull distribution, denoted by $\mathcal{W}(\alpha, \gamma)$ as defined by (1.4.6). It is often more convenient to write the model in terms of the parameterization $\lambda = \log(\gamma)$, leading to

$$f(y|\alpha, \lambda) = \alpha y^{\alpha-1} \exp(\lambda - \exp(\lambda) y^{\alpha}). \tag{2.2.1}$$

Throughout, we will use the notation $\mathcal{W}(\alpha, \lambda)$, where $\lambda = \log(\gamma)$ to denote the density in (2.2.1). Let $S(y|\alpha, \lambda) = \exp(-\exp(\lambda) y^{\alpha})$ denote the survival function. We can write the likelihood function of (α, λ) as

$$\begin{aligned} L(\alpha, \lambda|D) &= \prod_{i=1}^{n} f(y_i|\alpha, \lambda)^{\nu_i} S(y_i|\alpha, \lambda)^{(1-\nu_i)} \\ &= \alpha^d \exp\left\{ d\lambda + \sum_{i=1}^{n} (\nu_i(\alpha - 1)\log(y_i) - \exp(\lambda) y_i^{\alpha}) \right\}. \end{aligned} \tag{2.2.2}$$

When α is assumed known, the conjugate prior for $\exp(\lambda)$ is the gamma prior. No joint conjugate prior is available when (α, λ) are both assumed unknown. In this case, a typical joint prior specification is to take α and λ to be independent, where α has a gamma distribution and λ has a normal distribution. Letting $\mathcal{G}(\alpha_0, \kappa_0)$ denote a gamma prior for α, and $N(\mu_0, \sigma_0^2)$ denote the normal prior for λ, the joint posterior distribution of (α, λ) is given by

$$\pi(\alpha, \lambda|D) \propto L(\alpha, \lambda|D)\pi(\alpha|\alpha_0, \kappa_0)\pi(\lambda|\mu_0, \sigma_0^2)$$
$$\propto \prod_{i=1}^{n} f(y_i|\alpha, \lambda)^{\nu_i} S(y_i|\alpha, \lambda)^{(1-\nu_i)}$$
$$= \alpha^{\alpha_0+d-1} \exp\left\{ d\lambda + \sum_{i=1}^{n} (\nu_i(\alpha - 1)\log(y_i) - \exp(\lambda)y_i^{\alpha}) - \kappa_0\alpha - \frac{1}{2\sigma_0^2}(\lambda - \mu_0)^2 \right\}. \quad (2.2.3)$$

The joint posterior distribution of (α, λ) does not have a closed form, but it can be shown that the conditional posterior distributions $[\alpha|\lambda, D]$ and $[\lambda|\alpha, D]$ are log-concave (see Exercise 2.4), and thus Gibbs sampling is straightforward for this model.

To build the Weibull regression model, we introduce covariates through λ, and write $\lambda_i = \boldsymbol{x}_i'\boldsymbol{\beta}$. Common prior distributions for $\boldsymbol{\beta}$ include the uniform improper prior, i.e., $\pi(\boldsymbol{\beta}) \propto 1$, and a normal prior. Assuming a $N_p(\boldsymbol{\mu}_0, \Sigma_0)$ prior for $\boldsymbol{\beta}$ and a gamma prior for α, we are led to the joint posterior

$$\pi(\boldsymbol{\beta}, \alpha|D) \propto \alpha^{\alpha_0+d-1} \exp\left\{ \sum_{i=1}^{n} (\nu_i \boldsymbol{x}_i'\boldsymbol{\beta} + \nu_i(\alpha - 1)\log(y_i) - y_i^{\alpha}\exp(\boldsymbol{x}_i'\boldsymbol{\beta})) - \kappa_0\alpha - \frac{1}{2}(\boldsymbol{\beta} - \boldsymbol{\mu}_0)\Sigma_0^{-1}(\boldsymbol{\beta} - \boldsymbol{\mu}_0) \right\}. \quad (2.2.4)$$

Closed forms for the posterior distribution of $\boldsymbol{\beta}$ are generally not available, and therefore one needs to use numerical integration or MCMC methods. Due to the availability of statistical packages such as BUGS, the Weibull regression model in (2.2.4) can easily be fitted using MCMC techniques. For the Weibull regression model, it can be easily shown that the conditional posterior distributions of $[\alpha|\boldsymbol{\beta}, D]$ and $[\boldsymbol{\beta}|\alpha, D]$ are log-concave (see Exercise 2.5), so implementation of the Gibbs sampler is straightforward.

Example 2.2. Melanoma data (Example 2.1 continued). We now consider the melanoma data using a Weibull model, whose likelihood is given by (2.2.2). We use the same priors for $\boldsymbol{\beta}$ as in Example 2.1, and use a $\mathcal{G}(1, 0.001)$ prior for α. The Weibull model with right censored data can be easily fit in BUGS. The BUGS code for this model is given in the website "`http://www.springer-ny.com`." Using a burn-in of 1000 samples, and then an additional 3000 Gibbs samples, we get the following posterior summaries of $\boldsymbol{\beta}$ from BUGS.

From Table 2.2, we see that the posterior estimates of β_0 and β_1 are similar to the exponential model, and the 95% credible interval for β_1 is

TABLE 2.2. Posterior Summaries of $\boldsymbol{\beta}$ and α Using the Weibull Model for E1684 Data.

Parameter	Mean	SD	2.5%	97.5%	Median
β_0	-1.08	0.25	-1.57	-0.60	-1.07
β_1	-0.27	0.15	-0.56	0.03	-0.27
α	0.79	0.06	0.69	0.90	0.79

$(-0.56, 0.03)$. Figure 2.2 shows trace plots and marginal posterior densities of β_0, β_1, and α. Again, the trace plots in Figure 2.2 show good mixing of the Gibbs samples for all three parameters.

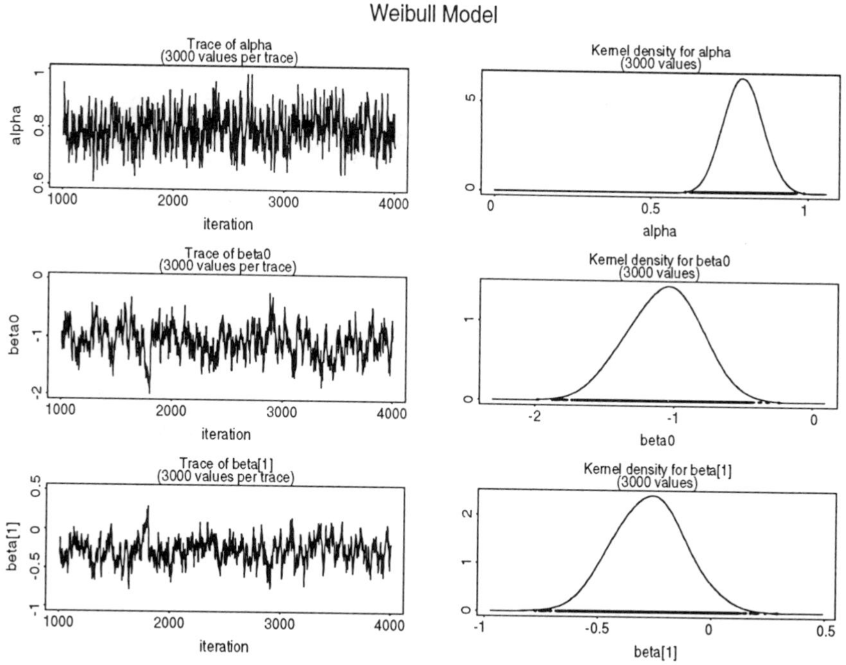

FIGURE 2.2. Trace plots and marginal posterior densities of $\boldsymbol{\beta}$ and α using the Weibull model for E1684 data.

2.3 Extreme Value Model

The extreme value model is another widely used parametric survival model. Suppose we have independent identically distributed survival times $\boldsymbol{y} = (y_1, y_2, \ldots, y_n)'$, each having an extreme value distribution, denoted by

$\mathcal{V}(\alpha, \lambda)$, with density

$$f(y|\alpha, \lambda) = \alpha \exp(\alpha y) \exp(\lambda - \exp(\lambda + \alpha y))$$

for $-\infty < y < \infty$. We note here that the extreme value distribution can be viewed as another parametrization of a Weibull distribution. To see this, let $T \sim \mathcal{W}(\alpha, \lambda)$. Then it can be shown that $Y = \log(T) \sim \mathcal{V}(\alpha, \lambda)$.

Now let $S(y|\alpha, \lambda) = \exp(-\exp(\lambda + \alpha y))$ denote the survival function. We can write the likelihood function of (α, λ) as

$$\begin{aligned} L(\alpha, \lambda|D) &= \prod_{i=1}^{n} f(y_i|\alpha, \lambda)^{\nu_i} S(y_i|\alpha, \lambda)^{(1-\nu_i)} \\ &= \alpha^d \exp\left\{ \sum_{i=1}^{n} (\alpha y_i \nu_i + \lambda \nu_i - \exp(\lambda + \alpha y_i)) \right\}. \end{aligned} \tag{2.3.1}$$

No joint conjugate prior is available when (α, λ) are both assumed unknown. In this case, a typical joint prior specification is to take α and λ to be independent, where α has a gamma distribution and λ has a normal distribution. Letting $\mathcal{G}(\alpha_0, \kappa_0)$ denote the prior for λ, and $N(\mu_0, \sigma_0^2)$ denote the prior for λ, the joint posterior distribution of (α, λ) is given by

$$\begin{aligned} \pi(\alpha, \lambda|D) &\propto L(\alpha, \lambda|D)\pi(\alpha|\alpha_0, \kappa_0)\pi(\lambda|\mu_0, \sigma_0^2) \\ &\propto \prod_{i=1}^{n} f(y_i|\alpha, \lambda)^{\nu_i} S(y_i|\alpha, \lambda)^{(1-\nu_i)} \\ &= \alpha^{\alpha_0+d-1} \exp\left\{ d\lambda + \sum_{i=1}^{n} (\alpha \nu_i y_i - \exp(\lambda + \alpha y_i)) \right. \\ &\qquad \left. - \kappa_0 \alpha - \frac{1}{2\sigma_0^2}(\lambda - \mu_0)^2 \right\}. \end{aligned} \tag{2.3.2}$$

The joint posterior distribution of (α, λ) does not have a closed form, but it can be shown that the conditional posterior distributions of $[\alpha|\lambda, D]$ $[\lambda|\alpha, D]$ are both log-concave, and thus Gibbs sampling is straightforward for this model.

To build the extreme value regression model, we introduce covariates through λ, and write $\lambda_i = \boldsymbol{x}_i'\boldsymbol{\beta}$. Common prior distributions for $\boldsymbol{\beta}$ include the uniform improper prior, i.e., $\pi(\boldsymbol{\beta}) \propto 1$, and a normal prior. Assuming a $N_p(\boldsymbol{\mu}_0, \Sigma_0)$ prior for $\boldsymbol{\beta}$ and a gamma prior for α, we are led to the joint posterior

$$\begin{aligned} \pi(\boldsymbol{\beta}, \alpha|D) \propto \alpha^{\alpha_0+d-1} \exp &\left\{ \sum_{i=1}^{n} [\nu_i(\boldsymbol{x}_i'\boldsymbol{\beta} + \alpha y_i) - \exp(\boldsymbol{x}_i'\boldsymbol{\beta} + \alpha y_i)] \right. \\ &\left. - \kappa_0 \alpha - \frac{1}{2}(\boldsymbol{\beta} - \boldsymbol{\mu}_0)\Sigma_0^{-1}(\boldsymbol{\beta} - \boldsymbol{\mu}_0) \right\}. \end{aligned} \tag{2.3.3}$$

Closed forms for the posterior distribution of $\boldsymbol{\beta}$ are generally not available, and therefore one needs to use numerical integration or MCMC methods. Due to the availability of statistical packages such as BUGS, the extreme value regression model in (2.3.3) can easily be fitted using MCMC techniques. For the extreme value regression model, it can be easily shown that the conditional posterior distributions of $[\alpha|\boldsymbol{\beta}, D]$ and $[\boldsymbol{\beta}|\alpha, D]$ are log-concave (see Exercise 2.6), so implementation of the Gibbs sampler is straightforward.

2.4 Log-Normal Model

Another commonly used parametric survival model is the the log-normal model. For this model, we assume that the logarithms of the survival times are normally distributed. If y_i has a log-normal distribution with parameters (μ, σ^2), denoted by $\mathcal{LN}(\mu, \sigma^2)$, then

$$f(y_i|\mu, \sigma) = (2\pi)^{-1/2}(y_i\sigma)^{-1} \exp\left\{-\frac{1}{2\sigma^2}(\log(y_i) - \mu)^2\right\}.$$

The survival function is given by

$$S(y_i|\mu, \sigma) = 1 - \Phi\left(\frac{\log(y_i) - \mu}{\sigma}\right).$$

We can thus write the likelihood function of (μ, σ) as

$$\begin{aligned} L(\mu, \sigma|D) &= \prod_{i=1}^{n} f(y_i|\mu, \sigma)^{\nu_i} S(y_i|\mu, \sigma)^{(1-\nu_i)} \\ &= (2\pi\sigma^2)^{-d/2} \exp\left\{-\frac{1}{2\sigma^2}\sum_{i=1}^{n} \nu_i(\log(y_i) - \mu)^2\right\} \\ &\quad \times \prod_{i=1}^{n} y_i^{-\nu_i}\left(1 - \Phi\left(\frac{\log(y_i) - \mu}{\sigma}\right)\right)^{1-\nu_i}. \end{aligned} \tag{2.4.1}$$

Let $\tau = 1/\sigma^2$. No joint conjugate prior is available when (μ, τ) are both assumed unknown. In this case, a typical joint prior specification is to take $\mu|\tau \sim N\left(\mu_0, \frac{1}{\tau\tau_0}\right)$ and $\tau \sim \mathcal{G}(\alpha_0/2, \lambda_0/2)$. Under this formulation, the joint posterior distribution of (μ, τ) is given by

$$\begin{aligned} \pi(\mu, \tau|D) &\propto L(\mu, \sigma|D)\pi(\mu, \tau|\mu_0, \tau_0, \alpha_0, \lambda_0) \\ &\propto \tau^{\frac{\alpha_0+d}{2}-1} \exp\left\{-\frac{\tau}{2}\left[\sum_{i=1}^{n} \nu_i(\log(y_i) - \mu)^2 + \tau_0(\mu - \mu_0)^2 + \lambda_0\right]\right\} \\ &\quad \times \prod_{i=1}^{n} y_i^{-\nu_i}\left(1 - \Phi\left(\tau^{1/2}(\log(y_i) - \mu)\right)\right)^{1-\nu_i}. \end{aligned} \tag{2.4.2}$$

The joint posterior distribution of (μ, τ) does not have a closed form. Unlike the parametric models discussed in the earlier sections, sampling from $\pi(\mu, \tau|D)$ is not easy. It can be shown that the conditional posterior distribution of $[\mu|\tau, D]$ is log-concave. However, the conditional posterior distribution of $[\tau|\mu, D]$ is not log-concave in general. Thus, we need to use the Metropolis algorithm to sample from $[\tau|\mu, D]$.

To construct the regression model, we introduce covariates through μ, and write $\mu_i = \boldsymbol{x}_i'\boldsymbol{\beta}$. Common prior distributions for $\boldsymbol{\beta}$ include the uniform improper prior, i.e., $\pi(\boldsymbol{\beta}) \propto 1$, and a normal prior. Assuming $\boldsymbol{\beta}|\tau \sim N_p(\boldsymbol{\mu}_0, \tau^{-1}\Sigma_0)$, the joint posterior for $(\boldsymbol{\beta}, \tau)$ is given by

$$\pi(\boldsymbol{\beta}, \tau|D) \propto \tau^{\frac{\alpha_0+d}{2}-1} \exp\left\{ -\frac{\tau}{2}\Big[\sum_{i=1}^{n} \nu_i(\log(y_i) - \boldsymbol{x}_i'\boldsymbol{\beta})^2 + (\boldsymbol{\beta} - \boldsymbol{\mu}_0)'\Sigma_0^{-1}(\boldsymbol{\beta} - \boldsymbol{\mu}_0) + \lambda_0\Big]\right\} \times \prod_{i=1}^{n} y_i^{-\nu_i} \left(1 - \Phi\left(\tau^{1/2}(\log(y_i) - \boldsymbol{x}_i'\boldsymbol{\beta})\right)\right)^{1-\nu_i}. \quad (2.4.3)$$

Closed forms for the posterior distribution of $\boldsymbol{\beta}$ are generally not available, and therefore one needs to use numerical integration or MCMC methods. For the log-normal regression model, it can be shown that the conditional posterior distribution of $[\boldsymbol{\beta}|\alpha, D]$ is log-concave (see Exercise 2.8). Thus, sampling from $[\boldsymbol{\beta}|\alpha, D]$ is easy. However, to sample from $[\tau|\boldsymbol{\beta}, D]$, we need to use the Metropolis algorithm.

2.5 Gamma Model

The gamma model is a generalization of the exponential model. For this model, $y_i \sim \mathcal{G}(\alpha, \lambda)$ and

$$f(y_i|\alpha, \lambda) = \frac{1}{\Gamma(\alpha)} y_i^{\alpha-1} \exp(\alpha\lambda - y_i \exp(\lambda)).$$

The survival function is given by

$$S(y_i|\alpha, \lambda) = 1 - IG(\alpha, y_i \exp(\lambda)),$$

where

$$IG(\alpha, y_i \exp(\lambda)) = \frac{1}{\Gamma(\alpha)} \int_0^{y_i \exp(\lambda)} u^{\alpha-1} \exp(-u)\, du \quad (2.5.1)$$

is the incomplete gamma function. We can thus write the likelihood function of (α, λ) as

$$\begin{aligned} L(\alpha, \lambda|D) =& \prod_{i=1}^{n} f(y_i|\alpha, \lambda)^{\nu_i} S(y_i|\alpha, \lambda)^{(1-\nu_i)} \\ =& \frac{1}{(\Gamma(\alpha))^d} \exp\left\{ d\alpha\lambda + \sum_{i=1}^{n} \nu_i(\alpha \log(y_i) - y_i \exp(\lambda)) \right\} \\ & \times \prod_{i=1}^{n} y_i^{-\nu_i} (1 - IG(\alpha, y_i \exp(\lambda)))^{1-\nu_i}. \end{aligned} \tag{2.5.2}$$

No joint conjugate prior is available when (α, λ) are both assumed unknown. In this case, a typical joint prior specification is to take $\alpha \sim \mathcal{G}(\alpha_0, \kappa_0)$ and $\lambda \sim N(\mu_0, \sigma_0^2)$ independently. Under this formulation, the joint posterior distribution of (α, λ) is given by

$$\begin{aligned} \pi(\alpha, \lambda|D) \propto& L(\alpha, \lambda|D)\pi(\alpha, \lambda|\alpha_0, \kappa_0, \mu_0, \sigma_0) \\ =& \frac{\alpha^{\alpha_0 - 1}}{(\Gamma(\alpha))^d} \exp\left\{ d\alpha\lambda + \sum_{i=1}^{n} \nu_i(\alpha \log(y_i) - y_i \exp(\lambda)) \right\} \\ & \times \prod_{i=1}^{n} y_i^{-\nu_i} (1 - IG(\alpha, y_i \exp(\lambda)))^{1-\nu_i} \\ & \times \exp\left(-\kappa_0 \alpha - \frac{1}{2\sigma_0^2}(\lambda - \mu_0)^2 \right). \end{aligned} \tag{2.5.3}$$

The joint posterior distribution of (α, λ) does not have a closed form. To simplify the posterior computation, we introduce complete data $\boldsymbol{y}^* = (y_1^*, y_2^*, \ldots, y_n^*)'$ such that $y_i^* = y_i$ if $\nu_i = 1$ and $y_i^* > y_i$ if $\nu_i = 0$. Then the complete data likelihood of (α, λ) given $\boldsymbol{y}^*$ and D takes the form

$$\begin{aligned} L(\alpha, \lambda|\boldsymbol{y}^*, D) =& \prod_{i=1}^{n} f(y_i^*|\alpha, \lambda) \\ =& \frac{1}{(\Gamma(\alpha))^n} \exp\left\{ n\alpha\lambda + \sum_{i=1}^{n} ((\alpha - 1)\log(y_i^*) - y_i^* \exp(\lambda)) \right\}. \end{aligned} \tag{2.5.4}$$

Then, we have

$$L(\alpha, \lambda|D) = \int \cdots \int_{y_i^* > y_i:\ \nu_i = 0} L(\alpha, \lambda|\boldsymbol{y}^*, D) \prod_{i:\ \nu_i = 0} dy_i^*, \tag{2.5.5}$$

where $L(\alpha, \lambda|D)$ is given in (2.5.2) (see Exercise 2.11). Using (2.5.4), the joint posterior distribution of $(\alpha, \lambda, \boldsymbol{y}^*)$ is thus given by

$$\pi(\alpha, \lambda, \boldsymbol{y}^*|D) \propto L(\alpha, \lambda|\boldsymbol{y}^*, D) \times \alpha^{\alpha_0 - 1} \exp\left(-\kappa_0\alpha - \frac{1}{2\sigma_0^2}(\lambda - \mu_0)^2\right). \tag{2.5.6}$$

It can be shown that the conditional posterior distributions $[\alpha|\lambda, \boldsymbol{y}^*, D]$ and $[\lambda|\alpha, \boldsymbol{y}^*, D]$ are log-concave as long as $n \geq 2$ (see Exercise 2.13). In addition, given (α, λ, D), for $\nu_i = 0$, y_i^* has a truncated gamma distribution with density

$$\pi(y_i^*|\alpha, \lambda, D) = \frac{(y_i^*)^{\alpha-1}}{\Gamma(\alpha)(1 - IG(\alpha, y_i\exp(\lambda)))}\exp(\alpha\lambda - y_i^*\exp(\lambda)), \quad y_i^* > y_i, \tag{2.5.7}$$

where $IG(\alpha, y_i \exp(\lambda))$ is the incomplete gamma function defined by (2.5.1). Thus, implementation of the Gibbs sampler is relatively straightforward.

To construct the regression model, we introduce covariates through λ, and write $\lambda_i = \boldsymbol{x}_i'\boldsymbol{\beta}$. Common prior distributions for $\boldsymbol{\beta}$ include the uniform improper prior, i.e., $\pi(\boldsymbol{\beta}) \propto 1$, and a normal prior. Assuming $\boldsymbol{\beta} \sim N_p(\boldsymbol{\mu}_0, \Sigma_0)$, we are led to the joint posterior

$$\begin{aligned}
\pi(\boldsymbol{\beta}, \alpha|D) \propto & \frac{\alpha^{\alpha_0-1}}{(\Gamma(\alpha))^d}\exp\left\{\sum_{i=1}^n \nu_i[\alpha(\boldsymbol{x}_i'\boldsymbol{\beta} + \log(y_i)) - y_i\exp(\boldsymbol{x}_i'\boldsymbol{\beta})]\right\} \\
& \times \prod_{i=1}^n y_i^{-\nu_i}\left(1 - IG(\alpha, y_i\exp(\boldsymbol{x}_i'\boldsymbol{\beta}))\right)^{1-\nu_i} \\
& \times \exp\left(-\kappa_0\alpha - \frac{1}{2}(\boldsymbol{\beta} - \boldsymbol{\mu}_0)'\Sigma_0^{-1}(\boldsymbol{\beta} - \boldsymbol{\mu}_0)\right).
\end{aligned} \tag{2.5.8}$$

For the gamma regression model, by introducing the complete data $\boldsymbol{y}^*$ above, we can easily implement the Gibbs sampler to sample from the joint posterior distribution of $(\boldsymbol{\beta}, \alpha, \boldsymbol{y}^*)$. See Exercise 2.14 for details.

Exercises

2.1 Suppose $y_1, y_2, \ldots, y_n$ is a random sample from an exponential distribution with mean $1/\lambda$. Further suppose that $y_1, y_2, \ldots, y_k$ are fully observed, and $y_{k+1}, \ldots, y_n$ are right censored.

(a) Derive Jeffreys's prior for λ. Note: Jeffrey's prior is defined to be any prior which is proportional to the square root of the Fisher information in λ.

(b) Derive the posterior distribution of λ using the prior in part (a).

(c) Using part (b), derive the predictive distribution of a future observation z, where z is assumed to be fully observed, i.e., not censored.

(d) Consider including a single covariate x_i in the exponential model, so that $\lambda_i = \exp(x_i\beta)$. Suppose $\beta \sim N(\mu_0, \sigma_0^2)$ a priori. Show that the posterior density of β is log-concave.

(e) Let $\theta = \exp(\beta)$ from part (d), and suppose that $\theta \sim \mathcal{G}(\alpha_0, \lambda_0)$, a priori. Derive a closed form expression for the posterior distribution of θ.

2.2 For the exponential model in (2.1.5), suppose we take a p dimensional normal prior for $\boldsymbol{\beta}$ with density $\pi(\boldsymbol{\beta}|\boldsymbol{\mu}_0, \Sigma_0)$, where $\boldsymbol{\mu}_0$ denotes the mean and Σ_0 denotes the covariance matrix. Show that the posterior $\pi(\boldsymbol{\beta}|D)$ for the exponential model given in (2.1.6) is log-concave in $\boldsymbol{\beta}$, i.e., $\frac{\partial^2}{\partial\boldsymbol{\beta}\partial\boldsymbol{\beta}'}\log(\pi(\boldsymbol{\beta}|D))$ is a negative definite matrix.

2.3 Consider the Weibull model given in (2.2.2).

(a) Suppose α is known. Derive the posterior distribution of λ using $\pi(\lambda) \propto \lambda^{-1}$.

(b) Using part (a), derive the predictive distribution of a future observation z, where z is assumed to be fully observed, i.e., not censored.

2.4 Consider the Weibull model given in (2.2.2).

(a) Derive the full conditional distributions of (α, λ) given in (2.2.3).

(b) Show that the full conditional distributions of $[\alpha|\lambda, D]$ and $[\boldsymbol{\beta}|\lambda, D]$ are log-concave.

2.5 Consider the Weibull model given in (2.2.2).

(a) Derive the full conditional distributions of $(\alpha, \boldsymbol{\beta})$ in (2.2.4).

(b) Show that the full conditional distributions of $[\alpha|\boldsymbol{\beta}, D]$ and $[\boldsymbol{\beta}|\alpha, D]$ are log-concave.

(c) Using the melanoma dataset in Example 1.2, fit the Weibull model in `BUGS` with treatment, age, and Breslow thickness as covariates. Also, include an intercept in the model. Use a $N_4(0, 10^4 I)$ prior for $\boldsymbol{\beta}$ and a $\mathcal{G}(1, 0.001)$ prior for α.

(i) What are your conclusions about the effect of each covariate? Compare your analysis with Example 1.2, which used only the single covariate, treatment.

(ii) Do a sensitivity analysis for the prior on $\boldsymbol{\beta}$, by varying the prior variances of the β_j's. What are your conclusions about the robustness of the posterior estimates to the choice of the prior variances of the β_j's?

(iii) Do a sensitivity analysis for the prior on α, by varying the shape and scale hyperparameters. What are your conclusions about the robustness of the posterior estimates?

2.6 Consider the extreme value model in (2.3.1).

(a) Derive the full conditional distributions for all of the parameters for the extreme value model in (2.3.3), and show that the full conditional distributions of $\boldsymbol{\beta}$ are log-concave.
(b) Show that (2.3.3) is a proportional hazards model.
(c) Write a program in BUGS to analyze the melanoma data in Example 1.2 using the extreme value model with only the treatment covariate. Compare your inferences to the exponential and Weibull models.

2.7 Let $\Phi(t)$ denote the $N(0,1)$ cumulative distribution function. Show that $\Phi(t)$ and $1-\Phi(t)$ are log-concave, i.e.,

$$\frac{d^2}{dt^2}\log[\Phi(t)] \le 0 \text{ and } \frac{d^2}{dt^2}\log[1-\Phi(t)] \le 0.$$

2.8 Consider the log normal model and the log normal regression model.

(a) For the joint posterior distribution given in (2.4.2), derive the full conditional posterior distribution of μ and show that it is log-concave.
(b) For the log-normal regression model given in (2.4.3), derive the full conditional posterior distribution of $\boldsymbol{\beta}$ and show that it is log-concave in each component of $\boldsymbol{\beta}$.
Hint: Use the results from Exercise 2.7.

2.9 Consider the log normal model. Let $\boldsymbol{y}^* = (y_1^*, y_2^*, \ldots, y_n^*)'$ denote the complete data from the log-normal model. That is, if $\nu_i = 1$, then $y_i^* = y_i$, and if $\nu_i = 0$, then $y_i^* \ge y_i$.

(a) Derive the complete data likelihood $L(\mu, \tau|\boldsymbol{y}^*, D)$, where $\tau = 1/\sigma^2$.
(b) Show that

$$L(\mu,\tau|D) = \underset{y_i^*>y_i:\ \nu_i=0}{\int \cdots \int} L(\mu,\tau|\boldsymbol{y}^*,D) \prod_{i:\ \nu_i=0} dy_i^*,$$

where $L(\mu, \tau|D)$ is given by (2.4.1).
(c) Suppose we take the same priors for μ and τ as the ones used for the posterior distribution given in (2.4.2). Derive the joint posterior distribution $\pi(\mu, \tau, \boldsymbol{y}^*|D)$.
(d) Derive the full conditional distributions of $[\mu|\tau, \boldsymbol{y}^*, D]$, $[\tau|\mu, \boldsymbol{y}^*, D]$, and $[\boldsymbol{y}^*|\mu, \tau, D]$, and show $[\mu|\tau, \boldsymbol{y}^*, D]$ is normal, $[\tau|\mu, \boldsymbol{y}^*, D]$ is gamma, and for $\nu_i = 0$, $[\boldsymbol{y}_i^*|\mu, \tau, D]$ is a truncated log-normal distribution.

(e) Use part (d) to write a program in BUGS to analyze the melanoma data in Example 1.2 using the log-normal model without any covariates, and take a $\mathcal{G}(1, 0.001)$ prior for τ and a $N(0, \tau^{-1}10^4)$ prior for $[\mu|\tau]$.

2.10 Let $\boldsymbol{y}^* = (y_1^*, y_2^*, \ldots, y_n^*)'$ denote the complete data from the log-normal regression model. That is, $\mu_i = \boldsymbol{x}_i'\boldsymbol{\beta}$, $y_i^* = y_i$ if $\nu_i = 1$, and $y_i^* \geq y_i$ if $\nu_i = 0$.

(a) Suppose $\tau \sim \mathcal{G}(\alpha_0/2, \lambda_0/2)$ and $\boldsymbol{\beta}|\tau \sim N_p(\boldsymbol{\mu}_0, \tau^{-1}\Sigma_0)$. Derive the joint posterior distribution $\pi(\boldsymbol{\beta}, \tau, \boldsymbol{y}^*|D)$.

(b) Obtain the full conditional distributions of $[\boldsymbol{\beta}|\tau, \boldsymbol{y}^*, D]$, $[\tau|\boldsymbol{\beta}, \boldsymbol{y}^*, D]$, and $[\boldsymbol{y}^*|\boldsymbol{\beta}, \tau, D]$.

(c) Using the melanoma dataset in Example 1.2, fit the log-normal regression model in BUGS with treatment, age, and Breslow thickness as covariates. Also, include an intercept in the model. Use a $N_4(0, \tau^{-1}10^4 I)$ prior for $\boldsymbol{\beta}$ and a $\mathcal{G}(1, .001)$ prior for τ. Compare your inferences to the Weibull regression model (see Exercise 2.5).

2.11 Prove (2.5.5).

2.12 Let $\Gamma(\alpha)$ denote the complete gamma function, i.e.,

$$\Gamma(\alpha) = \int_0^\infty t^{\alpha-1}e^{-t}dt.$$

Show that $\log[\Gamma(\alpha)]$ is convex, i.e.,

$$\frac{d^2}{d\alpha^2}\log[\Gamma(\alpha)] > 0, \quad \alpha > 0.$$

2.13 Consider the joint posterior distribution $\pi(\alpha, \lambda, \boldsymbol{y}^*|D)$ given by (2.5.6).

(a) Derive the full conditional distributions of $[\alpha|\lambda, \boldsymbol{y}^*, D]$ and $[\lambda|\alpha, \boldsymbol{y}^*, D]$.

(b) Show that $[\alpha|\lambda, \boldsymbol{y}^*, D]$ is log-concave if $n \geq 2$.

(c) Show that $[\lambda|\alpha, \boldsymbol{y}^*, D]$ is log-concave.

2.14 Consider the gamma regression model in (2.5.8). Let $\boldsymbol{y}^* = (y_1^*, y_2^*, \ldots, y_n^*)'$ denote the complete data from a $\mathcal{G}(\alpha, \exp(\boldsymbol{x}_i\boldsymbol{\beta}))$ distribution, such that $y_i^* = y_i$ if $\nu_i = 1$ and $y_i^* \geq y_i$ if $\nu_i = 0$. Suppose we take a $\mathcal{G}(\alpha_0, \kappa_0)$ prior for α, and a $N_p(\boldsymbol{\mu}_0, \Sigma_0)$ prior for $\boldsymbol{\beta}$.

(a) Derive the complete data likelihood and the joint posterior distribution of $(\boldsymbol{\beta}, \alpha, \boldsymbol{y}^*)$.

(b) Derive the full conditional distributions of $[\boldsymbol{\beta}|\alpha, \boldsymbol{y}^*, D]$, $[\alpha|\boldsymbol{\beta}, \boldsymbol{y}^*, D]$, and $[y_i^*|\boldsymbol{\beta}, \alpha, \boldsymbol{y}^*, D]$ for $\nu_i = 0$.

(c) Show that $[\boldsymbol{\beta}|\alpha, \boldsymbol{y}^*, D]$ is log-concave and $[\alpha|\boldsymbol{\beta}, \boldsymbol{y}^*, D]$ is log-concave if $n \geq 2$.

2.15 For the exponential regression model in (2.1.5), let X^* denote the $n \times p$ matrix with i^{th} row $\nu_i \boldsymbol{x}_i'$. Let $\pi(\boldsymbol{\beta}) \propto 1$. Then, the posterior distribution is given by

$$\pi(\boldsymbol{\beta}|D) \propto L(\boldsymbol{\beta}|D),$$

where $L(\boldsymbol{\beta}|D)$ is given in (2.1.5). Show that the posterior distribution is proper, i.e.,

$$\int L(\boldsymbol{\beta}|D) d\boldsymbol{\beta} < \infty,$$

if and only if X^* is of full rank p.

2.16 Consider the Weibull regression model in (2.2.4). Let X^* denote the $n \times p$ matrix with i^{th} row $\nu_i \boldsymbol{x}_i'$ and $d = \sum_{i=1}^n \nu_i$. Assuming $\pi(\boldsymbol{\beta}) \propto 1$ and $\pi(\alpha|\alpha_0, \kappa_0) \propto \alpha^{\alpha_0 - 1} \exp(-\kappa_0 \alpha)$, show that the posterior distribution $\pi(\boldsymbol{\beta}, \alpha|D)$ is proper if X^* is of full rank p, $\alpha_0 > -d$, and $\kappa_0 > 0$. Note that to obtain a proper posterior, the prior $\pi(\alpha|\alpha_0, \kappa_0)$ need not be proper, since α_0 is allowed to take on negative values.

2.17 For the extreme value regression model, the log-normal regression model, and the gamma regression model, characterize the propriety of the posterior distributions under a uniform improper prior for $\boldsymbol{\beta}$.

2.18 For the predictive distribution in (2.1.3), find

(a) the cumulative distribution function of y_f.
(b) a 95% credible interval for y_f.
(c) a 95% Highest Posterior Density (HPD) interval for y_f, and compare your result to part (b).

3
Semiparametric Models

Nonparametric and semiparametric Bayesian methods in survival analysis have recently become quite popular due to recent advances in computing technology and the development of efficient computational algorithms for implementing these methods. Nonparametric Bayesian methods have now become quite common and well accepted in practice, since they offer a more general modeling strategy that contains fewer assumptions. The literature on nonparametric Bayesian methods has been recently surging, and all of the references are far too enormous to list here. In this chapter, we discuss several types of nonparametric prior processes for the baseline hazard or cumulative hazard, and focus our discussion primarily on the Cox proportional hazards model. Specifically, we examine piecewise constant hazard models, the gamma process, beta process, correlated prior processes, and the Dirichlet process. In each case, we give a development of the prior process, construct the likelihood function, derive the posterior distributions, and discuss MCMC sampling techniques for inference. Several applications involving case studies are given to demonstrate the various methods.

3.1 Piecewise Constant Hazard Model

One of the most convenient and popular models for semiparametric survival analysis is the piecewise constant hazard model. To construct this model, we first construct a finite partition of the time axis, $0 < s_1 < s_2 < \ldots < s_J$, with $s_J > y_i$ for all $i = 1, 2, \ldots, n$. Thus, we have the J intervals $(0, s_1]$,

$(s_1, s_2]$, ..., $(s_{J-1}, s_J]$. In the jth interval, we assume a constant baseline hazard $h_0(y) = \lambda_j$ for $y \in I_j = (s_{j-1}, s_j]$. Let $D = (n, \boldsymbol{y}, X, \nu)$ denote the observed data, where $\boldsymbol{y} = (y_1, y_2, \ldots, y_n)'$, $\boldsymbol{\nu} = (\nu_1, \nu_2, \ldots, \nu_n)'$ with $\nu_i = 1$ if the ith subject failed and 0 otherwise, and X is the $n \times p$ matrix of covariates with i^{th} row $\boldsymbol{x}_i'$. Letting $\boldsymbol{\lambda} = (\lambda_1, \lambda_2, \ldots, \lambda_J)'$, we can write the likelihood function of $(\boldsymbol{\beta}, \boldsymbol{\lambda})$ for the n subjects as

$$L(\boldsymbol{\beta}, \boldsymbol{\lambda}|D) = \prod_{i=1}^{n} \prod_{j=1}^{J} (\lambda_j \exp(\boldsymbol{x}_i'\boldsymbol{\beta}))^{\delta_{ij}\nu_i} \exp\left\{ -\delta_{ij} \left[\lambda_j (y_i - s_{j-1}) + \sum_{g=1}^{j-1} \lambda_g (s_g - s_{g-1}) \right] \exp(\boldsymbol{x}_i'\boldsymbol{\beta}) \right\}, \tag{3.1.1}$$

where $\delta_{ij} = 1$ if the ith subject failed or was censored in the jth interval, and 0 otherwise, $\boldsymbol{x}_i' = (x_{i1}, x_{i2}, \ldots, x_{ip})$ denotes the $p \times 1$ vector of covariates for the ith subject, and $\boldsymbol{\beta} = (\beta_1, \beta_2, \ldots, \beta_p)'$ is the corresponding vector of regression coefficients. The indicator δ_{ij} is needed to properly define the likelihood over the J intervals. The semiparametric model in (3.1.1), sometimes referred to as a *piecewise exponential model*, is quite general and can accommodate various shapes of the baseline hazard over the intervals. Moreover, we note that if $J = 1$, the model reduces to a parametric exponential model with failure rate parameter $\lambda \equiv \lambda_1$. The piecewise exponential model is a useful and simple model for modeling survival data. It serves as the benchmark for comparisons with other semiparametric or fully parametric models for survival data.

A common prior of the baseline hazard $\boldsymbol{\lambda}$ is the independent gamma prior $\lambda_j \sim \mathcal{G}(\alpha_{0j}, \lambda_{0j})$ for $j = 1, 2, \ldots, J$. Here α_{0j} and λ_{0j} are prior parameters which can be elicited through the prior mean and variance of λ_j. Another approach is to build a prior correlation among the λ_j's (Leonard, 1978; Sinha, 1993) using a correlated prior $\boldsymbol{\psi} \sim N(\boldsymbol{\psi}_0, \Sigma_\psi)$, where $\psi_j = \log(\lambda_j)$ for $j = 1, 2, \ldots, J$.

The likelihood in (3.1.1) is based on continuous survival data. The likelihood function based on grouped or discretized survival data is given by

$$L(\boldsymbol{\beta}, \boldsymbol{\lambda}|D) \propto \prod_{j=1}^{J} G_j^*, \tag{3.1.2}$$

where

$$G_j^* = \exp\left\{ -\lambda_j \Delta_j \sum_{k \in \mathcal{R}_j - \mathcal{D}_j} \exp(\boldsymbol{x}_k'\boldsymbol{\beta}) \right\} \times \prod_{l \in \mathcal{D}_j} \left[1 - \exp\{-\lambda_j \Delta_j \exp(\boldsymbol{x}_l'\boldsymbol{\beta})\}\right], \tag{3.1.3}$$

$\Delta_j = s_j - s_{j-1}$, $\mathcal{R}_j$ is the set of patients at risk, and $\mathcal{D}_j$ is the set of patients having failures in the jth interval.

Example 3.1. Melanoma data. We consider the E1690 melanoma data discussed in Example 1.2. We use E1684 as the historical data (D_0), and incorporate the E1684 data via the power prior in (1.7.1). The power prior for (β, λ) for model (3.1.1) is given by

$$\pi(\beta, \lambda, a_0|D_0) \propto L(\beta, \lambda|D_0)^{a_0}\ \pi_0(\beta, \lambda)\ a_0^{\kappa_0 - 1}(1 - a_0)^{\xi_0 - 1}\ , \tag{3.1.4}$$

where (κ_0, ξ_0) are specified hyperparameters for the prior distribution of a_0, $\pi_0(\beta, \lambda)$ is the initial prior distribution for (β, λ), and $L(\beta, \lambda|D_0)$ is the likelihood function in (3.1.1) with D_0 in place of D. For the initial prior $\pi_0(\beta, \lambda)$, we assume that β and λ are independent, where β has a uniform prior and λ has a Jeffreys's prior. This leads to the joint initial improper prior

$$\pi_0(\beta, \lambda) \propto \prod_{j=1}^{J} \lambda_j^{-1}. \tag{3.1.5}$$

We consider the treatment covariate alone in the example here, and thus β is one dimensional. Table 3.1 shows results based on several values of a_0 using the initial prior in (3.1.5). The value $a_0 = 0$ corresponds to a Bayesian analysis of E1690 using noninformative priors, that is, not using any historical data. A value of $a_0 = 1$ corresponds to giving the historical and current data equal weight. In Table 3.1, HR denotes the hazard ratio of OBS to IFN, SD denotes the posterior standard deviation, and 95% HPD denotes 95% Highest Posterior Density intervals. We see from Table 3.1 that as more weight is given to the historical data, the posterior hazard ratios increase and the HPD intervals become narrower and do not include 1. This is reasonable since the posterior hazard ratios based on the E1684 data alone were much larger than E1690 alone, and therefore as more weight is given to E1684, the greater the posterior hazard ratios and the narrower the HPD intervals. Thus, the incorporation of E1684 into the current analysis via the power prior sharpens the assessment between IFN and OBS and leads to more definitive conclusions about the effect of IFN. This example thus demonstrates the effect of incorporating historical data into an analysis.

TABLE 3.1. Posterior Estimates of Hazard Ratio for E1684 Data.

$E(a_0\|D, D_0)$	HR	SD	95% HPD
0	1.30	0.17	(0.99, 1.64)
0.05	1.30	0.16	(0.99, 1.63)
0.30	1.33	0.15	(1.03, 1.63)
1	1.36	0.13	(1.11, 1.62)

3.2 Models Using a Gamma Process

The gamma process is perhaps the most commonly used nonparametric prior process for the Cox model. The seminal paper by Kalbfleisch (1978) describes the gamma process prior for the baseline cumulative hazard function (see also Burridge, 1981). The gamma process can be described as follows. Let $\mathcal{G}(\alpha, \lambda)$ denote the gamma distribution with shape parameter $\alpha > 0$ and scale parameter $\lambda > 0$. Let $\alpha(t), t \geq 0$, be an increasing left continuous function such that $\alpha(0) = 0$, and let $Z(t), t \geq 0$, be a stochastic process with the properties:

(i) $Z(0) = 0$;

(ii) $Z(t)$ has independent increments in disjoint intervals; and

(iii) for $t > s$, $Z(t) - Z(s) \sim \mathcal{G}(c(\alpha(t) - \alpha(s)), c)$.

Then the process $\{Z(t) : t \geq 0\}$ is called a gamma process and is denoted by $Z(t) \sim \mathcal{GP}(c\alpha(t), c)$. We note here that $\alpha(t)$ is the mean of the process and c is a weight or confidence parameter about the mean. The sample paths of the gamma process are almost surely increasing functions. It is a special case of a Levy process whose characteristic function is given by

$$E[\exp\{iy(Z(t) - Z(s)\}] = (\phi(y))^{c(\alpha(t)-\alpha(s))},$$

where ϕ is the characteristic function of an infinitely divisible distribution function with unit mean. The gamma process is the special case $\phi(y) = \{c/(c - iy)\}^c$.

3.2.1 Gamma Process on Cumulative Hazard

Under the Cox model, the joint probability of survival of n subjects given the covariate matrix X is given by

$$P(\mathbf{Y} > y|\boldsymbol{\beta}, X, H_0) = \exp\left\{-\sum_{j=1}^{n} \exp(\boldsymbol{x}_j'\boldsymbol{\beta}) H_0(y_j)\right\}. \qquad (3.2.1)$$

The gamma process is often used as a prior for the cumulative baseline hazard function $H_0(y)$. In this case, we take

$$H_0 \sim \mathcal{GP}(c_0 H^*, c_0), \qquad (3.2.2)$$

where $H^*(y)$ is an increasing function with $H^*(0) = 0$. H^* is often assumed to be a known parametric function with hyperparameter vector γ_0. For example, if H^* corresponds to the exponential distribution, then $H^*(y) = \gamma_0 y$, where γ_0 is a specified hyperparameter. If $H^*(y)$ is taken as Weibull, then $H^*(y) = \eta_0 y^{\kappa_0}$, where $\boldsymbol{\gamma}_0 = (\eta_0, \kappa_0)'$ is a specified vector of hyperparameters. The marginal survival function is given by

$$P(\mathbf{Y} > y|\boldsymbol{\beta}, X, \gamma_0, c_0) = \prod_{j=1}^{n} [\phi(iV_j)]^{c_0(H^*(y_{(j)}) - H^*(y_{(j-1)}))}, \tag{3.2.3}$$

where $V_j = \sum_{l \in \mathcal{R}_j} \exp(x_l' \boldsymbol{\beta})$, $\mathcal{R}_j$ is the risk set at time $y_{(j)}$ and $y_{(1)} < y_{(2)} < \ldots, < y_{(n)}$ are distinct ordered times. For continuous data, when the ordered survival times are all distinct, the likelihood of $(\boldsymbol{\beta}, \gamma_0, c_0)$ can be obtained by differentiating (3.2.3). Note that this likelihood, used by Kalbfleisch (1978), Clayton (1991), and among others, is defined only when the observed survival times are distinct. In the next subsection, we present the likelihood and prior associated with grouped survival data using a gamma process prior for the baseline hazard.

3.2.2 *Gamma Process with Grouped-Data Likelihood*

Again, we construct a finite partition of the time axis, $0 < s_1 < s_2 < \ldots < s_J$, with $s_J > y_i$ for all $i = 1, \ldots, n$. Thus, we have the J disjoint intervals $(0, s_1]$, $(s_1, s_2]$, $\ldots$, $(s_{J-1}, s_J]$, and let $I_j = (s_{j-1}, s_j]$. The observed data D is assumed to be available as grouped within these intervals, such that $D = (X, \mathcal{R}_j, \mathcal{D}_j : j = 1, 2, \ldots, J)$, where $\mathcal{R}_j$ is the risk set and $\mathcal{D}_j$ is the failure set of the jth interval I_j. Let h_j denote the increment in the cumulative baseline hazard in the jth interval, that is

$$h_j = H_0(s_j) - H_0(s_{j-1}), \quad j = 1, 2, \ldots, J.$$

The gamma process prior in (3.2.2) implies that the h_j's are independent and

$$h_j \sim \mathcal{G}(\alpha_{0j} - \alpha_{0,j-1}, c_0), \tag{3.2.4}$$

where $\alpha_{0j} = c_0 H^*(s_j)$, and H^* and c_0 are defined in the previous subsection. Thus, the hyperparameters (H^*, c_0) for h_j consist of a specified parametric cumulative hazard function $H^*(y)$ evaluated at the endpoints of the time intervals, and a positive scalar c_0 quantifying the degree of prior confidence in $H^*(y)$. Now writing $H_0 \sim \mathcal{GP}(c_0 H^*, c_0)$ implies that every disjoint increment in H_0 has the prior given by (3.2.4). Thus, the grouped data representation can be obtained as

$$P(y_i \in I_j|\boldsymbol{h}) = \exp\left\{ -\exp(\boldsymbol{x}_i'\boldsymbol{\beta}) \sum_{k=1}^{j-1} h_k \right\} [1 - \exp\{-h_j \exp(\boldsymbol{x}_i'\boldsymbol{\beta})\}],$$

where $\boldsymbol{h} = (h_1, h_2, \ldots, h_J)'$. This leads to the grouped data likelihood function

$$L(\boldsymbol{\beta}, \boldsymbol{h}|D) \propto \prod_{j=1}^{J} G_j, \tag{3.2.5}$$

where

$$G_j = \exp\Big\{ - h_j \sum_{k \in \mathcal{R}_j - \mathcal{D}_j} \exp(\boldsymbol{x}_k' \boldsymbol{\beta}) \Big\} \prod_{l \in \mathcal{D}_j} \Big[1 - \exp\{-h_j \exp(\boldsymbol{x}_l' \boldsymbol{\beta})\} \Big]. \tag{3.2.6}$$

Note that the grouped data likelihood expression in (3.2.6) is very general and not limited to the case when the h_j's are realizations of a gamma process on H_0. Since the cumulative baseline hazard function H_0 enters the likelihood in (3.2.6) only through the h_j's, our parameters in the likelihood are $(\boldsymbol{\beta}, \boldsymbol{h})$ and thus we only need a joint prior distribution for $(\boldsymbol{\beta}, \boldsymbol{h})$. One important case is that when one considers the piecewise constant baseline hazard of the previous section with $h_j = \Delta_j \lambda_j$ and $\Delta_j = s_j - s_{j-1}$. In this case, we observe a great similarity between the likelihoods (3.1.3) and (3.2.6). In the absence of covariates, (3.2.6) reduces to

$$G_j = \exp\{-h_j(r_j - d_j)\}\{1 - \exp(-h_j)\}^{d_j},$$

where r_j and d_j are the numbers of subjects in the sets $\mathcal{R}_j$ and $\mathcal{D}_j$, respectively.

A typical prior for $\boldsymbol{\beta}$ is a $N_p(\boldsymbol{\mu}_0, \Sigma_0)$ distribution. Thus, the joint posterior of $(\boldsymbol{\beta}, \boldsymbol{h})$ can be written as

$$\pi(\boldsymbol{\beta}, \boldsymbol{h}|D) \propto \prod_{j=1}^{J} \Big[G_j h_j^{(\alpha_{0j} - \alpha_{0,j-1}) - 1} \exp(-c_0 h_j) \Big] \times \exp\left\{ -\frac{1}{2} (\boldsymbol{\beta} - \boldsymbol{\mu}_0) \Sigma_0^{-1} (\boldsymbol{\beta} - \boldsymbol{\mu}_0) \right\}. \tag{3.2.7}$$

To sample from the joint posterior distribution of $(\boldsymbol{\beta}, \boldsymbol{h})$, it can be shown that $[\boldsymbol{\beta}|\boldsymbol{h}, D]$ is log-concave in $\boldsymbol{\beta}$ and thus the adaptive rejection algorithm can be used efficiently to sample the components of $\boldsymbol{\beta}$. Moreover, $[\boldsymbol{h}|\boldsymbol{\beta}, D]$ is also log-concave in the components of $\boldsymbol{h}$. We can thus carry out the following Gibbs sampling scheme:

(i) Sample from

$$\pi(\beta_j|\boldsymbol{\beta}^{(-j)}, \boldsymbol{h}, D) \propto \prod_{j=1}^{J} G_j \exp\left\{ -\frac{1}{2} (\boldsymbol{\beta} - \boldsymbol{\mu}_0) \Sigma_0^{-1} (\boldsymbol{\beta} - \boldsymbol{\mu}_0) \right\},$$

using the adaptive rejection algorithm for $j = 1, 2, \ldots, p$.

(ii) Sample from

$$\pi(h_j|\boldsymbol{h}^{(-j)}, \boldsymbol{\beta}, D) \propto h_j^{(\alpha_{0j} - \alpha_{0,j-1}) - 1} \times \exp\Big\{ - h_j \Big(\sum_{k \in \mathcal{R}_j - \mathcal{D}_j} \exp(\boldsymbol{x}_k' \boldsymbol{\beta}) + c_0 \Big) \Big\}, \tag{3.2.8}$$

where $\boldsymbol{h}^{(-j)}$ denote the $\boldsymbol{h}$ vector without the jth component. The full conditional distribution in (3.2.8) can be well approximated by a gamma distribution, and thus a more efficient Gibbs sampling scheme would be to replace (3.2.8) by

(ii*) Sample from $[\boldsymbol{h}|\boldsymbol{\beta}, D]$ using independent samples from a conditional posterior approximated by

$$h_j \sim \mathcal{G}\left(\alpha_{0j} - \alpha_{0,j-1} + d_j, c_0 + \sum_{k \in \mathcal{R}_j - \mathcal{D}_j} \exp(\boldsymbol{x}_k' \boldsymbol{\beta})\right). \tag{3.2.9}$$

3.2.3 Relationship to Partial Likelihood

In this section, we show that the partial likelihood defined by Cox (1975) can be obtained as a limiting case of the marginal posterior of $\boldsymbol{\beta}$ in the Cox model under a gamma process prior for the cumulative baseline hazard. Towards this goal, discretize the time axis as $(0, s_1]$, $(s_1, s_2]$, ..., $(s_{J-1}, s_J]$, and suppose $H_0 \sim \mathcal{GP}(c_0 H^*, c_0)$. Let $y_{(1)} < y_{(2)} < \ldots < y_{(n)}$ denote the ordered failure or censoring times. Therefore, if $h_j = H_0(y_{(j)}) - H_0(y_{(j-1)})$, then

$$h_j \sim \mathcal{G}(c_0 h_{0j}, c_0), \tag{3.2.10}$$

where $h_{0j} = H^*(y_{(j)}) - H^*(y_{(j-1)})$. Let $A_j = \sum_{l \in \mathcal{R}_j} \exp(\boldsymbol{x}_l' \boldsymbol{\beta})$ and E_{GP} denote expectation with respect to the gamma process prior. Then, we have

$$\begin{aligned} P(\mathbf{Y} > \boldsymbol{y}|X, \boldsymbol{\beta}, H_0) &= \exp\left\{-\sum_{j=1}^{n} \exp(\boldsymbol{x}_j' \boldsymbol{\beta}) H_0(y_{(j)})\right\} \\ &= \exp\left\{-\sum_{j=1}^{n} h_j \sum_{l \in \mathcal{R}_j} \exp(\boldsymbol{x}_l' \boldsymbol{\beta})\right\} \end{aligned}$$

and

$$\begin{aligned} &E_{GP}\Big[P(\mathbf{Y} > \boldsymbol{y}|X, \boldsymbol{\beta}, H_0)|H^*\Big] \\ &\quad = \prod_{j=1}^{n} \left(\frac{c_0}{c_0 + A_j}\right)^{c_0 h_{0j}} \\ &\quad = \prod_{j=1}^{n} \exp\left\{c_0 H^*(y_{(j)}) \log\left(1 - \frac{\exp(\boldsymbol{x}_j' \boldsymbol{\beta})}{c_0 + A_j}\right)\right\}. \end{aligned}$$

Now let $\boldsymbol{\theta} = (\boldsymbol{\beta}', h_0, c_0)'$, where $h_0(y) = \frac{d}{dy}H^*(y)$. We can write the likelihood function as

$$\begin{aligned}
L(\boldsymbol{\theta}|D) &= \prod_{j=1}^{n} \exp\left\{c_0 H^*(y_{(j)}) \log\left(1 - \frac{\exp(\boldsymbol{x}_j'\boldsymbol{\beta})}{c_0 + A_j}\right)\right\} \\
&\quad \times \left\{-c_0 \frac{dH^*(y_{(j)})}{dy_{(j)}} \left(\log\left(1 - \frac{\exp(\boldsymbol{x}_j'\boldsymbol{\beta})}{c_0 + A_j}\right)\right)\right\}^{\nu_i} \\
&= \prod_{j=1}^{n} \exp\left\{H^*(y_{(j)}) \log\left(1 - \frac{\exp(\boldsymbol{x}_j'\boldsymbol{\beta})}{c_0 + A_j}\right)^{c_0}\right\} \\
&\quad \times \left\{-c_0 h_0(y_{(j)}) \log\left(1 - \frac{\exp(\boldsymbol{x}_j'\boldsymbol{\beta})}{c_0 + A_j}\right)\right\}^{\nu_i}.
\end{aligned}$$

Let $d = \sum_{i=1}^{n} \nu_i$ and $h^* = \prod_{j=1}^{n} [h_0(y_{(j)})]^{\nu_i}$. Now we have

$$\lim_{c_0 \to 0} \exp\left\{H_0(y_{(j)}) \log\left(1 - \frac{\exp(\boldsymbol{x}_j'\boldsymbol{\beta})}{c_0 + A_j}\right)^{c_0}\right\} = 0$$

for $j = 1, 2, \ldots, n$, and

$$\lim_{c_0 \to 0} \log\left(1 - \frac{\exp(\boldsymbol{x}_j'\boldsymbol{\beta})}{c_0 + A_j}\right) = \log\left(1 - \frac{\exp(\boldsymbol{x}_j'\boldsymbol{\beta})}{A_j}\right) \approx -\frac{\exp(\boldsymbol{x}_j'\boldsymbol{\beta})}{A_j}$$

for $j = 1, 2, \ldots, n-1$. Thus, we have

$$\lim_{c_0 \to 0} \frac{L(\boldsymbol{\theta}|D)}{c_0^d \log(c_0) h^*} \approx \prod_{j=1}^{n} \left[\frac{\exp(\boldsymbol{x}_j'\boldsymbol{\beta})}{A_j}\right]^{\nu_i}. \tag{3.2.11}$$

We see that the right-hand side of (3.2.11) is precisely Cox's partial likelihood.

Now if we let $c_0 \to \infty$, we get the likelihood function based on $(\boldsymbol{\beta}, h_0)$. To see this, note that

$$\begin{aligned}
&\lim_{c_0 \to \infty} \left[\exp\left\{H^*(y_{(j)}) \log\left(1 - \frac{\exp(\boldsymbol{x}_j'\boldsymbol{\beta})}{c_0 + A_j}\right)^{c_0}\right\}\right. \\
&\quad \left. \times \left\{-c_0 h_0(y_{(j)}) \log\left(1 - \frac{\exp(\boldsymbol{x}_j'\boldsymbol{\beta})}{c_0 + A_j}\right)\right\}^{\nu_j}\right] \\
&= \exp\left\{-H^*(y_{(j)}) \exp(\boldsymbol{x}_j'\boldsymbol{\beta})\right\} \left\{h_0(y_{(j)}) \exp(\boldsymbol{x}_j'\boldsymbol{\beta})\right\}^{\nu_j},
\end{aligned}$$

and therefore,

$$\begin{aligned}
&\lim_{c_0 \to \infty} L(\boldsymbol{\beta}, c_0, h_0|D) \\
&= \prod_{j=1}^{n} \left(\exp\left\{-H^*(y_{(j)}) \exp(\boldsymbol{x}_j'\boldsymbol{\beta})\right\}\right) \left\{h_0(y_{(j)}) \exp(\boldsymbol{x}_j'\boldsymbol{\beta})\right\}^{\nu_j}.
\end{aligned} \tag{3.2.12}$$

Thus, we see that (3.2.12) is the likelihood function of $(\boldsymbol{\beta}, h_0)$ based on the proportional hazards model.

3.2.4 Gamma Process on Baseline Hazard

An alternative specification of the semiparametric Cox model is to specify a gamma process prior on the hazard rate itself. Such a formulation is considered by Dykstra and Laud (1981) in their development of the extended gamma process. Here, we consider a discrete approximation of the extended gamma process. Specifically, we construct the likelihood by using a piecewise constant baseline hazard model and use only information about which interval the failure times fall into. Let $0 = s_0 < s_1 < \ldots < s_J$ be a finite partition of the time axis and let

$$\delta_j = h_0(s_j) - h_0(s_{j-1})$$

denote the increment in the baseline hazard in the interval $(s_{j-1}, s_j]$, $j = 1, 2, \ldots, J$, and $\boldsymbol{\delta} = (\delta_1, \delta_2, \ldots, \delta_J)'$. We follow Ibrahim, Chen, and MacEachern (1999) for constructing the approximate likelihood function of $(\boldsymbol{\beta}, \boldsymbol{\delta})$. For an arbitrary individual in the population, the survival function for the Cox model at time y is given by

$$\begin{aligned} S(y|\boldsymbol{x}) &= \exp\left\{-\eta \int_0^y h_0(u)\, du\right\} \\ &\approx \exp\left\{-\eta \left(\sum_{i=1}^{J} \delta_i (y - s_{i-1})^+\right)\right\}, \end{aligned} \tag{3.2.13}$$

where $h_0(0) = 0$, $(u)^+ = u$ if $u > 0$, 0 otherwise, and $\eta = \exp(\boldsymbol{x}'\boldsymbol{\beta})$. This first approximation arises since the specification of $\boldsymbol{\delta}$ does not specify the entire hazard rate, but only the δ_j. For purposes of approximation, we take the increment in the hazard rate, δ_j, to occur immediately after s_{j-1}. Let p_j denote the probability of a failure in the interval $(s_{j-1}, s_j]$, $j = 1, 2, \ldots, J$. Using (3.2.13), we have

$$\begin{aligned} p_j =& S(s_{j-1}) - S(s_j) \\ \approx& \exp\left\{-\eta \sum_{l=1}^{j-1} \delta_l (s_{j-1} - s_{l-1})\right\} \left[1 - \exp\left\{-\eta (s_j - s_{j-1}) \sum_{l=1}^{j} \delta_l\right\}\right]. \end{aligned}$$

Thus, in the jth interval $(s_{j-1}, s_j]$, the contribution to the likelihood function for a failure is p_j, and $S(s_j)$ for a right censored observation. For $j = 1, 2, \ldots, J$, let d_j be the number of failures, $\mathcal{D}_j$ be the set of subjects failing, c_j be the number of right censored observations and $\mathcal{C}_j$ is the set of subjects that are censored. Also, let $D = (n, \boldsymbol{y}, X, \boldsymbol{\nu})$ denote the data. The grouped data likelihood function is thus given by

$$L(\boldsymbol{\beta}, \boldsymbol{\delta}|D) = \prod_{j=1}^{J} \left\{ \exp\{-\delta_j(a_j + b_j)\} \prod_{k \in \mathcal{D}_j} [1 - \exp\{-\eta_k T_j\}] \right\}, \quad (3.2.14)$$

where $\eta_k = \exp(\boldsymbol{x}_k' \boldsymbol{\beta})$,

$$a_j = \sum_{l=j+1}^{J} \sum_{k \in \mathcal{D}_l} \eta_k (s_{l-1} - s_{j-1}), \ b_j = \sum_{l=j}^{M} \sum_{k \in \mathcal{C}_l} \eta_k (s_l - s_{j-1}), \quad (3.2.15)$$

and

$$T_j = (s_j - s_{j-1}) \sum_{l=1}^{j} \delta_l. \quad (3.2.16)$$

We note that this likelihood involves a second approximation. Instead of conditioning on exact event times, we condition on the set of failures and set of right censored events in each interval, and thus we approximate continuous right censored data by grouped data.

3.3 Prior Elicitation

In this section, we demonstrate the use of the power prior for the gamma process model of Subsection 3.2.4. Let $D_0 = (n_0, \boldsymbol{y}_0, X_0, \boldsymbol{\nu}_0)$ denote the data from the previous study (i.e., historical data), where n_0 denotes the sample size of the previous study, y_0 denotes a right censored vector of survival times with censoring indicators $\boldsymbol{\nu}_0$, and X_0 denotes the $n \times p$ matrix of covariates. In general, for most problems, there are no firm guidelines on the method of prior elicitation. Typically, one tries to balance sound theoretical ideas with practical and computationally feasible ones. The issue of how to use D_0 for the current study in survival analysis has no obvious solution since it depends in large part of how similar the two studies are. In most clinical trials, for example, no two studies will ever be identical. In many cancer clinical trials, the patient populations typically differ from study to study even when the same regimen is used to treat the same cancer. In addition, other factors may make the two studies heterogeneous. These include conducting the studies at different institutions or geographical locations, using different physicians, using different measurement instruments and so forth. Due to these differences, an analysis which combines the data from both studies may not be desirable. In this case, it may be more appropriate to "weight" the data from the previous study so as to control its impact on the current study. Thus, it is desirable for the investigators to have a prior distribution that summarizes the prior data in an efficient and useful manner and allows them to tune or weight D_0 as they see fit in order to control its impact on the current study.

Let $\pi_0(\boldsymbol{\beta}, \boldsymbol{\delta})$ denote the initial prior for $(\boldsymbol{\beta}, \boldsymbol{\delta})$. The power prior distribution for $(\boldsymbol{\beta}, \boldsymbol{\delta})$ is given by

$$\pi(\boldsymbol{\beta}, \boldsymbol{\delta}|D_0, a_0) \propto \{L(\boldsymbol{\beta}, \boldsymbol{\delta}|D_0)\}^{a_0} \pi_0(\boldsymbol{\beta}, \boldsymbol{\delta}), \tag{3.3.1}$$

where $L(\boldsymbol{\beta}, \boldsymbol{\delta}|D_0)$ is the likelihood function of $(\boldsymbol{\beta}, \boldsymbol{\delta})$ based on the historical data D_0 and thus, $L(\boldsymbol{\beta}, \boldsymbol{\delta}|D_0)$ is (3.2.14) with D replaced by $D_0 = (n_0, \boldsymbol{y}_0, X_0, \boldsymbol{\nu}_0)$. To simplify the prior specification, we take

$$\pi_0(\boldsymbol{\beta}, \boldsymbol{\delta}) = \pi_0(\boldsymbol{\beta}|c_0)\pi_0(\boldsymbol{\delta}|\boldsymbol{\theta}_0),$$

where c_0 and $\boldsymbol{\theta}_0$ are fixed hyperparameters. Specifically, we take $\pi_0(\boldsymbol{\beta}|c_0)$ to be a p dimensional multivariate normal density, $N_p(0, c_0W_0)$, with mean 0 and covariance matrix c_0W_0, where c_0 is a specified scalar and W_0 is a $p \times p$ diagonal matrix. We take $\pi_0(\boldsymbol{\delta}|\boldsymbol{\theta}_0)$ to be a product of M independent gamma densities, each with mean f_{0i}/g_{0i} and variance f_{0i}/g_{0i}^2, $i = 1, 2, \ldots, M$. So, we get

$$\pi_0(\boldsymbol{\delta}|\boldsymbol{\theta}_0) \propto \prod_{i=1}^{M} \delta_i^{f_{0i}-1} \exp\{-\delta_i g_{0i}\}, \tag{3.3.2}$$

where $\boldsymbol{\theta}_0 = (f_{01}, g_{01}, \ldots, f_{0M}, g_{0M})'$.

The motivation for (3.3.2) is that it is a discrete approximation to an underlying gamma process prior for $h_0(y)$. Ibrahim, Chen, and MacEachern (1999) show numerically that this type of prior elicitation is quite robust in many situations. It is important to note that this prior elicitation method can be also adapted to other popular discretized prior processes such as the discretized gamma process for the cumulative baseline hazard (Sinha and Dey, 1997), discretized beta processes for the baseline hazard rate (Sinha, 1997) and the discretized correlated process on $h_0(y)$ (Arjas and Gasbarra, 1994). The modification required for these other processes would be an appropriate selection of $\pi_0(\boldsymbol{\delta}|\theta_0)$. We note that in (3.3.1), $(\boldsymbol{\beta}, \boldsymbol{\delta})$ are not independent, and also the components of $\boldsymbol{\delta}$ are not independent a priori.

The prior specification is completed by specifying a prior for a_0. We specify a beta prior for a_0 $(0 \le a_0 \le 1)$, so that

$$\pi(a_0|\alpha_0, \lambda_0) \propto a_0^{\alpha_0-1}(1-a_0)^{\lambda_0-1}, \tag{3.3.3}$$

thus obtaining the joint prior

$$\pi(\boldsymbol{\beta}, \boldsymbol{\delta}, a_0|D_0) \propto L(\boldsymbol{\beta}, \boldsymbol{\delta}|D_0)^{a_0}\pi_0(\boldsymbol{\beta}|c_0)\pi_0(\boldsymbol{\delta}|\boldsymbol{\theta}_0)\pi(a_0|\alpha_0, \lambda_0). \tag{3.3.4}$$

3.3.1 Approximation of the Prior

One of the potential drawbacks of (3.3.4) is that it does not have a closed form and may be computationally intensive. For these reasons, it may be desirable to approximate (3.3.4) in a suitable way. We begin by assuming a priori independence between the baseline hazard rate and the regression coefficients, and thus the joint prior density of $(\boldsymbol{\beta}, \boldsymbol{\delta})$ is given by

$$\pi(\boldsymbol{\beta}, \boldsymbol{\delta}|D_0) = \pi(\boldsymbol{\beta}|D_0)\pi(\boldsymbol{\delta}|D_0). \tag{3.3.5}$$

The assumption of prior independence between $\boldsymbol{\beta}$ and $\boldsymbol{\delta}$ is a sensible specification, since the hazard rate may be a nuisance parameter for many problems. We consider a fully parametric p dimensional multivariate normal prior for $\boldsymbol{\beta}$ given by

$$(\boldsymbol{\beta}|a_0, \boldsymbol{\mu}_0, T_0) \sim N_p(\boldsymbol{\mu}_0, a_0^{-1}T_0^{-1}), \tag{3.3.6}$$

where $\boldsymbol{\mu}_0$ is the prior mean, $a_0 T_0$ is the precision matrix, and a_0 is defined in the previous subsection. We take $\boldsymbol{\mu}_0$ to be the solution to Cox's partial likelihood (Cox, 1975) equations for $\boldsymbol{\beta}$ using D_0 as data. Suppose there are d_0 failures and $n_0 - d_0$ right censored values in $\boldsymbol{y}_0$. Cox's partial likelihood for $\boldsymbol{\beta}$ based on D_0 is given by

$$\mathrm{PL}(\boldsymbol{\beta}|D_0) = \prod_{i=1}^{d_0} \left\{ \frac{\exp\{\boldsymbol{x}_{0i}'\boldsymbol{\beta}\}}{\sum_{\ell \in \mathcal{R}(y_{0i})} \exp\{\boldsymbol{x}_{0\ell}'\boldsymbol{\beta}\}} \right\}, \tag{3.3.7}$$

where $\boldsymbol{x}_{0i}'$ is the ith row of X_0, $(y_{01}, \dots, y_{0d_0})$ are the ordered failures and $\mathcal{R}(y_{0i})$ is the set of labels attached to the individuals at risk just prior to y_{0i}. Now we take $\boldsymbol{\mu}_0$ to be the solution to

$$\frac{\partial \log\left(\mathrm{PL}(\boldsymbol{\beta}|D_0)\right)}{\partial \boldsymbol{\beta}} = 0. \tag{3.3.8}$$

The matrix T_0 is taken to be the Fisher information matrix of $\boldsymbol{\beta}$ based on the partial likelihood in (3.3.7). Thus

$$T_0 = \left[\frac{-\partial^2}{\partial \boldsymbol{\beta} \partial \boldsymbol{\beta}'} \log(\mathrm{PL}(\boldsymbol{\beta}|D_0)) \right] \Bigg|_{\boldsymbol{\beta}=\boldsymbol{\mu}_0}. \tag{3.3.9}$$

These priors represent a summary of the historical data D_0 through $(\boldsymbol{\mu}_0, T_0)$ which are obtained via Cox's partial likelihood. This is a practical and useful summary of the data D_0 as indicated by many authors, including Cox (1972, 1975), and Tsiatis (1981a).

We emphasize that since our main goal is to make inferences on $\boldsymbol{\beta}$, the parameter $\boldsymbol{\delta}$ is viewed as a nuisance parameter, and thus a noninformative prior for $\pi(\boldsymbol{\delta}|a_0, D_0)$ is suitable for it. Otherwise we can take $\pi(\boldsymbol{\delta}|a_0, D_0)$ as a product of M independent gamma densities with mean ϕ_i and variance $a_0^{-1}\gamma_i$ for $i = 1, 2, \dots, M$. The ϕ_i's and γ_i's can be elicited from D_0 by taking $\phi_i = \hat{\delta}_{0i}$ and $\gamma_i = \mathrm{Var}(\hat{\delta}_{0i})$ where $\hat{\delta}_{0i}$ is the MLE and $\mathrm{Var}(\hat{\delta}_{0i})$ is the corresponding estimated variance of δ_i calculated from the profile likelihood $L_c(\boldsymbol{\delta}|D_0, \boldsymbol{\beta} = \boldsymbol{\mu}_0)$. Here $L_c(\boldsymbol{\delta}, \boldsymbol{\beta}|D_0)$ is proportional to the probability of observing the grouped survival data D_0 given $(\boldsymbol{\beta}, \boldsymbol{\delta})$.

The prior given by (3.3.5) and (3.3.6) has several advantages. First, it has a closed form and is easy to interpret. Second, the prior elicitation is straightforward in the sense that $(\boldsymbol{\mu}_0, T_0)$ and a_0 completely determine the prior for $\boldsymbol{\beta}$. Third, (3.3.5) and (3.3.6) are computationally feasible and relatively simple. Fourth, our prior assumes a priori independence between $(\boldsymbol{\beta}, \boldsymbol{\delta})$, which further simplifies interpretations as well as the elicitation scheme. In addition, as discussed in Ibrahim, Chen, and MacEachern (1999), (3.3.5) and (3.3.6) provide a reasonable asymptotic approximation to the prior for $(\boldsymbol{\beta}, \boldsymbol{\delta})$ given in (3.3.4).

3.3.2 Choices of Hyperparameters

The choices of the prior parameters play a crucial role in any Bayesian analysis. We first discuss (α_0, λ_0). For the purposes of prior elicitation, it is easier to work with $\mu_{a_0} = \alpha_0/(\alpha_0 + \lambda_0)$ and $\sigma^2_{a_0} = \mu_{a_0}(1 - \mu_{a_0})(\alpha_0 + \lambda_0 + 1)^{-1}$. A uniform prior (i.e., $\alpha_0 = \lambda_0 = 1$), which corresponds to $(\mu_{a_0}, \sigma^2_{a_0}) = (1/2, 1/12)$, may be a suitable noninformative starting point and facilitates a useful reference analysis for other choices. The investigator may choose μ_{a_0} to be small (say, $\mu_{a_0} \leq 0.1$), if he/she wishes to have low prior weight on the historical data. If a large prior weight is desired, then $\mu_{a_0} \geq 0.5$ may be desirable. In any case, in an actual analysis, we recommend that several choices of $(\mu_{a_0}, \sigma^2_{a_0})$ be used, including ones that give small and large weight to the historical data, and several sensitivity analyses conducted. The choices recommended here can be used as starting points from which sensitivity analyses can be based.

It is reasonable to specify a noninformative prior for $\pi_0(\boldsymbol{\beta}|c_0)$ since this is the prior for $\boldsymbol{\beta}$ corresponding to the previous study and contains no information about the historical data D_0. The quantity $c_0 \geq 0$ is a scalar variance parameter which serves to control the impact of $\pi_0(\boldsymbol{\beta}|c_0)$ on $\pi(\boldsymbol{\beta}, \boldsymbol{\delta}, a_0|D_0)$. To make $\pi_0(\boldsymbol{\beta}|c_0)$ noninformative, we take large values of c_0 so that $\pi_0(\boldsymbol{\beta}|c_0)$ is flat relative to $L(\boldsymbol{\beta}, \boldsymbol{\delta}|D_0)^{a_0}$. Small values of c_0 will let $\pi_0(\boldsymbol{\beta}|c_0)$ dominate (3.3.4). The actual size of c_0 used will depend on the structure of the data set and the prior parameters for a_0. In Example 3.2, reasonable choices of c_0 are $c_0 \geq 3$. In any case, we do not recommend an automatic one-time specification for c_0, but rather that several sensitivity analyses be conducted with several values of c_0 to examine the impact of $\pi_0(\boldsymbol{\beta}|c_0)$ on posterior quantities of interest. The matrix W_0 has a less crucial role than c_0. We take W_0 to be a diagonal matrix consisting of the sample variances of the covariates. The purpose of picking W_0 in this way is to adjust properly for the different scales of the measured covariates. If the covariates are all standardized or are measured on the same scale, then we take $W_0 = I$. In any case, W_0 plays a minimal role when c_0 is taken large.

For $\pi_0(\boldsymbol{\delta}|\boldsymbol{\theta}_0)$, the values of $\boldsymbol{\theta}_0$ should be chosen so that $\pi_0(\boldsymbol{\delta}|\boldsymbol{\theta}_0)$ is flat relative to $L(\boldsymbol{\beta}, \boldsymbol{\delta}|D_0)^{a_0}$. Here, we choose f_{0i} to be proportional to the interval widths, i.e., $f_{0i} \propto s_i - s_{i-1}$, and $g_{0i} \to 0$, for $i = 1, \ldots, M$. Choosing g_{0i} small creates a noninformative gamma prior, and choosing the shape parameters to be proportional to the length of each interval is useful in allowing the variance of each δ_i to depend on the interval length.

3.3.3 *Sampling from the Joint Posterior Distribution of* $(\boldsymbol{\beta}, \boldsymbol{\delta}, a_0)$

The joint posterior density of $(\boldsymbol{\beta}, \boldsymbol{\delta}, a_0|D)$ for the current study is given by

$$\pi(\boldsymbol{\beta}, \boldsymbol{\delta}, a_0|D) \propto \left[\prod_{j=1}^{M}\left\{\exp\{-\delta_j(a_j+b_j)\}\prod_{k\in\mathcal{D}_j}\{1-\exp(-\eta_k T_j)\}\right\}\right]$$
$$\times L(\boldsymbol{\beta}, \boldsymbol{\delta}|D_0)^{a_0}\pi_0(\boldsymbol{\beta}|c_0)\pi_0(\boldsymbol{\delta}|\theta_0)\pi(a_0|\alpha_0, \lambda_0), \qquad (3.3.10)$$

where η_k, a_i, b_i, and T_j are given by (3.2.14), (3.2.15), and (3.2.16), and $L(\boldsymbol{\beta}, \boldsymbol{\delta}|D_0)$ is the likelihood function given by (3.2.14) with D replaced by D_0. In (3.3.10), we use the same subintervals $(s_{i-1}, s_i]$ for both $L(\boldsymbol{\beta}, \boldsymbol{\delta}|D)$ and $L(\boldsymbol{\beta}, \boldsymbol{\delta}|D_0)$ so that $\boldsymbol{\delta}$ has the same meaning across both the historical data and the current data. To specify the s_i's, (i) we combine $\{y_i, i = 1, 2, \ldots, n\}$ and $\{y_{0i}, i = 1, 2, \ldots, n_0\}$ together, denoted by $\{y_i^*, i = 1, 2, \ldots, n+n_0\}$; (ii) the subintervals $(s_{i-1}, s_i]$ are chosen to have equal numbers of failure or censored observations for the combined survival or censored time y_i^*'s, i.e., s_i is chosen to be the $(i/M)^{th}$ quantile of the y_i^*'s, where M is the total number of subintervals. We take M so that in each subinterval $(s_{i-1}, s_i]$, there is at least one failure or censored observation from the y_{0i}'s and the y_i's.

We now describe a Gibbs sampling strategy for sampling from the full conditionals of $[\boldsymbol{\beta}|\boldsymbol{\delta}, a_0, D]$, $[\boldsymbol{\delta}|\boldsymbol{\beta}, a_0, D]$, and $[a_0|\boldsymbol{\beta}, \boldsymbol{\delta}, D]$.

To sample $\boldsymbol{\beta}$ from its full conditional distribution, we first observe that the density function of $[\boldsymbol{\beta}|\boldsymbol{\delta}, a_0, D]$ is log-concave in each component of $\boldsymbol{\beta}$. Therefore, we may directly use the adaptive rejection algorithm to sample from $[\boldsymbol{\beta}|\boldsymbol{\delta}, a_0, D]$. Let $\pi(\boldsymbol{\delta}|\boldsymbol{\beta}, a_0, D)$ denote the conditional posterior density of $\boldsymbol{\delta}$. As for $\pi(\boldsymbol{\beta}|\boldsymbol{\delta}, a_0, D)$, it can be shown that $\pi(\boldsymbol{\delta}|\boldsymbol{\beta}, a_0, D)$ is log-concave in each component of $\boldsymbol{\delta}$ as long as $\pi_0(\boldsymbol{\delta}|\boldsymbol{\theta}_0)$ is log-concave in each component of $\boldsymbol{\delta}$. Thus, we may use the adaptive rejection algorithm again to sample $\boldsymbol{\delta}$. Note that sampling $\boldsymbol{\delta}$ is slightly different from the one for $\boldsymbol{\beta}$, that is, $\delta_i \geq 0$ while $-\infty < \beta_r < \infty$. To ensure log-concavity of $\pi(\boldsymbol{\delta}|\boldsymbol{\theta}_0)$, we need to take $f_{0i} \geq 1$.

The generation of a_0 does not depend on the data D from the current study. The conditional posterior density of a_0 can be written as

$$\pi(a_0|\boldsymbol{\beta}, \boldsymbol{\delta}, D) \propto L(\boldsymbol{\beta}, \boldsymbol{\delta}|D_0)^{a_0}\ \pi(a_0|\alpha_0, \lambda_0). \qquad (3.3.11)$$

In general, the full conditional posterior density of a_0 is not log-concave. Therefore, we use an adaptive Metropolis algorithm to sample a_0 as proposed in Chen, Ibrahim, and Yiannoutsos (1999). The algorithm proceeds as follows. Consider

$$a_0 = \frac{\exp(\xi)}{1 + \exp(\xi)}, \tag{3.3.12}$$

then the conditional posterior distribution $[\xi|\boldsymbol{\beta}, \boldsymbol{\delta}, D]$ is

$$\pi(\xi|\boldsymbol{\beta}, \boldsymbol{\delta}, D) \propto \pi(a_0|\boldsymbol{\beta}, \boldsymbol{\delta}, D)\frac{\exp(\xi)}{(1 + \exp(\xi))^2}, \tag{3.3.13}$$

where $\pi(a_0|\boldsymbol{\beta}, \boldsymbol{\delta}, D)$ is given by (3.3.11) and a_0 is evaluated at $a_0 = \exp(\xi)/(1 + \exp(\xi))$. Instead of directly generating a_0 from (3.3.11), we first generate ξ from (3.3.13) and then use (3.3.12) to obtain a_0. To generate ξ, we use a normal proposal $N(\hat{\xi}, \hat{\tau}^2_{\hat{\xi}})$, where $\hat{\xi}$ is the maximizer of the logarithm of the right side of (3.3.13), which can be obtained by the Nelder-Mead algorithm implemented by O'Neill (1971). Also, $\hat{\tau}^2_{\hat{\xi}}$ is minus the inverse of the second derivative of $\log \pi(\xi|\boldsymbol{\beta}, \boldsymbol{\delta}, D)$ evaluated at $\xi = \hat{\xi}$, given by

$$\hat{\tau}^{-2}_{\hat{\xi}} = -\left.\frac{d^2 \log \pi(\xi|\boldsymbol{\beta}, \boldsymbol{\delta}, D)}{d\xi^2}\right|_{\xi=\hat{\xi}}.$$

The algorithm to generate ξ operates as follows:

Step 1. Let ξ be the current value.

Step 2. Generate a proposal value ξ^* from $N(\hat{\xi}, \hat{\tau}^2_{\hat{\xi}})$.

Step 3. A move from ξ to ξ^* is made with probability

$$\min\left\{\frac{\pi(\xi^*|\boldsymbol{\beta}, \boldsymbol{\delta}, D)\phi\left(\frac{\xi-\hat{\xi}}{\hat{\tau}_{\hat{\xi}}}\right)}{\pi(\xi|\boldsymbol{\beta}, \boldsymbol{\delta}, D)\phi\left(\frac{\xi^*-\hat{\xi}}{\hat{\tau}_{\hat{\xi}}}\right)}, 1\right\},$$

where ϕ is the standard normal probability density function.

After we obtain ξ, we compute a_0 by using (3.3.12).

Example 3.2. Myeloma data. We examine in more detail the multiple myeloma study E2479 discussed in Example 1.1. Our main goal in this example is to illustrate the behavior of the power prior and the sensitivity of the posterior estimates to the choice of a_0 and c_0. The current dataset is E2479 (Study 2) and the historical dataset, which consists of a similar study in multiple myeloma conducted several years earlier, is labeled Study 1. Two superimposed Kaplan-Meier plots for the two studies are displayed in Figure 3.1. A total of $n = 339$ observations were available from E2479, with 8 observations being right censored. Our analysis used $p = 8$ covariates. These are blood urea nitrogen (x_1), hemoglobin (x_2), platelet count (x_3) (1 if normal, 0 if abnormal), age (x_4), white blood cell count (x_5), bone fractures (x_6), percentage of the plasma cells in bone marrow (x_7), and serum calcium (x_8). To ease the computational burden, we standardized all of the variables. The standardization helped the numerical stability in the implementation of the adaptive rejection algorithm for sampling the regression coefficients from the posterior distribution. Study 1 consisted of $n_0 = 65$ observations of which 17 were right censored.

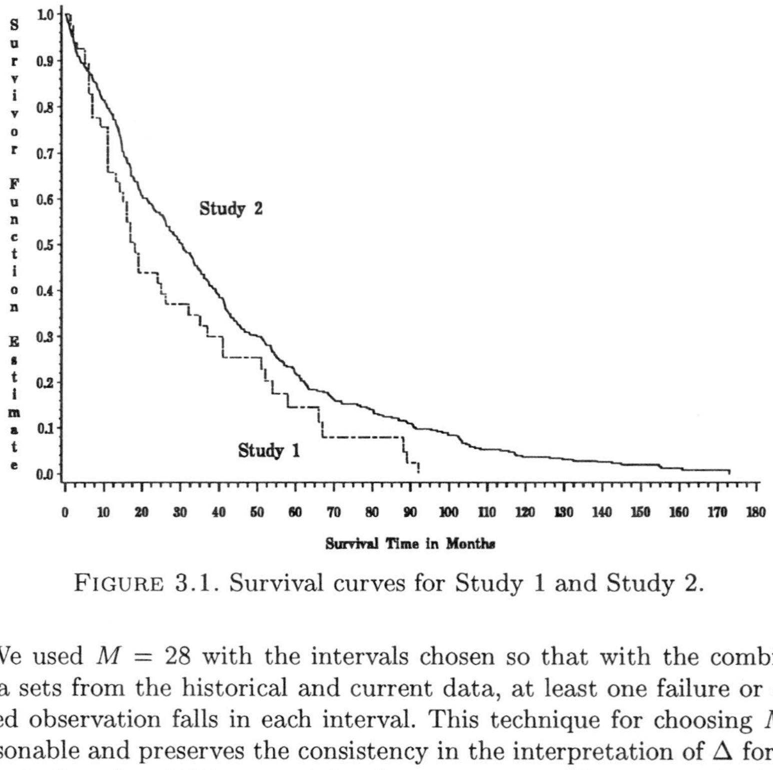

FIGURE 3.1. Survival curves for Study 1 and Study 2.

We used $M = 28$ with the intervals chosen so that with the combined data sets from the historical and current data, at least one failure or censored observation falls in each interval. This technique for choosing M is reasonable and preserves the consistency in the interpretation of Δ for the two studies. In addition, we take $f_{0i} = s_i - s_{i-1}$ if $s_i - s_{i-1} \geq 1$ and $f_{0i} = 1.1$ if $s_i - s_{i-1} < 1$, and $g_{0i} = 0.001$. For the last interval, we take $g_{0i} = 10$ for $i = M$ since very little information in the data is available for this last interval. The above choices of f_{0i} and g_{0i} ensure the log-concavity of $\pi_0(\delta \mid \theta_0)$, as this is required in sampling δ from its conditional prior and posterior distributions. To obtain the posterior estimates of $\boldsymbol{\beta}$, 50,000 Gibbs iterations were used after convergence. Tables 3.2 and 3.3 show the posterior estimates of $\boldsymbol{\beta}$ under the prior parameters $(\mu_{a_0}, \sigma_{a_0}) = (0,0)$, $(0.5, 0.06)$, $(1,0)$, and $c_0 = 3$, 10. We note that $\sigma_{a_0} = 0$ implies $a_0 = 0$ or $a_0 = 1$ with probability 1. From these two tables, it can be seen that the posterior means of $\boldsymbol{\beta}$ are very similar, and the posterior standard deviations are slightly smaller when μ_{a_0} is getting larger. Thus, the results are not too sensitive to these values of $(\mu_{a_0}, \sigma_{a_0})$ and c_0. This may be partially explained by the relatively small sample size of the historical data. From Table 3.2, we see that as a_0 increases, the posterior standard deviations of the regression coefficients decrease, and thus the precision in the estimates is improved. Thus, one of the advantages in incorporating historical data is that it increases the precision in estimation. Another feature in Table 3.2 is that the estimates are remarkably robust with respect to the choices of $(\mu_{a_0}, \sigma_{a_0})$ and c_0, thus revealing a desirable property in the power prior.

TABLE 3.2. Posterior Estimates of $\boldsymbol{\beta}$ for Myeloma Data with $c_0 = 3$.

Variable	Posterior Mean			Posterior Std. Error		
	$a_0 = 0$	$\mu_{a_0} = 0.5$	$a_0 = 1$	$a_0 = 0$	$\mu_{a_0} = 0.5$	$a_0 = 1$
x_1	0.133	0.134	0.153	0.046	0.046	0.044
x_2	−0.178	−0.182	−0.186	0.046	0.045	0.043
x_3	−0.069	−0.076	−0.087	0.049	0.049	0.046
x_4	0.406	0.411	0.371	0.049	0.049	0.046
x_5	0.229	0.236	0.218	0.054	0.054	0.051
x_6	−0.004	−0.002	0.015	0.051	0.051	0.048
x_7	0.440	0.446	0.399	0.058	0.058	0.054
x_8	0.235	0.241	0.224	0.057	0.057	0.053

TABLE 3.3. Posterior Estimates of $\boldsymbol{\beta}$ for Myeloma Data with $c_0 = 10$.

Variable	Posterior Mean			Posterior Std. Error		
	$a_0 = 0$	$\mu_{a_0} = 0.5$	$a_0 = 1$	$a_0 = 0$	$\mu_{a_0} = 0.5$	$a_0 = 1$
x_1	0.131	0.134	0.153	0.047	0.046	0.045
x_2	−0.180	−0.182	−0.189	0.046	0.046	0.044
x_3	−0.072	−0.076	−0.090	0.049	0.049	0.046
x_4	0.413	0.411	0.377	0.050	0.050	0.047
x_5	0.226	0.236	0.224	0.055	0.054	0.052
x_6	−0.005	−0.002	0.015	0.052	0.051	0.048
x_7	0.449	0.446	0.406	0.058	0.058	0.055
x_8	0.241	0.241	0.228	0.057	0.057	0.054

3.4 A Generalization of the Cox Model

We consider a model which can be viewed as a generalization of the piecewise constant model discussed in Section 3.1. For ease of exposition, we assume only one covariate, as the extension to multiple covariates is straightforward. Sinha, Chen, and Ghosh (1999) consider a piecewise constant hazard function with $h(y|x) = \lambda_k \exp(x\beta_k)$ for $y \in I_k$, where

$I_k = (s_{k-1}, s_k]$ for $k = 1, 2, \ldots, J$, and x is one-dimensional. This discretized hazard model allows for time-dependent regression coefficients, and therefore is not a proportional hazards model. It may be viewed as a generalized Cox model. More precisely, these features are captured through the prior specifications as follows:

(i) $\lambda_k \overset{indep}{\sim} \mathcal{G}(\eta_k, \gamma_k)$ for $k = 1, \ldots, J$;

(ii) $\beta_{k+1}|\beta_0, \beta_1, \ldots, \beta_k \sim N(\beta_k, w_k^2)$ for $k = 0, \ldots, J-1$; and

(iii) β_k's are a priori independent of $\boldsymbol{\lambda} = (\lambda_1, \lambda_2, \ldots, \lambda_J)'$.

We assume that the hyperparameters η_k, γ_k, w_k, and β_0 are known in advance. This model is a discretized version of a nonproportional hazards model with a discretized version of the gamma process prior for the baseline hazard $h_0(\cdot)$, where η_k/γ_k is the prior mean and η_k/γ_k^2 is the prior variance of λ_k. When the grid intervals are sufficiently small, this discretized version will be indistinguishable from the actual time-continuous gamma process. The discretized autocorrelated prior process for the β_k's allows the covariate effect to change over time, but also incorporates the prior information that the values of the regression coefficient $\boldsymbol{\beta}$ in adjacent intervals are expected to be somewhat close and the dependence among the β_k's decreases as the intervals become further apart. This assumption is relevant in situations where the covariate effect may change over time, but is not expected to change too much over time. The w_k's can be used as a tuning device to determine prior opinion about the possible change in the magnitude of the regression coefficient over time. For example, a priori, we may expect β_{k+1} to be within $1.96w_k$ from β_k with 95% confidence. The w_k's should depend on the lengths of the I_k's, thus allowing the coefficient to change more for larger grid intervals. Alternatively, it is possible to use an autocorrelated prior process for the baseline hazard, as discussed in Section 3.6. For further details on the use and properties of an autocorrelated process, see Sinha and Dey (1997), Sahu, Dey, Aslanidou, and Sinha (1997), Aslanidou, Dey, and Sinha (1998), and Sinha (1998).

Data from previous studies may be used to elicit the prior mean (η_k/γ_k) and variance (η_k/γ_k^2) of λ_k. Otherwise, one may select the hyperparameters to represent prior opinions that are nearly noninformative (flat) in the subset of the parameter space supported by the likelihood. Sinha, Chen, and Ghosh (1999) do this for the breast cancer data given in Example 1.3 by taking the common prior mean of the λ_k's as 0.5 and a common prior variance of 1.25. This results in a noninformative prior, and it is clear that for this example, we expect the λ_k's to be less than 1. This prior implies the following hyperparameters:

η_k	γ_k	β_0	w_k	w_0
0.2	0.4	0	1.0	2.0

These values are obtained using some simplifications, for example, by taking all the λ_k's to be identically distributed, which corresponds to having no prior belief that the hazard of deterioration changes over time.

Denote the observed data from n patients by $D_{obs} = \{x_i, (a_{l_i}, a_{r_i}];\ i = 1, 2, \ldots, n\}$, where the survival time y_i for the ith patient is known to be within $(a_{l_i}, a_{r_i}]$ and $a_{l_i} < a_{r_i}$ are two of the grid points $(a_1, a_2, \ldots, a_J)$ but not necessarily consecutive, and x_i is the covariate for the ith subject. Let $D = \{y_i, x_i, i = 1, 2, \ldots, n\}$ denote the complete data, and $\boldsymbol{\theta} = (\boldsymbol{\beta}', \boldsymbol{\lambda}')'$, where $\boldsymbol{\beta} = (\beta_1, \beta_2, \ldots, \beta_J)'$.

The distribution of $(y_i|x_i, \boldsymbol{\theta})$ is piecewise exponential, and thus the complete-data likelihood is given by

$$L(\boldsymbol{\theta}|D) \propto \prod_{k=1}^{J} \left[\lambda_k^{d_k} \exp\left\{ -\lambda_k \sum_{j \in \mathcal{R}_k} \exp(x_j \beta_k) \Delta_{jk} \right\} \exp\left(\sum_{j \in D_k} x_j \beta_k \right) \right], \tag{3.4.1}$$

where $\mathcal{R}_k$ is the set of patients at risk at a_{k-1}, $\Delta_{jk} = \min(y_j, a_k) - a_{k-1}$, $\mathcal{D}_k$ is the set of patients failing in $I_k = (a_{k-1}, a_k]$, and d_k is the number of patients in $\mathcal{D}_k$.

We use the Gibbs sampler for sampling from the posterior distribution of $\boldsymbol{\theta}$. For the full conditionals, we obtain

$$(\lambda_k|\boldsymbol{\lambda}^{(-k)}, \boldsymbol{\beta}, D) \sim \mathcal{G}\left(\eta_k + d_k,\ \gamma_k + \sum_{j \in \mathcal{R}_k} \exp(x_j \beta_k) \Delta_{jk} \right), \tag{3.4.2}$$

$$\pi(\beta_k|\boldsymbol{\beta}^{(-k)}, \boldsymbol{\lambda}, D) \propto \phi(\beta_k|\mu_k, \sigma_k^2) \exp\left(-\lambda_k \sum_{j \in \mathcal{R}_k} \exp(x_j \beta_k) \Delta_{jk} \right), \tag{3.4.3}$$

where $\phi(\cdot|\mu_k, \sigma_k^2)$ is the $N(\mu_k, \sigma_k^2)$ density function,

$$\mu_k = \frac{\left(\sum_{j \in \mathcal{D}_k} x_j\right) w_k^2 w_{k-1}^2 + \beta_{k-1} w_k^2 + \beta_{k+1} w_{k-1}^2}{w_k^2 + w_{k-1}^2},$$

$$\sigma_k^2 = \frac{w_k^2 w_{k-1}^2}{w_k^2 + w_{k-1}^2},$$

for $k = 1, 2, \ldots, J$, and $\beta_{J+1} = 0$. To obtain the Cox model, we set all of the β_k's equal to a common β.

The conditional distribution of $[y_i|\boldsymbol{\theta}, D_{obs}]$ is a truncated piecewise exponential with parameters $\exp(x_i\beta_k)\lambda_k$, given by

$$f(y_i|\boldsymbol{\theta}, D_{obs}) = \left[1 - \exp\left\{-\sum_{l=l_i+1}^{r_i} \lambda_l \exp(x_i\beta_l)\tilde{\Delta}_l\right\}\right]^{-1} \lambda_k \exp(x_i\beta_k)$$
$$\times \exp\left\{-\sum_{l=l_i+1}^{k-1} \left[\lambda_l \exp(x_i\beta_l)\tilde{\Delta}_l \right.\right.$$
$$\left.\left. - \lambda_k \exp(x_i\beta_k)(y_i - a_{k-1})\right]\right\}, \quad (3.4.4)$$

for $y_i \in I_k$, $l_i + 1 \leq k \leq r_i$, and $\tilde{\Delta}_l = a_l - a_{l-1}$. Note that (3.4.4) is a product of multinomial and truncated exponential densities. The conditional distribution of $(y_i|\theta)$ for the Cox model (i.e., $\beta_k \equiv \beta$) is similar to (3.4.4).

From above, we see that $\boldsymbol{\lambda}$ and the y_i's can be sampled straightforwardly using standard statistical subroutines. The full conditionals for the β_k's are log-concave and thus require the adaptive rejection algorithm.

Example 3.3. Breast cancer data. In this example, we examine the breast cancer data in Example 1.3. We implement the Gibbs sampler for the above generalized Cox model. The convergence of the Gibbs sampler is checked by using several diagnostic procedures as recommended by Cowles and Carlin (1996). After convergence, we generated 50,000 Gibbs iterates to calculate the posterior means and 95% HPD intervals for all of the β_k's. A plot of β_k versus k, $k = 1, 2, \ldots, J$ is given in Figure 3.2. From Figure 3.2, we observe that all of the 95% HPD intervals of the β_k's contain 0 and the HPD intervals change over time. We also notice that the HPD intervals are wider for large k. This suggests that the covariate effect may change over time.

3.5 Beta Process Models

3.5.1 Beta Process Priors

We first discuss time-continuous right censored survival data without covariates. Kalbfleisch (1978) and Ferguson and Phadia (1979) used the definition of the cumulative hazard $H(t)$ as

$$H(t) = -\log(S(t)), \quad (3.5.1)$$

where $S(t)$ is the survival function. The gamma process can be defined on $H(t)$ when this definition of the cumulative hazard is appropriate. A more general way of defining the hazard function, which is valid even when the

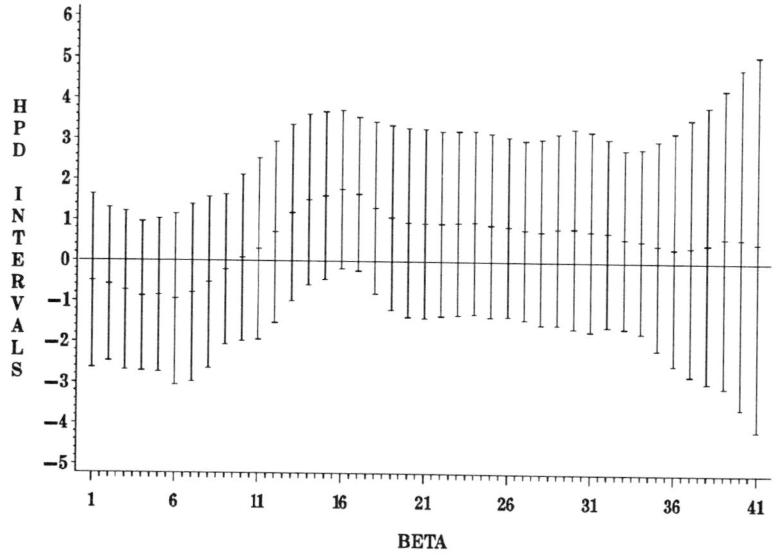

FIGURE 3.2. Plot of posterior means and 95% HPD intervals for the β_k's.

survival time distribution is not continuous, is to use the definition of Hjort (1990). General formulae for the cumulative hazard function $H(t)$ are

$$H(t) = \int_{[0,t]} \frac{dF(u)}{S(u)}, \tag{3.5.2}$$

where

$$F(t) = 1 - S(t) = 1 - \prod_{[0,t]} \{1 - dH(t)\}. \tag{3.5.3}$$

The cumulative hazard function $H(t)$ defined here is equal to (3.5.1) when the survival distribution is absolutely continuous. Hjort (1990) presents what he calls a beta process with independent increments as a prior for $H(.)$. A beta process generates a proper cdf $F(t)$, as defined in (3.5.2), and has independent increments of the form

$$dH(s) \sim \mathcal{B}(c(s)dH^*(s), c(s)(1 - dH^*(s))), \tag{3.5.4}$$

where $\mathcal{B}(a, b)$ denotes the beta distribution with parameters (a, b). Due to the complicated convolution property of independent beta distributions, the exact distribution of the increment $H(s)$ is only approximately beta over any finite interval, regardless of how small the length of the interval might be. See Hjort (1990) for formal definitions of the beta process prior and for properties of the posterior with right censored time-continuous data. It is possible to deal with the beta process for the baseline cumulative hazard appropriately defined under a Cox model with time continuous data, but survival data in practice is commonly grouped within some grid intervals, where the grid size is determined by the data and trial design. So for practical purposes, it is more convenient and often sufficient to use a discretized version of the beta process (Hjort, 1990; Sinha, 1997) along with grouped survival data. The beta process prior for the cumulative baseline hazard in (3.5.4) has been discussed by many authors, including Hjort (1990), Damien, Laud, and Smith (1996), Laud, Smith, and Damien (1996), Sinha (1997), and Florens, Mouchart, and Rolin (1999). Here we will focus only on the discretized beta process prior with a grouped data likelihood.

Within the spirit of the definition of the cumulative hazard function $H(t)$ defined in (3.5.2), a discretized version of the Cox model can be defined as

$$S(s_j|\boldsymbol{x}) = P(T > s_j|\boldsymbol{x}) = \prod_{k=1}^{j}(1 - h_k)^{\exp(\boldsymbol{x}'\boldsymbol{\beta})},$$

where h_k is the discretized baseline hazard rate in the interval $I_k = (s_{k-1}, s_k]$. The likelihood can thus be written as

$$L(\boldsymbol{\beta}, \boldsymbol{h}) = \prod_{j=1}^{J}\left((1 - h_j)^{\sum_{i\in\mathcal{R}_j-\mathcal{D}_j}\exp(\boldsymbol{x}_i'\boldsymbol{\beta})}\right)\prod_{l\in D_j}\left(1 - (1 - h_j)^{\exp(x_l'\boldsymbol{\beta})}\right), \tag{3.5.5}$$

where $\boldsymbol{h} = (h_1, h_2, \ldots, h_J)'$. To complete the discretized beta process model, we specify independent beta priors for the h_k's. Specifically, we take $h_k \sim \mathcal{B}(c_{0k}\alpha_{0k}, c_{0k}(1 - \alpha_{0k}))$, and independent for $k = 1, 2, \ldots, J$. Though it is reasonable to assume that the h_k's are independent from each other a priori, the assumption of an exact beta distribution of the h_k's is only due to an approximation to the true time-continuous beta process. Thus, according to the time-continuous beta process, the distribution of the h_k's is not exactly beta, but it can be well approximated by a beta distribution only when the width of I_k is small.

Under the discretized beta process defined here, the joint prior density of $\boldsymbol{h}$ is thus given by

$$\pi(\boldsymbol{h}) \propto \prod_{j=1}^{J} h_j^{c_{0j}\alpha_{0j}-1}(1 - h_j)^{c_{0j}(1-\alpha_{0j})-1}.$$

A typical prior for $\boldsymbol{\beta}$ is a $N_p(\boldsymbol{\mu}_0, \Sigma_0)$ prior, which is independent of $\boldsymbol{h}$. Assuming an arbitrary prior for $\boldsymbol{\beta}$, the joint posterior of $(\boldsymbol{\beta}, h)$ can be written as

$$\begin{aligned}\pi(\boldsymbol{\beta},\boldsymbol{h}|D) &\propto L(\boldsymbol{\beta},\boldsymbol{h}|D)\pi(\boldsymbol{h})\pi(\boldsymbol{\beta})\\ &= \prod_{j=1}^{J}\left((1-h_j)^{\sum_{i\in\mathcal{R}_j-\mathcal{D}_j}\exp(\boldsymbol{x}_i'\boldsymbol{\beta})}\right)\prod_{l\in\mathcal{D}_j}\left(1-(1-h_j)^{\exp(\boldsymbol{x}_i'\boldsymbol{\beta})}\right)\\ &\quad\times\prod_{j=1}^{J}h_j^{c_{0j}\alpha_{0j}-1}(1-h_j)^{c_{0j}(1-\alpha_{0j})-1}\pi(\boldsymbol{\beta}). \end{aligned} \tag{3.5.6}$$

We use the Gibbs sampler to sample from the joint posterior distribution of $(\boldsymbol{\beta},\boldsymbol{h})$. To sample from the full conditional of $[\boldsymbol{h}|\boldsymbol{\beta},D]$, we introduce auxiliary variables to create posterior independence between the components of $\boldsymbol{h}$. To this end, define the transformation $q_j = -\log(1-h_j)$, $j = 1,2,\ldots,J$, which leads to the full conditional

$$\begin{aligned}\pi(\boldsymbol{q}|\boldsymbol{\beta},D) \propto \prod_{j=1}^{J}\Bigg[&\exp(-q_j g_j(\boldsymbol{\beta}))\left(\prod_{i\in D_j}(1-\exp(-\theta_i q_j))\right)\\ &\times(1-\exp(-q_j))^{c_{0j}\alpha_{0j}-1}\Bigg], \end{aligned} \tag{3.5.7}$$

where $\boldsymbol{q} = (q_1, q_2, \ldots, q_J)'$, $\theta_i = \exp(\boldsymbol{x}_i'\boldsymbol{\beta})$ and $g_j(\boldsymbol{\beta}) = c_{0j}(1-\alpha_{0j}) + \sum_{i\in\mathcal{R}_j-\mathcal{D}_j}\theta_i$. From (3.5.7), we can see that

$$\begin{aligned}\pi(q_j|\boldsymbol{\beta},D) \propto \exp(-q_j g_j(\boldsymbol{\beta}))&\left(\prod_{i\in D_j}(1-\exp(-\theta_i q_j))\right)\\ &\times(1-\exp(-q_j))^{c_{0j}\alpha_{0j}-1}.\end{aligned}$$

Let $d_{0j} = c_{0j}\alpha_{0j} - 1$ and assume d_{0j} is a nonnegative integer. Now define auxiliary variables γ_{ij} and η_{lj} which are exponential random variables truncated at 1. Specifically,

$$\pi(\gamma_{ij}|q_j,\boldsymbol{\beta},D) \propto \theta_i q_j\exp(-\gamma_{ij}\theta_i q_j)(1-\exp(-\theta_i q_j))^{-1}I(0<\gamma_{ij}<1) \tag{3.5.8}$$

and

$$\pi(\eta_{lj}|q_j,\boldsymbol{\beta},D) \propto q_j\exp(-\eta_{lj}q_j)(1-\exp(-q_j))^{-d_{0j}}I(0<\eta_{lj}<1) \tag{3.5.9}$$

for $l = 1,2,\ldots,d_{0j}$. From (3.5.8) and (3.5.9), we see that

$$E(\eta_{ij}|q_j,\boldsymbol{\beta},D) = q_j^{-1} \text{ and } E(\gamma_{ij}|q_j,\boldsymbol{\beta},D) = (\theta_i q_j)^{-1}.$$

Let $\boldsymbol{\eta}_j = (\eta_{1j}, \eta_{2j}, \ldots, \eta_{d_{0j},j})$, and $\boldsymbol{\gamma}_j = (\gamma_{ij})$ for $i \in \mathcal{D}_j$. Using these latent variables, we now have

$$\pi(q_j|\eta_j, \gamma_j, \boldsymbol{\beta}, D) \propto q_j^{d_j+d_{0j}} \exp\left(-q_j\left(g_j(\beta) + \sum_{i\in\mathcal{D}_j} \gamma_{ij}\theta_i + \sum_{l=1}^{d_{0j}} \eta_{lj}\right)\right), \tag{3.5.10}$$

which can be recognized as

$$\mathcal{G}\left(d_j + d_{0j} + 1, g_j(\beta) + \sum_{i\in D_j} \gamma_{ij}\theta_i + \sum_{l=1}^{d_{0j}} \eta_{lj}\right).$$

Thus, a posteriori, the q_j's are independent gamma variates. The full conditional of $\boldsymbol{\beta}$ is given by

$$\pi(\boldsymbol{\beta}|\boldsymbol{q}, \boldsymbol{\gamma}, \boldsymbol{\eta}, D) \propto \prod_{j=1}^{J} \exp\left(-q_j\left(g_j(\beta) + \sum_{i\in\mathcal{D}_j} \gamma_{ij}\theta_i\right)\right)\pi(\boldsymbol{\beta}). \tag{3.5.11}$$

If $\pi(\boldsymbol{\beta})$ is log-concave, then so is $\pi(\boldsymbol{\beta}|\boldsymbol{q}, \boldsymbol{\gamma}, \eta, D)$, and therefore the adaptive rejection algorithm can be used to sample $\boldsymbol{\beta}$ from $[\boldsymbol{\beta}|\boldsymbol{q}, \boldsymbol{\gamma}, \boldsymbol{\eta}, D]$. Thus, the Gibbs sampler can be summarized as follows:

(i) Sample q_j from $[q_j|\eta_j, \gamma_j, \boldsymbol{\beta}, D]$ in (3.5.10)

(ii) Sample $\boldsymbol{\beta}$ from $[\boldsymbol{\beta}|\boldsymbol{q}, \boldsymbol{\gamma}, \boldsymbol{\eta}, D]$ in (3.5.11).

We note here that the likelihood and posterior are greatly simplified if we do not include covariates. By setting $\exp(\boldsymbol{x}_i'\boldsymbol{\beta}) = 1$, $[\boldsymbol{h}|D]$ is given by

$$\pi(\boldsymbol{h}|D) \propto \prod_{j=1}^{J} h_j^{c_{0j}\alpha_{0j}+d_j-1}(1-h_j)^{c_{0j}(1-\alpha_{0j})+r_j-d_j-1}$$

so that

$$h_j|D \sim \mathcal{B}(c_{0j}\alpha_{0j} + d_j, c_{0j}(1-\alpha_{0j}) + r_j - d_j).$$

Now define $h_j = P[s_{j-1} < Y \le s_j|Y > s_{j-1})]$. Therefore, the survival curve is given by $S(s_j) = \prod_{k=1}^{j}(1-h_k)$. The prior distribution of h_j is given by

$$h_j \sim \mathcal{B}\Big(c_{0j}\alpha_{0j},\ c_{0j}(1-\alpha_{0j})\Big) \quad \text{for} \quad j = 1, 2, \ldots, J, \tag{3.5.12}$$

where the h_j's are independent, and each with mean α_{0j} and variance $\alpha_{0j}(1-\alpha_{0j})/(c_{0j}+1)$. Therefore, c_{0j} is the measure of confidence around the prior mean α_{0j} of the hazard rate h_j in I_j. Given the prior structure of (3.5.12), the posterior distribution of the h_j's given grouped survival data is also independent beta with

$$h_j|D \sim \mathcal{B}\Big(c_{0j}\alpha_{0j} + d_j,\ c_{0j}(1 - \alpha_{0j}) + r_j - d_j\Big), \tag{3.5.13}$$

where $D = \{(d_j, r_j), j = 1, 2, \ldots, J\}$ denotes the complete grouped data. The joint posterior of the h_j's given interval-censored data is not as straightforward as (3.5.13), and is discussed in the next section. The next section shows how the natural conjugacy of the time-discrete beta process with the complete grouped data likelihood makes way for implementing the Gibbs sampling algorithm for interval censored data.

3.5.2 Interval Censored Data

We consider the beta process prior with interval censored data. The natural conjugacy of the time-discrete beta process with the complete grouped data likelihood facilitates the implementation of the Gibbs sampler for interval censored data. Let $D_{obs} = \{(a_{l_i}, a_{r_i}), i = 1, 2, \ldots, n\}$ denote the interval censored data and let D denote the completely grouped data. The Gibbs sampling steps for the ith iteration are given as follows:

Step 1. Sample $\boldsymbol{y}_i$ from $\pi(\boldsymbol{y}|\boldsymbol{h}_{i-1}, D_{obs})$.

Step 2. Sample $h_{j,i}$ from $\pi(h_j|\boldsymbol{y}_i, D_{obs})$ for $j = 1, 2, \ldots, J+1$.

Step 3. Use $\boldsymbol{h}_i$ as the starting value for the $(i+1)^{th}$ iteration.

After a sufficient number of iterations (or after satisfying some convergence criterion) the final value of $\boldsymbol{h}_i$ can be taken as a sample from the posterior distribution of $\boldsymbol{h}|D_{obs}$. For Step 2, the conditional posterior of h_j given D is of course the same as in (3.5.13) (which is a beta distribution) only with completely grouped data replaced by the augmented grouped data $(\mathbf{z}, \mathbf{y})$. To obtain the predictive distribution of $\mathbf{z}$ given $(\boldsymbol{h}, \mathbf{y})$ in Step 1, we first use the fact that given $(\boldsymbol{h}, \mathbf{y})$, each subject is independent. Let I_k be the set of grouping intervals such that the kth subject may fall into one of these groups (according to the incomplete data $\mathbf{y}$, we may not have the exact group). Let Ψ_k be the vector of indicators (augmented data for the kth subject) which indicates the specific grouping interval where the failure of the kth subject is augmented. So, ψ_{kj} is 1 for j being the augmented grouping interval of failure and 0 for other $j \in I_k$. One can generate $\mathbf{z}$ from $p(\mathbf{z}|\boldsymbol{h}, \mathbf{y})$ by independently sampling Ψ_k for the kth subject using

$$P(\psi_{kj} = 1|\boldsymbol{h}, \mathbf{y}) \ = \ \frac{h_j \displaystyle\prod_{u \in I_k}^{j-1} (1 - h_u)}{1 - \displaystyle\prod_{l \in I_k} (1 - h_l)}$$

for $j \in I_k$ and $k = 1, 2, \ldots, n$. So, Ψ_k is a multinomial random vector of indicators taking value 1 at exactly one cell; and the conditional posterior distributions used in Steps 1 and 2 of each Gibbs sampling iteration are standard distributions for sampling purposes. The major issues regarding the hyperparameters are the dimension of $\boldsymbol{h}$, the number of grid intervals required, the widths of the grid intervals, and the selection of the s_j's. Ideally, these choices should not depend on the data but rather on the design of the study. Depending on how precisely we want to estimate the survival function, the experimenter should decide the times of the scheduled appointments for the patients and each grid interval should be equal to the intended gap between two consecutive scheduled appointments. For the breast cancer data in Example 1.3, assuming that the intended gap between two consecutive appointments is one month, each prespecified grid interval is of length one month except for the interval starting after the 60th month. Hence, the total number of prespecified time intervals should be 61. But, in practice, if we have too many grid intervals and each patient misses the appointments too often, there may not be enough data from a moderate sized dataset with comparatively wide censoring intervals. In that case, we may have to adopt wider grid intervals than what were intended. But clearly the partition induced by the grid intervals should be finer than the partition induced by the censoring intervals.

3.6 Correlated Gamma Processes

Using ideas from Nieto-Barajas and Walker (2000), Mezzetti and Ibrahim (2000) develop a class of correlated gamma process priors for the Cox model. Again, consider the piecewise constant hazard model of Section 3.1. We construct a finite partition of the time axis, $0 < s_1 < s_2 < \ldots < s_J$, with $s_J > y_i$ for all $i = 1, 2, \ldots, n$, and assume a constant baseline hazard $h_0(y) = \lambda_k$ for $y \in I_k = (s_{k-1}, s_k]$. The likelihood function of $(\boldsymbol{\lambda}, \boldsymbol{\beta})$ can now be written as

$$L(\boldsymbol{\beta}, \boldsymbol{\lambda}|D) = \prod_{k=1}^{J} \left[\lambda_k^{d_k} \left(\prod_{i \in \mathcal{D}_k} \exp(\boldsymbol{x}_i'\boldsymbol{\beta}) \right) \right.$$
$$\left. \times \exp\left\{ -\lambda_k (s_k - s_{k-1}) \left(\sum_{i \in \mathcal{R}_k} \exp(\boldsymbol{x}_i'\boldsymbol{\beta}) \right) \right\} \right], \quad (3.6.1)$$

where $\mathcal{R}_k$ is the set of patients at risk and $\mathcal{D}_k$ is the set of patients having failures in the kth interval.

The gamma process prior of Kalbfleisch (1978) discussed in Section 3.2 assumes independent cumulative hazard increments. This is unrealistic in most applied settings, and does not allow for borrowing of strength between adjacent intervals. A correlated gamma process for the cumulative hazard yields a natural smoothing of the survival curve. Although the idea of smoothing is not new (Arjas and Gasbarra, 1994; Aslanidou, Dey, and Sinha, 1998; Sinha, 1998; Gamerman, 1991; Berzuini and Clayton, 1994), its potential has not been totally explored in the presence of covariates. Modeling dependence between hazard increments has been discussed by Gamerman (1991) and Arjas and Gasbarra (1994). Gamerman (1991) proposes a Markov prior process for the $\{\log(\lambda_k)\}$, by modeling

$$\log(\lambda_k) = \log(\lambda_{k-1}) + \epsilon_k, \qquad E(\epsilon_k) = 0, \quad \text{and} \quad \text{Var}(\epsilon_k) = \sigma_k^2.$$

Arjas and Gasbarra (1994) introduced a first-order autoregressive structure on the increment of the hazards by taking

$$\lambda_k | \lambda_{k-1} \sim \mathcal{G}(\alpha_k, \alpha/\lambda_{k-1})$$

for $k > 1$. Nieto-Barajas and Walker (2000) propose dependent hazard rates with a Markovian relation, given by

$$\lambda_1 \sim \mathcal{G}(\alpha_1, \gamma_1), \quad u_k | \lambda_k, v_k \sim \mathcal{P}(v_k \lambda_k), \quad v_k | \xi_k \sim \mathcal{E}\left(1/\xi_k\right), \tag{3.6.2}$$

$$\lambda_{k+1} | u_k, v_k \sim \mathcal{G}(\alpha_{k+1} + u_k, \gamma_{k+1} + v_k), \tag{3.6.3}$$

and

$$\boldsymbol{\beta} \sim \pi(\boldsymbol{\beta}),$$

for $k \geq 1$, where $\pi(\boldsymbol{\beta})$ denotes the prior for $\boldsymbol{\beta}$, which can be taken to be a normal distribution, for example.

Following Nieto-Barajas and Walker (2000), the innovation in the process is the introduction of two latent processes $\{u_k\}$ and $\{v_k\}$, which are not observable and cannot be expressed in terms of observable quantities. The role of the processes $\{u_k\}$ and $\{v_k\}$ is fundamental, since they allow us to control the strength of the correlation between different hazard increments. These processes can be characterized by examining the following conditional expectations:

$$E(\lambda_{k+1} | u_k, v_k) = \frac{\alpha_{k+1} + u_k}{\gamma_{k+1} + v_k},$$

$$E(u_k | \lambda_k, v_k) = v_k \lambda_k, \quad E(u_k | \lambda_k) = \lambda_k \xi_k,$$

and

$$E(\lambda_{k+1} | \lambda_k, v_k) = \frac{\alpha_{k+1} + v_k \lambda_k}{\gamma_{k+1} + v_k},$$

where α_k, γ_k, and ξ_k are specified hyperparameters. Note that we do not force the process to have the same first-order autoregressive structure, in that the dependence can vary along the time axis. We also note that the higher the value of v_k, the closer the process is to first-order autoregressive. When $v_k = 0$, we have the independence model of Kalbfleisch (1978).

Allowing v_k to be a random process in (3.6.2) enables us to learn about the degree of correlation between increments in adjacent time intervals through the data. This alternative Markov prior process generalizes the independent gamma process, but keeps the convenient conjugacy property of a gamma prior with the Poisson distribution. When α_k, γ_k and v_k are constant over time, the hazard increments λ_k are marginally distributed as independent and identically distributed (i.i.d.) gamma variates, and the process is a stationary process. Let $\boldsymbol{\lambda} = (\lambda_1, \lambda_2, \ldots, \lambda_J)'$, $\boldsymbol{u} = (u_0, u_1, \ldots, u_J)'$, $\boldsymbol{v} = (v_0, v_1, \ldots, v_J)'$, and define $u_0 = v_0 = 0$. With this process, the joint posterior is given by

$$\begin{aligned}\pi(\boldsymbol{\lambda}, \boldsymbol{\beta}, \boldsymbol{u}, \boldsymbol{v}|D) \propto & L(\boldsymbol{\lambda}, \boldsymbol{\beta}|D) \prod_{k=1}^{J} \{\pi(\lambda_k|u_{k-1}, v_k)\pi(u_k|\lambda_k, v_k)\pi(v_k|\xi_k)\}\, \pi(\boldsymbol{\beta}) \\ \propto & \prod_{k=1}^{J} \left[\lambda_k^{d_k} A_k \frac{\lambda_k^{\alpha_k + u_{k-1} - 1} \exp\left(-\lambda_k(\gamma_k + v_{k-1})\right)}{\Gamma(\alpha_k + u_{k-1})} \right. \\ & \left. \times (\gamma_k + v_{k-1})^{\alpha_k + u_{k-1}} \frac{\exp(-v_k\lambda_k)(v_k\lambda_k)^{u_k}}{u_k!} \exp(-\frac{v_k}{\xi_k}) \right] \\ & \times \exp\left(-\sum_{k=1}^{J} \lambda_k(s_k - s_{k-1})B_k \right) \pi(\boldsymbol{\beta}),\end{aligned}$$

where d_k is the number of events in $I_k = (s_{k-1}, s_k]$,

$$A_k = \prod_{i \in \mathcal{D}_k} \exp(\boldsymbol{x}_i'\boldsymbol{\beta}), \quad \text{and} \quad B_k = \sum_{i \in \mathcal{R}_k} \exp(\boldsymbol{x}_i'\boldsymbol{\beta}).$$

After some algebra, we obtain

$$\begin{aligned}\pi(\boldsymbol{\lambda}, \boldsymbol{\beta}, \boldsymbol{u}, \boldsymbol{v}|D) \propto & \exp\left\{ -\sum_{k=1}^{J} \lambda_k \left((s_k - s_{k-1})B_k + \gamma_k + v_{k-1} + v_k\right) \right\} \\ & \times \prod_{k=1}^{J} \frac{A_k(\gamma_k + v_{k-1})^{\alpha_k + u_{k-1}} v_k^{u_k} \exp(-\frac{v_k}{\xi_k})}{\Gamma(\alpha_k + u_{k-1}) u_k!} \\ & \times \prod_{k=1}^{J} \lambda_k^{d_k + \alpha_k + u_{k-1} + u_k - 1} \times \pi(\boldsymbol{\beta}).\end{aligned}$$

The full conditionals are given by

$$\lambda_k|D, \boldsymbol{\beta}, \boldsymbol{u}, \boldsymbol{v} \sim \mathcal{G}(d_k + \alpha_k + u_k + u_{k-1}, \gamma_k + v_k + v_{k-1} + (s_k - s_{k-1})B_k), \tag{3.6.4}$$

$$P\left[u_k = j|D, \boldsymbol{\lambda}, \boldsymbol{\beta}, \boldsymbol{v}\right] \propto \frac{\left(\lambda_k\lambda_{k+1}v_k(\gamma_{k+1} + v_k)\right)^j}{\Gamma(\alpha_{k+1} + j)\Gamma(j+1)}, \quad k < J, \tag{3.6.5}$$

$$(u_J|D, \boldsymbol{\lambda}, \boldsymbol{\beta}, \boldsymbol{v}) \sim \mathcal{P}(v_J\lambda_J),$$

$$\pi(v_k|D, \boldsymbol{\lambda}, \boldsymbol{u}, \boldsymbol{v}^{(-k)}) \propto (\gamma_{k+1} + v_k)^{\alpha_{k+1}+u_k} v_k^{u_k} \exp\left\{-v_k\left(\lambda_{k+1} + \lambda_k + \frac{1}{\xi_k}\right)\right\} \tag{3.6.6}$$

for $k < J$, where $\boldsymbol{v}^{(-k)}$ denotes the $\boldsymbol{v}$ vector without the kth component, and

$$v_J|D, \boldsymbol{\lambda}, \boldsymbol{u} \sim \mathcal{G}(u_J + 1, \lambda_J + 1/\xi_J).$$

To obtain the full conditional distribution for $\boldsymbol{\beta}$, we utilize the collapsed Gibbs technique of Liu (1994) by integrating over $\boldsymbol{\lambda}$. This step provides a key simplification to the Gibbs sampling algorithm, and leads to a tractable full conditional for $\boldsymbol{\beta}$. Thus, after integration over $\boldsymbol{\lambda}$, we get the conditional posterior distribution of $\boldsymbol{\beta}$,

$$\pi(\boldsymbol{\beta}|D, \boldsymbol{v}, \boldsymbol{u}) \propto \frac{\left(\prod_{i=1}^{n} \exp(\nu_i \boldsymbol{x}_i'\boldsymbol{\beta})\right) \pi(\boldsymbol{\beta})}{\prod_{k=1}^{J} \left((s_k - s_{k-1})B_k + \gamma_k + v_k + v_{k-1}\right)^{(\alpha_k + d_k + u_k + u_{k-1})}}. \tag{3.6.7}$$

We observe that (3.6.7) has the same kernel as Cox's partial likelihood,

$$\text{PL}(\boldsymbol{\beta}) = \prod_{i=1}^{n} \frac{\exp(\nu_i \boldsymbol{x}_i'\boldsymbol{\beta})}{\left(\sum_{l \in \mathcal{R}_i} \exp(\boldsymbol{x}_l'\boldsymbol{\beta})\right)^{\nu_i}}, \tag{3.6.8}$$

when $\pi(\boldsymbol{\beta})$ is vague and $\alpha_k = \gamma_k = v_k = v_{k-1} = u_k = u_{k-1} = 0$. In this case, equations (3.6.7) and (3.6.8) give the same estimates of $\boldsymbol{\beta}$. We demonstrate this methodology using a well-known published dataset.

Example 3.4. Leukemia data. We consider the well-known leukemia data analyzed by Cox (1972) and Kalbfleisch (1978). The data shown in Table 3.4 consists of remission times (in weeks) of leukemia patients assigned to treatment with a drug (6-MP) or a placebo during remission maintenance therapy. In Table 3.4, an asterisk (*) indicates censoring. We investigate the properties of the model and conduct sensitivity analyses for the hyperparameters. The partition of the time axis was based on the observed remission times, in which each interval corresponded to an observed remission time.

TABLE 3.4. Leukemia Data.

Group 0 (drug)	6*, 6, 6, 6, 7, 9*, 10*, 10, 11*, 13, 16, 17*, 19*, 20*, 22, 23, 25*, 32*, 32*, 34*, 35*
Group 1 (placebo)	1, 1, 2, 2, 3, 4, 4, 5, 5, 8, 8, 8, 11, 11, 12, 12, 15, 17, 22, 23

Figures 3.3 and 3.4 show the posterior estimates of the survival curves for various choices of the hyperparameters using the prior distributions

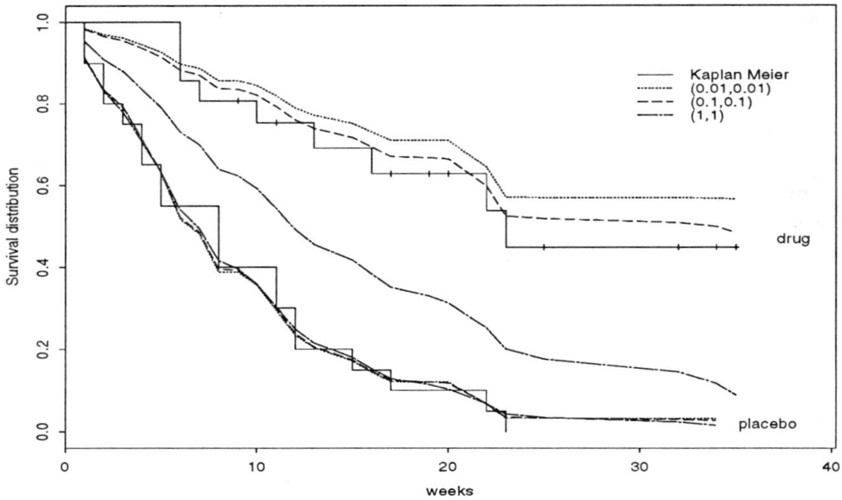

FIGURE 3.3. Posterior estimation of survival curves with $\xi_k = 5$ and varying values of (α_k, γ_k) for leukemia data.

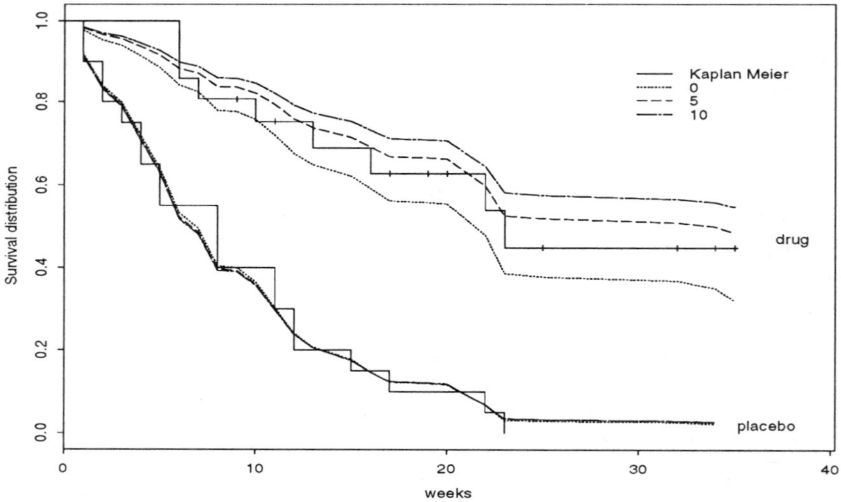

FIGURE 3.4. Posterior estimation of survival curves with $(\alpha_k = 0.1, \gamma_k = 0.1)$ and varying values of ξ_k for leukemia data.

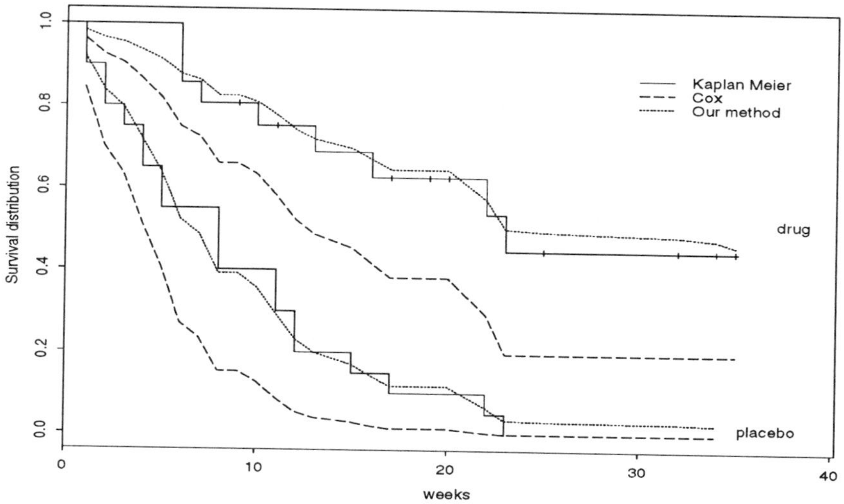

FIGURE 3.5. Posterior estimation of survival curves with hyperparameters defined in (3.6.9) for leukemia data.

in (3.6.2) and (3.6.3), and taking $\pi(\beta)$ to be a $N(0, 10^6)$ distribution, where β corresponds to the regression coefficient of the treatment covariate. More specifically, in Figure 3.3, we vary the values of the hyperparameters α_k, γ_k, $k = 1, \ldots, M$, and we fix $\xi_k = 5$, assuming a high prior correlation between adjacent hazard increments. In Figure 3.4, we fix the values of α_k, γ_k, $k = 1, \ldots, M$, to 0.01 and vary the prior correlation. Figure 3.5 gives posterior estimates of the survival distribution, obtained by varying the hyperparameters α_k, γ_k, and ξ_k through the time axis. The larger the prior precision, the closer the results are to the prior mean; the smaller the prior precision, the closer the results to the Cox's partial likelihood estimates. It also follows that the smaller the hyperparameters α_k and γ_k, the closer the final estimates (both the baseline hazard and the regression coefficient β) to the Cox's partial likelihood estimates. The higher the values of γ_k, the higher the prior precision, and thus the closer the final estimates of λ_k are to the prior mean, $\frac{\alpha_k + u_{k-1}}{\gamma_k + v_{k-1}}$, of the distribution in (3.6.3). In Figure 3.3, we observe that a bad fit is obtained when $\alpha_k = \gamma_k = 1$ and $\xi_k = 5$, since these hyperparameters give a very informative prior with mean 1 for the hazard increments.

The posterior estimates of β are not sensitive unless we assume a very informative prior distribution for $\pi(\beta)$. The parameters α_k, γ_k, and ξ_k are defined as functions of time, and the estimates shown in Figure 3.5 are obtained with the following choices of hyperparameters,

$$\alpha_k = \begin{cases} 0.01 & t_k \le 20 \\ 0.1 & t_k > 20 \end{cases} \quad \gamma_k = \begin{cases} 0.1 & t_k \le 10 \\ 0.01 & 10 < t_k \le 20 \\ 0.2 & t_k > 20 \end{cases} \quad \xi_k = \begin{cases} 1 & t_k \le 20 \\ 5 & t_k > 20. \end{cases} \tag{3.6.9}$$

Most of the events are observed before the 10th week, so at the beginning of the time axis, we let the prior distribution of the hazard increments be flat. As time increases, the number of observed events is smaller, and thus we let the prior correlation between adjacent hazard increments be stronger. The posterior mean of β is 1.587 with 95% credible interval (0.800, 2.412), which agrees with the Cox's partial likelihood model estimate of 1.506 with 95% confidence interval (0.702, 2.318). Furthermore, the model does not seem to be too sensitive to the choice of partition of the time axis. The lower the value of ξ_k, the more sensitive the estimates are to the choice of the time intervals.

3.7 Dirichlet Process Models

3.7.1 Dirichlet Process Prior

The Dirichlet process is perhaps the most celebrated and popular prior process in nonparametric Bayesian inference. The introduction of the Dirichlet process by Ferguson (1973, 1974) initiated modern day Bayesian nonparametric methods, and today this prior process is perhaps the most important and widely used nonparametric prior. Notable articles using the Dirichlet process in various applications include Susarla and Van Ryzin (1976), Gelfand and Kuo (1991), Kuo and Smith (1992), Ramgopal, Laud, and Smith (1993), Doss (1994), MacEachern (1994), Bush and MacEachern (1996), Escobar (1994), Escobar and West (1995), Doss and Narasimhan (1998), Kleinman and Ibrahim (1998a,b), Gelfand and Kuo (1991), Newton, Czado, and Chapell (1996), Gelfand and Mallick (1995), and Walker, Damien, Laud, and Smith (1999). The Dirichlet process provides the Bayesian with a nonparametric prior specification over the class of possible distribution functions $F(y)$ for a random variable Y, where $F(y) = P(Y \le y)$. To further discuss the Dirichlet process, we first define the Dirichlet distribution.

The random vector $(y_1, y_2, \ldots, y_{k-1})$ is said to have a $k-1$ dimensional Dirichlet distribution, denoted by $y \sim D_{k-1}(\alpha_1, \alpha_2, \ldots, \alpha_k)$, if

$(y_1, y_2, \ldots, y_{k-1})$ have joint density

$$f(y_1, y_2, \ldots, y_{k-1}) \\ = \left(\frac{\Gamma(\alpha)}{\prod_{j=1}^{k} \Gamma(\alpha_j)} \right) \left(\prod_{j=1}^{k-1} y_j^{\alpha_j - 1} \right) \left(1 - \sum_{j=1}^{k-1} y_j \right)^{\alpha_k - 1},$$

where

$$y_j \geq 0,\ j = 1, 2, \ldots, k-1,\ \sum_{j=1}^{k-1} y_j \leq 1,\ \ \alpha = \sum_{j=1}^{k} \alpha_j.$$

The joint density of $(y_1, y_2, \ldots, y_{k-1})$ above is called the $k-1$ dimensional Dirichlet density.

In Bayesian nonparametric inference with the Dirichlet process prior, the typical approach is to specify a prior distribution over the space all possible cumulative distribution functions, $F(t) = 1 - S(t)$. To define the Dirichlet process formally, let the sample space be denoted by Ω, and suppose $\Omega = B_1 \cup B_2 \cup \ldots \cup B_k$, where the B_j's are disjoint. The B_j's, for example, can be disjoint intervals. Then a stochastic process P indexed by elements of a particular partition $B = \{B_1, B_2, \ldots, B_k\}$ is said to be a Dirichlet process on (Ω, B) with parameter vector α, if for any partition of Ω, the random vector $(P(B_1), P(B_2), \ldots, P(B_k))$ has a Dirichlet distribution with parameter $(\alpha(B_1), \alpha(B_2), \ldots, \alpha(B_k))$. We can make the definition a bit less general and write for any partition $\Omega = B_1 \cup B_2 \cup \ldots \cup B_k$, $(F(B_1), F(B_2), \ldots, F(B_k))$ has a Dirichlet distribution with parameters $(\alpha(B_1), \alpha(B_2), \ldots, \alpha(B_k))$. The B_j's for example might be disjoint intervals of the form $B_j = (b_{1j}, b_{2j}]$, so that $F(B_j) = F(b_{2j}) - F(b_{1j})$ and $\alpha(B_j) = \alpha(b_{2j}) - \alpha(b_{1j})$.

The parameter vector α is a probability measure, i.e., a distribution function itself so that we can write $\alpha = F_0(\cdot)$, where $F_0(\cdot)$ is the prior hyperparameter for $F(\cdot)$, and thus $\alpha(B_j) = F_0(b_{2j}) - F_0(b_{1j})$. The hyperparameter $F_0(.)$ is often called the base measure of the Dirichlet process prior. Finally, we can define a weight parameter c_0 ($c_0 > 0$) that gives prior weight to $F_0(\cdot)$, so that $(F(B_1), F(B_2), \ldots, F(B_k))$ has a Dirichlet distribution with parameters $(c_0 F_0(B_1), c_0 F_0(B_2), \ldots, c_0 F_0(B_k))$. Finally, we say that F has a Dirichlet process prior with parameter $c_0 F_0$ if $(F(B_1), F(B_2), \ldots, F(B_k))$ has a Dirichlet distribution with parameters $(c_0 F_0(B_1), c_0 F_0(B_2), \ldots, c_0 F_0(B_k))$ for every possible partition of the sample space $\Omega = B_1 \cup B_2 \cup \ldots \cup B_k$. We denote the Dirichlet process prior by $\mathcal{DP}$.

We are now led to the following theorem which characterizes the posterior process with the Dirichlet process prior. A proof of the theorem can be found in Ferguson (1973).

Theorem 3.7.1 *Suppose $y_1, y_2, \ldots, y_n$ is a sample from $F(\cdot)$, and suppose a priori $F \sim \mathcal{DP}(c_0 F_0)$. Then*

$$F \mid \boldsymbol{y} \sim \mathcal{DP}\left(c_0 F_0 + \sum_{i=1}^{n} \delta_{y_i}\right),$$

where $\boldsymbol{y} = (y_1, y_2, \ldots, y_n)'$, and δ_{y_i} is a point mass giving probability 1 to y_i.

Thus, a posteriori, the distribution of F is a Dirichlet process with parameter

$$c_0 F_0 + \sum_{i=1}^{n} \delta_{y_i} = c_0 F_0 + n F_n \ ,$$

where F_n is the empirical cdf, defined as

$$F_n(y) = \frac{\#\ of\ y_j\ in\ the\ sample \le y}{n}.$$

In an actual data analysis, we can construct the intervals $B_1, B_2, \ldots, B_k$ after looking at the data $y_1, y_2, \ldots, y_n$, and construct them such that B_j has at least one of the y_i's.

Example 3.5. Simple example of Dirichlet process. Suppose y_1, y_2, $\ldots$, y_5 are i.i.d. from $F(\cdot)$ and suppose $F \sim \mathcal{DP}(c_0 F_0)$, where $c_0 = .1$ and $F_0 = \mathcal{E}(1)$, i.e., $F_0(y) = 1 - e^{-y}$. Further, suppose $y_1 = 1$, $y_2 = 0.7$, $y_3 = 0.8$, $y_4 = 1.2$, $y_5 = 1.3$, and let $B_1 = \{y : \ 0 < y \le 1\}$, $B_2 = \{y : \ 1 < y \le 1.25\}$, and $B_3 = \{y : \ 1.25 < y < \infty\}$. We note that the intervals are always left open and right closed. Thus, $F(B_1) = F(1) - F(0) = p_1$, $F(B_2) = F(1.25) - F(1) = p_2$, $F(B_3) = F(\infty) - F(1.25) = p_3$, and $p_1 + p_2 + p_3 = 1$. Furthermore,

$$(p_1, p_2, p_3) \sim D_2(c_0 F_0(B_1), c_0 F_0(B_2), c_0 F_0(B_3)),$$

where $F_0(B_1) = F_0(1) - F_0(0) = 1 - e^{-1} = 0.632$, $F_0(B_2) = F_0(1.25) - F_0(1) = e^{-1} - e^{-1.25} = 0.081$, and $F_0(B_3) = F_0(\infty) - F_0(1.25) = e^{-1.25} = 0.287$. Thus, a priori

$$(p_1, p_2) \sim D_2(0.0632, 0.0081, 0.0287).$$

By Theorem 3.7.1,

$$(p_1, p_2)|\boldsymbol{y} \sim D_2\Big(a_0 F_0(B_1) + n\ F_n(B_1), a_0 F_0(B_2) + n\ F_n(B_2),$$
$$a_0 F_0(B_3) + n\ F_n(B_3)\Big).$$

A different partition of the sample space leads to different definition of the p_j's and hence a Dirichlet prior with different prior parameters. Now $F_n(y) = (\#\text{ of } y_j \le y)/5$, so $F_n(B_1) = F_n(1) - F_n(0) = \frac{3}{5} - 0 = 3/5$,

$F_n(B_2) = F_n(1.25) - F_n(1) = \frac{4}{5} - \frac{3}{5} = \frac{1}{5}$, $F_n(B_3) = F_n(\infty) - F_n(1.25) = 1 - \frac{4}{5} = \frac{1}{5}$, and $F_n(B_1) + F_n(B_2) + F_n(B_3) = 1$. Thus,

$$\begin{aligned}(p_1, p_2) \mid y \sim & D_2\left(0.0632 + 5\left(\frac{3}{5}\right), 0.0081 + 5\left(\frac{1}{5}\right), 0.0287 + 5\left(\frac{1}{5}\right)\right) \\ = & D_2(3.0632, 1.0081, 1.0287).\end{aligned}$$

3.7.2 Dirichlet Process in Survival Analysis

Some of the earliest work on Dirichlet processes in the context of survival analysis is based on the work of Ferguson and Phadia (1979) and Susarla and Van Ryzin (1976). Susarla and Van Ryzin derive the Bayes estimator of the survival function under the Dirichlet process prior and also derive the posterior distribution of the cumulative distribution function with right censored data. In this section, we summarize the fundamental results of Ferguson and Phadia (1979) and Susarla and Van Ryzin (1976). Letting $S(t)$ denote the survival function, Susarla and Van Ryzin (1976) derive the Bayes estimator of $S(t)$ under the squared error loss

$$L(\hat{S}, S) = \int_0^\infty (\hat{S}(t) - S(t))^2 dw(t),$$

where w is a weight function, i.e., a nonnegative decreasing function on $(0, \infty)$ and $\hat{S}(t)$ is an estimator of $S(t)$. Suppose we have n observations, $y_1, y_2, \ldots, y_n$, and for ease of exposition, let $y_1, y_2, \ldots, y_k$ denote the uncensored observations and $y_{k+1}, y_{k+2}, \ldots, y_n$ denote the censored observations. Further, let $y_{(k+1)}, y_{(k+2)}, \ldots, y_{(m)}$ denote the distinct observations among the censored observations $y_{k+1}, y_{k+2}, \ldots, y_n$. Further let λ_j denote the number of censored observations that are equal to $y_{(j)}$, for $j = k+1, k+2, \ldots, m$, and let $N(t)$ and $N^+(t)$ denote the number of observations (censored or not) greater than or equal to t and the number greater than t, respectively, and let F_0 denote the base measure. Then Susarla and Van Ryzin (1976) show that the Bayes estimator $\hat{S}(u)$ under squared error loss is given by

$$\begin{aligned}\hat{S}(u) = & \frac{c_0(1 - F_0(u)) + N^+(u)}{c_0 + n} \\ & \times \prod_{j=k+1}^{l} \left(\frac{(c_0(1 - F_0(y_{(j)})) + N(y_{(j)})}{c_0(1 - F_0(y_{(j)})) + N(y_{(j)}) - \lambda_j} \right)\end{aligned} \tag{3.7.1}$$

in the interval $y_{(j)} \le u \le y_{(l+1)}$, $l = k, k+1, \ldots, m$, with $y_{(k)} = 0$, $y_{(m+1)} = \infty$. The Kaplan-Meier estimator of $S(u)$ (Kaplan and Meier, 1958) is a limiting case of (3.7.1) and is obtained when $F_0 \to 0$, as shown by Susarla and Van Ryzin (1976). To see this, list and label the n observed y values in order of increasing magnitude so that one has $0 \le y'_1 \le y'_2 \le \cdots \le y'_n$. Then from Kaplan and Meier (1958), the product–limit estimator of $S(u)$ is

$$\hat{S}(u) = \prod_r (n-r)/(n-r+1), \tag{3.7.2}$$

where r assumes those values for which $y'_r \leq u$ and y'_r is an uncensored observation. Kaplan and Meier (1958) show that this estimate maximizes the likelihood of the observations. Note that (3.7.1) converges to

$$(N^+(u)/n) \prod_{j=k+1}^{l} (N(y_{(j)})/(N(y_{(j)}) - \lambda_j)) \tag{3.7.3}$$

as $F_0 \to 0$. Now let $i(u)$ be the largest integer such that $y'_{(i(u))} \leq u$. Then

$$N^+(u)/n = \prod_{j \leq i(u)} N^+(y'_j)/N(y'_j). \tag{3.7.4}$$

Substituting (3.7.4) into (3.7.3) and canceling the ratios common to both the products (note that these ratios correspond to the censored observations) yields precisely (3.7.2) provided that the uncensored observations are distinct.

In order to calculate the Kaplan-Meier (KM) estimate in (3.7.2), one need not know the actual observations but only the number of censored observations between two uncensored observations, while one needs all of the observations, both censored and uncensored in order to calculate the Bayes estimator in (3.7.1). So in this way, the Bayes estimator uses all of the data and may be preferable to the KM estimator. Also, define $F_0(u,\infty) = c_0(1 - F_0(u))$. Then, if $F_0(u,\infty)$ is continuous in u, the data $D = (\boldsymbol{y}, \boldsymbol{\nu})$ can be recovered exactly given the estimator, which is not true for the KM estimator. In this sense, the Bayes estimator is a function of the sufficient statistic, i.e., the data D, whereas the KM estimator is not. Both the KM estimator and Bayes estimator are discontinuous at uncensored observations provided F_0 is assumed to be continuous in the latter case. But the Bayes estimator smoothes the estimator at censored observations, resulting in the continuity at the censored observations provided F_0 does not have point masses. More generally, Susarla and Van Ryzin (1976) also derived the posterior p^{th} moment of $S(u)$ as

$$E(S(u)^p|D) = \prod_{s=0}^{p-1} \left(\frac{F_0(u,\infty) + s + N^+(u)}{F_0(R^+) + s + n} \right) \times \prod_{j=k+1}^{l} \left(\frac{F_0(y_{(j)},\infty) + s + N(y_{(j)})}{F_0(y_{(j)},\infty) + s + N(y_{(j)}) - \lambda_j} \right). \tag{3.7.5}$$

From (3.7.5), we see that when $p = 1$, we get (3.7.1).

Example 3.6. Kaplan-Meier data. We consider the example from Susarla and Van Ryzin (1976) to obtain the Bayes estimator of S, where $F_0(u) = 1 - \exp(-\theta u)$. We consider values of $\theta = 0.12$ and $c_0 = 4, 8, 16$. The data consists of deaths at 0.8, 3.1, 5.4, and 9.2 months, and censored observations at 1.0, 2.7, 7.0, and 12.1 months. Table 3.5 gives ranges of u and the Bayes estimator of $S(u)$.

From Table 3.5, it can be observed that all of the uncensored observation points—namely, 0.8, 3.1, 5.4, and 9.2—are points of discontinuity of $\hat{S}(.)$. At the censored observation points—namely, 1.0, 2.7, 7.0, and 12.1, $\hat{S}(.)$ is continuous but the left and right derivatives of $\hat{S}(.)$ are different at these points. Further, in the plot given by Susarla and Van Ryzin (1976), the Bayes estimator is seen to be smoother then the KM estimator in that the jumps at the uncensored observations are not as large for the Bayes estimator.

TABLE 3.5. Bayes and KM Estimates of $S(y)$.

u in	Bayes estimate, $F_0 = 1 - \exp(-\theta u)$	KM Estimate
$[0, 0.8)$	$\frac{c_0 \exp(-\theta u)+8}{c_0+8}$	1.0
$[0.8, 1.0)$	$\frac{c_0 \exp(-\theta u)+7}{c_0+8}$	7/8
$[1.0, 2.7)$	$\frac{c_0 \exp(-\theta u)+6}{c_0+8} \frac{c_0 \exp(-\theta u)+7}{c_0+6}$	7/8
$[2.7, 3.1)$	$\frac{c_0 \exp(-\theta u)+6}{c_0+8} \frac{c_0 \exp(-\theta u)+7}{c_0+6} \frac{c_0 \exp(-\theta u)+6}{c_0+5}$	7/8
$[3.1, 5.4)$	$\frac{c_0 \exp(-\theta u)+6}{c_0+8} \frac{c_0 \exp(-\theta u)+7}{c_0+6} \frac{c_0 \exp(-\theta u)+6}{c_0+5}$	7/8
$[5.4, 7.0)$	$\frac{c_0 \exp(-\theta u)+6}{c_0+8} \frac{c_0 \exp(-\theta u)+7}{c_0+6} \frac{c_0 \exp(-\theta u)+6}{c_0+5}$	7/8
$[7.0, 9.2)$	$\frac{c_0 \exp(-\theta u)+6}{c_0+8} \frac{c_0 \exp(-\theta u)+7}{c_0+6} \frac{c_0 \exp(-\theta u)+6}{c_0+5}$	7/8
$[9.2, 12.1)$	$\frac{c_0 \exp(-\theta u)+6}{c_0+8} \frac{c_0 \exp(-\theta u)+7}{c_0+6} \frac{c_0 \exp(-\theta u)+6}{c_0+5}$	7/8
$[12.1, \infty)$	$\frac{c_0 \exp(-\theta u)+6}{c_0+8} \frac{c_0 \exp(-\theta u)+7}{c_0+6} \frac{c_0 \exp(-\theta u)+6}{c_0+5}$	7/8

Source: Susarla and Van Ryzin (1976).

On the other hand, if F_0 has positive mass at any of the censored observation points, then the Bayes estimate $\hat{S}(.)$ would be discontinuous there. The question of the choice of the parameter F_0 of the Dirichlet process prior is, of course, important. In the example with $F_0 = 1 - \exp(-\theta u)$, θ was taken as 0.12 based on the following heuristic argument. If one were to consider the Bayes estimator in the no data problem, the Bayes estimator would be $\hat{S}(u) = \exp(-\theta u)$. If we force this estimator to satisfy $\exp(-\theta M) = 0.525$ where M is the 0.525 quartile from the KM curve, then the θ so derived is approximately 0.12. We choose the 0.525 (21/40) quartile as the closest jump to the 50th quartile on the observed KM curve. Thus, by double use of the data, a reasonable value of θ is 0.12. Other values of θ tend to pull the Bayes estimator away from the KM estimator. Roughly speaking, they are taking the prior mass on $(0, \infty)$ to be 1/2, equal, and twice the sample size $n = 8$. The larger the c_0, the smoother the Bayes estimator becomes in comparison to the KM estimator. That is, the jumps at the death points are smaller. The extreme case of this would be to let $c_0 \to \infty$ in which case the Bayes estimator reduces to $\exp(-\theta u)$ with $\theta = 0.12$ in the example.

3.7.3 Dirichlet Process with Doubly Censored Data

Kuo and Smith (1992) discuss Dirichlet process priors with doubly censored survival data. The data structure and the model can be described as follows. Let $t_1, t_2, \ldots, t_n$ denote the true survival times of n subjects that could be observed precisely if no censoring were present. The t_i are independent and identically distributed with cumulative distribution function $F(t) = P(T \leq t)$. For each i, we assume that there are windows of observations (a_i, b_i), $a_i \leq b_i$, that are either fixed constants or random variables independent of the t_i. Thus, we observe

$$y_i = \max(\min(t_i, b_i), a_i).$$

Moreover, for each i, we also know whether it is left censored with $y_i = a_i$, or right censored with $y_i = b_i$, or a precisely observed time with $y_i = t_i$. We assume that the subjects are examined at discrete times, $0 < s_1 < s_2 < \ldots, s_J$, with observed events classified into one of J intervals $(0, s_1], (s_1, s_2], \ldots, (s_{J-1}, s_J]$. Let d_j denote the number of precise observations in the interval $(s_{j-1}, s_j]$; m_j, the number of left censored observations at s_j; and l_j, the number of right censored observations at s_j. It is assumed that the left censored subjects m_j all occur at the end of the period $(s_{j-1}, s_j]$, and the right censored subjects l_j all occur at the beginning of $(s_j, s_{j+1}]$. Let $p_j = p(s_j) = 1 - F(s_j)$ denote the survival function evaluated at s_j, so that the likelihood is given by

$$L(\boldsymbol{p}|D) \propto \prod_{j=1}^{J} (p_{j-1} - p_j)^{d_j} (1 - p_j)^{m_j} p_j^{l_j}, \tag{3.7.6}$$

where $\boldsymbol{p} = (p_1, p_2, \ldots, p_J)'$. Let $\theta_j = p_{j-1} - p_j$ for $j = 1, 2, \ldots, J$, and let $\theta_{J+1} = p_J$. We specify a Dirichlet process prior for the θ_j's, and thus the prior distribution for $\boldsymbol{\theta} = (\theta_1, \ldots, \theta_{J+1})'$ is a Dirichlet distribution, given by

$$\pi(\boldsymbol{\theta}) \propto \prod_{j=1}^{J+1} \theta_j^{\alpha_j - 1}, \tag{3.7.7}$$

where

$$\alpha_j = c_0(F_0(s_j) - F_0(s_{j-1})),$$

and F_0 is the base measure, $j = 1, 2, \ldots, J+1$, and $F_0(s_{J+1}) = 1$. The posterior distribution of $\boldsymbol{\theta}$ is known to be a mixture of Dirichlet distributions (see Antoniak, 1974). We discuss mixtures of Dirichlet processes in the next subsection. The Gibbs sampler for sampling the posterior of $\boldsymbol{\theta}$ avoids the direct computation of this mixture. To employ the Gibbs sampler, we introduce latent variables that decompose the numbers of censored entries into the numbers of observations belonging to individual intervals. Let $z_{ij}, \ldots, z_{jj}$ denote the random variables that count the number of observations in m_j that might fall in the intervals $(0, s_1]$, $(s_1, s_2]$, $\ldots$, $(s_{j-1}, s_J]$, respectively, so that $m_j = \sum_{l=1}^{j} z_{1j}$. Further, let $z_{j+1,j}, \ldots, z_{m+1,j}$ denote the number of observations l_j that might fall in the intervals $(s_j, s_{j+1}]$, $\ldots$, $(s_{J-1}, s_J]$, $[s_J, \infty)$, respectively, so that $l_j = \sum_{l=j+1}^{J+1} z_{1j}$. The posterior full conditionals for $\boldsymbol{\theta}$ given the z's and the data, is easily seen to be an updated Dirichlet distribution depending only on the z's, and the posterior full conditional for the z's given $\boldsymbol{\theta}$ and the data is easily seen to be a product of multinomial distributions. Thus, suppose at the ith iteration step of the Gibbs sampler, we have the realization $\boldsymbol{\theta}^{(i)} = (\theta_1^{(i)}, \ldots, \theta_{J+1}^{(i)})'$, with $\sum_{j=1}^{J+1} \theta_j^{(i)} = 1$. We then update the z variables from the multinomial distributions as follows. For each j, $j = 1, 2, \ldots, J$, we sample $z_{1j}^{(i+1)}, \ldots, z_{jj}^{(i+1)}$ from the multinomial distribution with sample size m_j and parameters $r_{1j}^{(i)}, \ldots, r_{jj}^{(i)}$, where $r_{\ell j}^{(i)} = \theta_{\ell j}^{(i)} / \sum_{\ell=1}^{j} \theta_\ell^{(i)}$ for $\ell = 1, \ldots, j$. Similarly, we sample $z_{j+1,j}^{(i+1)}, \ldots, z_{J+1,j}^{(i+1)}$ from the multinomial distribution with sample size l_j and parameters $r_{j+1,j}, \ldots, r_{J+1,j}$, where $r_\ell^{(i)} = \theta_\ell^{(i)} / \sum_{\ell=j+1}^{J+1} \theta_\ell^{(i)}$ for $\ell = j+1, j+2, \ldots, J+1$.

Having sampled the z random variables, we then generate new θ variables from the Dirichlet distribution as follows. We compute, for each ℓ, $\ell = 1, 2, \ldots, J+1$,

$$\lambda_\ell^{(i+1)} = \alpha_\ell + d_\ell + \sum_{j=1}^{J} z_{\ell j}^{(i+1)}$$

and then sample $(\theta_1^{(i+1)}, \ldots, \theta_J^{(i+1)}, \theta_{J+1}^{(i+1)})$ from the Dirichlet distribution with parameters $(\lambda_1^{(i+1)}, \ldots, \lambda_{J+1}^{(i+1)})$. By running M parallel chains, after the ith iteration, we have $\theta_{1m}^{(i)}, \theta_{2m}^{(i)}, \ldots, \theta_{J+1,m}^{(i)}$ and $\lambda_{1m}^{(i)}, \ldots, \lambda_{J+1,m}^{(i)}$ for $m = 1, 2, \ldots, M$. The posterior distribution of θ_j for $j = 1, \ldots, J+1$ can thus be approximated by

$$\hat{F}(\theta_j | D) = M^{-1} \sum_{m=1}^{M} \mathcal{B}(\lambda_{\ell m}^{(i)}, \sum_{k \neq 1}^{J+1} \lambda_{km}^{(i)}),$$

where $\mathcal{B}(a, b)$ denotes the beta distribution with parameters a and b. The posterior mean of θ_j is thus estimated by

$$\hat{\theta}_j = M^{-1} \sum_{m=1}^{M} \frac{\lambda_{1m}^{(i)}}{\sum_{l=1}^{J+1} \lambda_{lm}^{(i)}}.$$

Other posterior summaries can be computed similarly from the replicated samples, i and M having been selected to achieve convergence to smooth estimates.

Example 3.7. Kaplan-Meier data (Example 3.6 continued). We consider the data set by Kaplan and Meier (1958), which was also discussed in the previous subsection. For comparison purposes, we consider the same prior specifications as those of the previous subsection, and thus take $F_0(1) = 1-\exp(-\phi t)$ with $\phi = .12$ and $c_0 = 4, 8, 16$. We discretized the time axis into the intervals $(0, 0.8]$, $(0.8^-, 0.8]$, $(0.8, 1]$, $(1, 2.7]$, $(2.7, 3.1^-]$, $(3.1^-, 3.1]$, $(3.1, 5.4^-]$, $(5.4^-, 5.4]$, $(5.4, 7]$, $(7, 9.2^-]$, $(9.2^-, 9.2]$, $(9.2, 12.1]$, and $(12.1, \infty)$. Here, the notation 0.8^- means a time value just below 0.8, and so forth. We label these intervals by $(0, s_1]$, $(s_1, s_2]$, ..., $(s_{12}, s_{13}]$, and let θ_1, θ_2, ..., θ_{13}, respectively, denote the probabilities assigned to the intervals. The likelihood function of $\boldsymbol{\theta} = (\theta_1, \theta_2, \ldots, \theta_{13})'$ given the data D is given by

$$L(\boldsymbol{\theta}|D) = \theta_2\theta_6\theta_8\theta_{11}(\theta_4 + \ldots + \theta_{13})(\theta_5 + \ldots + \theta_{13})(\theta_{10} + \ldots + \theta_{13})\theta_{13}. \tag{3.7.8}$$

Let $\alpha_j = c_0(\exp(-\phi s_{j-1}) - \exp(-\phi s_j))$, so that the prior distribution of θ is

$$\pi(\theta) \propto \prod_{j=1}^{13} \theta_j^{\alpha_j - 1}.$$

We note that θ_2, θ_6, θ_8, and θ_{13} in the likelihood combine simply with the corresponding θ variables in the prior distribution, so that the parameters θ_2, θ_6, θ_8, θ_{11}, and θ_{13} are each updated by 1 in the posterior distribution. Therefore, we need only introduce three z variables for the incomplete data, namely, $\boldsymbol{z}_1 = (z_{41}, z_{51}, \ldots, z_{13,1})'$, $\boldsymbol{z}_2 = (z_{52}, z_{62}, \ldots, z_{13,2})'$, and $\boldsymbol{z}_3 = (z_{10,3}, z_{11,3}, z_{12,3}, z_{13,3})'$. We then sample $\boldsymbol{z}_j$ for $j = 1, 2, 3$ from

TABLE 3.6. Bayes Estimates for $c_0 = 4$.

Estimates	0.8^-	0.8	1	2.7	3.1^-	3.1
$\hat{p}_s$ with $M = 1000$	0.970	0.886	0.879	0.819	0.805	0.702
$\hat{p}_s$ with $M = 4000$	0.970	0.886	0.879	0.819	0.805	0.701
Exact Bayes	0.970	0.886	0.879	0.819	0.805	0.701
Estimates	5.4^-	5.4	7	9.2^-	9.2	12.1
$\hat{p}_s$ with $M = 1000$	0.632	0.529	0.491	0.437	0.305	0.253
$\hat{p}_s$ with $M = 4000$	0.632	0.529	0.491	0.438	0.307	0.256
Exact Bayes	0.632	0.528	0.490	0.438	0.306	0.255

Source: Kuo and Smith (1992).

the appropriate multinomial distribution with sample size 1 and rescaled probabilities.

To estimate the survival function at s_j, we accumulate the θ_k for $k > j$. For s between s_j and s_{j+1}, an interpolation formula that connects the survival function at the two endpoints according to the prior shape can be used. Tables 3.6 and 3.7 exhibit the Gibbs sampler results for the survival function evaluated at s_j with $M = 1000$ and $M = 4000$, both with $i = 10$. The exact Bayes solutions given by Susarla and Van Ryzin (1976) are also listed for comparison. The tables show that the Gibbs sampler results for $M = 1000$ are already very accurate in approximating the exact Bayes rule. Similar results are obtained for $c_0 = 16$.

TABLE 3.7. Bayes Estimates for $c_0 = 8$.

Estimates	0.8^-	0.8	1	2.7	3.1^-	3.1
$\hat{p}_s$ with $M = 1000$	0.954	0.892	0.881	0.792	0.773	0.698
$\hat{p}_s$ with $M = 4000$	0.954	0.892	0.881	0.792	0.773	0.700
Exact Bayes	0.954	0.892	0.881	0.793	0.773	0.699
Estimates	5.4^-	5.4	7	9.2^-	9.2	12.1
$\hat{p}_s$ with $M = 1000$	0.600	0.527	0.474	0.405	0.316	0.249
$\hat{p}_s$ with $M = 4000$	0.602	0.529	0.474	0.405	0.318	0.250
Exact Bayes	0.602	0.528	0.474	0.405	0.318	0.250

Source: Kuo and Smith (1992).

3.7.4 Mixtures of Dirichlet Process Models

The Mixture of Dirichlet Process (MDP) model arises in cases of the following general situation. Suppose an $n_i \times 1$ random vector $\boldsymbol{y}_i$ has a parametric distribution indexed by the $w \times 1$ vector $\boldsymbol{\theta}_i$, $i = 1, \ldots, N$. Then suppose the $\boldsymbol{\theta}_i$ themselves have a prior distribution with known hyperparameters $\boldsymbol{\psi}_0$. Thus

$$\begin{aligned} &\text{Stage 1:} \quad [\boldsymbol{y}_i|\boldsymbol{\theta}_i] \sim \Pi_{n_i}(h_1(\boldsymbol{\theta}_i)), \\ &\text{Stage 2:} \quad [\boldsymbol{\theta}_i|\boldsymbol{\psi}_0] \sim \Pi_w(h_2(\boldsymbol{\psi}_0)), \end{aligned} \tag{3.7.9}$$

where $\Pi_w(\cdot)$ is a generic label for a w-dimensional parametric multivariate distribution and $h_1(\cdot)$ and $h_2(\cdot)$ are functions. The MDP model (Escobar, 1994; MacEachern, 1994) removes the assumption of a parametric prior at the second stage, and replaces it with a general distribution G. The distribution G then in turn has a Dirichlet process prior (Ferguson, 1973), leading to

$$\begin{aligned} &\text{Stage 1:} \quad [\boldsymbol{y}_i|\boldsymbol{\theta}_i] \sim \Pi_{n_i}(h_1(\boldsymbol{\theta}_i)), \\ &\text{Stage 2:} \quad \boldsymbol{\theta}_i|G \overset{i.i.d.}{\sim} G, \\ &\text{Stage 3:} \quad [G|c_0, \boldsymbol{\psi}_0] \sim \mathcal{DP}(c_0 \cdot G_0(h_2(\boldsymbol{\psi}_0))), \end{aligned} \tag{3.7.10}$$

where G_0 is a w-dimensional parametric distribution, often called the *base measure*, and c_0 is a positive scalar. The parameters of a Dirichlet process are $G_0(\cdot)$, a probability measure, and c_0, a positive scalar. The parameter $c_0G_0(\cdot)$ contains a distribution, $G_0(\cdot)$, which approximates the true nonparametric shape of G, and the scalar c_0, which reflects our prior belief about how similar the nonparametric distribution G is to the base measure $G_0(\cdot)$.

There are two special cases in which the MDP model leads to the fully parametric case. As $c_0 \to \infty$, $G \to G_0(\cdot)$, so that the base measure is the prior distribution for $\boldsymbol{\theta}_i$. Also, if $\boldsymbol{\theta}_i \equiv \boldsymbol{\theta}$ for all i, the same is true. For a more hierarchical modeling approach, it is possible to place prior distributions on (c_0, ψ_0). The specification in (3.7.10) results in a semiparametric specification in that a fully parametric distribution is given in Stage 1 and a nonparametric distribution is given in Stages 2 and 3.

The Polya urn representation of the Dirichlet process was developed by Blackwell and MacQueen (1973) and is useful for sampling purposes. We describe it as follows. The draw of $\boldsymbol{\theta}_1$ is always from the base measure, $G_0(\cdot)$. The draw of $\boldsymbol{\theta}_2$ is equal to θ_1 with probability p_1 and is from the base measure with probability $p_0 = 1 - p_1$. The draw of $\boldsymbol{\theta}_3$ is equal to θ_1 with probability p_1, equal to θ_2 with probability p_2; and is a draw from the base measure with probability $p_0 = 1 - (p_1 + p_2)$. The values of the p_i's change with each new draw. This process continues until $\boldsymbol{\theta}_N$ is equal to each of the preceding $\boldsymbol{\theta}$'s with probability p_i, $i \in \{1, \ldots, N-1\}$ and is a draw from the base measure with probability $p_0 = 1 - \sum_{i=1}^{N-1} p_i$. We determine the values of p_i, $i = 0, \ldots, N-1$, from the Dirichlet process parameters. In other words, the $\boldsymbol{\theta}$'s are actually drawn from a mixture distribution, where the mixing probabilities are determined by the Dirichlet process of Stage 3, thus giving rise to the MDP label. From this representation, it is clear that if all of the $\boldsymbol{\theta}_i \equiv \boldsymbol{\theta}$ for all i, then we draw $\boldsymbol{\theta}$ from the base measure with probability 1 and thus the base measure is the prior.

The MDP model is simplified in practice by the Polya urn representation, using the fact that, marginally, the $\boldsymbol{\theta}_i$ are distributed as the base measure along with the added property that $P(\boldsymbol{\theta}_i = \boldsymbol{\theta}_j, i \neq j) > 0$. The Dirichlet process prior results in what MacEachern (1994) calls a "cluster structure" among the $\boldsymbol{\theta}_i$'s. This cluster structure partitions the N $\boldsymbol{\theta}_i$'s into k sets or clusters, $0 < k \leq N$. All of the observations in a cluster share an identical value of $\boldsymbol{\theta}$ and subjects in different clusters have differing values of $\boldsymbol{\theta}$.

As described by Escobar (1994), conditional on, $\boldsymbol{\theta}^{(-i)}$, $\boldsymbol{\theta}_i$ has the following mixture distribution:

$$\pi(\boldsymbol{\theta}_i | \boldsymbol{y}, \boldsymbol{\theta}^{(-i)}) \propto \sum_{j \neq i} q_j \delta_{\boldsymbol{\theta}_j} + c_0 q_0 g_0(\boldsymbol{\theta}_i) f(\boldsymbol{y}_i | \boldsymbol{\theta}_i), \tag{3.7.11}$$

where $\boldsymbol{y} = (\boldsymbol{y}_1', \boldsymbol{y}_2', \ldots, \boldsymbol{y}_N')'$, $\boldsymbol{\theta}^{(-i)} = (\boldsymbol{\theta}_1', \ldots, \boldsymbol{\theta}_{i-1}', \boldsymbol{\theta}_{i+1}', \ldots, \boldsymbol{\theta}_N')'$, and $f(\boldsymbol{y}_i | \boldsymbol{\theta}_i)$ is the sampling distribution of $\boldsymbol{y}_i$. We normalize the values q_j and

$c_0 q_0$ to obtain the selection probabilities $p_i, i = 0, \ldots, N-1$, in the Polya urn scheme described above. In addition, δ_s is a degenerate distribution with point mass at s, and $g_0(\cdot)$ is the density corresponding to the probability measure $G_0(\cdot)$. Finally, $q_j = f(\boldsymbol{y}_i|\boldsymbol{\theta}_j)$, $j = 1, \ldots, i-1, i+1, \ldots, N$, and $q_0 = \int f(\boldsymbol{y}_i|\boldsymbol{\theta}) g_0(\boldsymbol{\theta})\, d\boldsymbol{\theta}$.

To demonstrate the MDP model, we consider the seminal example of Escobar (1994) and Escobar and West (1995). Suppose that y_i has a univariate normal distribution with unknown mean θ_i and known variance σ_y^2. In this case, we have $n_i = 1$, $i = 1, \ldots, N$. Also assume that each θ_i has a univariate normal distribution. Then (3.7.9) becomes

$$\begin{aligned} &\text{Stage 1:} \quad [y_i|\theta_i, \sigma_y] \sim N(\theta_i, \sigma_y^2), \\ &\text{Stage 2:} \quad [\theta_i|\mu, \sigma_\theta] \sim N(\mu, \sigma_\theta^2). \end{aligned}$$

The MDP model removes the assumption of normality at the second stage, resulting in

$$\begin{aligned} &\text{Stage 1:} \quad [y_i|\theta_i, \sigma_y] \sim N(\theta_i, \sigma_y^2), \\ &\text{Stage 2:} \quad \theta_i|G \sim G, \\ &\text{Stage 3:} \quad [G|M, \psi_0] \sim \mathcal{DP}(c_0 \cdot G_0(h_2(\psi_0))). \end{aligned} \tag{3.7.12}$$

3.7.5 Conjugate MDP Models

Suppose $G_0 = N(\mu, \sigma_\theta^2)$ in (3.7.12) so that $\psi_0 = (\mu, \sigma_\theta^2)$. In this case, the unnormalized selection probability q_j is equal to $f(y_i|\theta_j) = \phi(y_i|\theta_j, \sigma_y^2)$, where $\phi(\cdot|\mu, \sigma^2)$ denotes the normal density with mean μ and variance σ^2. With probability proportional to q_j, $\theta_i \sim \delta_{\theta_j}$, which means that $\theta_i = \theta_j$ with probability 1. The unnormalized selection probability q_0 is given by

$$q_0 = \int f(y_i|\theta, \sigma_y^2) g_0(\theta|\psi_0)\, d\theta = \int \phi(y_i|\theta, \sigma_y^2)\phi(\theta|\mu, \sigma_\theta^2)\, d\theta.$$

With probability proportional to $c_0 q_0$,

$$[\theta_i|y_i] \sim g_0(\theta) f(y_i|\theta) = N(\theta|\mu, \sigma_\theta^2) N(y_i|\theta, \sigma_y^2),$$

where $N(y_i|\mu, \sigma^2)$ indicates that y_i has a normal distribution with mean μ and variance σ^2. Then

$$[\theta_i|y_i] \sim N\left([(\sigma_\theta^2 + \sigma_y^2)^{-1}\sigma_\theta^2\sigma_y^2] \left(\frac{\mu}{\sigma_\theta^2} + \frac{y_i}{\sigma_y^2}\right), (\sigma_\theta^2 + \sigma_y^2)^{-1}\sigma_\theta^2\sigma_y^2 \right).$$

In the example above, selecting G_0 to be normal when the sampling distribution of the data is normal emulates the conjugate relationship between sampling distribution and prior in the usual Bayesian hierarchy. In the MDP case, the sampling distribution is conjugate to the base measure. MacEachern (1994) calls MDP models with base measures and sampling distributions that are conjugate in this fashion "conjugate MDP models." The computational advantages of the conjugate MDP model are clear from the example. First, q_0 has a closed form. Second, the distribution of θ_i corresponding to q_0 is from the same exponential family as the base measure. As a result, Gibbs sampling in the conjugate model described above can proceed in a relatively straightforward fashion, as described in detail in Kleinman and Ibrahim (1998a).

3.7.6 Nonconjugate MDP Models

When we do not assume conjugacy, the integral needed for q_0 typically has no closed-form solution. Since we must evaluate this integral N times within each Gibbs cycle, the cost in time of numerical integration is compounded, as is the cost in accuracy of approximations. Several attempts to avoid this integration have been made. For example, West, Müller, and Escobar (1994) approximate q_0 with $f(\boldsymbol{y}_i|\boldsymbol{\theta}_i)$, where $\boldsymbol{\theta}_i \sim G_0(\cdot)$. This is certainly simple, but unfortunately, the stationary distribution underlying the Gibbs sampler is no longer the posterior distribution we desire. In fact, the stationary distribution may be quite different from the posterior. Fortunately, MacEachern and Müller (1998) describe a technique whereby one can fit nonconjugate MDP models without numerical integration or approximation. Other techniques have also been suggested by Walker, Damien, Laud, and Smith (1999) and Damien, Wakefield, and Walker (1999). Some additional notation is necessary for the exposition of the MacEachern and Müller method. Recall that when the $\boldsymbol{\theta}_i$'s are known, the observations are grouped into clusters which have equal $\boldsymbol{\theta}_i$'s. There will be some number k, $0 < k \leq N$, of unique values among the $\boldsymbol{\theta}_i$'s. Denote these unique values by γ_l, $l = 1, \ldots, k$, and recall from the Polya urn scheme that the γ_l are independent observations from $G(\cdot)$. Let n_l be the number of observations that share the value γ_l. Additionally, let l represent the set of subjects with common random effect γ_l. Note that knowing the $\boldsymbol{\theta}_i$'s is equivalent to knowing k, γ_l, n_l, and the cluster memberships l, $l = 1, \ldots, k$.

The routine of MacEachern and Müller (1998) is closely intertwined with the Gibbs sampler it generates. The method relies on the augmentation of the k independent γ_l's with an additional $N - k$ independent samples from $G_0(\cdot)$ at the start of each loop of the Gibbs sampler. Label these additional draws $\gamma_{k+1}, \ldots, \gamma_N$. Then the routine proceeds in the following fashion. If $n_l > 1$, $i \in l$, meaning that at least one other subject has the same value of $\boldsymbol{\theta}$ as subject i, then $\boldsymbol{\theta}_i$ has the distribution

$$\pi(\boldsymbol{\theta}_i|\boldsymbol{\theta}^{(-i)}, \boldsymbol{y}) \propto \sum_{l=1}^{k} n_l^- q_l \delta_{\gamma_l} + \frac{c_0}{k^* + 1} q_{k+1} \delta_{\gamma_{k+1}}, \tag{3.7.13}$$

where $k^* = k$ and n_l^- is the number of observations sharing γ_l when we exclude observation i. Note that this means $n_l^- = n_l$, except when $i \in l$, in which case $n_l^- = n_l - 1$. Also, $q_l = f(\boldsymbol{y}_i|\gamma_l), l = 1, \ldots, k+1$. In other words, with probability proportional to $n_l^- f(\boldsymbol{y}_i|\gamma_l)$, $\boldsymbol{\theta}_i$ is equal to γ_l with probability 1, $l = 1, \ldots, k$. With probability proportional to $[c_0/(k^*+1)]f(\boldsymbol{y}_i|\gamma_{k+1})$, $\boldsymbol{\theta}_i$ is distributed $\delta_{\gamma_{k+1}}$, meaning that $\boldsymbol{\theta}_i = \gamma_{k+1}$ with probability 1. If $n_l = 1$, $i \in l$, then only subject i has the value $\boldsymbol{\theta}_i$. In this case, we do the following. With probability $k^*/(k^* + 1)$, leave $\boldsymbol{\theta}_i$ unchanged. Otherwise, with probability $1/(k^* + 1)$, $\boldsymbol{\theta}_i$ is distributed according to (3.7.13), with the modification that $k^* = k - 1$.

If it should occur that this routine causes a cluster to disappear, meaning that $n_{l'} = 0$ for some $l' \leq k$, switch the cluster labels of l' and k. Notice that k decreases as a result of this process. Another important point is that the value $\gamma_{l'}$ is not removed, but becomes γ_{k+1} in the distribution of $\boldsymbol{\theta}_{i+1}$. Once we have completed an iteration of the Gibbs sampler, we discard the augmentary values $\gamma_{k+1}, \ldots, \gamma_N$.

3.7.7 MDP Priors with Censored Data

Doss (1994), Doss and Huffer (1998), and Doss and Narasimhan (1998) discuss the implementation of MDP priors for $F(t) = 1 - S(t)$ in the presence of right censored data using the Gibbs sampler. A Bayesian nonparametric approach based on mixtures of Dirichlet priors (Ferguson, 1973, 1974; Antoniak, 1974) offers a reasonable compromise between purely parametric and purely nonparametric models. The MDP prior for F can be defined as follows. If ν is some prior distribution for $\boldsymbol{\theta}$, where $\boldsymbol{\theta} \in \Theta$, and $c_0 > 0$ for each $\boldsymbol{\theta}$, then if $F|\boldsymbol{\theta} \sim \mathcal{DP}(c_0 F_{0\boldsymbol{\theta}})$, then F, unconditional on $\boldsymbol{\theta}$ is a mixture of Dirichlet's. The weight parameter may depend on $\boldsymbol{\theta}$, but in most applications it does not. For the case where the data are not censored, there is a closed form expression for the posterior distribution of F, given by Theorem 3.7.1. From this result, it can be easily seen that in the uncensored case, estimators based on mixtures of Dirichlet priors continuously interpolate between those based on the purely parametric and nonparametric models. For large values of c_0, the estimators are close to the Bayes estimator based on the parametric model. On the other hand, for small values of c_0, the estimators are essentially equal to the nonparametric maximum likelihood estimator. One therefore expects that the same will be true for the case where the data are censored.

For the censored case, there is no closed form expression for the posterior distribution of F given the data, and one has to use Monte Carlo methods. We review a Gibbs sampling scheme described by Doss and Huffer (1998) and Doss and Narasimhan (1998) that will enable us to estimate the posterior distributions of interest. Using the notation of Doss (1994), we write the MDP prior on F as

$$F \sim \int \mathcal{DP}_{\alpha_{\boldsymbol{\theta}}} \nu(d\boldsymbol{\theta}), \tag{3.7.14}$$

where for each $\boldsymbol{\theta} \in \Theta$, $\alpha_{\boldsymbol{\theta}} = c_0 F_{0\boldsymbol{\theta}}$ and $c_0 > 0$. We note here that c_0 may also depend on $\boldsymbol{\theta}$. Let D_{obs} denote the observed data, that is, the event

$\{y_i \in A_i : i = 1, \ldots, n\}$ where A_i is some set. We use the convention that subscripting a probability measure signifies conditioning. For example if U and V are random variables, then $\pi_V(U)$ denotes $\pi(U|V)$, the conditional distribution of U given V.

We are interested in the posterior distribution of F, denoted by $\pi(F|D)$. We use the fact that conditional on $\boldsymbol{\theta}$, the posterior of F is Dirichlet, so that

$$\begin{aligned}\pi(F|D_{obs}) &= \int \pi(F|\boldsymbol{\theta}, D) L(\boldsymbol{\theta}|D_{obs}) d\boldsymbol{\theta} \\ &= \int \mathcal{DP}_{\alpha_{\boldsymbol{\theta}} + \sum_{i=1}^n \delta_{y_i}} L(\boldsymbol{\theta}|D_{obs}) d\boldsymbol{\theta}, \qquad (3.7.15)\end{aligned}$$

where δ_i is a point mass at y_i, D_{obs} denotes the observed data, and $L(\boldsymbol{\theta}|D)$ denotes the complete data likelihood. From (3.7.15), we see that if we can generate an i.i.d. sample $(y^{(1)}, \boldsymbol{\theta}^{(1)}), \ldots, (y^{(G)}, \boldsymbol{\theta}^{(G)}) \sim \pi(y, \boldsymbol{\theta}|D_{obs})$, then we can estimate $\pi(F|D_{obs})$. Thus, for example, for any fixed t, we can estimate $\pi(F(t)|D_{obs})$ by the average of beta distributions

$$\frac{1}{G}\sum_{g=1}^{G} \mathcal{B}\Big(\alpha_{\boldsymbol{\theta}^{(g)}}(t) + \textstyle\sum_{i=1}^n \delta_{y_i^{(g)}}(t),\ c_{0\boldsymbol{\theta}^{(g)}} + n - \alpha_{\boldsymbol{\theta}^{(g)}}(t) - \sum_{i=1}^n \delta_{y_i^{(g)}}(t)\Big). \qquad (3.7.16)$$

So we need to be able to generate samples from $\pi(y_1, y_2, \ldots, y_n, \boldsymbol{\theta}|D_{obs})$. We will do this using the Gibbs sampler and using two well-known facts about the Dirichlet prior. The first of these states that if $F \sim \mathcal{DP}_\beta$, where $\beta = c_0 G_0$, and if $y_1, y_2, \ldots, y_n$ are i.i.d. with distribution F, then for any j,

$$\pi(y_j|y_i,\, i \neq j) = \frac{G_0 + \sum_{i\neq j} \delta_{y_i}}{c_0 + n - 1}. \qquad (3.7.17)$$

The second fact is the proposition below stated and proved by Doss and Narasimhan (1998), which enables the generation of $\boldsymbol{\theta}$ in the Gibbs sampler. Several versions of this result are already known (e.g., Lemma 1 of Antoniak (1974), Lemmas 2.1 and 3.1 of Diaconis and Freedman (1986)) and a number of different proofs are possible. We refer the reader to these articles for proofs of Theorem 3.7.2. If $\boldsymbol{v}$ is a vector in R^n, we use $\#(\boldsymbol{v})$ to denote the number of distinct values of $v_1, v_2, \ldots, v_n$.

Theorem 3.7.2 *Assume that for each $\boldsymbol{\theta} \in \Theta$, $G_{\boldsymbol{\theta}}$ is absolutely continuous, with a density $g_{\boldsymbol{\theta}}$ that is continuous on R. If the prior on F is given by (3.7.14), then the posterior distribution of F given $y_1, y_2, \ldots, y_n$ is*

$$\int \mathcal{DP}_{\alpha_{\boldsymbol{\theta}} + \sum_{i=1}^n \delta_{y_i}} \nu_{\boldsymbol{y}}(d\boldsymbol{\theta}), \qquad (3.7.18)$$

where $\nu_{\boldsymbol{y}}$ is the measure which is absolutely continuous with respect to ν and is defined by

$$\nu_{\boldsymbol{y}}(d\boldsymbol{\theta}) = c(\boldsymbol{y})\left(\prod_{\text{dist. } j} g_{\boldsymbol{\theta}}(y_i)\right)\left[\frac{(c_{0\boldsymbol{\theta}})^{\#(\boldsymbol{y})}\Gamma(c_{0\boldsymbol{\theta}})}{\Gamma(c_{0\boldsymbol{\theta}}+n)}\right]\nu(d\boldsymbol{\theta}). \tag{3.7.19}$$

In (3.7.19) the "dist." in the product indicates that the product is taken over distinct values only, Γ is the gamma function, and $c(\boldsymbol{y})$ is a normalizing constant.

The Gibbs sampler described in Doss and Huffer (1998) is straightforward, and relies on (3.7.17) and (3.7.19). We pick starting values $y_1^{(0)}, \ldots, y_n^{(0)}, \boldsymbol{\theta}^{(0)}$, with $y_i^{(0)} \in A_i$. Suppose that the current value of this vector is $\left(y_1^{(k-1)}, \ldots, y_n^{(k-1)}, \boldsymbol{\theta}^{(k-1)}\right)$. Generate

$$y_i^{(k)} \sim \left(\frac{\alpha_{\boldsymbol{\theta}^{(k-1)}} + \sum_{j<i}\delta_{y_j^{(k)}} + \sum_{j>i}\delta_{y_j^{(k-1)}}}{c_{0\boldsymbol{\theta}}+n-1}\right)_{A_i} \quad i = 1, \ldots, n, \tag{3.7.20}$$

and then

$$\boldsymbol{\theta}^{(k)} \sim \nu_{\boldsymbol{y}^{(k)}}, \tag{3.7.21}$$

where $\nu_{\boldsymbol{y}^{(k)}}$ is given by (3.7.19). In (3.7.20) we are using the following notation. If Λ is a probability measure, B is a set, and U is distributed according to Λ, then Λ_B denotes the conditional distribution function of U given that $U \in B$; that is,

$$\Lambda_B(A) = \Lambda(A \cap B)/\Lambda(B) \tag{3.7.22}$$

when $\Lambda(B) > 0$. Thus, in (3.7.20), the subscript A_i indicates that y_i is generated from the probability distribution inside the parentheses, conditional on its lying in the set A_i. (When conditioning on y_j $(j \neq i)$ and the data, conditioning on the events $y_j \in A_j$ $(j \neq i)$ is superfluous.)

Consider now (3.7.19). If $c_{0\boldsymbol{\theta}}$ is constant in $\boldsymbol{\theta}$, then the expression in brackets in (3.7.19) is a constant, and can be absorbed into $c(\boldsymbol{y})$. In this case, except for the fact that the product is over the distinct y_j's only, (3.7.19) is the posterior distribution of $\boldsymbol{\theta}$ in the standard finite-dimensional parametric model in which $\boldsymbol{\theta} \sim \nu$, and the y_i are i.i.d. with $y_i \sim G_{\boldsymbol{\theta}}$ given $\boldsymbol{\theta}$, for which the posterior is proportional to the likelihood times the prior.

This Gibbs sampler is distinct from the one described in Doss (1994). Doss and Huffer (1998) describe other Monte Carlo schemes for generating observations from $(y_1, y_2, \ldots, y_n, \boldsymbol{\theta}|D_{obs})$. For the case where the data are right censored, one of these schemes, based on the idea of sequential importance sampling (Kong, Liu, and Wong 1994), yields independent observations that are distributed exactly according to $(y_1, y_2, \ldots, y_n, \boldsymbol{\theta}|D_{obs})$.

Finally, we mention that Doss and Narasimhan (1998) implement this methodology in `Lisp-Stat` and have written software for carrying out

Bayesian survival analysis with MDP priors. The software for dynamic graphics is written in the lisp dialect implemented in `Lisp-Stat`. It is available in Doss and Narasimhan (1998) as a literate program in the sense of Knuth (1992). Briefly, a literate program is a program written in a style that makes it easy for humans to read, understand and modify. All their programs can be obtained electronically from `http://www-stat.stanford.edu/~naras/dirichlet-dynamic-graphics`. It must be noted that the software only does sensitivity analysis—no general facility is provided for generating observations from Markov chains. However, they have included the Fortran programs used in generating the output for their examples so that others may use it for running Gibbs samplers in models similar to theirs. The software can handle output from any source provided the samples are available in the proper format. The specific requirements are spelled out in detail in the literate program.

3.7.8 Inclusion of Covariates

Throughout this chapter, we have discussed Dirichlet process priors in survival analysis without the inclusion of covariates in the model. Dirichlet processes are quite difficult to work with in the presence of covariates, since they have no direct representation through either the hazard or cumulative hazard function. As explained by Hjort (1990), there are two ways of representing the hazard function. When the survival time distribution is continuous, we can write $S(y) = \exp(-H(y))$ and otherwise we can use the more general product-integral representation of Hjort (as given in Section 3.6). With the Dirichlet process prior on $S_0(y)$ and using the Cox model $S(y|x) = \{S_0(y)\}^{\exp(\boldsymbol{x}'\boldsymbol{\beta})}$, we cannot use either of these hazard representations. Another way of using the Dirichlet process would be to specify a Dirichlet process prior for the cdf of the random error ϵ in the accelerated life-time model $\log(y) = \alpha + \mathbf{x}'\boldsymbol{\beta} + \epsilon$. But, for such a model, we need to center the cdf of ϵ at 0 to assure identifiability. With the Dirichlet process, such a centering is not possible. To obtain a cdf for ϵ which is centered at 0, we can use a Polya tree process prior as discussed in Lavine, (1992). We discuss Polya tree priors in Chapter 10.

Exercises

3.1 Derive the likelihoods in (3.1.1), (3.1.2), and (3.2.5).

3.2 Consider the likelihood in (3.1.1) and the joint prior $\pi(\boldsymbol{\beta}, \boldsymbol{\lambda}) \propto \prod_{j=1}^{J} \lambda_j^{-1}$.

(a) Give necessary and sufficient conditions for the joint propriety of the joint posterior $(\boldsymbol{\beta}, \boldsymbol{\lambda})$.

(b) Derive the full conditional distributions for $\boldsymbol{\beta}$ and $\boldsymbol{\lambda}$.
(c) Using the myeloma data, write a Gibbs sampling algorithm to sample from the joint posterior of $(\boldsymbol{\beta}, \boldsymbol{\lambda})$.
(d) Obtain the posterior mean and covariance matrix of $\boldsymbol{\beta}$.
(e) Plot the marginal posterior density of β_1.

3.3 (a) Derive the likelihood function given by (3.2.14) using the equation in (3.2.13).
(b) Write a Gibbs sampling algorithm to sample from the joint posterior of $(\boldsymbol{\beta}, \boldsymbol{\delta})$ using $a_0 = 0$, 0.1, 0.5 for the myeloma data.

3.4 Using the myeloma data (E2479), write a program in BUGS to sample from the joint posterior of $(\boldsymbol{\beta}, \boldsymbol{h})$ in (3.2.7):

(a) Using noninformative choices for (μ_0, Σ_0) and c_0.
(b) Using Study 1 as the historical data, construct an informative normal prior for $\boldsymbol{\beta}$ using (3.3.7)–(3.3.9).
(c) Using the informative prior in part (b), and a noninformative prior for $\boldsymbol{h}$, carry out a data analysis using $a_0 = 0.01$, 0.1, 0.5, and 1. What are your conclusions about the impact of the historical data?

3.5 From (3.3.4), show that

$$\frac{\partial^2 \log \pi(\boldsymbol{\beta}|\boldsymbol{\delta}, a_0, D)}{\partial \beta_r^2} \leq 0,$$

and hence deduce that $[\boldsymbol{\beta}|\boldsymbol{\delta}, a_0, D]$ is log-concave in the components of $\boldsymbol{\beta}$.

3.6 (a) Using (3.2.1), show that for a gamma process prior on the cumulative baseline hazard $H_0(t)$, the likelihood function $L(\boldsymbol{\beta}, \gamma_0, c_0)$ for time continuous survival data is given by

$$L(\boldsymbol{\beta}, c_0, h_0|D) \propto \prod_{j=1}^{n} \left[\left(\frac{c_0}{c_0 + A_j} \right)^{c_0(H^*(y_{(j)}) - H^*(y_{(j-1)}))} \times \left(-c_0 h_0(y_{(j)}) \log \left(\frac{c_0 + A_{j+1}}{c_0 + A_j} \right) \right)^{\nu_{(j)}} \right],$$

where $\nu_{(j)}$ is the censoring indicator for the jth ordered survival time $y_{(j)}$.
(b) Using part (a), show that as $c_0 \to \infty$, $L(\boldsymbol{\beta}, c_0, h_0|D)$ is proportional to the likelihood function of $\boldsymbol{\beta}$ given in (3.2.12).

3.7 (a) Given time-continuous survival data, $D = \{y_{(j)}, \nu_{(j)}, x_{(j)},\ j = 1, 2, \ldots, n\}$, for $y_{(1)} < y_{(2)} < \cdots < y_{(n)}$, show that the full likelihood is given by

$$L(\boldsymbol{\beta}, H_0|D) \propto \prod_{j=1}^{n} \left[\left(\exp(x'_{(j)}\boldsymbol{\beta}) dH_0(y_{(j)}) \right)^{\nu_{(j)}} \right.$$
$$\left. \times \exp\left(- \exp(x'_{(j)}\boldsymbol{\beta} H_0(y_{(j)}) \right) \right],$$

where $H_0(t)$ is the cumulative baseline hazard.

(b) A popular naive approach is to assume $dH_0(t) = 0$ everywhere except at observed failure times, that is, at the $y_{(j)}$'s where $\nu_{(j)} = 1$. Suppose we assume i.i.d. improper priors

$$\pi(dH_0(y_{(j)})) \propto \left(dH_0(y_{(j)})\right)^{-1}$$

for $\nu_{(j)} = 1$. Show that the conditional posteriors for the Gibbs steps are

(i) $dH_0(y_{(j)}) \sim \mathcal{E}(A_j)$, where $A_j = \sum_{l \in \mathcal{R}_j} \exp(\boldsymbol{x}'_l\boldsymbol{\beta})$.

(ii) $\pi(\boldsymbol{\beta}|dH_0, D) \propto \left(- \sum_{j=1}^{n} A_j dH_0(y_{(j)}) \nu_{(j)} + \sum \nu_{(j)} \boldsymbol{x}'_{(j)}\boldsymbol{\beta} \right)$ where $\pi(\boldsymbol{\beta})$ is the prior for $\boldsymbol{\beta}$, which is independent of the prior for dH_0.

(c) Show that $\pi(\boldsymbol{\beta}|dH_0, D)$ is log-concave in $\boldsymbol{\beta}$ whenever $\pi(\boldsymbol{\beta})$ is log-concave.

(d) Show that the Cox's partial likelihood of $\boldsymbol{\beta}$ is given by

$$\text{PL}(\boldsymbol{\beta}|D) \propto \prod_{j=1}^{n} \left\{ \int_0^{\infty} \exp(\boldsymbol{x}'_{(j)}\boldsymbol{\beta}) \exp(-A_j u) du \right\}^{\nu_j},$$

and thus the Gibbs steps applied to part (d) would give us a method of sampling from the posterior of $\boldsymbol{\beta}$ based on partial likelihood.

3.8 For the generalized Cox model,

(a) derive the likelihood of the model in (3.4.1).

(b) derive the full conditionals given by (3.4.2)–(3.4.4).

3.9 Redo Exercise 3.8 for p covariates, i.e., when $\boldsymbol{x}_k$ is a $p \times 1$ vector of covariates.

3.10 Instead of the prior specifications given in Section 3.4 for the generalized Cox model with a univariate covariate, consider

(i) $\alpha_{k+1}|\alpha_1, \alpha_2, \ldots, \alpha_k \sim N(\alpha_k, v_k^2)$, where $\alpha_k = \log(\lambda_k)$, for $k = 0, 1, \ldots, J-1$;

(ii) $\beta_{k+1}|\beta_0, \beta_1, \ldots, \beta_k \sim N(\beta_k, w_k^2)$ for $k = 0, 1, \ldots, J-1$; and

(iii) the β_k's are a priori independent of $\boldsymbol{\alpha} = (\alpha_1, \alpha_2, \ldots, \alpha_J)'$.

The values of the hyperparameters are given as follows:

β_0	w_k	w_0	v_k	α_0	v_0
0	1.0	2.0	1.0	−0.1	2.0

(a) Using the complete-data likelihood given by (3.4.1), derive all necessary full conditionals for the Gibbs sampler under the above model.

(b) Reanalyze the breast cancer data given in Example 1.3 under this new model, and compare the posterior estimates of the β_k's to those displayed in Figure 3.2.

3.11 Consider the E1684 melanoma data from Example 1.2 using treatment and age as covariates.

(a) Write a Gibbs sampling program to sample from the correlated gamma process model in Section 3.6, using a normal prior for $\boldsymbol{\beta}$.

(b) Compare your inferences about $\boldsymbol{\beta}$ to those of the gamma process prior of Kalbfleisch (1978).

(c) Do a sensitivity analysis on the choices of α_k, ξ_k and γ_k. Are posterior inferences for $\boldsymbol{\beta}$ sensitive to the choices of α_k, ξ_k and γ_k?

3.12 In Exercise 3.7, let us define

$$\tilde{H}(t) = \sum_{j:y_{(j)} \le t} \nu_{(j)} dH_0(y_{(j)}).$$

(a) Prove that

$$E(\tilde{H}(t)|D) \approx \sum_{j:y_{(j)} \le t} \nu_{(j)} \frac{1}{\sum_{l \in \mathcal{R}_j} \exp(\boldsymbol{x}_l' \hat{\boldsymbol{\beta}})},$$

where $\hat{\boldsymbol{\beta}}$ is the posterior mean of $\boldsymbol{\beta}$ based on Cox's partial likelihood $\mathrm{PL}(\boldsymbol{\beta}|D)$.

(b) Use the Gibbs steps in Exercise 3.7 on a dataset using BUGS and compare the quantities $E(\boldsymbol{\beta}|D)$, $\mathrm{Var}(\boldsymbol{\beta}|D)$, and $E((\tilde{H}(t)|D))$ with classical estimators.

3.13 Consider the grouped data likelihood of Section 3.2.2. Suppose we have a "data-driven" prior for the r_j's such that $r_j = 0$ if $\mathcal{D}_j = \phi$ (i.e., the null set). For non-null $\mathcal{D}_j$'s assume an improper uniform prior for the r_j's, and assume that the r_j's are independent a priori. Show that

$$\pi(\boldsymbol{\beta}|D) \propto \sum_{R_c} \mathrm{PL}(\pi(\boldsymbol{\beta}|D_c)),$$

where the sum is over all possible complete rankings (R_c's) of the survival times which are compatible with the incomplete ranking induced by the observed data D.

3.14 Consider the beta process model in Section 3.5.1.

(a) Show that $\pi(\boldsymbol{q}|\boldsymbol{\beta}, D)$ given by (3.5.7) is log-concave in the components of $\boldsymbol{q}$ provided $c_{0j}\alpha_{0j} \geq 1$.

(b) Using (3.5.7), show that the conditional distribution $\pi(\boldsymbol{\beta}|\boldsymbol{q}, D)$ is log-concave in the components of $\boldsymbol{\beta}$ whenever $\pi(\boldsymbol{\beta})$ is log-concave.

(c) Use parts (a) and (b) to develop a Gibbs sampling algorithm to sample from $\pi(\boldsymbol{\beta}, \boldsymbol{q}|D)$. Compared to the Gibbs sampler using the auxiliary variables $\boldsymbol{\eta}$ and $\boldsymbol{\gamma}$, which version of the Gibbs algorithm is more efficient? Justify your answer.

3.15 Let $z_1, z_2, \ldots, z_k$ be independent random variables, with

$$z_j \sim \mathcal{G}(\alpha_j, 1), \quad j = 1, 2, \ldots k.$$

Define

$$u = \sum_{j=1}^{k} z_j,$$

and

$$y_j = \frac{z_j}{u} = \frac{z_j}{\sum_{j=1}^{k} z_j}.$$

Show that $(y_1, y_2, \ldots, y_{k-1})$ defined above have a $k-1$ dimensional Dirichlet distribution, denoted by $D_{k-1}(\alpha_1, \alpha_2, \ldots, \alpha_k)$.

3.16 Suppose

$$(x_1, x_2, \ldots, x_k) \sim \text{Multinomial}(p_1, p_2, \ldots, p_k),$$

where $\sum_{j=1}^{k} p_j = 1$. Show that the conjugate prior for $(p_1, p_2, \ldots, p_{k-1})$ is

$$(p_1, p_2, \ldots, p_{k-1}) \sim D_{k-1}(\alpha_1, \alpha_2, \ldots, \alpha_k),$$

and therefore show that a posteriori,

$$(p_1, p_2, \ldots, p_{k-1}|\boldsymbol{x}) \sim D_{k-1}(\alpha_1 + x_1, \alpha_2 + x_2, \ldots, \alpha_k + x_k),$$

where $\boldsymbol{x} = (x_1, x_2, \ldots, x_k)'$.

3.17 Suppose that $(y_1, y_2, \ldots, y_{k-1}) \sim D_{k-1}(\alpha_1, \alpha_2, \ldots, \alpha_k)$.

(a) Show that univariate marginal distributions are beta distributions. That is,

$$y_j \sim \mathcal{B}(\alpha_j, \alpha - \alpha_j)$$

for $j = 1, 2, \ldots, k-1$.

(b) Show that $(y_j, y_k) \sim D_2(\alpha_j, \alpha_k, \alpha - \alpha_j - \alpha_k)$.

(c) Using parts (a) and (b), show that all marginal distributions of Dirichlet are Dirichlet.

(d) Show that

$$(y_1+y_2, y_3+y_4+y_5) \sim D_2\left(\alpha_1+\alpha_2, \alpha_3+\alpha_4+\alpha_5, \alpha-\sum_{j=1}^{5}\alpha_j\right).$$

(e) Show that

$$E(y_j)=\frac{\alpha_j}{\alpha},\quad E(y_j^2)=\frac{\alpha_j(\alpha_j+1)}{\alpha(\alpha+1)},\quad E(y_iy_j)=\frac{\alpha_i\alpha_j}{\alpha(\alpha+1)},$$
$$\text{Var}(y_j)=\frac{\alpha_j(\alpha-\alpha_j)}{\alpha^2(\alpha+1)},\ \text{ and }\ \text{Cov}(y_i,y_j)=-\frac{\alpha_i\alpha_j}{\alpha^2(\alpha+1)}.$$

3.18 The breast cancer data of Example 1.3 can be used to demonstrate the discretized beta process prior. We will consider every grouping interval as the unit of the recorded data (length is one month).

(a) Impute the survival times as the left endpoint of the observed interval. Combine both groups and find the median (η) of the imputed survival times.

(b) Compute μ_0 from $(1-\mu_0)^\eta = .5$ and use this μ_0 as the prior mean of the discretized hazards (h_j's).

(c) Compute the c_{0j}'s as $c_0 = 4$, and $c_{0,j+1} = c_{0j} - \mu_0 c_{0j}$ for $j = 1, 2, \ldots, J$. We note that this implies that our confidence around the prior is equivalent to the confidence obtained by observing four complete observations from previous data.

(d) Use this common beta process prior as interval-censored data from two groups to obtain Gibbs samples from two independent posterior distributions.

(e) Plot $E(S_0(y)|D)$, $E(S_1(y)|D)$, and $E(S_c(t)|D)$ versus time on the same graph, where $E(S_c(t)|D)$ is the common prior estimate of the survival function.

(f) Comment on the equality of the survival distributions for the two groups.

4
Frailty Models

In studies of survival, the hazard function for each individual may depend on a set of risk factors or explanatory variables but usually not all such variables are known or measurable. This unknown and unobservable risk factor of the hazard function is often termed as the individual's heterogeneity or *frailty*—a term coined by Vaupel, Manton, and Stallard (1979). Frailty models are becoming increasing popular in multivariate survival analysis since they allow us to model the association between the individual survival times within subgroups or *clusters* of subjects. With recent advances in computing technology, Bayesian approaches to frailty models are now computationally feasible, and several approaches have been discussed in the literature. The various approaches differ in the modeling of the baseline hazard or in the distribution of the frailty. Fully parametric approaches to frailty models are examined in Sahu, Dey, Aslanidou, and Sinha (1997), where they consider a frailty model with a Weibull baseline hazard. Semiparametric approaches have also been examined. Clayton (1991) and Sinha (1993, 1997) consider a gamma process prior on the cumulative baseline hazard in the frailty model. Sahu, Dey, Aslanidou, and Sinha (1997), Sinha and Dey (1997), Aslanidou, Dey, and Sinha (1998), and Sinha (1998) discuss frailty models with piecewise exponential baseline hazards. Qiou, Ravishanker, and Dey (1999) examine a positive stable frailty distribution, and Gustafson (1997) and Sargent (1998) examine frailty models using Cox's partial likelihood. In this chapter, we present an overview of these various approaches to frailty models, and discuss Bayesian inference as well as computational implementation of these methods.

4.1 Proportional Hazards Model with Frailty

The most common type of frailty model is called the shared-frailty model, which is an extension of the Cox proportional hazards model (Cox, 1972). This model can be derived as follows. Let y_{ij} denote the survival time for the jth subject in the ith cluster, $i = 1, 2, \ldots, n$, and $j = 1, 2, \ldots, m_i$. Thus, m_i represents the number of subjects in the ith cluster, and therefore we have a total of $N = \sum_{i=1}^n m_i$ subjects. In the shared frailty model, we assume that the conditional hazard function of y_{ij} given the unobserved frailty random variable w_i for the ith cluster and the fixed covariate vector x_{ij} is given by

$$h(y|w_i, x_{ij}) = h_0(y) w_i \exp(\boldsymbol{x}'_{ij}\boldsymbol{\beta}), \tag{4.1.1}$$

$i = 1, 2, \ldots n$, $j = 1, 2, \ldots, m_i$, where $\boldsymbol{\beta}$ is the $p \times 1$ vector of unknown regression coefficients, $h_0(.)$ is an unknown baseline hazard function common to every subject, and $\boldsymbol{x}_{ij}$ is the $p \times 1$ covariate vector for the jth subject in the ith cluster, and may be time dependent. The model in (4.1.1) was first motivated by Vaupel, Manton, and Stallard (1979), and further developed by Clayton and Cuzick (1985), and Oakes (1986, 1989). A common method is to use a parametric distribution for the frailty w_i. Finite mean frailty distributions such as the gamma and log-normal are very popular in the literature in spite of their theoretical limitations (see Hougaard, 1995). Other alternatives include using infinite mean distributions such as the positive stable distribution.

The gamma distribution is the most commonly used finite mean distribution to model the frailty term w_i. For finite mean frailty distributions, we need the mean of the frailty distribution to be unity in order for the parameters of the model to be identifiable. Thus, for the gamma frailty model, the w_i's are assumed to be i.i.d. with

$$w_i \sim \mathcal{G}(\kappa^{-1}, \kappa^{-1}), \tag{4.1.2}$$

where κ is the (unknown) variance of the w_i's. Thus, larger values of κ imply greater heterogeneity among clusters.

Letting $\boldsymbol{w} = (w_1, w_2, \ldots, w_n)'$, we have

$$\pi(\boldsymbol{w}|\kappa) \propto \prod_{i=1}^n w_i^{\kappa^{-1}-1} \exp(-\kappa^{-1} w_i). \tag{4.1.3}$$

The frequentist literature on estimation with the gamma frailty model given by (4.1.1) and (4.1.3) is enormous. We refer the reader to Klein and Moeschberger (1997, Chapter 13), and Andersen, Borgan, Gill, and Keiding (1993) and the references therein for a review of likelihood based frequentist inferences for the gamma frailty model. As mentioned in Anderson et al. (1993), in spite of promising results by several authors, formal and completely satisfactory justifications of these likelihood-based methods await more results on their asymptotic and convergence properties.

Moreover, none of these likelihood-based methods directly maximize the full likelihood given the data, and the small sample properties of these estimators have yet to be studied. Thus, Bayesian methods are attractive for these models since they easily allow an analysis using the full likelihood and inference does not rely on asymptotics. In the following subsections, we present several Bayesian approaches to the frailty model.

4.1.1 Weibull Model with Gamma Frailties

Let ν_{ij} denote the censoring indicator variable, taking value 1 if the jth subject ($j = 1, 2, \ldots, m_i$) of the ith cluster ($i = 1, 2, \ldots, n$) fails and 0 otherwise. Hence, y_{ij} is a failure time if $\nu_{ij} = 1$ and a censoring time otherwise. Further, let $\boldsymbol{\nu} = (\nu_{11}, \nu_{12}, \ldots, \nu_{nm_n})'$, $\boldsymbol{y} = (y_{11}, y_{12}, \ldots, y_{nm_n})'$, and $X = (X_1, X_2, \ldots, X_n)$, where X_i is the $m \times p$ matrix of covariates for the ith cluster. Let $D = (X, \boldsymbol{\nu}, \boldsymbol{y}, \boldsymbol{w})$ denote the complete data and let $D_{obs} = (X, \boldsymbol{\nu}, \boldsymbol{y})$ denote the observed data. Here, we only allow right censored survival data and assume that the censoring is noninformative. Let the Weibull baseline hazard function be given by

$$h_0(y_{ij}) = \gamma \alpha y_{ij}^{\alpha - 1},$$

where (γ, α) are the parameters of the Weibull distribution. The hazard function is given by

$$h(y_{ij} | \boldsymbol{x}_{ij}, w_i) = \gamma \alpha w_i y_{ij}^{\alpha - 1} \theta_{ij}, \tag{4.1.4}$$

where $\theta_{ij} = \exp(\boldsymbol{x}_{ij}' \boldsymbol{\beta})$, and the complete data likelihood is given by

$$L(\boldsymbol{\beta}, \gamma, \alpha | D) = \prod_{i=1}^{n} \prod_{j=1}^{m_i} \left(\gamma \alpha y_{ij}^{\alpha - 1} w_i \theta_{ij} \right)^{\nu_{ij}} \exp\left\{ -\gamma y_{ij}^{\alpha} \theta_{ij} w_i \right\}. \tag{4.1.5}$$

The likelihood function of $(\boldsymbol{\beta}, \gamma, \alpha)$ based on the observed data D_{obs} can be obtained by integrating out the w_i's from (4.1.5) with respect to the density $\pi(\boldsymbol{w}|\kappa)$ given in (4.1.3). The observed data likelihood is far too complicated to work with, and thus it is difficult to evaluate the joint posterior distribution of $(\boldsymbol{\beta}, \gamma, \alpha)$ analytically. To circumvent this problem, we use the Gibbs sampler to generate samples from the joint posterior distribution.

Let $\pi(\cdot)$ denote the prior density of its argument and let

$$S = \sum_{i=1}^{n} \sum_{j=1}^{m_i} y_{ij}^{\alpha} \theta_{ij} w_i. \tag{4.1.6}$$

The full conditional distribution of each w_i is a gamma distribution, i.e.,

$$(w_i | \boldsymbol{\beta}, \alpha, \gamma, D_{obs}) \sim \mathcal{G}\left\{ \kappa^{-1} + \sum_{j=1}^{m_i} \nu_{ij}, \kappa^{-1} + \gamma \sum_{j=1}^{m_i} y_{ij}^{\alpha} \theta_{ij} \right\}, \tag{4.1.7}$$

for $i = 1, 2, \ldots, n$. Letting $\eta = \kappa^{-1}$, the full conditional distribution of η is given by

$$\pi(\eta|\boldsymbol{\beta}, \boldsymbol{w}, D_{obs}) \propto \prod_{i=1}^{n} w_i^{\eta-1} \eta^{-n\eta} \frac{\exp\{-\eta \sum_{i=1}^{n} w_i\}}{[\Gamma(\eta)]^n} \pi(\eta). \tag{4.1.8}$$

The full conditional of $\boldsymbol{\beta}$ is given by

$$\pi(\boldsymbol{\beta}|\eta, \boldsymbol{w}, D_{obs}) \propto \exp\left\{\boldsymbol{\beta}' \sum_{i=1}^{n} \sum_{j=1}^{m_i} \nu_{ij} \boldsymbol{x}_{ij} - \gamma S\right\} \pi(\boldsymbol{\beta}). \tag{4.1.9}$$

If a priori $\gamma \sim \mathcal{G}(\rho_1, \rho_2)$, then the full conditional of γ is a gamma distribution given by

$$(\gamma|\boldsymbol{\beta}, \alpha, \boldsymbol{w}, D_{obs}) \sim \mathcal{G}\left\{\rho_1 + \sum_{i=1}^{n} \sum_{j=1}^{m_i} \nu_{ij},\ \rho_2 + S\right\}. \tag{4.1.10}$$

Finally, the full conditional of α is given by

$$\pi(\alpha|\boldsymbol{\beta}, \gamma, \boldsymbol{w}, D_{obs}) \propto \left(\prod_{i=1}^{n} \prod_{j=1}^{m_i} y_{ij}^{\nu_{ij}}\right)^{\alpha-1} \alpha^{\sum_{i=1}^{n} \sum_{j=1}^{m_i} \nu_{ij}} \exp\{-\gamma S\} \pi(\alpha). \tag{4.1.11}$$

A priori, it is common to take $\alpha \sim \mathcal{G}(a_1, a_2)$, so that $\pi(\alpha) \propto \alpha^{a_1-1} \exp(-\alpha a_2)$. With these choices of priors, it can be shown that each of the above full conditional densities in (4.1.7)–(4.1.11) is log-concave. The frailty model in (4.1.4) is a multiplicative frailty model. The modeling strategy used in the BUGS (Spiegelhalter, Thomas, Best, and Gilks, 1995) software manual is based on an additive frailty model,

$$h(y_{ij}|\boldsymbol{x}_{ij}, b_i) = \xi_{ij} \alpha y_{ij}^{\alpha-1}, \tag{4.1.12}$$

where

$$\log(\xi_{ij}) = \zeta + \boldsymbol{x}_{ij}'\boldsymbol{\beta} + b_i, \tag{4.1.13}$$

the b_i's are assumed i.i.d. $N\left(0, \kappa^{-1}\right)$, and κ is given a $\mathcal{G}(\phi_1, \phi_2)$ prior. The prior for α is $\mathcal{G}(a_1, a_2)$, and ζ is given a normal prior. It is expected that both the multiplicative and additive hazard Weibull model formulations will yield similar inferences since this additive frailty model is actually a multiplicative frailty model with a log-normal frailty distribution.

Example 4.1. Kidney infection data. We consider the kidney infection data discussed in Example 1.5, and fit a Weibull frailty model. Spiegelhalter, Thomas, Best, and Gilks (1995) fit a Weibull frailty model in BUGS to these data using an additive frailty model for the hazard. Spiegelhalter, Thomas, Best, and Gilks (1995) consider five covariates in the model (age$_{ij}$, sex$_i$, disease$_{i1}$, disease$_{i2}$, disease$_{i3}$). For ease of exposition, we label these covariates as $(x_{ij1}, x_{ij2}, x_{ij3}, x_{ij4}, x_{ij5})$, respectively. The model is thus given by

$$y_{ij} \sim \mathcal{W}(\alpha, \gamma_{ij}), \tag{4.1.14}$$

where $i = 1, 2, \ldots, 38$, $j = 1, 2$, and

$$\log(\gamma_{ij}) = \beta_0 + \beta_1 x_{ij1} + \beta_2 x_{ij2} + \beta_3 x_{ij3} + \beta_4 x_{ij4} + \beta_5 x_{ij5} + b_i,$$

where $b_i \sim N(0, \sigma^2)$. Following Spiegelhalter, Thomas, Best, and Gilks (1995), we take $\boldsymbol{\beta} \sim N(0, 10^5 I)$, and $\sigma^{-2} \sim \mathcal{G}(10^{-4}, 10^{-4})$. Further, we take $\alpha \sim \mathcal{G}(1, 10^{-4})$, which yields a prior that is slowly decreasing on the positive real line. The `BUGS` code and datasets for this model are given at "`http://www.springer-ny.com`." Table 4.1 below gives posterior estimates of the regression parameters based on 2500 iterations after a 2500 iteration burn-in. The results of McGilchrist and Aisbett (1991) (labeled M & A) using an iterative Newton-Raphson estimation procedure are also shown in Table 4.1 for comparison.

TABLE 4.1. Posterior Estimates of $\boldsymbol{\beta}$ for Kidney Infection Data.

Parameter	`BUGS` Estimate (SE)	M & A Estimate (SE)
β_1	0.003 (0.015)	0.006 (0.013)
β_2	−1.866 (0.496)	−1.795 (0.434)
β_3	−0.055 (0.592)	0.206 (0.484)
β_4	0.586 (0.582)	0.410 (0.494)
β_5	−1.269 (0.822)	−1.296 (0.712)
σ	0.496 (0.378)	0.382 –

Source: Spiegelhalter, Thomas, Best, and Gilks (1995).

4.1.2 Gamma Process Prior for $H_0(t)$

Clayton (1991) derives the Bayesian frailty model assuming a gamma process for the cumulative baseline hazard $H_0(t)$. When the gamma process prior on $H_0(t)$ as developed in Chapter 3 is used, the complete (time-continuous) data likelihood of $(\boldsymbol{\beta}, H_0)$ is given by

$$\begin{aligned} L(\boldsymbol{\beta}, H_0) &= \prod_{i=1}^{n} \prod_{j=1}^{m_i} \left(w_i \exp(\boldsymbol{x}_{ij}'\boldsymbol{\beta}) dH_0(y_{ij})\right)^{\nu_{ij}} \exp\left\{-\exp(\boldsymbol{x}_{ij}'\boldsymbol{\beta}) w_i H_0(y_{ij})\right\} \\ &\propto \left(\prod_{i=1}^{n} w_i^{d_i}\right) \exp\left(\sum_{i=1}^{n}\sum_{j=1}^{m_i} \nu_{ij}\boldsymbol{x}_{ij}'\boldsymbol{\beta}\right) \left(\prod_{i=1}^{n}\prod_{j=1}^{m_i} (dH_0(y_{ij}))^{\nu_{ij}}\right) \\ &\times \exp\left(-\sum_{i=1}^{n}\sum_{j=1}^{m_i} \exp(\boldsymbol{x}_{ij}'\boldsymbol{\beta}) w_i H_0(y_{ij})\right). \end{aligned} \tag{4.1.15}$$

As discussed in Chapter 3, a grouped data version of (4.1.15) is given by

$$L(\boldsymbol{\beta}, \boldsymbol{h}|w, D) = \prod_{j=1}^{J} G_j h_j^{c_0\alpha_{0j}-1} \exp(-c_0 h_j), \tag{4.1.16}$$

where

$$\begin{aligned} G_j &= \exp(-h_j \sum_{k\in\mathcal{R}_j} \exp(\boldsymbol{x}_k'\boldsymbol{\beta} + \log(w_k))) \\ &\times \prod_{l\in\mathcal{D}_j} (1 - \exp(-\exp(h_j \boldsymbol{x}_l'\boldsymbol{\beta} + \log(w_l)))), \end{aligned}$$

$\boldsymbol{h} = (h_1, h_2, \ldots, h_J)'$, the h_j's are independent, $h_j \sim \mathcal{G}(\alpha_{0j} - \alpha_{0,j-1}, c_0)$ as defined in (3.2.4), $\mathcal{R}_j$ is the set of patients at risk, and $\mathcal{D}_j$ is the set of patients having failures in the jth interval as defined in Section 3.2.2. The joint posterior distribution of $(\boldsymbol{\beta}, \boldsymbol{h}, \boldsymbol{w}|D)$ is given by

$$\pi(\boldsymbol{\beta}, \boldsymbol{h}, \boldsymbol{w}|D_{obs}) \propto L(\boldsymbol{\beta}, \boldsymbol{h}|D)\pi(\boldsymbol{w})\pi(\boldsymbol{\beta}), \tag{4.1.17}$$

where $\pi(\boldsymbol{\beta})$ is the prior distribution for $\boldsymbol{\beta}$, and $\pi(\boldsymbol{w})$ is given by (4.1.2). Clayton (1991) assumes a normal prior for $\boldsymbol{\beta}$, although other priors can certainly be used. To sample from (4.1.17), we need to sample from the following full conditional distributions:

$$(i) \quad [\boldsymbol{\beta}|\boldsymbol{w}, \boldsymbol{h}, D_{obs}]; \tag{4.1.18}$$

$$(ii) \quad [\boldsymbol{w}|\boldsymbol{\beta}, \boldsymbol{h}, D_{obs}]; \text{ and} \tag{4.1.19}$$

$$(iii) \quad [\boldsymbol{h}|\boldsymbol{\beta}, \boldsymbol{w}, D_{obs}]. \tag{4.1.20}$$

We leave the derivations of these full conditional distributions as an exercise (see Exercise 4.2).

Example 4.2. Carcinogenesis data. We consider the animal carcinogenesis dataset described by Clayton (1991). The dataset is given in Table 1.5. The experiment was litter-matched and used three rats from each of 50 litters, one rat being treated with putative carcinogen and the other two serving as controls. The time to tumor occurrence or censoring is recorded to the nearest week. The data set contains a single covariate x, with value 1 or 0 indicating exposure or nonexposure to carcinogen. As a reference, a Cox regression analysis was carried out using the SAS PHREG procedure. This yielded a maximum partial likelihood estimate of β of 0.907 and an estimate of the asymptotic variance of 0.1008. For the Bayesian analysis, vague priors were used throughout. The posterior mode of (β, κ) is $(0.919, 0.502)$. Table 4.2 is based on 2000 Gibbs iterations.

TABLE 4.2. Posterior Summaries of (β, κ) for Carcinogenesis Data.

	β	κ
Mean	0.986	0.685
Variance	0.114	0.170
Quantiles		
0.05	0.436	0.147
0.25	0.759	0.366
0.50	0.981	0.618
0.75	1.211	0.935
0.95	1.548	1.494

Source: Clayton (1991).

4.1.3 Piecewise Exponential Model for $h_0(t)$

Piecewise exponential models provide a very flexible framework for modeling univariate survival data. Although in a strict sense it is a parametric model, a piecewise exponential hazard can approximate any shape of a nonparametric baseline hazard. Therefore, in practice, defining a prior for a piecewise exponential hazard is the same as defining a prior process for the nonparametric hazard function.

We construct the piecewise exponential model in the same way discussed in Chapter 3. We first divide the time axis into J prespecified intervals $I_k = (s_{k-1}, s_k]$ for $k = 1, 2, \ldots, J$, where $0 = s_0 < s_1 < \ldots < s_J < \infty$, s_J being the last survival or censored time and assume the baseline hazard to be constant within intervals. That is,

$$h_0(y) = \lambda_k, \quad \text{for} \quad y \in I_k. \tag{4.1.21}$$

Breslow (1974) used distinct failure times as the endpoints of the intervals. Kalbfleisch and Prentice (1973) suggested that the selection of the grid $\{s_1, s_2, \ldots, s_J\}$ should be made independent of the data. We will discuss the choice of grid later in this subsection.

Although modeling the baseline hazard through an independent increment prior process such as the gamma or beta process is attractive and common, in many applications, it often turns out that prior information is available on the smoothness of the hazard rather than the actual baseline hazard itself (see, e.g., Leonard, 1978; Gamerman, 1991). For the frailty model, the marginal hazard is always a nonlinear function of the baseline hazard involving covariates and frailty parameters as well. However, the ratio of marginal hazards at the nearby time-points given the same covariates is approximately equal to the ratio of baseline hazards at these points. Thus, in these cases, it is more attractive to consider some type of correlated prior process for the baseline hazard. Such models have been considered by Aslanidou, Dey, and Sinha (1998) and Sahu, Dey, Aslanidou, and Sinha (1997), and we follow their approach in this subsection.

To correlate the λ_k's in adjacent intervals, a discrete-time martingale process is used, similar to that of Arjas and Gasbarra (1994) for the univariate survival model. Given $(\lambda_1, \lambda_2, \ldots, \lambda_{k-1})$ we specify that

$$\lambda_k | \lambda_1, \lambda_2, \ldots, \lambda_{k-1} \sim \mathcal{G}\left(\alpha_k, \frac{\alpha_k}{\lambda_{k-1}}\right), \quad k = 1, 2 \ldots, J, \tag{4.1.22}$$

where $\lambda_0 = 1$, and $E(\lambda_k | \lambda_1, \lambda_2, \ldots, \lambda_{k-1}) = \lambda_{k-1}$. The parameter α_k in (4.1.22) controls the amount of smoothness available and a small α_k indicates less information on the smoothing of the λ_k's. If $\alpha_k = 0$, then λ_k and λ_{k-1} are independent as in Chapter 3. When $\alpha_k \to \infty$, then the baseline hazard is the same in the intervals I_k and I_{k-1}: i.e., $\lambda_k = \lambda_{k-1}$. Thus, in the limiting case, we get a constant baseline hazard. Notice, though, that if prior information is available, the shape and scale of the marginal prior of λ_k will change accordingly and the prior will be more informative. A version of the above process which can also be used, was given by Leonard (1978) and Gamerman (1991). They modeled $\log(\lambda_k) = \xi_k$ and took

$$\xi_k | \xi_{k-1} \sim N(\xi_{k-1}, \tau^2), \quad k = 1, 2, \ldots, J \tag{4.1.23}$$

with $\xi_0 = 0$. Taking this further, we can assume a second difference prior process for ξ_k, i.e., $\xi_k | \xi_{k-2}, \xi_{k-1} \sim N(2\xi_{k-1} - \xi_{k-2}, \tau^2), k = 3, 4, \ldots, J$, and so forth.

A few remarks are in order on the choice of the number of grid intervals J. It is clear that a very large J will make the model nonparametric. However, too large a J will produce unstable estimators of the λ_k's and too small a choice may lead to poor model fitting. Hence, a robust choice of J should be considered here. Note that the maximum likelihood estimate of λ_k depends on the number of failures, d_k, in the kth interval I_k and is 0 if d_k is zero. One advantage of the Bayesian approach with the correlated prior process described here is to smooth out such jumps to zero. A random choice of J will make the posterior distribution have a varying dimension and sampling techniques other than the Gibbs sampler, e.g., reversible jump MCMC (see Green, 1995), can be used to compute the posterior distribution. The above models can also be easily altered to accommodate monotone baseline hazard functions. Suppose that one intends to model the λ_k's with the constraint $\lambda_1 \le \lambda_2 \le \ldots \le \lambda_J$. Following Arjas and Gasbarra (1994), we can assume that

$$\lambda_k - \lambda_{k-1} \sim \mathcal{G}(\alpha_k, \alpha_k),\ k = 1, 2, \ldots, J$$

instead of (4.1.22) or (4.1.23).

The likelihood can now be derived as follows. The jth subject of the ith cluster has a constant hazard of $h_{ij} = \lambda_k \theta_{ij} w_i$ in the kth interval ($k = 1, 2, \ldots, J$) given the unobserved frailty w_i. If the subject has survived beyond the kth interval, i.e., $y_{ij} > s_k$, the likelihood contribution is $\exp\{-\lambda_k \Delta_k \theta_{ij} w_i\}$, where $\Delta_k = s_k - s_{k-1}$. If the subject has failed or is censored in the kth interval, i.e., $s_{k-1} < y_{ij} \le s_k$, then the likelihood contribution is $(\lambda_k \theta_{ij} w_i)^{\delta_{ij}} \exp\left\{-\lambda_k (y_{ij} - s_{k-1})\theta_{ij} w_i\right\}$. Hence, the complete data likelihood is given by

$$L(\boldsymbol{\beta}, \boldsymbol{\lambda}|D) \propto \prod_{i=1}^{n} \prod_{j=1}^{m_i} \left[\left\{ \prod_{k=1}^{g_{ij}} \exp(-\lambda_k \Delta_k \theta_{ij} w_i) \right\} \left(\lambda_{g_{ij}+1} \theta_{ij} w_i\right)^{\delta_{ij}} \right.$$
$$\left. \times \exp\left\{-\lambda_{g_{ij}+1}(y_{ij} - s_{g_{ij}})\theta_{ij} w_i\right\} \right], \qquad (4.1.24)$$

where g_{ij} is such that $y_{ij} \in (s_{g_{ij}}, s_{g_{ij}+1}] = I_{g_{ij}+1}$.

The full conditionals needed for Gibbs sampling are given as follows. A posteriori, the w_i's are independent, and the full conditional of each w_i is a gamma distribution given by

$$(w_i|\boldsymbol{\beta}, \boldsymbol{\lambda}, D) \sim \mathcal{G}\left\{ \kappa^{-1} + \sum_{j=1}^{m_i} \nu_{ij},\ \kappa^{-1} + S_i \right\},\ i = 1, 2, \ldots, n, \qquad (4.1.25)$$

where

$$S_i = \sum_{j=1}^{m_i} \theta_{ij} \left(\sum_{k=1}^{g_{ij}} \lambda_k \Delta_k + \lambda_{g_{ij}+1}(s_{ij} - s_{g_{ij}}) \right). \qquad (4.1.26)$$

The full conditional density of $\eta = \kappa^{-1}$ is given by

$$\pi(\eta|\boldsymbol{\beta}, \boldsymbol{\lambda}, D) \propto \prod_{i=1}^{n} w_i^{\eta-1} \eta^{-n\eta} \frac{\exp\left\{-\eta \sum_{i=1}^{n} w_i\right\}}{[\Gamma(\eta)]^n} \pi(\eta). \qquad (4.1.27)$$

The full conditional density of $\boldsymbol{\beta}$ is

$$\pi(\boldsymbol{\beta}|\boldsymbol{\lambda}, D) \propto \exp\left\{ \sum_{i=1}^{n} \sum_{j=1}^{m_i} \nu_{ij} \boldsymbol{x}_{ij}' \boldsymbol{\beta} - \sum_{i=1}^{n} w_i S_i \right\} \pi(\boldsymbol{\beta}). \qquad (4.1.28)$$

Let

$$V_k = \sum_{(i,j) \in \mathcal{R}_k} \Delta_k \theta_{ij} w_i + \sum_{(i,j) \in \mathcal{D}_k} (y_{ij} - s_{k-1})\theta_{ij} w_i,$$

where $\mathcal{R}_k = \{(i,j); y_{ij} > s_k\}$ is the risk set at s_k, and $\mathcal{D}_k = \mathcal{R}_{k-1} - \mathcal{R}_k$. The full conditional density of λ_k, $k = 1, 2, \ldots, J$, is thus given by

$$\pi(\lambda_k|\beta, D) \propto \lambda_k^{d_k} \exp\{-\lambda_k V_k\}\, \pi(\lambda_k|\boldsymbol{\lambda}^{(-k)}), \tag{4.1.29}$$

where d_k is the number of failure times that occurred in the interval I_k, $\boldsymbol{\lambda}^{(-k)}$ is $\boldsymbol{\lambda} = (\lambda_1, \lambda_2, \ldots, \lambda_J)'$ without the kth component, and $\pi(\lambda_k|\boldsymbol{\lambda}^{(-k)})$ is the conditional prior, given by

$$\pi(\lambda_1|\lambda^{(-1)}) \propto \lambda_1^{-\alpha_0} \exp(-\alpha_0 \lambda_1^{-1}\lambda_2),$$

$$\pi(\lambda_k|\lambda^{(-k)}) \propto \lambda_k^{-1} \exp\left\{-\alpha_0\left(\frac{\lambda_k}{\lambda_{k-1}} + \frac{\lambda_{k+1}}{\lambda_k}\right)\right\},\quad k = 2, 3, \ldots, J-1,$$

and

$$\pi(\lambda_J|\lambda^{(-J)}) \propto \lambda_J^{\alpha_0 - 1} \exp\left(-\alpha_0 \frac{\lambda_J}{\lambda_{J-1}}\right).$$

It can be shown that, except for the density of λ_k in (4.1.29), each of the above conditional densities in (4.1.25)–(4.1.27) is log-concave. A sufficient condition for the density of λ_k in (4.1.29) to be log-concave is $d_k \geq 1$, that is, there is at least one failure in the kth interval. Log-concavity of each λ_k will be ensured if there is at least one failure in each of the intervals $(s_{k-1}, s_k]$.

Example 4.3. Kidney infection data (Example 4.1 continued). We again consider the kidney infection data given in Chapter 1. Here, we consider the two covariates: age of the patient at the time of each infection and sex of the patient. The dataset has been reanalyzed by Spiegelhalter, Thomas, Best, and Gilks (1995) using an additive frailty model as described in Section 4.1.1. Following Sahu, Dey, Aslanidou, and Sinha (1997), we consider the following four models for this dataset.

Model I: Piecewise exponential model with gamma priors for the λ_k's as in (4.1.22).

Model II: Weibull baseline hazard with multiplicative gamma frailties.

Model III: Piecewise exponential baseline hazard with normal priors for the $\log(\lambda_k)$'s as in (4.1.23).

Model IV: Weibull baseline hazard with additive frailties.

The proportional hazard's component of each of the above models is

$$\theta_{ij} = \exp(\boldsymbol{x}_{ij}'\boldsymbol{\beta}) = \exp(\beta_{sex}\, sex_i + \beta_{age}\, age_{ij}),$$

where $sex_i = 1$ if the ith patient is a female and 0 otherwise, age_{ij} is the age at the jth infection of the ith patient.

Each of the above models was fitted using Gibbs sampling, and the BUGS software (Spiegelhalter Thomas, Best, and Gilks, 1995) was used to fit Model IV. Similar to other data examples in this book,

(i) the adaptive rejection sampling of Gilks and Wild (1992) was used whenever the full conditional distribution did not have an analytic closed form and was log-concave;

(ii) a Metropolis step was implemented if the distribution was not log-concave; and

(iii) several convergence diagnostics measures were calculated, e.g., Geweke (1992) and Raftery and Lewis (1992) to monitor convergence.

The first 1000 iterates in each case were discarded and the subsequent 10,000 iterates were used to make inference.

The following values of hyperparameters were used in this example. For the prior on κ^{-1}, we take $\phi_1 = \phi_2 = 0.001$. Each component of $\boldsymbol{\beta}$ was assumed a priori normal with 0 mean and variance 10^3. The same prior was assumed for ζ in Model IV. For Model I all the α_k's were assumed to be 0.01. For Model III, τ^2 was taken as 10^4 to make it comparable with the corresponding prior precision for the λ_k's in Model I. For Models III and IV we took $\rho_1 = \rho_2 = 0.001$, and $a_1 = a_2 = 0.001$. We first investigated different choices of the grid size J for Models I and III. We experimented with three choices of $J = 5$, 10, and 20. The $J = 5$ case seemed to give worse model fitting than the $J = 10$ case and the last choice of J did not provide substantially better results than the $J = 10$ case. Hence we decided to use $J = 10$ throughout. Model fitting and/or model choice were not very sensitive to small variations on the values of the other hyperparameters as given above. Widely different α_k's in Model I did change the estimates a little bit. However, that did not alter the model choice ordering as reported below.

Table 4.3 shows the posterior mean, standard deviation, and 95% credible intervals for $\beta_{sex}, \beta_{age}, \kappa$. We show the estimates of α and γ (for Model IV $\mu = \exp\{\zeta\}$) in Table 4.4. In both Tables 4.3 and 4.4, posterior means are followed by standard deviations in the first row, and 95% credible intervals are shown in the second row. The estimates of β_{sex} show that the female patients have a slightly lower risk for infection. The estimates of κ from different models show that there is strong posterior evidence of a high degree of heterogeneity in the population of patients. Some patients are expected to be very prone to infection compared to others with the same covariate value. This is not very surprising, as in the dataset there is a male patient with infection times 8 and 16, and there is also another male patient with infection times 152 and 562. The high posterior means of κ also provide evidence of a strong positive association between two infection times for the same patient. The above analysis suggests that Models I and

III are very close to each other while Models II and IV are also somewhat similar.

TABLE 4.3. Parameter Estimates from Different Models for Kidney Infection Data.

	β_{sex}	β_{age}	κ
Model I	−1.493 (0.468)	0.006 (0.013)	0.499 (0.283)
	(−2.430, −0.600)	(−0.018, 0.032)	(0.061, 1.160)
Model II	−1.888 (0.564)	0.007 (0.013)	0.585 (0.307)
	(−3.034, −0.846)	(−0.018, 0.032)	(0.115, 1.317)
Model III	−1.500 (0.480)	0.007 (0.013)	0.523 (0.285)
	(−2.467, −0.624)	(−0.018, 0.036)	(0.089, 1.195)
Model IV	−1.69 (0.529)	0.006 (0.014)	0.816 (0.507)
	(−2.780, −0.699)	(−0.019, 0.036)	(0.079, 2.050)

Source: Sahu, Dey, Aslanidou, and Sinha (1997).

TABLE 4.4. Parameter Estimates of α and μ for Weibull Models II and IV for Kidney Infection Data.

	Model II	Model IV
α	1.278 (0.190)	1.22 (0.160)
	(0.937, 1.692)	(0.916, 1.540)
μ	0.016 (0.015)	0.013 (0.014)
	(0.001, 0.058)	(0.001, 0.053)

Source: Sahu, Dey, Aslanidou, and Sinha (1997).

Finally, we mention that in Sections 4.1.1–4.1.3, we have used a gamma distribution for the frailty. The choice of the gamma distribution for frailties arises partly for mathematical convenience, since this distribution is conjugate to the likelihood for w_i and partly because, in the case of bivariate survival data without covariates, integration over the unknown frailty yields a class of bivariate survival time distributions with appealing properties (see Clayton, 1978; Cox and Oakes, 1984). This choice also has some less desirable consequences, however. The marginal relationship between the hazard and covariates no longer follows the proportional hazards model, since the marginal hazard function $h(y|\boldsymbol{x})$ is given by

$$h(y|\boldsymbol{x}) = \frac{h_0(y)\exp(\boldsymbol{x}'\boldsymbol{\beta})}{\kappa H_0(y)\exp(\boldsymbol{x}'\boldsymbol{\beta}) + 1}. \tag{4.1.30}$$

Instead, we see in (4.1.30) that there is a convergence of hazards, at a rate determined by κ. Clayton and Cuzick (1985) exploit this fact in a univariate semiparametric survival model. In the multivariate case with covariates, this property has the consequence that information for estimation of κ comes partly from the coincidence of failure within clusters, and partly from the marginal convergence of hazards in relation to the covariates. Hougaard (1986a,b) has pointed out that this is not a desirable property for the multivariate model since it renders interpretation of κ difficult. This problem persists with any other finite mean frailty distribution, such as the log-normal. The assumption of a positive stable distribution of the w_i's avoids this problem, since the proportional hazards assumption for covariates then remains true marginally. We discuss alternative frailty distributions in the next subsection.

4.1.4 Positive Stable Frailties

The gamma frailty model discussed in the previous subsections has several advantages, but as noted in the previous subsection, it also has several drawbacks. In addition, Oakes (1994, p. 279) notes that in the finite mean frailty model, the unconditional effect of a covariate, which is measured by the hazard ratio between unrelated subjects (i.e., with different frailties) is always less than its conditional effect, measured by the hazard ratio among subjects with the same frailty. In particular, suppose we consider two subjects from different clusters with respective covariates $\boldsymbol{x}_1$ and $\boldsymbol{x}_2$. Let $S_1(y)$ and $S_2(y)$ denote the corresponding unconditional survivor functions derived under a frailty specification. Oakes showed that the covariate effects, as measured by the hazard ratio, are always attenuated and further, $S_1(y)$ and $S_2(y)$ are usually not related via a proportional hazards model. The degree of attenuation of the core effect is not easy to quantify unless both conditional and marginal specifications have a proportional hazards structure. However, if the frailty distribution is an infinite variance positive stable distribution with Laplace transform

$$E(\exp(-sw)) = \exp(-s^{\alpha})$$

for $0 < \alpha < 1$ (Feller 1971), then $S_1(y)$ and $S_2(y)$ will have proportional hazards (Hougaard 1986a,b) since

$$S(y|\boldsymbol{x}) = \exp\left\{-(\theta H_0(y))^{\alpha}\right\} = \exp\left\{-\exp(\alpha \boldsymbol{x}'\boldsymbol{\beta})H_0^{\alpha}(y)\right\}, \qquad (4.1.31)$$

where $H_0(y)$ denotes the cumulative baseline hazard function.

From (4.1.31), it is easy to quantify the attenuation of the covariate effect, since the parameter α is clearly the attenuation of $\boldsymbol{\beta}$ in the marginal hazard. Thus, we no longer need to choose between conditional and unconditional Cox model specifications, since a single specification can be interpreted either way. Although the positive stable frailty model is conceptually simple, estimation of the resulting model parameters is complicated due to the lack of a closed form expression for the density function of a stable random variable. Frequentist methods of estimation for positive stable frailty models have been described by Hougaard (1986a,b, 1995), Lam and Kuk

(1997), Manatunga (1989), and Oakes (1994). Bayesian methods have been discussed by Qiou, Ravishanker, and Dey (1999).

The Bayesian framework for this model using MCMC methods offers an attractive alternative to frequentist methods. Specifically, it greatly reduces the difficulty in estimating α, the parameter that characterizes the tail behavior in stable distributions. Bayesian estimation of stable law parameters using the Gibbs sampler was investigated by Buckle (1995) with application to portfolio analysis. Qiou (1996) described an alternate approach by approximating the posterior density of the parameters by a mixture of distributions.

The frailty parameters w_i, $i = 1, 2, \ldots, n$, are assumed to be independent and identically distributed for every cluster, according to a positive α-stable distribution. The stable distribution $S_\alpha(\sigma, \gamma, \delta)$ (Samorodnitsky and Taqqu, 1994) is characterized by four parameters. The stability parameter α, which lies in $(0, 2]$ measures the degree of peakedness of the density function and heaviness of its tails. When α approaches 2, the distribution approaches the (symmetric) normal distribution, regardless of the value of γ, while for $\alpha = 1$, it is the Cauchy distribution. When $0 < \alpha < 1$, the tails of the distribution are so heavy that even the mean does not exist. In survival analysis, we concentrate on the special case $\gamma = 1$, $0 < \alpha < 1$ and $\delta = 0$ since the support of the distribution $S_\alpha(\sigma, 1, 0)$ is the positive half of the real line. In this case, the distribution is totally skewed to the right and is called the positive stable distribution. Here, the scale parameter σ is also fixed at the value of 1 for identifiability purposes. The density function of a positive stable random variable w_i is not available in closed form. However, its characteristic function is available and has the form

$$E(e^{i\vartheta w_j}) = \exp\{-|\vartheta|^\alpha(1 - i\ \mathrm{sign}(\vartheta)\tan(\pi\alpha/2))\}, \tag{4.1.32}$$

where ϑ is a real number, $i = \sqrt{-1}$, $\mathrm{sign}(\vartheta) = 1$ if $\vartheta > 0$, $\mathrm{sign}(\vartheta) = 0$ if $\vartheta = 0$ and $\mathrm{sign}(\vartheta) = -1$ if $\vartheta < 0$.

Buckle (1995) provides an expression for the joint density of n independent and identically distributed observations from a stable distribution by utilizing a bivariate density function $f(w_j, z|\alpha)$ whose marginal density with respect to one of the two variables gives exactly a stable density. This approach facilitates implementation of sampling based Bayesian inference. Let $f(w_j, z|\alpha)$ be a bivariate function such that it projects $(-\infty, 0) \times (-1/2, l_\alpha) \cup (0, \infty) \times (l_\alpha, 1/2)$ to $(0, \infty)$:

$$f(w_j, z|\alpha) = \frac{\alpha}{|\alpha - 1|} \exp\left\{-\left|\frac{w_j}{\tau_\alpha(z)}\right|^{\alpha/(\alpha-1)}\right\} \left|\frac{w_j}{\tau_\alpha(z)}\right|^{\alpha/(\alpha-1)} \frac{1}{|w_j|}, \tag{4.1.33}$$

where

$$\tau_\alpha(z) = \frac{\sin(\pi\alpha z + \psi_\alpha)}{\cos \pi z} \left[\frac{\cos \pi z}{\cos\{\pi(\alpha - 1)z + \psi_\alpha\}}\right]^{(\alpha-1)/\alpha},$$

$w_j \in (-\infty, \infty)$, $z \in (-1/2, 1/2)$, and $\psi_\alpha = \min(\alpha, 2-\alpha)\pi/2$ and $l_\alpha = -\psi_\alpha/\pi\alpha$. Then

$$f(w_j|\alpha) = \frac{\alpha|w_j|^{1/(\alpha-1)}}{|\alpha-1|} \int_{-1/2}^{1/2} \exp\left\{-\left|\frac{w_j}{\tau_\alpha(z)}\right|^{\alpha/(\alpha-1)}\right\} \left|\frac{1}{\tau_\alpha(z)}\right|^{\alpha/(\alpha-1)} dz. \tag{4.1.34}$$

As in the previous subsection, we use the same piecewise exponential model with the correlated prior process in (4.1.22) on the baseline hazard to construct the likelihood function. Specifically, given $(\lambda_1, \lambda_2, \cdots, \lambda_{k-1})$, we take

$$\lambda_k|\lambda_1, \lambda_2, \cdots, \lambda_{k-1} \sim \mathcal{G}\left(\alpha_k, \frac{\alpha_k}{\lambda_{k-1}}\right), k = 1, 2, \cdots, J, \tag{4.1.35}$$

where $\lambda_0 = 1$, so that $E(\lambda_k|\lambda_1, \lambda_2, \cdots, \lambda_{k-1}) = \lambda_{k-1}$. The complete data likelihood is thus given by (4.1.24). The observed data likelihood is obtained by integrating out the w_i's from (4.1.24) using the stable density expression in (4.1.34), and thus

$$\begin{aligned} L(\boldsymbol{\beta}, \boldsymbol{\lambda}, \alpha|D) \propto \prod_{i=1}^{n}\prod_{j=1}^{m_i} \Bigg[&\left\{\prod_{k=1}^{g_{ij}} \exp(-\lambda_k \Delta_k \theta_{ij} w_i)\right\} (\lambda_{g_{ij}+1}\theta_{ij}w_i)^{\delta_{ij}} \\ &\times \exp\left\{-\lambda_{g_{ij}+1}(y_{ij} - s_{g_{ij}})\theta_{ij}w_i\right\} \times \frac{\alpha}{(1-\alpha)w_i} \\ &\times \int_{-1/2}^{1/2} \left(\frac{\tau_\alpha(z_i)}{w_i}\right)^{\alpha/(1-\alpha)} \exp\left\{-\left(\frac{\tau_\alpha(z_i)}{w_i}\right)^{\alpha/(1-\alpha)}\right\}\Bigg] dz_i. \end{aligned} \tag{4.1.36}$$

Let $\pi(\alpha), \pi(\boldsymbol{\beta})$, and $\pi(\boldsymbol{\lambda})$ denote the priors for the model parameters α, $\boldsymbol{\beta}$, and $\boldsymbol{\lambda}$. Since α is on (0,1), a common prior for alpha is a beta prior. Qiou, Ravishanker, and Dey (1999) use a $\mathcal{B}(1,1)$ prior for α, i.e., a uniform prior, and a normal prior for $\boldsymbol{\beta}$. The full conditional distributions for $\boldsymbol{\beta}$ and $\boldsymbol{\lambda}$ are exactly the same as those in (4.1.28) and (4.1.29). However, the full conditional distribution of $\boldsymbol{w}$ is given by

$$\begin{aligned} \pi(w_i|\boldsymbol{\beta}, \boldsymbol{\lambda}, \alpha, D_{obs}) \propto w_i^{\sum_{j=1}^{m_i}\nu_{ij} - \frac{1}{1-\alpha}} \\ \times \exp\Bigg\{ &-\frac{\sum_{k=1}^{J}\lambda_k}{J}\sum_{j=1}^{m_i} y_{ij}\exp(w_i \boldsymbol{x}_{ij}'\boldsymbol{\beta}) \\ &- \left(\frac{\tau_\alpha(z_i)}{w_i}\right)^{\alpha/(1-\alpha)}\Bigg\}. \end{aligned} \tag{4.1.37}$$

The ratio-of-uniforms method (Wakefield, Gelfand, and Smith, 1991; Chen, Ibrahim, and Sinha, 2001; Devroye, 1986, pp. 40–65, 194–205) is used to generate samples of $\boldsymbol{\lambda}$, $\boldsymbol{\beta}$ and w_i. The full conditional distribution of α is

$$\pi(\alpha|\boldsymbol{\lambda},\boldsymbol{\beta},\boldsymbol{z},D) \propto \left(\frac{\alpha}{1-\alpha}\right)^n \prod_{i=1}^{n} \left(\frac{\tau_\alpha(z_i)}{w_i}\right)^{\frac{\alpha}{1-\alpha}} \exp\left\{-\sum_{i=1}^{n}\left(\frac{\tau_\alpha(z_i)}{w_i}\right)^{\frac{\alpha}{1-\alpha}}\right\}, \tag{4.1.38}$$

where $\boldsymbol{z} = (z_1, z_2, \ldots, z_n)'$, which does not have a standard form. Qiou, Ravishanker, and Dey (1999) use the Metropolis-Hastings algorithm with a beta distribution as the proposal density to generate α (see Buckle, 1995). The full conditional distribution for $\boldsymbol{z}$ is thus given by

$$p(z_i|\boldsymbol{\beta},\alpha,D) \propto \left(\frac{\tau_\alpha(z_i)}{w_i}\right)^{\frac{\alpha}{1-\alpha}} \exp\left\{1-\left(\frac{\tau_\alpha(z_i)}{w_i}\right)^{\frac{\alpha}{1-\alpha}}\right\}. \tag{4.1.39}$$

A rejection algorithm (Devroye, 1986) is then applied here. Details of all the sampling algorithms are given in Qiou, Ravishanker, and Dey (1999).

Example 4.4. Kidney infection data (Example 4.3 continued). We illustrate the positive stable frailty model on the kidney infection data using a single covariate *sex*. This analysis is also presented in Qiou, Ravishanker, and Dey (1999). First, we model the frailty w_i by a positive α-stable specification. The baseline hazard is modeled by a piecewise exponential model with a correlated prior process given in (4.1.35). We choose a uniform prior for α, i.e., $\pi(\alpha) = 1, 0 < \alpha < 1$, while the prior on $\eta = \kappa^{-1}$ is $N(0, 10^3)$. Model fits were compared after an initial investigation with grid sizes $J = 5$, 10, and 20. Since the choice of $J = 10$ produced a model fit that was at least as good as the other two, we assume $J = 10$ in the following analysis. The choice of each a_k to be 0.01 was made similarly. This choice of a_k is crucial if a goal is to make inferences about the baseline hazard function; in theory, one could put a prior on a_k and estimate it.

Using Gibbs sampling, we generated 10,000 iterates from the posterior distribution, and then selected every fifth of the last 5000 iterates to make inference, after monitoring convergence (Geweke, 1992). Column 2 and Column 3 of Table 4.5 present the estimated posterior means and standard deviations of each parameter under the positive stable frailty model. The negative estimate of η shows that the female patients have a slightly lower risk for infection. Since the cases $\alpha = 0$ and $\alpha = 1$ correspond to maximal dependence and independence respectively, the estimate of α of 0.86 confirms some degree of dependence between infection times for each patient. See Section 4.4 for more on bivariate measures of dependence.

We also present a comparison between the positive stable frailty model and the alternative and more widely used gamma frailty model, again using a piecewise exponential baseline hazard model (see Sahu, Dey, Aslanidou, and Sinha, 1997). To do this, we assume $w_i \sim \mathcal{G}(\kappa^{-1}, \kappa^{-1})$, where the prior for the hyperparameter κ^{-1} is chosen to be $\mathcal{G}(0.001, 0.001)$ in order to allow for a sufficiently large variance. Columns 4 and 5 of Table 4.5 present the estimated posterior mean and standard deviation of each parameter under the gamma frailty model. Figure 4.1 (a) presents box plots of the posterior means of w_i; the plot on the left corresponds to the stable frailties and the plot on the right to the gamma frailties. Figure 4.1 (b) presents similar plots of the posterior standard deviations of w_j under the two frailty models. The posterior means of the stable frailties and gamma frailties are similar while the posterior standard deviations appear to be somewhat smaller under the gamma frailty model.

TABLE 4.5. Posterior Estimates of the Model Parameters for the Kidney Infection Data under Piecewise Exponential Hazard Model.

	Stable Frailty		Gamma Frailty	
Parameter	Mean	Std Dev	Mean	Std Dev
η	−1.060	0.3613	−1.6200	0.4186
α	0.860	0.0745	-	-
$1/\kappa$	-	-	0.3268	0.1737
λ_1	0.0010	0.0016	0.0015	0.0030
λ_2	0.0029	0.0067	0.0012	0.0025
λ_3	0.0038	0.0094	0.0011	0.0020
λ_4	0.0039	0.0089	0.0010	0.0020
λ_5	0.0034	0.0077	0.0012	0.0023
λ_6	0.0042	0.0114	0.0011	0.0024
λ_7	0.0044	0.0101	0.0011	0.0024
λ_8	0.0063	0.0135	0.0014	0.0028
λ_9	0.0265	0.0225	0.0056	0.0051
λ_{10}	1.8409	0.6039	0.3667	0.1495

Source: Qiou, Ravishanker, and Dey (1999).

It is interesting to look at the estimated frailties by sex (see also McGilchrist and Aisbett, 1991; Walker and Mallick, 1996). Table 4.6 shows the estimated posterior means and the corresponding standard deviations of the w_j by sex under the stable frailty model and the gamma frailty model.

It appears that the frailties for the female patients are centered around one, and are nicely distributed. The frailties for the males are rather irregularly distributed, with larger posterior standard deviations. From Table 4.5, it may also be seen that the parameter estimates are quite different in the two models. In particular, the λ_i values are substantially higher under the positive stable frailty model. The positive stable frailty model also gives a larger baseline hazard in general. In retrospect, it seems that we may get a better fit with different κ's for males and females, though this would increase the dimension of the model. We would also like to caution the reader about trying to infer too much from the posterior distributions of individual frailties. For example, $\pi(w_i|D_{obs})$ is bound to have high mean with males, $\nu_{ij} = 1$, and y_{ij} small, though it is unclear whether we should attribute the small y_{ij} values to gender effects or to a frailty. Investigation of a more versatile model, involving a suitable mixture of positive α stable distributions is discussed in Ravishanker and Dey (2000).

TABLE 4.6. Posterior Means and Standard Deviations of Stable and Gamma frailties by Sex.

Stable Frailty				Gamma Frailty			
Female		Male		Female		Male	
Mean	SD	Mean	SD	Mean	SD	Mean	SD
1.52	1.52	3.58	4.01	1.25	0.73	1.40	0.70
0.98	0.49	1.83	1.53	0.99	0.48	1.10	0.55
0.75	0.27	0.58	0.17	0.66	0.33	0.19	0.13
1.06	0.61	1.98	1.85	1.00	0.48	1.18	0.57
0.89	0.44	5.34	5.62	0.88	0.42	1.54	0.74
0.96	0.57	0.94	0.49	0.95	0.52	0.59	0.32
1.40	0.92	1.72	1.95	1.25	0.60	1.10	0.59
0.73	0.28	0.58	0.18	0.58	0.38	0.20	0.14
0.68	0.23	1.87	1.77	0.51	0.30	1.08	0.51
0.94	0.45	2.73	2.83	0.92	0.44	1.32	0.67
0.84	0.36	1.01	0.60	0.79	0.40	0.74	0.43
0.78	0.36			0.63	0.42		
1.07	0.86			1.02	0.54		
0.73	0.29			0.60	0.34		
2.15	1.85			1.47	0.77		
1.29	1.13			1.15	0.62		
0.79	0.34			0.68	0.38		
0.96	0.48			0.94	0.48		
2.48	2.06			1.56	0.80		
1.26	0.84			1.21	0.60		
2.00	1.65			1.45	0.72		
1.57	1.58			1.20	0.65		
1.21	0.75			1.18	0.57		
0.94	0.51			0.87	0.50		
1.46	1.12			1.33	0.68		
0.92	0.61			0.79	0.52		
1.17	0.96			1.09	0.65		

Source: Qiou, Ravishanker, and Dey (1999).

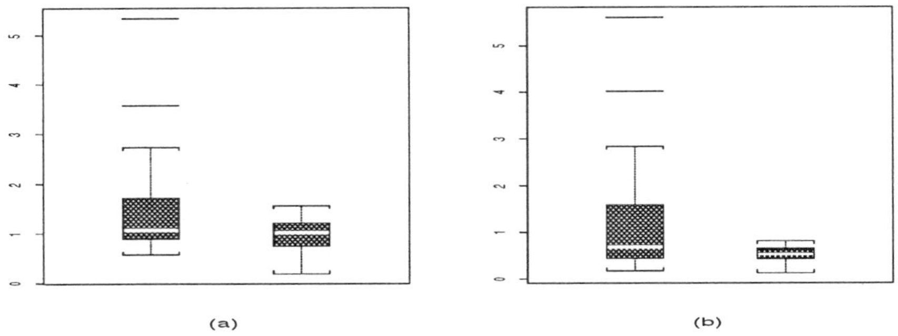

FIGURE 4.1. Box-plots of posterior means in (a) and posterior standard deviations in (b) for frailties under the stable model (left) and the gamma model (right).

4.1.5 A Bayesian Model for Institutional Effects

In randomized clinical trials comparing treatments for diseases such as cancer, it is often necessary to include patients from a number of separate centers to accrue the required sample size in a reasonable period of time. The contributing institutions vary from comprehensive cancer centers to other university hospitals to community hospitals to private practice clinics. Although treatment protocols are tightly structured, there can be variations in interpretation of dose modifications, type of supportive care given, and so on, which lead to differences in how treatment is given at different institutions. There are several reasons why it might be important to examine institutional variation. If there are substantial institutional effects, then a conventional analysis ignoring institution would be based on an incorrect model, so the inferences from the analysis may not be valid. A second reason is that the objective of a clinical trial is to try and draw conclusions about the overall effect of therapy in the general population. Since the institutions participating in trials are not selected at random, and generally only a small nonrandom subset of the patients from these institutions are entered on trials, with substantial institutional variation it would not be clear exactly what effect would be seen in the general population, and in particular whether it would be more like that seen in the institutions with poor outcomes or that seen in those with good outcomes. Another reason to examine institutional variation is that it might make it possible to learn more about how the therapy should be given or to whom it should be given. That is, if institutional variation is seen in the effect of therapy, then a detailed comparison of the characteristics of patients and of how therapy was given among institutions with different results might identify reasons why the results are different, and lead to modifications in the recommendations for treatment.

Gray (1994) proposes a Bayesian frailty model for assessing institutional effects in cancer clinical trials. He specifies a proportional hazards model for the covariates and institutional effects. Institutional differences in log hazards and the regression coefficients are then assumed to have a joint normal distribution, which leads to a hierarchical Bayesian model with a Wishart prior on the covariance matrix of the normal distribution. By examining the posterior distributions for the institutional effects, it is possible to quantify the amount of institutional variation, and to identify the institutions with more extreme effects, so they could be studied further. Related papers discussing institutional effects in clinical trials include Skene and Wakefield (1990), who used the same type of hierarchical Bayesian structure to model data from multicenter binary response trials. They also consider heavier tailed alternatives to the normal-Wishart model proposed by Gray (1994).

The model of Gray (1994) is motivated from a clinical trial conducted by the Eastern Cooperative Oncology Group (E1582), which investigated therapies for small cell lung cancer. This trial included 579 patients from 31 distinct institutions, and the number of patients per institution varied from 1 to 56, with a median of 17. Each "institution" is an administrative grouping consisting of a main institution and several affiliated centers, but if each affiliate is regarded as a separate institution then the number of patients per institution becomes too small for there to be much information about institutional variation. There was an initial randomization to two different chemotherapy regimens, a standard therapy (CAV, consisting of cytoxan, adriamycin and vincristine), and an alternating regimen (CAV-HEM), where cycles of CAV are alternated with HEM (hexamethylmelamine, etoposide and methotrexate). The effect of the initial induction therapy on overall survival is the focus of this clinical trial. Further details on the trial and the main clinical results are given in Ettinger et al. (1990). The conventional analysis of E1582 reported by Ettinger et al. (1990) identified four factors besides treatment which were significantly associated with survival. These factors with longer survival were ambulatory performance status at entry, absence of liver metastases, absence of bone metastases, and no weight loss prior to entry. A proportional hazards model is assumed for the effects of treatment and these four factors. Let x_{ijk} be covariate k for subject j from institution i, and in general assume there are

n institutions with m_i cases from institution i and $p-1$ covariates, with x_{ijp} denoting the treatment variable. A piecewise constant model is used for the underlying hazard. Let $0 = s_0 < s_1 < \cdots < s_J$ be the boundaries of time intervals, and set $I_l(s) = I(s_{l-1} < s \leq s_l)$. The full model for the hazard for subject ij can then be written

$$h(y|\boldsymbol{x}_{ij}, \boldsymbol{\alpha}, \boldsymbol{\beta}, \boldsymbol{\theta}_i) = \exp\left\{\sum_{l=1}^{J} \alpha_l I_l(s) + \theta_{i0} + \sum_{k=1}^{p} x_{ijk}\beta_k + \theta_{i1}x_{ijp}\right\},$$

where $\boldsymbol{\alpha} = (\alpha_1, \ldots, \alpha_J)'$, $\boldsymbol{\beta} = (\beta_1, \ldots, \beta_p)'$ and $\boldsymbol{\theta}_i = (\theta_{i0}, \theta_{i1})'$ are unknown parameters, and $\boldsymbol{x}_{ij} = (x_{ij1}, \ldots, x_{ijp})'$. In the analysis of E1582, the four covariates and the treatment variable are binary, coded as contrasts summing to 0. Treatment is coded so that $\beta_p + \theta_{i1}$ is the log hazard ratio for CAV-HEM/CAV. Also, $J = 30$ intervals are used in the underlying hazard, with the boundaries chosen to give roughly equal numbers of failures in each interval. Choosing $J = 30$ is arbitrary, but inferences are not sensitive to this parameter unless it is too small.

A convenient family of priors for this model is

$$\boldsymbol{\beta} \sim N_p(0, \sigma_\beta^2 I), \quad \boldsymbol{\theta}_i|\Sigma \overset{i.i.d.}{\sim} N_2(0, \Sigma), \quad \Sigma^{-1} \sim W(d, V),$$

$$\alpha_l - \alpha_{l-1}|\psi \overset{i.i.d.}{\sim} N(0, \psi^{-1}), \; l = 2, 3, \ldots, J, \quad \psi \sim \mathcal{G}(a, r),$$

and with α_1 given an essentially flat prior, where the distributions α, $\boldsymbol{\beta}$, and the $\boldsymbol{\theta}_i$ are assumed to be mutually independent. Here $W(d, V)$ is the Wishart distribution with density (for a $k \times k$ matrix S)

$$f(S) \propto |S|^{(d-k-1)/2} \exp\{-\text{trace}(SV^{-1})/2\}.$$

The prior for α restricts the magnitude of the jump between adjacent intervals in the piecewise constant model, which is done to reflect the fact that this model is really an approximation to an underlying continuous function.

Conditional on β_p and Σ, the institutional log hazard ratios for CAV-HEM/CAV, $\beta_p + \theta_{i1}$, are i.i.d. $N(\beta_p, \Sigma_{11})$, so θ_{i1} is the deviation in the ith institution from an overall effect β_p. Similarly, the θ_{i0} are institutional deviations from an overall underlying log hazard $\sum_{l=1}^{J} \alpha_l I_l(s)$. The quantities $\exp(\beta_p)$ and $\exp\left\{\sum_{l=1}^{J} \alpha_l I_l(s)\right\}$ are referred to as the overall treatment hazard ratio and the overall underlying hazard. The logs of these overall quantities are means over the distribution of random institutional effects. However, these overall effects need not be the actual hazard ratio or underlying hazard from any institution, nor are they a true average of these quantities over the institutions in the trial.

Let $\nu_{ijl} = 1$ if subject ij is observed to fail in time interval l, and $= 0$ otherwise, and let u_{ijl} be the total follow-up time of subject ij in interval l ($u_{ijl} = 0$ if the failure or censoring time is $< s_{l-1}$). The likelihood corresponding to (4.1.40) is $\prod_i L_i(\boldsymbol{\alpha}, \boldsymbol{\beta}, \boldsymbol{\theta}_i)$, where

$$L_i(\boldsymbol{\alpha}, \boldsymbol{\beta}, \boldsymbol{\theta}_i) = \exp\left\{ \sum_{j=1}^{m_i} \sum_{l=1}^{J} [\nu_{ijl}\zeta_{ijl} - u_{ijl}\exp(\zeta_{ijl})] \right\},$$

with

$$\zeta_{ijl} = \alpha_l + \theta_{i0} + \sum_{k=1}^{p} x_{ijk}\beta_k + x_{ijp}\theta_{i1}.$$

Noting the independence assumptions in the prior, the joint posterior is then proportional to

$$\left[\prod_{i=1}^{n} L_i(\boldsymbol{\alpha}, \boldsymbol{\beta}, \boldsymbol{\theta}_i)\pi(\boldsymbol{\theta}_i|\Sigma)\right] \pi(\Sigma)\pi(\boldsymbol{\alpha}|\psi)\pi(\psi)\pi(\boldsymbol{\beta}). \tag{4.1.40}$$

Samples from the joint posterior of (4.1.40) can be obtained via the Gibbs sampler. Two generalizations of the model are worth noting. First, instead of limiting institutional variation to a single covariate, institutional effects can be included for an arbitrary subset of the covariates. The only change needed in the prior when this is done is that the prior for $\boldsymbol{\theta}_i$ has more components, so Σ is of larger dimension. The same type of Wishart prior can still be used. Second, multiple strata can be included, with different underlying hazards in the different strata. This is done by having a separate set of parameters α_s for each stratum. The boundaries of the time intervals can also be stratum dependent. For the prior, one possibility is to take the priors on the underlying hazard for the different strata to be independent, each with the same form given above, with a different ψ for each stratum.

TABLE 4.7. $P(|Z| > \log(R))$ (in %), Where Z is a Normal Deviate with Mean 0 and Variance σ_Z^2.

		R			
σ_Z^2	$\psi = 1/\sigma_Z^2$	1.1	1.25	1.5	2.0
0.364	2.75	87	71	50	25
0.1	10	76	48	20	2.8
0.04	25	63	26	4.3	0.05
0.02	50	50	11	0.4	0
0.01	100	34	2.6	0.005	0
0.00334	299	9.9	0.01	0	0

Source: Gray (1994).

The parameters which need to be specified are σ_β^2 in the prior for $\boldsymbol{\beta}$, d, and V in the Wishart prior for Σ^{-1}, and a and r in the prior for ψ. Proper priors are used, but the parameters are chosen to keep the priors fairly weak. For the institutional effects this is justified since there is little prior information on the magnitude of these parameters. The variance σ_β^2 was set equal to 10,000, which gives an essentially flat prior. This prior was applied after transforming all the covariates to have variance 1. For the prior for ψ, note that $1/\psi$ is the variance of the jumps in the log of the underlying hazard at the boundaries of the time intervals. A jump of ± 0.405 corresponds to a 50% increase or 33% decrease in the hazard from one interval to the next. The first 29 intervals span a period of 2.3 years, so the interval widths are small, and a 50% jump would be very unlikely. Table 4.7 gives $P(|Z| > \log(R))$ for various values R, where $Z \sim N(0, \sigma_Z^2)$. Here $\psi = 1/\sigma_Z^2$, so the probability of a jump of 50% or larger is 20% when $\psi = 10$. Since frequent large jumps are not realistic, while an exponential (constant hazard) model might be, the values in Table 4.7 suggest taking a prior for ψ centered towards larger values. For example, taking $a = 3$ and $r = 0.05$ gives $P(\psi > \psi_0)$ =98.6%, 87%, 54%, 12%, for ψ_0 =10, 25, 50, and 100. These were the values used in the analysis, and although they are centered on larger values, the prior is still diffuse enough to allow a considerable range of values for ψ.

To specify V in the prior for Σ, first note that the off-diagonal elements relate to how much correlation we expect in the components of $\boldsymbol{\theta}_i$. Since the treatment variable is coded to be a contrast (summing to 0), there is no particular reason here to assume that institutions with high baseline risks (θ_{i0}) should also have large treatment differences (θ_{i1}), or to assume the opposite, so the prior for Σ is centered on distributions with no correlation, by setting the off-diagonal elements of V to 0.

The diagonal elements of V relate to the magnitude of the variation in the components of $\boldsymbol{\theta}_i$. The parameter θ_{i0} is the log of the ratio of the underlying hazard in the ith institution to the overall underlying hazard. The most significant prognostic factor in the conventional analysis of the E1582 data was performance status, with an estimated hazard ratio of 1.8 (nonambulatory/ambulatory). Thus a ratio of baseline hazards between two institutions of 1.5–2.0 would be possible, for example, if one institution only put on ambulatory patients and another only put on nonambulatory, and if this factor were not taken into account in the model. Since such an extreme difference in accrual patterns is quite unlikely, and the major known prognostic variables are being incorporated in the model, differences this large among institutions would not be expected. On the other hand, smaller ratios, say, on the order of 1.1 or less, are difficult to detect and may be commonplace. For θ_{i1}, the treatment hazard ratio in the ith institution is $\exp(\beta_p + \theta_{i1})$. Given the limited effectiveness of most therapies for lung cancer, it is unlikely that a treatment hazard ratio of CAV-HEM/CAV would be $< 1/1.5$, and institutional differences of this magnitude would be very surprising. These considerations suggest that plausible magnitudes of institutional effects are similar for both parameters, so for convenience the diagonal elements of V are chosen to have the same value v. From

Table 4.7, if the variances of the institutional effect parameters (diagonal elements of Σ) are 0.364, a great deal of institutional variation is allowed, since then $P(\exp(|\theta_{ih}|) > 2|\Sigma) = 0.25$, while a variance of 0.00334 is quite small, since this gives $P(\exp(|\theta_{ih}|) > 1.1|\Sigma) = 0.10$. With two components, the Wishart distribution must have at least 2 degrees of freedom for the density to exist. When generating Σ^{-1} from the $W(2, vI)$ distribution, trial and error showed that with $v = 125$ the probability that the variances (diagonal elements of Σ) were > 0.364 was 12.1%, and the probability they were < 0.00334 was 12.5% (estimates based on 5000 samples). Thus this gives a weak prior which allows a wide range for the amount of institutional variation.

Since these priors are fairly arbitrary, the sensitivity of the results to changes in the prior also needs to be examined. Here the part of the prior specification which is most important for inferences about institutional variability are the diagonal elements v of V. To examine sensitivity to the value of v, posteriors are also evaluated using alternate priors with $v = 125/2$, which gives that the prior probability that the variances are > 0.364 and < 0.00334 are 16.5% and 2.6%, and $v = 2 \times 125$, where these prior probabilities are 8.3% and 28.3%. Thus these changes shift the prior distribution, but still leave considerable prior probability of extreme values for the variances. If the model is generalized to include institutional variation in the effects of several covariates, it would often still be appropriate to set the off-diagonal elements of V to 0, provided there is no reason to expect institutional effects for different variables to be in the same direction. Even for a factor with more than two categories, this might still be reasonable if the factor is coded using orthogonal contrasts. Using orthogonal contrasts is also important because collinearity in the covariates greatly increases the serial correlation in the Gibbs sampling algorithm.

For the E1582 study, we have 570 patients from 26 institutions, with the number of patients per institution varying from 5 to 56 (median= 18.5). This number excludes patients from institutions contributing three or fewer patients. Of the 570 patients, 560 had died. The posterior distribution of $\exp(\beta_p)$ had a median of 0.76 and (0.025, 0.975) percentiles of (0.62, 0.93). For comparison, fitting the Cox proportional hazards model ignoring institutional effects gave an estimated treatment hazard ratio of 0.77 with 95% confidence interval (0.65, 0.92). Coefficients of the other covariates showed similar agreement. This indicates that incorporating institutional effects was not critical for drawing conclusions on the overall effects.

Figure 4.2 summarizes the posterior distributions of the institutional effects, where (a) denotes the distribution of the ratio of the hazard in each institution to the overall hazard, for patients on CAV, (b) denotes the distribution of this ratio for patients on CAV-HEM, and (c) denotes the distribution of $\exp(\theta_{i1})$, the ratio of the treatment hazard ratio in the ith institution to the overall treatment hazard ratio. Plots (a) and (b) in Figure 4.2 give percentiles of the posterior distribution of $\exp(\theta_{i0} + \theta_{i1}C)$

and $\exp(\theta_{i0} + \theta_{i1}[C + 1])$, where C is the value of the treatment covariate for patients on CAV. These are the ratios of the institutional hazards to the overall hazard for patients treated on CAV (Figure 4.2 (a)) and CAV-HEM (Figure 4.2 (b)). Figure 4.2 (c) gives percentiles for the distribution of $\exp(\theta_{i1})$, the institutional deviations from the overall treatment hazard ratio. The actual values plotted are the 0.025, 0.5, and 0.975 percentiles from the marginal posterior for each institution. Institutions are coded from 1 to 26, with 1 being the smallest institution (5 patients entered) and 26 the largest (56 patients entered).

Plot (c) in Figure 4.2 appears to indicate substantial variation in the effect of treatment across institutions. From Plots (a) and (b) in Figure 4.2 it also appears this is mostly due to variation in the CAV-HEM arm, with little variation in the CAV arm. To further quantify the institutional variation in the effect of therapy, Figure 4.3 gives the posterior probabilities that the institutional effects $\exp(\theta_{i1})$ are $> R$ and $< 1/R$ for $R = 1.1$, 1.25, and 1.5. From Figure 4.3 (a) there are three institutions (14, 23, 24) with posterior probability $> 40\%$ that the institutional treatment hazard ratio $\exp(\beta_p + \theta_{i1})$ is at least 1.25 times larger than the overall ratio $\exp(\beta_p)$, and the probabilities that $\exp(\theta_{i1}) > 1.1$ for these three institutions are in the range 63%-66%. From Figure 4.3 (b), there are two institutions (11, 13) with about a 35% chance that $\exp(\theta_{i1}) < 1/1.25$, and about a 57% chance that $\exp(\theta_{i1}) < 1/1.1$. In both panels, there are a number of institutions with smaller but still substantial probabilities of deviations of this magnitude, and the probabilities of at least one institution having $\exp(|\theta_{i1}|) > R$ are 98%, 79%, and 56% for $R = 1.1$, 1.25, and 1.5.

Standard errors for the estimated posterior probabilities are also computed, and most of these are $< 1.5\%$, although a few are around 2%. This indicates that for the most part, the estimated probabilities have reasonably good precision, but this uncertainty in the estimates should be kept in mind when examining the results. To get some idea what these differences in hazard rates mean in terms of survival, note that under the proportional hazards assumption, if for a certain combination of covariates and for a given institution the probability of survival beyond y is q when CAV is used, then the probability of survival beyond y is q^R on CAV-HEM, where $R = \exp(\beta_p + \theta_{i1})$. For a rough point of reference, the Kaplan-Meier estimates of survival probabilities from the pooled sample of patients treated with CAV are 78%, 35%, 10%, and 4% at 0.5, 1, 1.5, and 2 years. If the one-year survival rate on CAV is $q = 35\%$, and if the treatment hazard ratio is 0.76 (the overall estimate), then the one-year survival rate on CAV-HEM is about 45%. Similar differences (5% to 10%) hold for baseline survival probabilities q from 5% to 75%, so CAV-HEM has a substantial advantage over a considerable range of follow-up. However, if for some institutions $\theta_{i1} \approx \log(1.25)$, which is consistent with some of the more extreme institutions, then the treatment hazard ratio would be 0.95 in these institutions, and the difference in survival from these institutions is probably of no im-

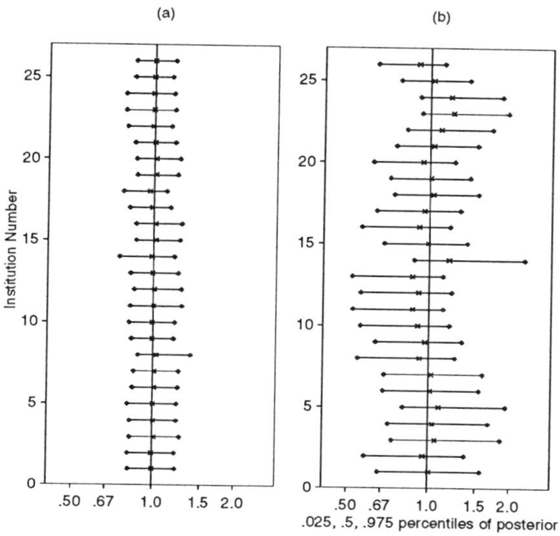

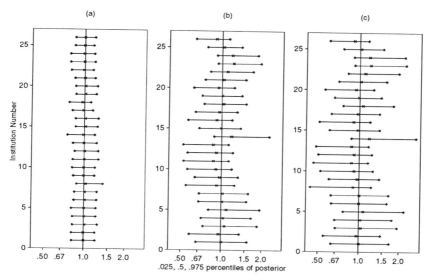

FIGURE 4.2. Posterior percentiles for institutional effects.

portance. On the other hand, $\theta_{i1} \approx -\log(1.25)$ would give a treatment hazard ratio of $< 1/1.5$, giving differences in survival rates of 9% to 18% over a considerable range of follow-up.

The question of the magnitude of survival differences can also be addressed using predictive distributions. The survival curve for the predictive distribution for a new case from institution i is given by the integral of

$$\exp\left[-\int_0^t h(u|\boldsymbol{x}_{ij}, \boldsymbol{\alpha}, \boldsymbol{\beta}, \boldsymbol{\theta}_i)\, du\right]$$

over the posterior distribution given by (4.1.40). With Gibbs sampling, the integral over the posterior is calculated by averaging over the generated parameter values.

When the posterior distributions were recomputed using other priors obtained by changing the magnitude of the diagonal v of V, the patterns remained similar, but the exact magnitude of the variation is quite sensitive to the prior. For example, the probabilities of at least one institution having $\exp(\theta_{i1}) > 1.25$ or $< 1/1.25$ were 91%, 79%, and 71% for $v = 62.5$, 125, and 250. However, in all three cases the magnitudes remain large enough so that the main conclusions are not affected. Both treatments here are toxic chemotherapy regimens, so the main criteria for choosing between them is overall survival. Thus, on the basis of these results, for the most part CAV-HEM would be preferred. However, there is somewhat higher toxicity on CAV-HEM (see Ettinger et al., 1990), so at institutions where the results indicate little evidence for improved survival with CAV-HEM,

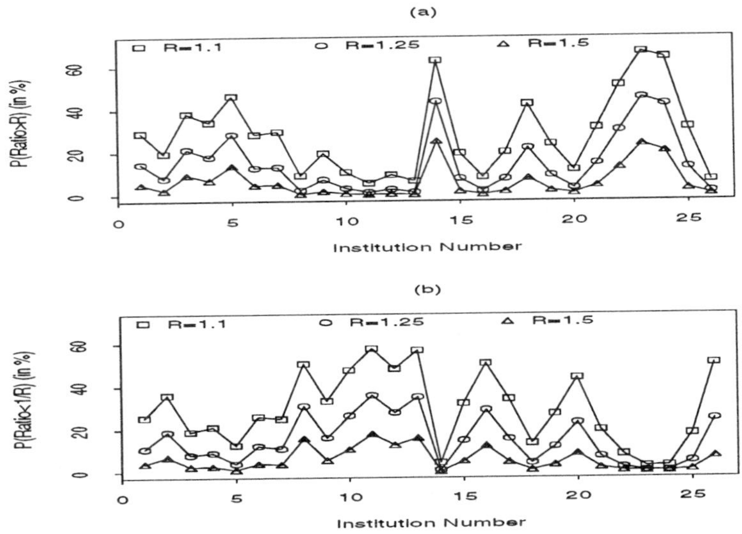

FIGURE 4.3. Posterior probabilities of large institutional effects in the effect of therapy for each institution ratio = $\exp(\theta_{i1})$.

there might still be a preference for CAV. Without further information on why different results are seen in different institutions, there is some doubt as to which therapy to use in the general population, since details of therapy administration to guarantee a better outcome on CAV-HEM, or the characteristics of patients to indicate a preference for one therapy over the other, remain unknown.

4.1.6 Posterior Likelihood Methods

Posterior likelihood methods for frailty models are based on the notion that prior information about H_0 is represented by a prior likelihood and this is multiplied with the complete data likelihood to obtain the posterior likelihood (Leonard, 1973, 1978). The main difference between posterior likelihood and posterior distributions is that the prior process can be improper, that is, the stochastic process may not be specified completely. Sinha (1998) discusses posterior likelihood methods for frailty models using correlated prior processes different from the ones discussed thus far. We first describe the correlated prior process of Sinha (1998), and then give the induced posterior distribution using a prior likelihood on H_0.

Consider the hazard for the jth subject in the ith cluster to be of the form

$$h(y|\boldsymbol{x}_{ij}, \boldsymbol{\beta}, w_i) = h_0(y) w_i \exp(\boldsymbol{x}'_{ij}\boldsymbol{\beta}), \qquad (4.1.41)$$

where $w_i \sim \mathcal{G}(\kappa^{-1}, \kappa^{-1})$. Following Sinha (1998), we model the smoothness in the hazard by an independent increment process for the discretized log-hazard $\alpha_j = \log \lambda_j$:

$$\alpha_{j+1} = \alpha_j + e_{j+1} \qquad (4.1.42)$$

for $j = 1, 2, \cdots, J-1$, and the e_j's are i.i.d. normal random variables with mean 0 and variance σ^2. Ideally, the variance of the e_j's should depend on the length of the jth interval, but when the grid intervals are sufficiently small, then equal variances may be a reasonable assumption. The process in (4.1.42) is not a stationary process as the marginal prior variance of α_j increases with j. For survival models, increasing uncertainty about the hazard as time increases is not an unreasonable assumption. To make the prior process stationary, we need to slightly modify (4.1.42) to $\alpha_{j+1} = \rho\alpha_j + e_{j+1}$ $(0 < \rho < 1)$. Higher values of σ give rise to less prior smoothing of the λ_j's and in the limiting case, when σ approaches ∞, the λ_j's will be almost independent, which means absence of prior smoothness in the baseline hazard. For the other limiting case, $\sigma = 0$, the baseline hazard is constant.

One may also incorporate prior knowledge (if available) about the beginning value α_1 and a prior distribution for α_1 gives prior information on the mean of the entire baseline hazard function. However, (4.1.42) alone (without any prior distribution for α_1) may be good enough to incorporate the only available prior belief (the degree of smoothness) about the baseline hazard. Leonard (1978) gives a more formal justification for using this kind of incomplete prior information and prior process in theoretical nonparametric Bayesian inference. The prior process in (4.1.42) is a discrete approximation of the Brownian process on the log-hazard. The prior process of (4.1.42) can be also viewed as a simple Kalman-Filter model for the baseline hazard. See Meinhold and Singpurwalla (1983) for a review of Kalman-Filter models and their frequentist and Bayesian interpretations.

The prior information about the discretized baseline hazard $\boldsymbol{\lambda} = (\lambda_1, \lambda_2, \cdots, \lambda_J)$ described above does not lead to a proper prior distribution for $\boldsymbol{\lambda}$, since no specification of the marginal distribution of λ_1 has been given and the prior mean of the baseline hazard is also left unspecified. This is a likely advantage in frailty models.

Once the posterior likelihood is obtained, it can be maximized to obtain point estimates of the parameters $(\boldsymbol{\lambda}, \boldsymbol{\beta}, \kappa)$. Following the arguments of Leonard (1978), it may be possible to show that these estimates also correspond to the Bayes estimates of corresponding parameters under vague priors for α_1, $\boldsymbol{\beta}$ and κ and a zero-one loss function, though the rigorous proofs of these properties are not attempted in the current literature. The posterior likelihood (denoted by L_p) has two components as

$$L_p(\boldsymbol{\lambda}, \boldsymbol{\beta}, \kappa | D) \propto L_1(H_0, \boldsymbol{\beta}, \kappa | D_{obs}) \times L_0(H_0),$$

where L_1 is the likelihood given D and L_0 is the prior likelihood. Sinha (1998) derives a closed form for L_1, and the logarithm of L_1, i.e., the sample log-likelihood (denoted by l_1), which is given by

$$\begin{aligned} l_1(\boldsymbol{\lambda}, \boldsymbol{\beta}, \kappa | D_{obs}) = & \sum_{i=1}^{n} \sum_{j=1}^{m_i} \sum_{k=1}^{J} \nu_{ijk} \boldsymbol{x}'_{ij} \boldsymbol{\beta} \\ & + \sum_{i=1}^{n} \Big[\log\Big\{ \Gamma(\eta + d_i) \Big\} - (\eta + d_i) \log(S_i + \eta) \Big] \\ & + n\ \eta \log(\eta) - n \log\{\Gamma(\eta)\} + \sum_{k=1}^{J} d_k^* \log(\lambda_k), \end{aligned} \tag{4.1.43}$$

where $\eta = \kappa^{-1}$, d_i is the total number of events from the ith cluster and d_k^* is the number of events in the kth interval. The other quantity S_i in (4.1.43) is given by $\sum_{j=1}^{m_i} \sum_{k=1}^{J} \lambda_j \exp(\boldsymbol{x}'_{ij}\boldsymbol{\beta})(y_{ijk} - s_{k-1})^+$, where $(u)^+ = u$ if $u > 0$ and 0 otherwise. Using the prior process of (4.1.42), we obtain the prior log-likelihood as

$$l_0(\boldsymbol{\lambda}) = -\sum_{k=1}^{J-1} \frac{1}{2\sigma^2} (\alpha_k - \alpha_{k+1})^2, \tag{4.1.44}$$

where α_k is the logarithm of λ_k. If the prior mean of λ_{k+1}/λ_k is different from 1, we might use the prior process $\alpha_{k+1} = \rho \alpha_j + e_{k+1}$. In that case, we need to modify $l_0(H_0)$ accordingly.

To find the maximizer of the posterior log-likelihood, given by $l_p = l_1 + l_0$ with l_1 and l_0 given, respectively, in (4.1.43) and (4.1.44), one may use the Newton-Raphson algorithm since the first and second derivative matrices of l_p with respect to $\Psi = (\boldsymbol{\beta}, \kappa, \boldsymbol{\lambda})$ are not difficult to compute. Because l_0 is a concave function, l_p is concave whenever l_1 is concave, and sometimes l_p can be concave even when l_1 is not concave. This is an additional advantage of the posterior (or penalized) likelihood method because we need fewer Newton-Raphson iterations when the likelihood is concave. The elements of the first and second derivative matrices of l_p involve di-gamma and tri-gamma functions. One may also use an EM algorithm (Dempster, Laird, and Rubin, 1977) to maximize l_p using $\boldsymbol{w} = (w_1, w_2, \ldots, w_n)'$ as the augmented data in the E-step. The maximizer of l_p with respect to Ψ, say, $\hat{\Psi}$, is the point estimate of the parameter vector Ψ.

The prior log-likelihood l_0 incorporates the information about the smoothness of the discretized baseline hazard $h_0(y)$. Other references on the discretized posterior likelihood include Leonard (1973) and Hickman and Miller (1981). This kind of prior likelihood approach is closely related to the penalized likelihood method of Good and Gaskins (1971) for estimating smooth functions in semiparametric regression. In the penalized likelihood method, a roughness penalty, which is $-l_0(\boldsymbol{\lambda})$ here, is deducted from the data likelihood to get a smooth estimate of the function. The estimate $\hat{\boldsymbol{\lambda}} = (\hat{\lambda}_1, \hat{\lambda}_2, \cdots, \hat{\lambda}_J)'$ is considered as a *discretized maximum penalized likelihood estimator* (DMPLE) introduced by Scott, Tapia, and Thompson (1980). Hastie and Tibshirani (1993) introduced the penalized likelihood in survival analysis in a varying-coefficient model. Here, the penalty term incorporates the smoothing of the baseline hazard function (a nuisance function) instead of smoothing of the covariate effect over time (found in Hastie and Tibshirani, 1993; Gamerman, 1991). The aim is to incorporate the smoothness information on the nuisance function to reduce the effective dimension of this high-dimensional parameter space. See Bacchetti (1990) for more discussion on the analysis of univariate survival data under the proportional hazards model using baseline hazard smoothing via penalized likelihood.

Different kinds of prior smoothing for the α_j's lead to different penalty functions. The process in (4.1.42) gives approximate linear spline smoothing (Scott, Tapia, and Thompson, 1980) for the baseline hazard. Fahrmeir (1994) showed how to introduce other available prior information on hazards, such as a local linear trend or a decaying effect, through the prior likelihood. The methodology discussed here is compatible with all of these higher order smooth prior processes. The prior process in (4.1.42) converges to a constant hazard as σ approaches 0 and does not require specification of the prior mean of the process. If we have some information about this, we may instead use the prior process

$$\alpha_{j+1} - \mu_{j+1} = \gamma(\alpha_j - \mu_j) + e_{j+1}, \tag{4.1.45}$$

where μ_j is the known prior mean of α_j and γ is the hyperparameter quantifying the amount of the pull of the process towards its prior mean.

The amount of smoothing of $h_0(y)$ depends on σ, and hence σ is a tuning constant in the penalized likelihood methodology to determine the relative importance of the information from the observed data to the prior information on the smoothness of the baseline hazard. It is possible to incorporate a hyperprior for σ, which can be elicited via prior knowledge about the smoothness of $h(y|\boldsymbol{x})$. There are many other suggestions in the existing literature on penalized likelihood for determining σ including a cross-validation method (Silverman, 1985), eyeball-cross-validation (Anderson and Blair, 1982), Akaike's (1973) information criterion, maximization of the marginal likelihood of σ using the EM algorithm (Leonard, 1982), Fahrmeir's (1994) EM-type algorithm, and the effective degrees-of-freedom method (Hastie and Tibshirani, 1993). Some of these methods are very costly in computer time. Adaptation of these methods for this particular problem and the asymptotic behavior of the estimates are yet to be explored. For selecting σ, we recommend using the available prior information about the amount of change in hazards from one time interval to the next. Say, if we have a prior belief that the ratio of baseline hazards in adjacent intervals should be within a factor of 2 (with 95% prior probability), then the corresponding σ should be 0.353 (because, $1.9645 \times 0.353 \approx \log(2)$). This method of selection of σ is called the "rule of thumb based on experience" by Wahba (1993). We recommend trying several choices for σ in order to explore the effects of different degrees of prior smoothing on the estimates.

Example 4.5. Kidney infection data (Example 4.4 continued). We consider the kidney infection data of Example 4.4. There are statistical methods available to explore the possible relationship of the sequence of survival times on the same patients (related through the frailty) and to compute the effect of the covariate on the intensity of the point process for each patient (Gamerman, 1991; Sinha, 1993; Yau and McGilchrist, 1994). For this study, the interest is not on the sequence of infection times for the patient. A fixed number of (only two) possibly right censored infection times have been considered; and before the second insertion of the tube, sufficient time was allowed for each patient to recover from the wound of the first insertion. The infection times of the patients are related because they are infections to the same patient, but occurrence of any infection at a particular time-point does not directly contribute to the increase of risk for the second infection following that time-point.

The prior process of (4.1.42) is used to model prior belief on the baseline hazard. The grids have been set with 16 intervals with endpoints at $(0, 10, 20, 30, \cdots, 100, 150, 200, 300, 400, 500, \infty)$. The interval length was increased after 100 days because there are fewer persons at risk beyond 100 days. The prior likelihood of (4.1.44) has been used for the analysis with different degrees of smoothing to assess the effect of smoothing on the estimates of κ and β. A `FORTRAN` program with `IMSL` subroutines has been used to maximize the posterior likelihoods (i.e., the penalized likelihoods) with different values of the smoothness parameter (σ).

Table 4.8 gives the values of the estimates of β and κ, their standard errors, values of $q = \exp(1.9645\sigma)$, and the weight of the penalty function ($\xi = \frac{1}{2\sigma^2}$) for 10 different values of the smoothing parameter, σ. The estimate of the regression parameter (β) did not change much with changes in smoothing. But the frailty parameter κ (i.e., the variance of the random effects due to the association of the times to infection within the same patient) increased for some time with an increase in smoothing and then remained steady for higher values of smoothing. The estimates of β in these analyses are between -1.243 and -1.301, which signify a lower infection rate for female patients. This value is smaller in magnitude than -1.8, which is the estimate of β by MA's original analysis of their model. But they used all the covariates for their analysis. From Table 4.8, $\sigma = 3.17$

corresponds approximately to $q = 500$, which means that the prior belief allows the ratio of hazards in two adjacent intervals to be within 500 with a 95% probability. So we can say that there is virtually no penalty imposed by the prior-likelihood for $\sigma > 3.17$. The estimates of the parameters for higher values of σ have to be taken with the caution that the data likelihood for this example is based on only 38 patients, and may not have enough information to give a reliable estimate of a very high dimensional parameter (dimension = 18). This actually supports the idea of using available prior belief on smoothness of the baseline hazard to make better inferences for this kind of situation. The estimates of κ are between 2.06 and 3.10, indicating a high degree of heterogeneity among patients. Comparison of the estimates of κ and the corresponding standard errors reveal that κ is significant for all degrees of smoothing.

Estimation of the dispersion matrix of the posterior likelihood estimates is obtained by inverting the observed information matrix by treating the log-posterior likelihood as the usual log-likelihood (Wahba, 1983; Silverman, 1985). To compute this, we need to evaluate the double derivative matrix of the log-posterior likelihood at the DMPLE (see Sinha, 1998). In practice, Table 4.8 shows that the standard errors of β and κ decrease with higher degree of smoothing, as expected. The comparison between the estimates of β and the corresponding standard errors show a significant gender effect for all degrees of smoothing. The estimates and standard errors do not change much once q (degree of prior smoothing) stays within a moderate range (between 2 and 10 for this example).

TABLE 4.8. Estimates of (κ, β) and Their Standard Errors for Different Values of σ.

q	σ	ξ	$\hat{\beta}$	s.e.$(\hat{\beta})$	$\hat{\kappa}$	s.e.$(\hat{\kappa})$
500	3.170	0.05	−1.277	0.621	2.06	1.833
7.1	1.000	0.50	−1.243	0.436	2.94	1.402
6	0.912	0.60	−1.243	0.557	2.98	1.617
5	0.819	0.74	−1.270	0.478	3.07	1.288
4	0.706	1.00	−1.298	0.406	3.10	1.125
3	0.559	1.60	−1.301	0.397	3.10	1.114
2.75	0.515	1.89	−1.298	0.396	3.01	1.113
2.50	0.466	2.30	−1.297	0.396	2.91	1.113
2.25	0.413	2.93	−1.297	0.396	2.89	1.113
2	0.353	4.02	−1.297	0.397	2.89	1.114

Source: Sinha (1998).

4.1.7 *Methods Based on Partial Likelihood*

Sargent (1998) and Gustafson (1997) discuss Bayesian frailty models based on Cox's partial likelihood (Cox, 1975) and thus leave the baseline hazard function totally unspecified. In their development, Cox's partial likelihood is used as the likelihood component for the model parameters. One possible disadvantage of this approach is that it does not use the full likelihood as do the other methods discussed in the previous subsections. However, one advantage of this method is that by eliminating the hazard, the prior specification and computations are simplified. The semiparametric Bayesian justification for using partial likelihood is due to Kalbfleisch (1978), demonstrating the convergence of the marginal likelihood of the regression parameters in the univariate case to the partial likelihood when the gamma process is used on the cumulative baseline hazard of the Cox model, and the variance of the gamma process goes to infinity. For details, see Section 3.2.3.

Sargent (1998) works with the reparameterized frailty model. Following Sargent (1998), define $b_i = \log(w_i)$ and write the hazard function for subject j in cluster i as

$$h(y|\boldsymbol{x}_{ij}, b_i) = h_0(y)\exp(\boldsymbol{x}_{ij}'\boldsymbol{\beta} + b_i),$$

which is the same model as the shared frailty model with a slightly different parameterization. A typical Cox regression analysis consists of maximizing the partial likelihood $C(\boldsymbol{\beta}, \boldsymbol{b}|\boldsymbol{x})$ for the parameter vector $\boldsymbol{b} = (b_1, b_2, \ldots, b_n)'$ and $\boldsymbol{\beta}$. For concreteness, if there are no ties, then Cox's partial likelihood of $\boldsymbol{\beta}$ and $\boldsymbol{b}$ is given by

$$C(\boldsymbol{\beta}, \boldsymbol{b}|D) = \prod_{i=1}^{k} \left(\frac{\exp(\boldsymbol{x}_i'\boldsymbol{\beta} + b_{(i)})}{\sum_{l \in \mathcal{R}_i} \exp(\boldsymbol{x}_l\boldsymbol{\beta} + b_{(l)})} \right),$$

where D denotes the data; k denotes the distinct observed survival times ordered as $y_{(1)} < y_{(2)} < \ldots < y_{(k)}$; $b_{(i)}$ is the log-frailty for the subject with survival time $y_{(i)}$; and $\mathcal{R}_i$ is the risk set just before time $y_{(i)}$. We assume throughout this subsection that the b_i's are i.i.d. with a $N(0, \xi)$ distribution. Let $\pi(\xi, \boldsymbol{\beta})$ denote the joint prior distribution for the model parameters $\boldsymbol{\beta}$ and ξ. Some degree of prior specification for the variance component is essential for computational stability. The posterior distribution for the model parameters, $\pi(\boldsymbol{\beta}, \boldsymbol{b}, \xi|D)$, is thus given by

$$\pi(\boldsymbol{\beta}, \boldsymbol{b}, \xi|D) \propto C(\boldsymbol{\beta}, \boldsymbol{b}|D)\pi(\boldsymbol{b}|\xi)\pi(\xi, \boldsymbol{\beta}).$$

To sample $\boldsymbol{\beta}$ and $\boldsymbol{b}$, Sargent (1998) uses a univariate Metropolis algorithm. Due to the use of the partial likelihood $C(\boldsymbol{\beta}, \boldsymbol{b}|D)$ instead of the full likelihood, the conditional posteriors of the b_i's and $\boldsymbol{\beta}$ are nonstandard distributions. In particular, normal candidate distributions are used with mean equal to the current realization and variance chosen to give a Metropolis acceptance rate of between 25% and 50% (Gelman, Roberts, and Gilks, 1995). For ξ, we use Gibbs steps only when conjugacy is available; when it is not, we use univariate Hastings steps with gamma candidate distributions, again with an appropriately chosen mean and variance. We note that in Exercise 4.9, we outline an alternative Gibbs chain to perform the posterior simulation based on $C(\boldsymbol{\beta}, \boldsymbol{b}|D)$. This alternative algorithm requires a Metropolis step for $\boldsymbol{\beta}$ only.

Calculation of Cox's partial likelihood can be performed using the package of survival routines developed by Therneau (1994). These routines include the capability for stratified analyses, so this method can be extended trivially to this case. Any standard method for handling ties in the data can be used; we use the Efron approximation (Efron, 1977) defined as

$$
\begin{aligned}
&C_e(\boldsymbol{\beta}, \boldsymbol{b}|D)\\
&=\prod_{i=1}^{k} \frac{\exp(\boldsymbol{s}_i'\boldsymbol{\beta} + v_{(i)})}{\prod_{j=1}^{d_i}\left[\sum_{l\in\mathcal{R}_i}\exp(\boldsymbol{x}_l'\boldsymbol{\beta} + b_{(l)}) - \frac{j-1}{d_i}\sum_{l\in\mathcal{D}_i}\exp(\boldsymbol{x}_l'\boldsymbol{\beta} + b_l)\right]}, \qquad (4.1.46)
\end{aligned}
$$

where d_i is the size of the set $\mathcal{D}_i$ of survival times that have events at time $y_{(i)}$, $\mathcal{R}_i$ is the risk set just before $y_{(i)}$, $\boldsymbol{s}_i = \sum_{l\in\mathcal{D}_i}\boldsymbol{x}_l$, and $v_{(i)} = \sum_{l\in\mathcal{D}_i} b_{(l)}$. It is important to be aware that Efron's and other approximations for handling ties in the partial likelihood are merely approximations to be used in presence of only a moderate proportion of ties and there are no known Bayesian justifications for such approximations. Including time-dependent covariates in this framework should be straightforward, although we do not discuss this here.

Example 4.6. Mammary tumor data. We consider the data on mammary tumors in rats discussed in Example 1.6. In the dataset, $n = 48$ rats were injected with a carcinogen and then randomized to receive either treatment ($x = 1$) or control ($x = 0$), and thus β is one dimensional. The event times are times to tumor for each rat. The number of tumors per rat ranged from 0 to 13, with a total of 210 tumors, and all rats were censored six months (182 days) after randomization, resulting in $253 = \sum_{i=1}^{n} m_i$ event times. For comparative purposes, standard Cox regression using only the time to the first event gives an estimated treatment effect of -0.69 with a standard error of 0.31. A standard Cox regression using all of the events, but assuming that they are independent, gives an estimated treatment effect of -0.82 with standard error 0.15. Sinha (1993) presented a frailty model-based analysis of this dataset using an independent increment gamma process prior on the baseline cumulative intensity of the conditional Poisson process. For the present analysis, the ties in the data are handled through Efron's approximation. In contrast, the analysis of Sinha (1993) is based on a grouped data likelihood which automatically handles the ties due to scheduled follow-ups. In this particular dataset, there are heavy ties in the data due to regularly scheduled follow-ups and the data collection method.

We take $b_i \sim N(0, \xi)$, $i = 1, 2, \ldots, n$, where ξ is taken to have an inverse gamma (α, ϕ) distribution. Further, we take a uniform improper prior for β independent of ξ, and thus $\pi(\beta) \propto 1$. The joint posterior distribution of $(\beta, \boldsymbol{b}, \xi)$ is given by

$$\pi(\beta, \boldsymbol{b}, \xi | D) \propto C_e(\beta, \boldsymbol{b} | D) \left(\prod_{j=1}^{48} \xi^{1/2} \exp\left(-\frac{\xi b_j^2}{2} \right) \right) \xi^{(\alpha-1)} \exp(\xi\phi),$$

where $C_e(\boldsymbol{\beta}, \boldsymbol{b}|D)$ is defined as in (4.1.46). Following Sargent (1998), we set $\alpha = 1.1$, $\phi = 0.1$, resulting in a diffuse but still proper prior for ξ. Results with an improper prior for ξ are discussed in Sargent (1998). Convergence was monitored by the Gelman and Rubin (1992) diagnostic and by monitoring within-chain autocorrelations. Three chains were generated at overdispersed locations and monitored for convergence; for this dataset all chains reached an equilibrium distribution within 500 iterations. Each chain was run for a total of 3000 iterations, using the last 2000 from each chain for the summary statistics presented here.

The results of Sargent's analysis are consistent with the results of Sinha (1993). The posterior mean for β is -0.81, with posterior standard deviation 0.22, providing strong evidence of a beneficial treatment effect. The batched estimate of sampling error (Ripley, 1987) for $E(\beta|D)$ is < 0.02, indicating a small MCMC error. These results show the benefit of using all of the events in the analysis. Compared to the analysis that used only time to first event, the treatment effect's posterior standard deviation is approximately one-third smaller in the frailty analysis than in the standard (time to first event) analysis, and the estimated treatment effect is larger. The standard deviation is approximately one-third larger than in the analysis that used all of the events as if they were independent.

Sinha (1993) finds strong evidence of heterogeneity and Sargent's results concur: the posterior mean for ξ is 3.9 with a posterior standard deviation of 1.8. These results provide strong evidence that the variance of the log-frailties is greater than 0.10 and thus effectively nonzero. The *prior* probability that $\xi > 10$ is approximately 0.31, while the *posterior* probability that $\xi > 10$ is approximately 0.01, giving strong evidence that the data are consistent with heterogeneity. In addition, 13 of the 48 posterior distributions for the log-frailties have greater than 90% of their posterior mass on one side of zero.

4.2 Multiple Event and Panel Count Data

The mammary tumor data of Gail, Santner, and Brown (1980) is a classic data example where every subject (rat) may experience a recurrent and possibly non-fatal event. The data are recorded over an interval of observation and the data on each subject contain a moderate number of events within its interval of observation. A close inspection of the data recording method of this study reveals another additional feature of this dataset—each rat has been inspected for tumors only at regular intervals of time. So, for each rat, only the number of new events (tumors) in between two inspection times is recorded in the data and the inspection schedules followed roughly a routine of two inspections in each week. This is the reason for heavy ties of crudely recorded event times among the rats as well as within each rat. Such data are often called panel count data (Sun and Wei, 2000). We mention that there are many models available for multiple event time data. We refer the reader to Oakes (1992) for a survey.

The model we discuss here is based on grouped panel count data. Suppose, $N_i(y)$ denotes the number of events by time y for the ith subject and $0 = s_0 < s_1 < \cdots < s_J$ are the ordered inspection times. We assume that the inspection times are the same for all subjects. The methodology can be extended easily to data with different inspection schedules for different subjects. Let $N_{ij} = N_i(s_j) - N_i(s_{j-1})$ be the number of events for the ith subject in the jth inspection interval $I_j = (s_{j-1}, s_j]$. The observed data is given by $D = \{N_{ij}, \boldsymbol{x}_{ij} : i = 1, 2, \cdots, n; j = 1, 2, \cdots, J\}$. We note that for the mammary tumor data, the $\boldsymbol{x}_{ij}$'s are all equal to $\boldsymbol{x}_i$. A model for such data is given by Sinha (1993) using the assumption that each $N_i(t)$ is a realization of a conditional non-homogeneous Poisson process (Karlin and Taylor, 1981) given the frailty random effect w_i. A conditional proportional intensity function for the model is given by

$$\lim_{\Delta \to 0} \frac{P(N_i(y+\Delta) - N_i(y) = 1 | \boldsymbol{x}_i, w_i)}{\Delta} = h_0(y) w_i \exp(\boldsymbol{x}_i' \boldsymbol{\beta}),$$

where $h_0(y)$ is the baseline intensity and the subject specific random effects w_i $(i = 1, 2, \ldots, n)$, are assumed to be i.i.d. $\mathcal{G}(\kappa^{-1}, \kappa^{-1})$. We note that for this model, we must use a finite mean frailty distribution, otherwise the expected number of events within any finite time interval would be infinite. The frailty random effect takes care of the heterogenity among subjects with respect to their proneness to events. A quick look at the mammary tumor data clearly suggests that there are some rats within the control group that are more prone to tumors compared to some other rats from the same group.

The conditional distributions of the observables, N_{ij} (for $j = 1, 2, \cdots, J$) given w_i, are independent poisson and can be expressed as

$$N_{ij} \sim \mathcal{P}(H_{0j} w_i \exp(\boldsymbol{x}_{ij}' \boldsymbol{\beta})),$$

where, $\mathcal{P}(\mu)$ denotes the Poisson distribution with mean μ and $H_{0j} = \int_{I_j} h_0(u)du$. The likelihood $L(\boldsymbol{\beta}, H_0, \boldsymbol{w}|D)$ involves the baseline cumulative intensity only through its increments H_{0j}. The likelihood is given by

$$L(\boldsymbol{\beta}, H_0, \boldsymbol{w}|D) \propto \prod_{j=1}^{J}\prod_{i=1}^{n}[H_{0j}w_i \exp(\boldsymbol{x}'_{ij}\boldsymbol{\beta})]^{N_{ij}} \exp\left(-H_{0j}w_i \exp(\boldsymbol{x}'_{ij}\boldsymbol{\beta})\right).$$

A convenient prior process is the gamma process on the cumulative baseline intensity. From the above likelihood it is clear that we only need the joint distributions of the independent increments. Using this prior process, we get the prior distribution $H_{0j} \sim \mathcal{G}(c\gamma_j, c)$, independent for $j = 1, 2, \cdots, J$. The hyperparameters c and γ_j are determined by the prior mean and confidence of the baseline cumulative intensity, which are assumed to be known here.

The joint posterior is given by

$$\pi(\boldsymbol{\beta}, \boldsymbol{w}, \eta|D) \propto L(\boldsymbol{\beta}, H_0, \boldsymbol{w}|D)\pi(\boldsymbol{w}|\eta)\pi(\boldsymbol{\beta}, \eta) \prod_{j=1}^{J}[H_{0j}^{c\gamma_j - 1} \exp(-cH0j)],$$

where $\eta = \kappa^{-1}$,

$$\pi(\boldsymbol{w}|\eta) = \prod_{i=1}^{n}[\eta^{\eta} w_i^{\eta-1} \exp(-w_i\eta)/\Gamma(\eta)],$$

and $\pi(\boldsymbol{\beta}, \eta)$ is the joint prior for $(\boldsymbol{\beta}, \eta)$. The full conditionals are quite easy to sample for this model. The full conditional of H_{0j} is $\mathcal{G}(c\gamma_j + N_{+j}, c + V_j)$, where $N_{+j} = \sum_{i=1}^{n} N_{ij}$ and $V_j = \sum_{i=1}^{n} w_i \exp(\boldsymbol{x}'_{ij}\boldsymbol{\beta})$. The full conditional of w_i is again $\mathcal{G}(\eta + N_{i+}, \eta + U_i)$, where $N_{i+} = \sum_{j=1}^{J} N_{ij}$ and $U_i = \sum_{j=1}^{J} H_{0j} \exp(\boldsymbol{x}'_{ij}\boldsymbol{\beta})$. The full conditional of η is proportional to

$$\frac{\eta^{n\eta} \exp(-\eta G)}{(\Gamma(\eta))^n} \times \pi(\boldsymbol{\beta}, \eta),$$

where $G = \sum_{i=1}^{n}(w_i - \log(w_i))$. The full conditional of $\boldsymbol{\beta}$ is proportional to $\exp\left\{-\sum_{i=1}^{n} w_i U_i + \boldsymbol{\beta}' \sum_{i=1}^{n}\sum_{j=1}^{J} \boldsymbol{x}_{ij} N_{ij}\right\} \times \pi(\boldsymbol{\beta}, \eta)$. The full conditional posteriors of $\boldsymbol{\beta}$ and η clearly depend on the form of $\pi(\boldsymbol{\beta}, \eta)$, but when $\pi(\boldsymbol{\beta}, \eta)$ is log-concave with respect to $\boldsymbol{\beta}$ and η, it can be shown that the full conditionals of $\boldsymbol{\beta}$ and η are both log-concave. See Exercise 4.14 for details about the MCMC steps for the mammary tumor data using this model.

4.3 Multilevel Multivariate Survival Data

Gustafson (1997) discusses Bayesian hierarchical frailty models for multilevel multivariate survival data. The hierarchical modeling given by Gustafson (1997) has elements in common with the work of Clayton (1991), Gray (1994), Sinha (1993), Stangl (1995), and Stangl and Greenhouse (1995). All these authors discuss Bayesian hierarchical models for failure time data. Gustafson's approach differs in several respects, including the treatment of baseline hazards, and the shape of priors for variance components. As well, the previous work does not deal with multiple sets of units, or with computation in very high-dimensional parameter spaces.

For the sake of clarity, we decribe Gustafson's model for the specific structure of the data set exmained in Example 4.8. In particular, in Example 4.8, two sets of units are considered: patients and clinical centers. We expect the responses from patients within a clinical trial center to be associated with each other. Moreover, there is an association between two failure times within a patient. Because of the association of responses at two levels (within patient association at first level and within a center association at second level), we call such data multilevel multivariate survival data. It is clear, however, that similar models can be adopted in a wide variety of circumstances. The bivariate failure time pairs for the n patients are denoted (y_{1i}, y_{2i}) $i = 1, 2, \ldots, n$. In the example, y_1 is the time from entry in the clinical trial to disease progression, and y_2 is the time from disease progression to death. Relevant information about the ith patient is encapsulated in the explanatory variables $\mathbf{u}_i$ and $\boldsymbol{x}_i$. Here $\mathbf{u}_i$ is a vector of dummy variables indicating the clinical site at which the patient received treatment, while $\boldsymbol{x}_i$ is a covariate vector indicating the patient's treatment arm and health status upon enrollment.

The unit-specific parameters are patient-specific log-frailties b_i, $i = 1, 2, \ldots, n$ along with bivariate center-specific effects $(\alpha_{1j}, \alpha_{2j})$, $j = 1, 2, \ldots, n_c$. Here n_c is the number of centers participating in the clinical trial. The effects α_{1j} and α_{2j} act on y_1 and y_2 respectively. Thus it is not presumed that a center effect is the same for both responses. The common parameters needed to specify the first stage of the model are $\boldsymbol{\beta}_1$ and $\boldsymbol{\beta}_2$. These are vectors of regression parameters, which modulate the relationship between the covariate vector $\boldsymbol{x}_i$ and the response pairs (y_{1i}, y_{2i}). Given parameters $(\boldsymbol{b}, \boldsymbol{\alpha}, \boldsymbol{\beta})$, the responses (y_{1i}, y_{2i}), $i = 1, 2, \ldots, n$ are modeled as independent, both within and across the bivariate pairs, with respective hazard rates:

$$
\begin{aligned}
h_{1i}(y_1 | \mathbf{u}_i, \xi, b_i) &= \exp(\boldsymbol{u}_i' \boldsymbol{\alpha}_1 + \boldsymbol{x}_i' \boldsymbol{\beta}_1 + b_i) h_{01}(y_1), \\
h_{2i}(y_2 | \mathbf{u}_i, \xi, b_i) &= \exp(\boldsymbol{u}_i' \boldsymbol{\alpha}_2 + \boldsymbol{x}_i' \boldsymbol{\beta}_2 + b_i)\, h_{02}(y_2).
\end{aligned}
$$

Here h_{01} and h_{02} are unknown baseline hazards. The second stage of the hierarchy assigns distributions to the unit-specific parameters. Because all the unit-specific parameters act additively on the baseline log–hazards, normal specifications seem reasonable. That is, log-normal frailty distributions are specified. Particularly, b_i and $(\alpha_{1j}, \alpha_{2j})$ are taken to be mutually independent and identically distributed sequences, with

$$
b_i \sim N(0, \sigma_F^2),
$$

and

$$\begin{pmatrix}\alpha_{1j}\\ \alpha_{2j}\end{pmatrix} \sim N\left(\begin{pmatrix}0\\ 0\end{pmatrix}, \begin{pmatrix}\sigma_1^2 & \rho\sigma_1\sigma_2\\ \rho\sigma_1\sigma_2 & \sigma_2^2\end{pmatrix}\right). \tag{4.3.1}$$

The zero mean constraints are imposed so that the patient and center effects represent deviations from population averages.

Conventional wisdom dictates that it is comparatively safe to make parametric assumptions at the second stage of a hierarchical model, as is done above. There is some literature to support this claim, including the asymptotic investigations of Neuhaus, Hauck, and Kalbfleisch (1992), Neuhaus, Kalbfleisch, and Hauck (1994), and Gustafson (1996) in a few particular parametric models. As well, the simulation studies of Pickles and Crouchley (1995) lends support to the claim in a frailty model context. Thus inference is expected to be similar whether the frailty distribution is modeled as log-normal or some other parametric family. While the gamma and positive stable distributions are commonly used as frailty distributions, the log-normal distribution is selected for the ease with which bivariate effects can be incorporated via (4.3.1). The extra complication of specifying a non-normal bivariate distribution for $(\alpha_{1i}, \alpha_{2i})$ does not seem warranted, since the choice of parametric family is not expected to have a large effect on inference.

The third stage of the model involves specifications of the prior distribution for the common parameters. Improper uniform priors are specified for $\boldsymbol{\beta}_1$ and $\boldsymbol{\beta}_2$. Proper prior distributions are needed for the variance components to obtain proper posteriors. Let σ^2 be a generic label for a variance component. Although gamma priors can be used for σ^2, Gustafson (1997) argues for prior densities for σ^2 that are positive and finite at $\sigma^2 = 0$, and which decrease monotonically in σ^2. This choice of prior leads to a posterior marginal density for σ^2 that can have its mode at zero or to the right of zero, depending on the data observed. As well, a prior with a monotonically decreasing density can be regarded as a parsimonious prior; it favors simpler models involving random effects of smaller magnitude.

Thus, Gustafson (1997) recommends independent half-Cauchy priors for the variance components, with densities of the form $\pi(\sigma^2) = (2/\phi)[\pi\{1 + (\sigma^2/\phi)^2\}]^{-1}$. Each variance component requires the specification of a scale hyperparameter ϕ; these are denoted ϕ_F, ϕ_1, and ϕ_2 for σ_F^2, σ_1^2, and σ_2^2 respectively. As well as having an appealing shape, a half-Cauchy specification involves only one hyperparameter, as compared to two for an inverse gamma specification. This is a practical advantage, since well-formed prior information about variance components is typically not available. For the same reason, a uniform prior on $(-1, 1)$ is adopted for ρ, the correlation parameter.

The argument put forward by Kalbfleisch (1978) given in Section 3.2.3 can be used to justify the use of the partial likelihood provided the baseline cumulative hazards corresponding to h_{01} and h_{02} are assumed to have diffuse gamma process priors and h_{01} and h_{02} are *a priori* independent of the other parameters. After integrating out h_{01} and h_{02}, the marginal posterior density of the other parameters can be expressed as

$$\pi(\boldsymbol{\theta}, \boldsymbol{\alpha}, \boldsymbol{\beta}, \sigma^2, \rho | D) \propto C_1((\boldsymbol{\alpha}_1, \boldsymbol{\beta}_1, \boldsymbol{b} | D_1) C_2((\boldsymbol{\alpha}_2, \boldsymbol{\beta}_2, \boldsymbol{b} | D_2)$$
$$\times \pi(\boldsymbol{b} | \sigma_F^2) \pi(\boldsymbol{\alpha} | \sigma_1^2, \sigma_2^2, \rho) \pi(\boldsymbol{\beta}, \sigma_F^2, \sigma_1^2, \sigma_2^2, \rho),$$

where C_k is the partial likelihood,

$$C_k(\boldsymbol{\alpha}_k, \boldsymbol{\beta}_k, \boldsymbol{b} | D_k) = \prod_{i=1}^{n} \left[\frac{\exp(\boldsymbol{u}_i' \boldsymbol{\alpha}_k + \boldsymbol{x}_i' \boldsymbol{\beta}_k + b_i)}{\sum_{\{j \in \mathcal{R}_{ki}\}} \exp(\boldsymbol{u}_j' \boldsymbol{\alpha}_k + \boldsymbol{x}_j' \boldsymbol{\beta}_k + z_l)} \right]^{\nu_{ik}},$$

$D = (D_1, D_2)$, $D_k = \{\mathbf{y}_k, U, X, \boldsymbol{\nu}_k\}$ is the observed data ($k = 1, 2$), and U denotes the matrix of the $\boldsymbol{u}_i$'s. For each k, ($k = 1, 2$) we assume that there are no ties amongst the observed survival times y_{ki}, $i = 1, 2, \ldots, n$. Also, $\mathcal{R}_{ki}$ is the risk set for the kth survival component ($k = 1, 2$) at the time y_{ik} and $\nu_{ik} = 1$ ($\nu_{ik} = 0$) corresponds to y_{ik} being an actual survival time (right censoring time) for $k = 1, 2$.

Example 4.8. Colorectal cancer data. Gustafson (1997) presents data from a clinical trial of chemotherapies for advanced cases of colorectal cancer. The data were originally reported upon by Poon et al. (1989). As mentioned previously, the ith patient's responses are y_{1i}, the time from study entry to disease progression, and y_{2i}, the time from progression to death. A total of 419 patients participated in the trial, which was conducted at 16 North American clinical sites. Five new chemotherapy regimens were investigated, in addition to a standard therapy regimen. The standard regimen involved fluorouracil (5-FU), while the novel regimens involved a combination of 5-FU and a second agent, either cisplatin (CDDP), leucovorin (LV), or methotrexate (MTX). Leucovorin and methotrexate were each administered at two dosages, denoted here by $-$ and $+$ for low and high dosages, respectively. Thus, there are six treatment arms in all. To reflect patient's overall health and severity of disease at time of enrollment, four binary covariates are incorporated in the analysis. These are performance status (PS), grade of anaplasia (AN), measurable disease (MS), and presence of symptoms (SY). For further details of the trial, see Poon et al. (1989).

A total of 29 patients did not follow the prevalent pattern of progression prior to death. In particular, these patients are recorded either as simultaneously progressing and dying, or as dying before progression. In the former case, y_1 is observed and equal to the death time, while y_2 is not manifested. In the latter case, y_1 is right censored at the death time, and y_2 is not manifested. In either case, y_2 is missing. For the sake of analysis, it is simplest to treat y_2 as being right censored at time zero in this situation. Additionally, there are 15 cases for which y_2 is left censored (y_1 is right censored, but

$y_1 + y_2$ is observed), and 2 cases for which y_2 is subject to a complicated censoring whereby the censoring time depends on the unobserved value of y_1 (both y_1 and $y_1 + y_2$ are right censored). These 17 cases in total are also treated as instances of missing (right censored at zero) y_2 values. Gustafson (1997) notes that this is not a good practice from the point of view of data analysis, but it does permit illustration of the methodology. While the use of the partial likelihood at the first-stage of the model has advantages, it also has limitations in terms of dealing with types of censoring other than right censoring. On the contrary, a full likelihood based analysis is capable of dealing with such interval-censored data. Based on the above assumptions, there are 386 observed y_1 times and 33 right censored y_1 times. For y_2 there are 369 observed times and 50 right censored times.

A hierarchical Bayesian analysis of these data is carried out in Gustafson (1995). A drawback of this analysis is the parametric assumption of exponentially distributed response times. As well, dependence between progression and progression-to-death times is incorporated by using a residual based on y_{1i} as an explanatory variable for the distribution of $y_{2i}|y_{1i}$, rather than by using a frailty structure. The former approach has the advantage of allowing negative or positive dependence between y_1 and y_2, whereas the frailty approach permits only positive dependence. The difficulty in extending the former approach to the current setting is that the nonparametric first-stage modeling does not yield an obvious residual for y_1. Physically, negative dependence is unlikely to be present in this situation and fortunately, the previous analysis indicates a strong positive dependence between y_1 and y_2, so the frailty approach seems appropriate.

The data are coded so that $\boldsymbol{u}_i$ is a 0 or 1 dummy vector of length 16, which indicates the center at which the ith patient received treatment. Each $\boldsymbol{x}_i$ is a vector of length 9; the first five components are dummy variables, taking the value 1 if the patient is on the corresponding novel therapy and 0 otherwise. The last four components are the binary covariate levels, coded as 1 for the higher risk level and 0 for the lower risk level. To specify the requisite scale hyperparameters, it is helpful to contemplate the potential magnitudes of the random effect variances. These are difficult to think about, however, because the random effects act additively upon the log-hazard. Thus a transformation is employed. For instance, the patient-specific parameters have variance σ_F^2. The transformation is given by

$$\zeta_F = \exp\left(\Phi^{-1}(0.75)\sigma_F\right),$$

where Φ^{-1} is the inverse of the standard normal distribution function. This is interpretable in that ζ_F is the prior median of each $\exp(|b_i|)$. More specifically, the population distribution of frailties, $\exp(b_i)$, assigns probability 0.5 to the interval (ζ_F^{-1}, ζ_F). Moreover, symmetry on the log-scale implies that 1 is the median frailty. As an example, let $\zeta_F = 1.25$. Then with probability 0.5, a single patient's multiplicative effect on the baseline hazard lies between 0.8 and 1.25. Given background knowledge of cancer, chemotherapy, and modern hospitals, $\zeta = 1.1$ is selected as a plausible value for all three variance components. Thus, the scale hyperparameters $\phi_F = \phi_1 = \phi_2 = 0.02$ are specified, as this fixes the prior medians of ζ_F, ζ_1, and ζ_2 at 1.1.

It is convenient to work with Cox's partial likelihood when all observed failure times are distinct. The present dataset, however, contains a small number of tied failure times. Several approximations to the partial likelihood for ties have been suggested; see, for example, Efron (1977) and the discussion in Kalbfleisch and Prentice (1980), and the references therein. Also, see Subsection 4.1.7 on the use of Efron's approximation for ties. It would be possible to implement one of these approximations in the present context. It is even simpler, however, randomly to break the ties after every iteration in the MCMC scheme. This is the approach actually used to deal with ties in the analysis. Clearly, this avoids the potential pitfall of basing an analysis on a single random breaking of the ties. Because of the modest number of ties, the difference between results based on one of the partial likelihood approximations and results based on the iterative rebreaking of ties should be negligible. Again, the most logical way of analyzing such data is to use a grouped data based likelihood which permits the use of the data as it has been reported.

The algorithm used in the present application is the hybrid Monte Carlo method (Neal, 1993a,b 1994a,b) discussed in Chapter 1. In contrast to its competitors, the hybrid algorithm updates the whole parameter vector simultaneously, and relies on evaluations of the gradient of the log posterior density. The rationale for employing the hybrid Monte Carlo method, rather than a better known algorithm, is twofold. First, the form of the Cox's partial likelihood is not amenable to Gibbs sampling. In contrast, it is straightforward to evaluate the gradient of the log posterior density, as required by the hybrid algorithm. Second, while Neal (1993a, 1994b) reports experience with the algorithm for Bayesian analysis in neural network models, we are not aware of its implementation in more traditional statistical models. Thus it seems worthwhile to report on its performance in a hierarchical modeling setting. A gradient-based algorithm for computing posterior quantities would also be of interest in other statistical applications.

The computational analysis is undertaken by implementing five production runs of the hybrid algorithm. As starting values for each run, regression parameters are set to zero, variance components are sampled from their priors, and random effects are sampled from their conditional priors, given the variance components. Each run consists of 9000 iterations of the algorithm. The first 1000 iterations are discarded as burn-in variates, to allow the Markov chain to reach its stationary distribution. From its partially random starting point, the Markov chain consistently moves to the same region of the parameter space. This suggests that the posterior distribution has a well-defined peak. To be more specific about the implementation of the hybrid Monte Carlo method, ϵ is simulated from a normal distribution with mean zero and standard deviation 0.05 (step 1), and k is simulated from a binomial (15, 0.5) distribution (step 2). These specifications are arrived at via trial and error experimentation. The resulting rejection rate in step 4 is in the range of 15% to 20%. Note that it is not desirable to have a very small rejection rate, as this suggests that the chain is not moving very far at each iteration.

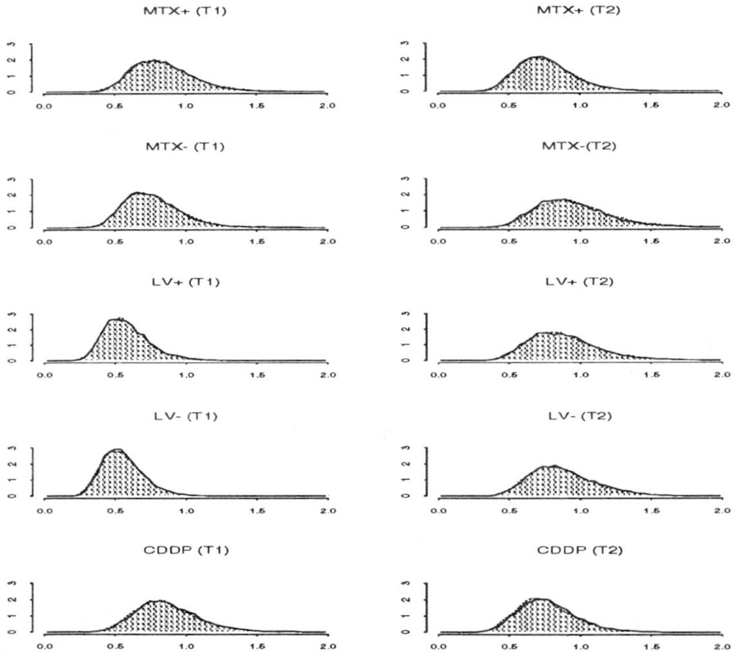

FIGURE 4.4. Posterior marginal densities of the treatment effects on the multiplicative scale.

Careful coding of the algorithm is required to ensure that proposed steps to very low probability regions of the parameter space are properly rejected, without leading to numerical overflow error. The computation is expensive; a single production run requires approximately 150 minutes of CPU time on a SPARC 20 workstation with a 50 MHz CPU. The expense is attributed to the high dimensionality of the problem; the dimension of the parameter vector is close to 500. Of course most of these parameter components are random effects, so there are far fewer parameters from a statistical point of view. But the random effects cannot be integrated out analytically, so the high dimensionality cannot be avoided. The unwieldy form of the Cox's partial likelihood also contributes to the computational expense.

There are many ways to summarize the results of a Bayesian analysis. Because the present analysis is illustrative, results are reported in the form of histograms of posterior marginal distributions of parameters. If quantitative summaries were desired, it would be straightforward to report MCMC estimates of posterior moments and/or quantiles of parameters. One way to assess the numerical accuracy of such estimates would be to report simulation standard errors based on the variability of independent estimates across the five production runs. If the analysis were to be used for decision-making on the part of the investigators, then the posterior analysis could be used to select actions which are optimal with respect to appropriate loss functions. Figure 4.4 displays posterior marginal histograms of the treatment effects on the multiplicative scale, that is, the posterior distributions of $\exp(\beta_{1j})$ and $\exp(\beta_{2j})$ are examined. These transformed parameters are multiplicative effects on the hazard rate associated with each novel therapy relative to the standard therapy. In terms of the progression time, y_1, the LV regimens appear to be the most beneficial, followed by the MTX regimens. For the progression-to-death times, y_2, the only regimens for which there is strong evidence of improvement relative to standard therapy are MTX+ and CDDP. In general, inferences for treatment effects are similar to those of Gustafson (1995).

Analogous histograms for covariate effects are given in Figure 4.5, again on the multiplicative scale. These indicate that the covariates are important predictors of both responses, with the only exception being for disease measurability combined with y_2. This is not surprising, as all four covariates are basic measures of health and disease severity. The covariate effects are generally much larger in magnitude than the treatment effects. The left column of Figure 4.6 displays posterior marginal histograms for the variance components, on the interpretable ζ-scale introduced previously. The estimated patient to patient variability is very large, suggesting strong association in the (y_{1i}, y_{2i}) pairs. That is, the observed covariates do not explain all of a patient's frailty. The center to center variability in the y_2 response appears to be greater than in the y_1 response, though the difference appears to be slightly less pronounced than that indicated by the analysis of Gustafson (1995).

Inference about the correlation parameter ρ is interesting in light of its physical interpretation. Under a positive value for ρ, a particular center's effects for y_1 and y_2 will tend to be simultaneously high or low. In this case, overall patient outcomes in terms of the death time $y_1 + y_2$ will tend to exhibit center-to-center heterogeneity. Conversely, under a negative value for ρ, the centers with positive effects for y_1 will tend to have negative effects for y_2, and vice versa. In this case, center differences can be explained by differential methods of progression determination across centers. This hypothesis is consistent with investigators' concerns about the accuracy

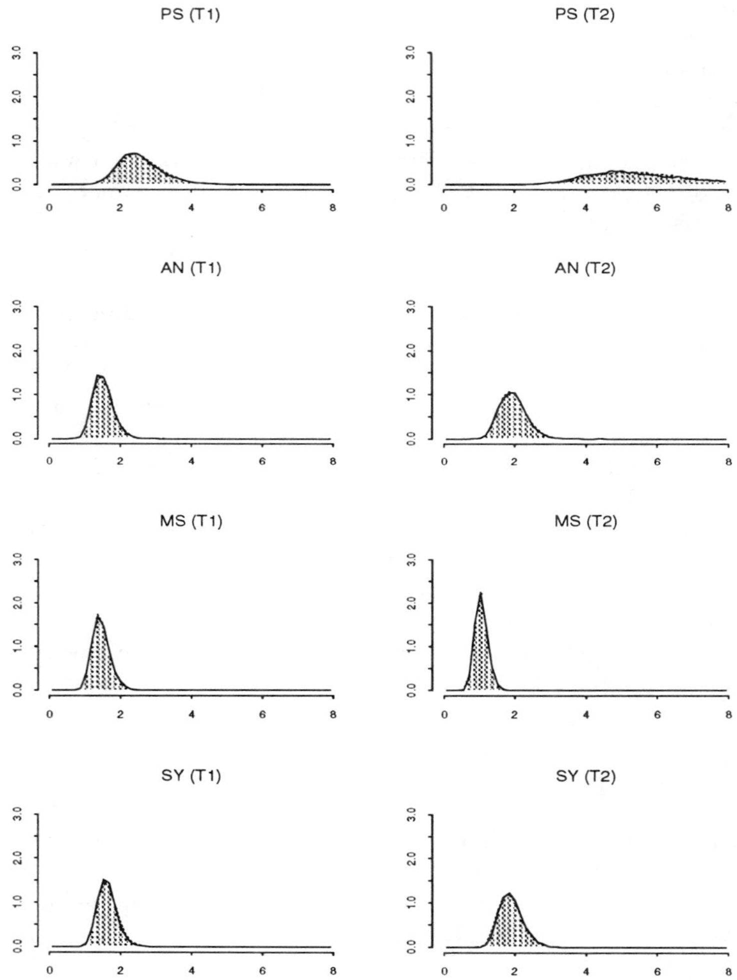

FIGURE 4.5. Posterior marginal histograms for covariate effects.

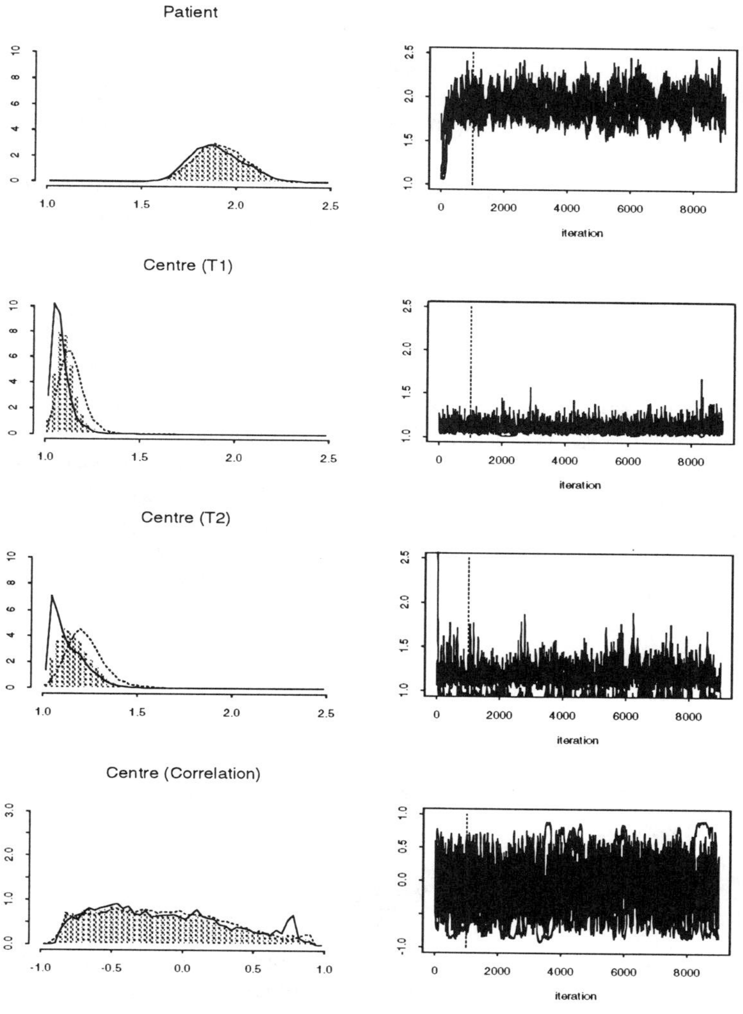

FIGURE 4.6. Posterior marginal histograms for the variance components.

with which progressions are identified clinically. In fact, Figure 4.6 shows that the data are decidedly noninformative about ρ. This is probably due to the small center to center variability in progression times. As σ_1^2 tends to zero with σ_2^2 fixed (or vice versa), the correlation ρ becomes meaningless.

To give an idea of the Markov chain performance, time-series plots of the chain evolution are presented, for each variance component and each production run, in the right-hand column of Figure 4.6. The dashed vertical lines indicate the end of the burn-in period. These plots are given for the variance components as these are typically the more difficult parameters for MCMC sampling. That is, the variance component marginals of the Markov chain mix more slowly than the treatment effect or covariate effect marginals. Though, as mentioned, the marginal histograms are consistent across production runs, even for the variance components. To investigate the sensitivity of inferences to the hyperparameter ϕ, several alternative specifications are considered. Recall that the specification $\phi = 0.020$ is chosen to make the prior median of each transformed variance component equal to 1.1. Instead one might choose $\phi = 0.062$ to make 1.1 the 0.2 quantile, or $\phi = 0.0065$ to make 1.1 the 0.8 quantile. Due to the computational burden, it is not desirable to rerun the analysis for every entertained hyperparameter value. Instead, the original posterior sample is reweighted according to the new hyperparameter value, as suggested by Smith and Gelfand (1992) and Besag, Green, Higdon, and Mengersen (1995), amongst others. Specifically, consider a sample $q^{(1)}, \ldots, q^{(m)}$ approximately from the posterior density $\pi(q|\phi_0)$ under hyperparameter ϕ_0. To approximate the posterior under an arbitrary hyperparameter ϕ, a weighted sample is formed, giving weight u_i to $q^{(i)}$, where

$$u_i \propto \frac{\pi(q^{(i)}|\phi)}{\pi(q^{(i)}|\phi_0)},$$

and $\sum_{i=1}^{m} u_i = 1$. The weights are easy to compute as only the final–stage contribution to the log–posterior need be considered. That is, before normalization,

$$u_i \propto \frac{\pi(\sigma_F^2, \sigma_1^2, \sigma_2^2|\phi)}{\pi(\sigma_F^2, \sigma_1^2, \sigma_2^2|\phi_0)}.$$

The reweighted posteriors are graphically represented in tandem with the posterior marginals displayed in Figures 4.4–4.6. Two sets of connected line segments are superimposed on each posterior marginal histogram. The height of the line segment at the midpoint of each histogram bin is the weight assigned to that bin by the weighted sample, appropriately normalized. Inference about most parameters is highly insensitive to the hyperparameter value. The exceptional cases are inference about the components of center to center variability, ζ_1 and ζ_2. For these parameters, the hyperparameter specification is quite important. In contrast, the specification is not influential for ζ_F, the parameter reflecting patient-to-patient variability.

4.4 Bivariate Measures of Dependence

The previous sections addressed how dependence is created between survival times. It is now of interest to quantify the dependence in some formal way. When the variables have a multivariate normal distribution, the Pearson product moment correlation is a common measure of linear dependence between two variables. In survival data, however, the marginal distributions are not normal and the dependence structure is nonlinear, and therefore we need more general measures of dependence for bivariate survival data. Towards this goal, Oakes (1989) proposes a local measure of dependence, given by

$$\theta^*(y_1, y_2) = \frac{S(y_1, y_2)\Delta_1\Delta_2 S(y_1, y_2)}{(\Delta_1 S(y_1, y_2))(\Delta_2 S(y_1, y_2))}, \tag{4.4.1}$$

where Δ_j denotes the operator $-\frac{\partial}{\partial y_j}$, $j = 1, 2$. This function, which was introduced by Clayton (1978), may be interpreted as the ratio of the conditional distribution of Y_1 given $Y_2 = y_2$, to that of Y_1 given $Y_2 > y_2$. This measure has proved to be a useful quantity for assessing the degree of association between survival times for a given frailty distribution.

For a model with gamma frailties, that is, $w \sim \mathcal{G}(\kappa^{-1}, \kappa^{-1})$, it can be shown that the joint survival function of (Y_1, Y_2) is given by

$$S(y_1, y_2) = \frac{1}{\left[\frac{1}{\kappa} - \log(S_1(y_1)) - \log(S_2(y_2))\right]^{1/\kappa}}, \tag{4.4.2}$$

where $S_1(y_1)$ and $S_2(y_2)$ are the baseline survival functions. For this model, it can be shown that $\theta^*(y_1, y_2) = 1 + \kappa$, so that $\theta^*(y_1, y_2)$ is free of (y_1, y_2) for this model (see Exercise 4.16). Also, $\theta^*(y_1, y_2) = 1 + \kappa$ implies that the larger the κ, the larger the correlation between Y_1 and Y_2, and thus this measure of dependence sheds light on the role of κ when using gamma frailties. In particular, when $\kappa \to 0$, $\theta^*(y_1, y_2) \to 1$, indicating independence between Y_1 and Y_2.

For a model with positive stable frailties, as discussed in Section 4.1.4, the Laplace transform of w is $E(\exp(-sw)) = \exp(-s^\alpha)$, and the bivariate survival function is

$$S(y_1, y_2) = \exp\left\{-\left[-\log(S_1(y_1)) - \log(S_2(y_2))\right]^\alpha\right\}. \tag{4.4.3}$$

The parameter α $(0 < \alpha < 1)$ is a scalar parameter that is a measure of association between (Y_1, Y_2). Small values of α indicate high association between (Y_1, Y_2). As $\alpha \to 1$, this implies less association between (Y_1, Y_2)

which can be seen from (4.4.3). For (4.4.3),

$$\theta^*(y_1, y_2) = \alpha^{-1}(1-\alpha)\left[-\log(S_1(y_1)) - \log(S_2(y_2))\right]^{-\alpha} + 1. \quad (4.4.4)$$

Thus (4.4.4) decreases in (y_1, y_2). Therefore, the association between (y_1, y_2) is greater when (y_1, y_2) are small and the association decreases over time. Such a property is desirable, for example, when y_1 denotes time to relapse and y_2 denotes relapse to death. This property is also desirable in twin studies. Genetic association will tend to have larger effects on early failures, but that effect may decrease as time increases due to random environmental exposure on the twins. As $\alpha \to 0$, Y_1 and Y_2 achieve maximal dependence. When $\alpha \to 1$, $\theta^*(y_1, y_2) \to 1$, indicating independence between Y_1 and Y_2.

Finally, we note here that other measures of dependence are also available for quantifying the degree of association between (Y_1, Y_2). These include Kendall's τ, the Pearson correlation coefficient, Spearman's correlation coefficient, or median concordance measures. These measures may be less attractive and more difficult to compute than (4.4.1) for some models. We have focused here on the local dependence measure (4.4.1) since it is widely used, is readily computed, and provides nice interpretations for gamma and positive stable frailties. We refer the reader to Hougaard (2000) for more details on local and global measures of dependence.

Exercises

4.1 Consider the Weibull frailty model.

(a) Derive the likelihood function in (4.1.5).

(b) Derive the full conditional distributions given by (4.1.7).

(c) Using the kidney infection data, write a Gibbs sampling program to sample from the joint posterior distribution of $(\alpha, \boldsymbol{\beta}, \gamma, \eta)$ given by (4.1.7)–(4.1.11).

(d) Write a program in BUGS to sample from the joint posterior distribution of $(\alpha, \boldsymbol{\beta}, \gamma, \eta)$ using the additive frailty model in (4.1.13).

(e) Write a brief summary of your inferences about $\boldsymbol{\beta}$ in parts (c) and (d).

4.2 Consider the gamma process prior for $H_0(t)$ in Section 4.1.2.

(a) Derive the likelihood function in (4.1.15).

(b) Derive the full conditional distributions in (4.1.18)–(4.1.20).

4.3 Prove (4.1.30).

4.4 Prove (4.1.31).

4.5 Consider the carcinogenesis data of Example 4.2.

(a) Using the same model and priors given by Clayton (1991) (i.e., (4.1.17)), write a program in BUGS to reproduce the results of Table 4.2.

(b) It is reasonable to assume in this dataset that the association between the control and experimental rats in the same litter will be different from the association between the experimental rats in the same litter. How would you modify the frailty model in (4.1.17) to take into account this association?

4.6 Show that for the digamma function given by $\Psi(x) = \frac{\Gamma'(x)}{\Gamma(x)}$,

$$\Psi'(x) = \sum_{j=0}^{\infty} \frac{1}{(x+j)^2}.$$

4.7 Consider the positive stable frailty model.

(a) Show that the joint survival function of (t_1, t_2) is given by

$$S(t_1, t_2|\boldsymbol{x}_1, \boldsymbol{x}_2) = \exp\left\{-\left(H_{01}\exp(\boldsymbol{x}_1'\boldsymbol{\beta}) + H_{02}\exp(\boldsymbol{x}_2'\boldsymbol{\beta})\right)^{\alpha}\right\}.$$

(b) Show that the marginal hazard in part (a) is given by

$$h_j(t|\boldsymbol{x}_j) = \alpha\exp(\alpha\boldsymbol{x}_j'\boldsymbol{\beta})h_{0j}(t)(H_{0j}(t))^{\alpha-1}, j = 1, 2.$$

(c) Using part (b), show that

$$\frac{h_1(t|\boldsymbol{x}_1)}{h_1(t+\Delta|\boldsymbol{x}_1)} \geq \frac{h_{01}(t)}{h_{01}(t+\Delta)}.$$

4.8 Let $S(t_1, t_2|\boldsymbol{x}_1, \boldsymbol{x}_2)$ denote the joint survival function based on the gamma frailty model.

(a) Show that

$$S_j(t|\boldsymbol{x}_j) = E(\exp(-w\exp(\boldsymbol{x}_j'\boldsymbol{\beta})H_{0j})) = \eta\frac{h_{0j}\exp(\boldsymbol{x}_j'\boldsymbol{\beta})}{\eta + \exp(\boldsymbol{x}_j'\boldsymbol{\beta})H_{0j}(t)}$$

for $j = 1, 2$.

(b) Show that

$$\begin{aligned}\frac{h_1(t|\boldsymbol{x}_1)}{h_1(t+\Delta|\boldsymbol{x}_1)} &= \frac{h_{01}(t_1)}{h_{01}(t_1+\Delta)}\left(1 + \frac{\exp(\boldsymbol{x}_1'\boldsymbol{\beta})H_{01}(t_1, t_1+\Delta)}{\eta + \exp(\boldsymbol{x}_1'\boldsymbol{\beta})H_{01}(t_1)}\right)\\ &\leq \frac{h_{01}(t_1)}{h_{01}(t_1+\Delta)}.\end{aligned} \tag{4.E.1}$$

4.9 (Continuation of Exercise 3.2.) Suppose we have continuously monitored multivariate survival data. From the model in (4.1.1) and the frailty distribution given by (4.1.2), we would like to explore a Gibbs sampling scheme for sampling from

$$\pi(\boldsymbol{\beta}, \eta|D) \propto \int C(\boldsymbol{\beta}, \boldsymbol{w}|D)\pi(\boldsymbol{w}|\boldsymbol{\beta})\, d\boldsymbol{w}, \tag{4.E.2}$$

where

$$C(\boldsymbol{\beta}, \boldsymbol{w}|D) \propto \prod_{i=1}^{n}\prod_{j=1}^{m_i}\left\{\frac{w_i \exp(\boldsymbol{x}_{ij}'\boldsymbol{\beta})}{\sum_{k\in\mathcal{D}_{ij}} w_k \exp(\boldsymbol{x}_k'\boldsymbol{\beta})}\right\}^{\nu_{ij}} \tag{4.E.3}$$

is the partial likelihood and $\mathcal{D}_{ij}$ is the risk set for the survival times t_{ij}.

(a) Show that

$$\frac{w_i \exp(\boldsymbol{x}_{ij}'\boldsymbol{\beta})}{\sum_{k\in\mathcal{D}_{ij}} w_k \exp(\boldsymbol{x}_k'\boldsymbol{\beta})} \propto \int_0^\infty \exp(\boldsymbol{x}_{ij}'\boldsymbol{\beta})\exp(-v_{ij}y)dy,$$

where $v_{ij} = \sum_{k\in\mathcal{D}_{ij}} w_k \exp(\boldsymbol{x}_k'\boldsymbol{\beta})$.

(b) Derive the full conditional posteriors of $dH(t_{ij})$ for $\nu_{ij} = 1$, $i = 1, 2, \ldots, n$, $j = 1, 2, \ldots, m_i$.

(c) Derive the full conditionals of the w_i's.

(d) Assuming a $N_p(\boldsymbol{\mu}_0, \Sigma_0)$ prior for $\boldsymbol{\beta}$ and an independent $\mathcal{G}(\gamma_0, \gamma_0)$ prior for η, derive the full conditional distributions of $\boldsymbol{\beta}$ and η.

(e) Show that the full conditional of η is log-concave.

(f) Using the kidney infection data, implement these Gibbs sampling steps in BUGS to sample from the joint posterior distribution of $(\boldsymbol{\beta}, \eta)$ in (4.E.2).

(g) How would the conditional posteriors change if we reparameterize the frailty model by taking $b_i = \log(w_i)$, where $b_i \sim N(0, \sigma^2)$, and $\sigma^{-2} \sim \mathcal{G}(\gamma_0, \gamma_0)$?

4.10 The M-site model discussed by Oakes (1992) is a potentially useful model for analyzing multiple event time data. Under this model, a fixed number (M) of independent competing events can occur to a subject with $F(t)$ denoting the common cdf of the promotion time to each event. Suppose $\mathcal{H}_i(t)$ is the history of the point-process $N_i(t)$. Here $N_i(t) = 1$ if the ith subject had an event by time t, and 0 otherwise.

(a) Show that the hazard for a new event at time t is given by

$$h_i(t|\mathcal{H}_i(t-)) = (M-k)h_F(t),$$

where $t-$ denotes the time right before t, k denotes the number of events by time t, and $h_F(t) = \frac{f(t)}{F(t)}$.

(b) A more realistic model may assume that M varies with i. Suppose that the M_i are i.i.d. and $M_i \sim \mathcal{P}(\theta)$, where $\mathcal{P}(\theta)$ denotes the Poisson distribution with mean θ. Show that in this case,

$$h_i(t|\mathcal{H}_i(t-)) = \theta f(t).$$

(c) Let $\boldsymbol{x}_i$ be a covariate vector, and suppose we model the covariates through θ via some link function $\theta_i \equiv \theta(\boldsymbol{x}_i'\boldsymbol{\beta})$. For what link θ_i will we get a Cox structure for the hazard $h_i(t|\mathcal{H}_i(t-)$?

(d) Derive the joint distribution of $(N_i(A), N_i(B))$, where A and B are two disjoint intervals.

(e) Would this model be appropriate for the mammary tumor data of Example 4.6? Justify your answer.

4.11 Consider a univariate survival model with conditional hazard

$$h(y|w) = \lambda w,$$

where $w \sim \mathcal{G}(\eta, \eta)$.

(a) Show that the unconditional survival function is given by

$$S(y) = \left(1 + \frac{\lambda y}{\eta}\right)^{-\eta}, \tag{4.E.4}$$

and note that this corresponds to a Pareto distribution.

(b) Show that $E(Y^a)$ exists only when $-1 < a < \eta$.

(c) Consider the inclusion of a vector of covariates $\boldsymbol{x}$. Show that $S(y|w, \boldsymbol{x})$ follows a proportional hazards model.

4.12 Consider the frailty model

$$h(y|w) = h_0(y)w, \tag{4.E.5}$$

where w has some frailty distribution. Derive the unconditional hazard $h(y)$, and show that

$$\begin{aligned} h(y) &= \lim_{\Delta \to 0} P(y \le Y \le y + \Delta | Y > y) \\ &= h_0(y) E_w(w|Y > y). \end{aligned}$$

4.13 For the random M-site model discussed in Exercise 4.10, let us assume a Weibull density

$$f(y) = \alpha \lambda y^{\alpha - 1} \exp(-\lambda y^{\alpha})$$

and let $\theta_i = \exp(\boldsymbol{x}_i' \boldsymbol{\beta})$.

(a) Using a $N_p(\boldsymbol{\mu}_0, \Sigma_0)$ prior for $\boldsymbol{\beta}$, and a prior $\pi(\alpha, \lambda)$ independent of $\boldsymbol{\beta}$, derive the full conditional distributions for $(\alpha, \lambda, \boldsymbol{\beta})$.

(b) What choices of $\pi(\alpha, \lambda)$ yield log-concave full conditionals for α and λ? Justify your answer.

(c) Using the mammary tumor data of Example 4.6, write a Gibbs sampler to fit the M-site model using a noninformative normal prior for $\boldsymbol{\beta}$ and independent noninformative gamma priors for α and λ.

4.14 Consider the panel count data model discussed in Section 4.2,

(a) Derive the full conditional distributions of $\boldsymbol{\beta}$ and η when $\pi(\boldsymbol{\beta}, \eta) = \pi(\boldsymbol{\beta})\pi(\eta)$, where $\pi(\boldsymbol{\beta})$ is a normal distribution and $\pi(\eta)$ is a gamma distribution.

(b) What can you say about the log-concavity of the full conditionals of $\boldsymbol{\beta}$ and η? Justify your answer.
(c) Suppose $\eta \sim \mathcal{G}(\alpha_0, \lambda_0)$ and $\boldsymbol{\beta} \sim N_p(\boldsymbol{\mu}_0, \sigma_0^2 I)$, where $\alpha_0 = .4$, $\lambda_0 = 0.2$, $\boldsymbol{\mu}_0 = 0$ and $\sigma_0^2 = 4$. Carry out an analysis of the carcinogenesis data using the Gibbs sampler. Carry out a sensitivity analysis using different choices of (α_0, λ_0) and $(\boldsymbol{\mu}_0, \sigma_0^2)$.
(d) What can you say about the full conditional of η using a $\mathcal{U}(0.1, 100)$ prior for η, which is independent of $\boldsymbol{\beta}$?
(e) Carry out an analysis using the Gibbs sampler with the prior for η in part (d). What procedure do you prefer for sampling η within the Gibbs sampler and why?

4.15 A generalization of the univariate Cox model has hazard function given by

$$h(y|\boldsymbol{x}, w) = h_0(y) w \exp(\boldsymbol{x}'\boldsymbol{\beta}), \tag{4.E.6}$$

where w is a frailty parameter with $E(w) = 1$. Suppose $\phi(u) = E(\exp(-wu))$ is the Laplace transform of the frailty distribution.

(a) Show that

$$h(y|\boldsymbol{x}) = -\frac{\phi'(H_0(y)\exp(\boldsymbol{x}'\boldsymbol{\beta}))}{\phi(H_0(y)\exp(\boldsymbol{x}'\boldsymbol{\beta}))} h_0(y)\exp(\boldsymbol{x}'\boldsymbol{\beta}), \tag{4.E.7}$$

where $\phi'(t)$ denotes the derivative of ϕ with respect to t.
(b) Use part (a) to show that $h(y|\boldsymbol{x}) = h_0(y)\exp(\boldsymbol{x}'\boldsymbol{\beta})$ only when $w = 1$.
(c) When w has support on $(0, \infty)$, show that $h(y|\boldsymbol{x}) \to 0$ as $y \to \infty$. Assume here that $H_0(y) \to \infty$ as $y \to \infty$, but $h_0(y) \leq M$ for some $M > 0$ for all y.
(d) Using the E1684 melanoma data of Example 1.2, write a Gibbs sampler based on model (4.E.7) using treatment as the only covariate and using a $N(0, \sigma_0^2)$ prior for β, where σ_0^2 is large.
(e) For $w \sim \mathcal{G}(\eta, \eta)$, show that

$$h(y|\boldsymbol{x}, \eta) = \frac{\eta}{\eta + H_0(y)\exp(\boldsymbol{x}'\boldsymbol{\beta})} h_0(y)\exp(\boldsymbol{x}'\boldsymbol{\beta}).$$

4.16 Consider the local measure of dependence for bivariate survival data given by (4.4.1).

(a) Show that for the gamma frailty model, $\theta^*(y_1, y_2)$ is constant.
(b) Find $\theta^*(y_1, y_2)$ for the positive stable law frailty model.
(c) Show that for both the gamma and positive stable frailty models, $\theta^*(y_1, y_2)$ can be expressed as a function of $S(y_1, y_2)$ alone.
(d) Show that for any frailty model for bivariate survival data, $\theta^*(y_1, y_2)$ can be expressed as a function of $S(y_1, y_2)$ alone.

(e) Consider the Gumbel distribution (Gumbel, 1961), in which

$$S(y_1, y_2) = \exp\Big\{ - H_1(y_1) - H_2(y_2) + a\left([H_1(y_1)]^{-1} + [H_2(y_2)]^{-1}\right)^{-1} \Big\},$$

where $a \geq 0$. Show that $\theta^*(y_1, y_2)$ cannot be expressed as a function of $S(y_1, y_2)$ alone.

4.17 Consider the breast cancer data of Example 1.3. Suppose we impute the interval censored data with the midpoint of the interval. For subjects with right censored data, we do not impute the survival time.

(a) With this imputed data, write a Gibbs sampler for the frailty model

$$h(y|x, w) = h_0(y) w \exp(x\beta),$$

where x is the binary covariate for treatment, $h_0(y) = \lambda\alpha y^{\alpha-1}$ and $w \sim \mathcal{G}(\eta, \eta)$.

(b) Compute a 95% HPD interval for β.

(c) Compute $E(S(y|x)|D)$, and plot $E(S(y|x)|D)$ and the Kaplan-Meier estimate of $S(y|x)$ on the same plot. Why kind of conclusions do you draw?

(d) Do you think that the implicit assumption of $h(y|x) \to 0$ as $y \to \infty$ is justified for this dataset? Justify your answer.

4.18 Consider the animal carcinogenesis data of Example 1.5. The association between the control rats within the same litter should be greater than the association between the control and exposed rat of the same litter. This can be modeled using the two frailty random variables w_1, w_2, such that

$$h(y|x, w_1, w_2) = h_0(y) w_1^{1-x} w_2 \exp(x\beta),$$

where $x = 1$ for an exposed rat and 0 for a control rat. Assume that $w_1 \sim \mathcal{G}(\eta_1, \eta_1)$, $w_2 \sim \mathcal{G}(\eta_2, \eta_2)$ and w_1 and w_2 are independent.

(a) Find an expression for $S(Y_e, Y_{c_1}, Y_{c_2})|x) = P(Y_e > y_e, Y_{c_1} > y_{c_1}, Y_{c_2} > y_{c_2}|x)$, where Y_e denotes the survival time for the exposed rats, and Y_{c_j} denotes the survival time for the control rats, $j = 1, 2$. Note that this expression is quite complicated and does not have a closed form.

(b) Suppose $\eta_1 \sim \mathcal{G}(a_1, a_2)$, $\eta_2 \sim \mathcal{G}(b_1, b_2)$, where η_1 and η_2 are independent, $\beta \sim N(\mu_0, \sigma_0^2)$, and $h_0(y) = \lambda\alpha y^{\alpha-1}$, where α and λ have independent gamma prior distributions. Take noninformative choices for all of the prior parameters. Write a Gibbs sampler for sampling from the joint posterior of $(\alpha, \lambda, \eta_1, \eta_2, \beta)$.

(c) Find 95% credible sets for α, λ, η_1, η_2, and β. What kinds of inferences do you draw about η_2? Note that η_2 is important in identifying whether the associations are similar for all three pairs within a litter.

(d) Plot $E(S(y|x = 1)|D)$ and compare it to the corresponding Kaplan-Meier plot. What do you conclude?

(e) Plot $E(S(y|x = 0)|D)$ and compare it to the corresponding Kaplan-Meier plot. What do you conclude?

5
Cure Rate Models

5.1 Introduction

Survival models incorporating a cure fraction, often referred to as *cure rate models*, are becoming increasingly popular in analyzing data from cancer clinical trials. The cure rate model has been used for modeling time-to-event data for various types of cancers, including breast cancer, non-Hodgkins lymphoma, leukemia, prostate cancer, melanoma, and head and neck cancer, where for these diseases, a significant proportion of patients are "cured." Perhaps the most popular type of cure rate model is the mixture model discussed by Berkson and Gage (1952). In this model, we assume a certain fraction π of the population is "cured," and the remaining $1 - \pi$ are not cured. The survivor function for the entire population, denoted by $S_1(t)$, for this model is given by

$$S_1(t) = \pi + (1 - \pi)S^*(t), \tag{5.1.1}$$

where $S^*(t)$ denotes the survivor function for the non-cured group in the population. Common choices for $S^*(t)$ are the exponential and Weibull distributions. We shall refer to the model in (5.1.1) as the *standard cure rate model.* The standard cure rate model has been extensively discussed in the statistical literature by several authors, including Farewell (1982, 1986), Goldman (1984), Greenhouse and Wolfe (1984), Halpern and Brown (1987a,b), Gray and Tsiatis (1989), Sposto, Sather, and Baker (1992), Laska and Meisner (1992), Kuk and Chen (1992), Yamaguchi (1992), Taylor (1995), Ewell and Ibrahim (1997), Stangl and Greenhouse (1998), and

Sy and Taylor (2000). The book by Maller and Zhou (1996) gives an extensive discussion of frequentist methods of inference for the standard cure rate model.

Although the standard cure rate model is attractive and widely used, it has some drawbacks. In the presence of covariates, it cannot have a proportional hazards structure if the covariates are modeled through π via a binomial regression model. Proportional hazards is a desirable property in survival models when doing covariate analyses. Also, when including covariates through the parameter π via a standard binomial regression model, (5.1.1) yields improper posterior distributions for many types of noninformative improper priors, including the uniform prior for the regression coefficients. This is a crucial drawback of (5.1.1) since it implies that Bayesian inference with (5.1.1) essentially requires a proper prior. We elaborate more on this issue in subsection 5.2.2. These drawbacks can be overcome with an alternative definition of a cure rate model, which we discuss in the next section.

5.2 Parametric Cure Rate Model

5.2.1 Models

We present a formulation of the parametric cure rate model discussed by Yakovlev et al. (1993), Yakovlev (1994), and Yakovlev and Tsodikov (1996). A Bayesian formulation of this model is given in Chen, Ibrahim, and Sinha (1999). The alternative model can be derived as follows. Suppose that for an individual in the population, let N denote the number of *metastasis-competent* tumor cells for that individual left active after the initial treatment. A metastatis-competent tumor cell is a tumor cell which has the potential of metastasizing. Further, assume that N has a Poisson distribution with mean θ. Let Z_i denote the random time for the ith metastatis-competent tumor cell to produce detectable metastatic disease. That is, Z_i can be viewed as a promotion time for the ith tumor cell. Given N, the random variables Z_i, $i = 1, 2, \ldots,$ are assumed to be independent and identically distributed with a common distribution function $F(y) = 1 - S(y)$ that does not depend on N. The time to relapse of cancer can be defined by the random variable $Y = \min\{Z_i, 0 \le i \le N\}$, where $P(Z_0 = \infty) = 1$. The survival function for Y, and hence the survival function for the population, is given by

$$\begin{aligned}S_{pop}(y) =& P(\text{no metastatic cancer by time y})\\ =& P(N = 0) + P(Z_1 > y, \ldots, Z_N > y, N \ge 1).\end{aligned}$$

After some algebra, we obtain

$$S_{pop}(y) = \exp(-\theta) + \sum_{k=1}^{\infty} S(y)^k \frac{\theta^k}{k!} \exp(-\theta)$$
$$= \exp(-\theta + \theta S(y)) = \exp(-\theta F(y)). \qquad (5.2.1)$$

Since $S_{pop}(\infty) = \exp(-\theta) > 0$, (5.2.1) is not a proper survival function. As Yakovlev and Tsodikov (1996) point out, (5.2.1) shows explicitly the contribution to the failure time of two distinct characteristics of tumor growth: the initial number of metastatis-competent tumor cells and the rate of their progression. Thus the model incorporates parameters bearing clear biological meaning. Aside from the biological motivation, the model in (5.2.1) is suitable for any type of survival data which has a surviving fraction. Thus, survival data which do not "fit" the biological definition given above can still certainly be modeled by (5.2.1) as long as the data has a surviving fraction and can be thought of as being generated by an unknown number N of latent competing risks (Z_i's). Thus the model can be useful for modeling various types of survival data, including time to relapse, time to death, time to first infection, and so forth.

We also see from (5.2.1) that the cure fraction (i.e., cure rate) is given by

$$S_{pop}(\infty) \equiv P(N = 0) = \exp(-\theta). \qquad (5.2.2)$$

As $\theta \to \infty$, the cure fraction tends to 0, whereas as $\theta \to 0$, the cure fraction tends to 1. The subdensity corresponding to (5.2.1) is given by

$$f_{pop}(y) = \theta f(y) \exp(-\theta F(y)), \qquad (5.2.3)$$

where $f(y) = \frac{d}{dy}F(y)$. We note here that $f_{pop}(y)$ is not a proper probability density since $S_{pop}(y)$ is not a proper survival function . However, $f(y)$ appearing on the right side of (5.2.3) is a proper probability density function. The hazard function is given by

$$h_{pop}(y) = \theta f(y). \qquad (5.2.4)$$

We note that $h_{pop}(y) \to 0$ at a fast rate as $y \to \infty$ and $\int_o^\infty h_{pop}(y)\, dy < \infty$.

The cure rate model (5.2.1) yields an attractive form for the hazard in (5.2.4). Specifically, we see that $h_{pop}(y)$ is multiplicative in θ and $f(y)$, and thus has the proportional hazards structure when the covariates are modeled through θ. This form of the hazard is more appealing than the one from the standard cure rate model in (5.1.1), which does not have the proportional hazards structure if π is modeled as a function of covariates. The proportional hazards property in (5.2.4) is also computationally attractive, as Markov chain Monte Carlo (MCMC) sampling methods are relatively easy to implement. The survival function for the "non-cured" population is given by

$$S^*(y) = P(Y > y | N \geq 1) = \frac{\exp(-\theta F(y)) - \exp(-\theta)}{1 - \exp(-\theta)}. \tag{5.2.5}$$

We note that $S^*(0) = 1$ and $S^*(\infty) = 0$ so that $S^*(t)$ is a proper survival function. The survival density for the non-cured population (a proper density function) is given by

$$f^*(y) = -\frac{d}{dy} S^*(y) = \left(\frac{\exp(-\theta F(y))}{1 - \exp(-\theta)} \right) \theta f(y),$$

and the hazard function for the non-cured population is given by

$$\begin{aligned} h^*(y) =& \frac{f^*(y)}{S^*(y)} = \left(\frac{\exp(-\theta F(y))}{\exp(-\theta F(y)) - \exp(-\theta)} \right) h_{pop}(y) \\ =& \left(\frac{1}{P(Y < \infty | Y > y)} \right) h_{pop}(y). \end{aligned} \tag{5.2.6}$$

Thus, (5.2.6) is magnified by the factor $\frac{1}{P(Y<\infty|Y>y)} > 1$ compared to the hazard function $h_{pop}(t)$ of the entire population. Clearly, (5.2.6) does not have a proportional hazards structure since $\frac{1}{P(Y<\infty|Y>y)}$ can never be free of y for any $f(y)$ with support on $(0, \infty)$. Furthermore, it can be shown that $h^*(y) \to \frac{f(y)}{S(y)}$ as $y \to \infty$, and thus $h^*(y)$ converges to the hazard function of the promotion time random variable Z as $y \to \infty$. Finally, it can be shown that the hazard function $h^*(y)$ is an increasing function of θ.

There is a mathematical relationship between the model in (5.1.1) and (5.2.1). We can write

$$S_{pop}(y) = \exp(-\theta) + (1 - \exp(-\theta)) S^*(y),$$

where $S^*(y)$ is given by (5.2.5). Thus $S_{pop}(y)$ is a standard cure rate model with cure rate equal to $\pi = \exp(-\theta)$ and survival function for the non-cured population given by $S^*(y)$. This shows that every model defined by (5.2.1) can be written as a standard cure rate model with a specific family of survival functions $S^*(y)$ given by (5.2.5). This result also implies that every standard cure rate model corresponds to some model of the form (5.2.1) for some θ and $F(.)$. If the covariates enter through θ, $S_{pop}(y)$ can be taken to have a Cox proportional hazards structure, but then in this case $h^*(y)$ will not have a proportional hazards structure. Equations (5.2.1)–(5.2.6) and the discussion above summarize the fundamental modeling differences between (5.2.1) and the standard cure rate model. In model (5.2.1), we model the entire population as a proportional hazards model, whereas in the standard cure rate model, only the non-cured group can be modeled with a proportional hazards structure.

In model (5.2.1), we let the covariates depend on θ through the relationship $\theta = \exp(\boldsymbol{x}'\boldsymbol{\beta})$, where $\boldsymbol{x}$ is a $p \times 1$ vector of covariates and $\boldsymbol{\beta}$ is a $p \times 1$ vector of regression coefficients. We demonstrate in the next section that entering the covariates in this fashion corresponds to a canonical link in a Poisson regression model. Using $\theta = \exp(\boldsymbol{x}'\boldsymbol{\beta})$, (5.2.2), and (5.2.6), we can interpret the role of the regression coefficients for the cured and non-cured group. For the cured group, the sign of the regression coefficients affects the cure fraction. Thus a negative regression coefficient, for example, leads to a larger cure fraction, when the corresponding covariate takes a positive value. For the non-cured group, the regression coefficients affect the hazard function in (5.2.6). Specifically, a negative regression coefficient, for example, leads to a larger hazard, whereas a positive regression coefficient leads to a smaller hazard, when the corresponding covariate takes a positive value.

Following, Chen, Ibrahim, and Sinha (1999), we can now construct the likelihood function as follows. Suppose we have n subjects, and let N_i denote the number of metastasis-competent tumor cells for the ith subject. Further, we assume that the N_i's are i.i.d. Poisson random variables with mean θ, $i = 1, 2, \ldots, n$. We emphasize here that the N_i's are not observed, and can be viewed as latent variables in the model formulation. Further, suppose $Z_{i1}, Z_{i2}, \ldots, Z_{i,N_i}$ are the i.i.d. promotion times for the N_i metastasis-competent cells for the ith subject, which are unobserved, and all have proper cumulative distribution function $F(.)$, $i = 1, 2, \ldots, n$. In this subsection, we will specify a parametric form for $F(.)$, such as a Weibull or gamma distribution. We denote the indexing parameter (possibly vector valued) by $\boldsymbol{\psi}$, and thus write $F(.|\boldsymbol{\psi})$ and $S(.|\boldsymbol{\psi})$. For example, if $F(.|\boldsymbol{\psi})$ corresponds to a Weibull distribution, then $\boldsymbol{\psi} = (\alpha, \lambda)'$, where α is the shape parameter and λ is the scale parameter. Let y_i denote the survival time for subject i, which may be right censored, and let ν_i denote the censoring indicator, which equals 1 if y_i is a failure time and 0 if it is right censored. The observed data is $D_{obs} = (n, \boldsymbol{y}, \boldsymbol{\nu})$, where $\boldsymbol{y} = (y_1, y_2, \ldots, y_n)'$, and $\boldsymbol{\nu} = (\nu_1, \nu_2, \ldots, \nu_n)'$. Also, let $\boldsymbol{N} = (N_1, N_2, \ldots, N_n)'$. The complete data is given by $D = (n, \boldsymbol{y}, \boldsymbol{\nu}, \boldsymbol{N})$, where $\boldsymbol{N}$ is an unobserved vector of latent variables. The complete data likelihood function of the parameters $(\boldsymbol{\psi}, \theta)$ can then be written as

$$L(\theta, \boldsymbol{\psi}|D) = \left(\prod_{i=1}^{n} S(y_i|\boldsymbol{\psi})^{N_i - \nu_i} \left(N_i f(y_i|\boldsymbol{\psi})\right)^{\nu_i}\right) \times \exp\left\{\sum_{i=1}^{n}(N_i \log(\theta) - \log(N_i!)) - n\theta\right\}. \tag{5.2.7}$$

Throughout the remainder of this subsection, we will assume a Weibull density for $f(y_i|\boldsymbol{\psi})$, so that

$$f(y|\boldsymbol{\psi}) = \alpha y^{\alpha-1} \exp\left\{\lambda - y^{\alpha}\exp(\lambda)\right\}.$$

We incorporate covariates for the parametric cure rate model (5.2.1) through the cure rate parameter θ. When covariates are included, we have a different cure rate parameter, θ_i, for each subject, $i = 1, 2, \ldots, n$. Let

$\boldsymbol{x}_i' = (x_{i1}, \dots, x_{ip})$ denote the $p \times 1$ vector of covariates for the ith subject, and let $\boldsymbol{\beta} = (\beta_1, \dots, \beta_p)'$ denote the corresponding vector of regression coefficients. We relate θ to the covariates by $\theta_i \equiv \theta(\boldsymbol{x}_i'\boldsymbol{\beta}) = \exp(\boldsymbol{x}_i'\boldsymbol{\beta})$, so that the cure rate for subject i is $\exp(-\theta_i) = \exp(-\exp(\boldsymbol{x}_i'\boldsymbol{\beta}))$, $i = 1, 2, \dots, n$. This relationship between θ_i and $\boldsymbol{\beta}$ is equivalent to a canonical link for θ_i in the setting of generalized linear models. With this relation, we can write the complete data likelihood of $(\boldsymbol{\beta}, \boldsymbol{\psi})$ as

$$L(\boldsymbol{\beta}, \boldsymbol{\psi}|D) = \left(\prod_{i=1}^{n} S(y_i|\boldsymbol{\psi})^{N_i - \nu_i} (N_i f(y_i|\boldsymbol{\psi}))^{\nu_i} \right) \times \exp\left\{ \sum_{i=1}^{n} [N_i \boldsymbol{x}_i'\boldsymbol{\beta} - \log(N_i!) - \exp(\boldsymbol{x}_i'\boldsymbol{\beta})] \right\}, \tag{5.2.8}$$

where $D = (n, \boldsymbol{y}, X, \boldsymbol{\nu}, \boldsymbol{N})$, X is the $n \times p$ matrix of covariates, $f(y_i|\boldsymbol{\psi})$ is the Weibull density given above, and $S(y_i|\boldsymbol{\psi}) = \exp(-y_i^{\alpha} \exp(\lambda))$. If we assume independent priors for $(\boldsymbol{\beta}, \boldsymbol{\psi})$, then the posterior distributions of $(\boldsymbol{\beta}, \boldsymbol{\psi})$ are also independent. We mention that the part of the complete data likelihood in (5.2.8) involving $\boldsymbol{\beta}$ looks exactly like a Poisson generalized linear model with a canonical link, with the N_i's being the observables.

5.2.2 *Prior and Posterior Distributions*

We discuss classes of noninformative priors as well as the power priors for $(\boldsymbol{\beta}, \boldsymbol{\psi})$, and examine some of their properties. Consider the joint noninformative prior $\pi(\boldsymbol{\beta}, \boldsymbol{\psi}) \propto \pi(\boldsymbol{\psi})$ where $\boldsymbol{\psi} = (\alpha, \lambda)'$ are the Weibull parameters in $f(y|\boldsymbol{\psi})$. This noninformative prior implies that $\boldsymbol{\beta}$ and $\boldsymbol{\psi}$ are independent a priori and $\pi(\boldsymbol{\beta}) \propto 1$ is a uniform improper prior. We will assume throughout this section that

$$\pi(\boldsymbol{\psi}) = \pi(\alpha|\delta_0, \tau_0)\pi(\lambda),$$

where

$$\pi(\alpha|\delta_0, \tau_0) \propto \alpha^{\delta_0 - 1} \exp(-\tau_0 \alpha),$$

and δ_0 and τ_0 are two specified hyperparameters. With these specifications, the posterior distribution of $(\boldsymbol{\beta}, \boldsymbol{\psi})$ based on the observed data $D_{obs} = (n, \boldsymbol{y}, X, \boldsymbol{\nu})$ is given by

$$\pi(\boldsymbol{\beta}, \boldsymbol{\psi}|D_{obs}) \propto \left(\sum_{\boldsymbol{N}} L(\boldsymbol{\beta}, \boldsymbol{\psi}|D) \right) \pi(\alpha|\delta_0, \tau_0)\pi(\lambda), \tag{5.2.9}$$

where the sum in (5.2.9) extends over all possible values of the vector $\boldsymbol{N}$. The following theorem, due to Chen, Ibrahim, and Sinha (1999), gives conditions concerning the propriety of the posterior distribution in (5.2.9) using the noninformative prior $\pi(\boldsymbol{\beta}, \boldsymbol{\psi}) \propto \pi(\boldsymbol{\psi})$.

Theorem 5.2.1 *Let* $d = \sum_{i=1}^{n} \nu_i$ *and* X^* *be an* $n \times p$ *matrix with rows* $\nu_i \boldsymbol{x}_i'$. *Then if (i)* X^* *is of full rank, (ii)* $\pi(\lambda)$ *is proper, and (iii)* $\tau_0 > 0$ *and* $\delta_0 > -d$, *the posterior given in* (5.2.9) *is proper.*

The proof of Theorem 5.2.1 is given in the Appendix. Note that the conditions given in Theorem 5.2.1 are sufficient but *not* necessary for the propriety of the posterior distribution. However, the conditions stated in the theorem are quite general and are typically satisfied for most datasets. We also note that a proper prior for α is not required in order to obtain a proper posterior. This can be observed from condition (iii) since $\pi(\alpha|\delta_0, \tau_0)$ is no longer proper when $\delta_0 < 0$. Based on condition (ii), $\pi(\lambda)$ is required to be proper. Although several choices can be made, we will use a normal density for $\pi(\lambda)$. Theorem 5.2.1 guarantees propriety of the posterior distribution of $\boldsymbol{\beta}$ using an improper uniform prior. This enables us to carry out Bayesian inference with improper priors for the regression coefficients and facilitates comparisons with maximum likelihood. However, under the improper priors $\pi(\boldsymbol{\beta}, \boldsymbol{\psi}) \propto \pi(\boldsymbol{\psi})$, the standard cure rate model in (5.1.1) always leads to an improper posterior distribution for $\boldsymbol{\beta}$. This result is stated in the following theorem.

Theorem 5.2.2 *For the standard cure rate model given in* (5.1.1), *suppose we relate the cure fraction* π *to the covariates via a standard binomial regression*

$$\pi_i = G(\boldsymbol{x}_i'\boldsymbol{\beta}), \tag{5.2.10}$$

where $G(\cdot)$ *is a continuous cumulative distribution function (cdf). Assume that the survival function* $S^*(.)$ *for the non-cured group depends on the parameter* $\boldsymbol{\psi}^*$. *Let* $L_1(\boldsymbol{\beta}, \boldsymbol{\psi}^*|D_{obs})$ *denote the resulting likelihood function based on the observed data. Then, if we take an improper uniform prior for* $\boldsymbol{\beta}$ *(i.e.,* $\pi(\boldsymbol{\beta}) \propto 1$*), the posterior distribution*

$$\pi_1(\boldsymbol{\beta}, \boldsymbol{\psi}^*|D_{obs}) \propto L_1(\boldsymbol{\beta}, \boldsymbol{\psi}^*|D_{obs})\pi(\boldsymbol{\psi}^*) \tag{5.2.11}$$

is always improper regardless of the propriety of $\pi(\boldsymbol{\psi}^*)$.

The proof of Theorem 5.2.2 is given in the Appendix.

We now examine the power priors for $(\boldsymbol{\beta}, \boldsymbol{\psi})$. Let n_0 denote the sample size for the historical data, $\boldsymbol{y}_0$ be an $n_0 \times 1$ of right censored failure times for the historical data with censoring indicators $\boldsymbol{\nu}_0$, $\boldsymbol{N}_0$ is the unobserved vector of latent counts of metastasis-competent cells, and X_0 is an $n_0 \times p$ matrix of covariates corresponding to $\boldsymbol{y}_0$. Let $D_0 = (n_0, \boldsymbol{y}_0, X_0, \boldsymbol{\nu}_0, \boldsymbol{N}_0)$ denote the complete historical data. Further, let $\pi_0(\boldsymbol{\beta}, \boldsymbol{\psi})$ denote the initial prior distribution for $(\boldsymbol{\beta}, \boldsymbol{\psi})$. The power prior for $(\boldsymbol{\beta}, \boldsymbol{\psi})$ takes the form

$$\pi(\boldsymbol{\beta}, \boldsymbol{\psi}|D_{0,obs}, a_0) \propto \left[\sum_{\boldsymbol{N}_0} L(\boldsymbol{\beta}, \boldsymbol{\psi}|D_0)\right]^{a_0} \pi_0(\boldsymbol{\beta}, \boldsymbol{\psi}), \tag{5.2.12}$$

where $L(\boldsymbol{\beta}, \boldsymbol{\psi}|D_0)$ is the complete data likelihood given in (5.2.8) with D being replaced by the historical data D_0, and $D_{0,obs} = (n_0, \boldsymbol{y}_0, X_0, \boldsymbol{\nu}_0)$. We take a noninformative prior for $\pi_0(\boldsymbol{\beta}, \boldsymbol{\psi})$, such as $\pi_0(\boldsymbol{\beta}, \boldsymbol{\psi}) \propto \pi_0(\boldsymbol{\psi})$, which implies $\pi_0(\boldsymbol{\beta}) \propto 1$. For $\boldsymbol{\psi} = (\alpha, \lambda)'$, we take a gamma prior for α with small shape and scale parameters, and an independent normal prior for λ with mean 0 and variance c_0. A beta prior is chosen for a_0 leading to the joint prior distribution

$$\pi(\boldsymbol{\beta}, \boldsymbol{\psi}, a_0|D_{0,obs}) \propto \left[\sum_{\boldsymbol{N}_0} L(\boldsymbol{\beta}, \boldsymbol{\psi}|D_0)\right]^{a_0} \pi_0(\boldsymbol{\beta}, \boldsymbol{\psi}) a_0^{\gamma_0 - 1}(1 - a_0)^{\lambda_0 - 1}, \tag{5.2.13}$$

where (γ_0, λ_0) are specified prior parameters. The prior in (5.2.13) does not have a closed form but has several attractive properties. First, we note that if $\pi_0(\boldsymbol{\beta}, \boldsymbol{\psi})$ is proper, then (5.2.13) is guaranteed to be proper. Further, (5.2.13) can be proper even if $\pi_0(\boldsymbol{\beta}, \boldsymbol{\psi})$ is improper. The following theorem characterizes the propriety of (5.2.13) when $\pi_0(\boldsymbol{\beta}, \boldsymbol{\psi})$ is improper.

Theorem 5.2.3 *Assume that*

$$\pi_0(\boldsymbol{\beta}, \boldsymbol{\psi}) \propto \pi_0(\boldsymbol{\psi}) \equiv \pi_0(\alpha|\delta_0, \tau_0)\pi_0(\lambda) \propto \alpha^{\delta_0 - 1} \exp(-\tau_0 \alpha)\pi_0(\lambda),$$

where δ_0 and τ_0 are specified hyperparameters. Let $d_0 = \sum_{i=1}^{n_0} \nu_{0i}$ and X_0^* be an $n_0 \times p$ matrix with rows $\nu_{0i}\boldsymbol{x}_{0i}'$. If (i) X_0^* is of full rank, (ii) $\delta_0 > 0$ and $\tau_0 > 0$, (iii) $\pi_0(\lambda)$ is proper, and (iv) $\gamma_0 > p$ and $\lambda_0 > 0$, then the joint prior given in (5.2.13) *is proper.*

The proof of Theorem 5.2.3 is given in the Appendix. We mention that the power prior for $\boldsymbol{\beta}$ based on the model (5.1.1) will lead to an improper prior as well as an improper posterior distribution. Thus, the power priors based on (5.1.1) will not work. This result can be summarized in the following theorem.

Theorem 5.2.4 *For the standard cure rate model given in* (5.1.1), *suppose we relate the cure fraction π to the covariates via a standard binomial regression given by* (5.2.10). *Assume that the survival function for the noncured group $S^*(.)$ depends on the parameter $\boldsymbol{\psi}^*$. Let $L_1(\boldsymbol{\beta}, \boldsymbol{\psi}^*|D_{0,obs})$ and $L_1(\boldsymbol{\beta}, \boldsymbol{\psi}^*|D_{obs})$ denote the likelihood functions based on the observed historical and current data. Suppose we use an improper uniform initial prior for $\boldsymbol{\beta}$ (i.e., $\pi_0(\boldsymbol{\beta}) \propto 1$) to construct the joint prior as*

$$\pi_1(\boldsymbol{\beta}, \boldsymbol{\psi}^*, a_0|D_{0,obs}) \propto [L_1(\boldsymbol{\beta}, \boldsymbol{\psi}^*|D_{0,obs})]^{a_0} \ \pi_0(\gamma^*) \ a_0^{\gamma_0 - 1}(1 - a_0)^{\lambda_0 - 1}, \tag{5.2.14}$$

where γ_0 and λ_0 are specified hyperparameters. Then, $\pi_1(\boldsymbol{\beta}, \boldsymbol{\psi}^*, a_0|D_{0,obs})$ is always improper regardless of the propriety of $\pi_0(\boldsymbol{\psi}^*)$. In addition, if we use $\pi_1(\boldsymbol{\beta}, \boldsymbol{\psi}^*, a_0|D_{0,obs})$ as a prior, the resulting posterior, given by

$$\pi_1(\boldsymbol{\beta}, \boldsymbol{\psi}^*, a_0|D_{obs}) \propto L_1(\boldsymbol{\beta}, \boldsymbol{\psi}^*|D_{obs})\pi_1(\boldsymbol{\beta}, \boldsymbol{\psi}^*, a_0|D_{0,obs}) \tag{5.2.15}$$

is also improper.

The proof of Theorem 5.2.4 is similar to that of Theorem 5.2.2, and therefore the details are omitted.

5.2.3 Posterior Computation

In this subsection, we describe a simple MCMC algorithm for sampling from the joint posterior density of $(\boldsymbol{\beta}, \boldsymbol{\psi}, a_0)$. The joint posterior of $(\boldsymbol{\beta}, \boldsymbol{\psi}, a_0)$ based on the observed data D_{obs} is given by

$$\pi(\boldsymbol{\beta}, \boldsymbol{\psi}, a_0|D_{0,obs}) \propto \sum_{\boldsymbol{N}} L(\boldsymbol{\beta}, \boldsymbol{\psi}|D) \left[\sum_{\boldsymbol{N}_0} L(\boldsymbol{\beta}, \boldsymbol{\psi}|D_0)\right]^{a_0} \times \pi_0(\boldsymbol{\beta}, \boldsymbol{\psi}) a_0^{\gamma_0 - 1}(1-a_0)^{\lambda_0 - 1}, \tag{5.2.16}$$

where $L(\boldsymbol{\beta}, \boldsymbol{\psi}|D)$ and $L(\boldsymbol{\beta}, \boldsymbol{\psi}|D_0)$ are the complete data likelihoods with data D and D_0, respectively. To sample from the posterior distribution $\pi(\boldsymbol{\beta}, \boldsymbol{\psi}, a_0|D_{obs})$, we introduce the latent variables $\boldsymbol{N} = (N_1, N_2, \ldots, N_n)'$ and $\boldsymbol{N}_0 = (N_{01}, N_{02}, \ldots, N_{0n})'$. It can be shown that

$$\begin{aligned}\left[\sum_{\boldsymbol{N}_0} L(\boldsymbol{\beta}, \boldsymbol{\psi}|D_0)\right]^{a_0} &= \prod_{i=1}^{n_0} \Big(\theta_{0i} f(y_{0i}|\boldsymbol{\psi})\Big)^{a_0 \nu_{0i}} \exp\{-a_0\theta_{0i}(1 - S(y_{0i}|\boldsymbol{\psi}))\} \\ &= \sum_{\boldsymbol{N}_0} \prod_{i=1}^{n_0} \Bigg[\Big(\theta_{0i} f(y_{0i}|\boldsymbol{\psi})\Big)^{a_0\nu_{0i}} \frac{1}{N_{0i}!} [a_0\theta_{0i} S(y_{0i}|\boldsymbol{\psi})]^{N_{0i}} \\ &\qquad \times \exp\{-a_0\theta_{0i}\}\Bigg].\end{aligned} \tag{5.2.17}$$

Thus, the joint posterior of $(\boldsymbol{\beta}, \boldsymbol{\psi}, a_0, \boldsymbol{N}, \boldsymbol{N}_0)$ can be written as

$$\begin{aligned}\pi(\boldsymbol{\beta}, \boldsymbol{\psi}, a_0, \boldsymbol{N}, \boldsymbol{N}_0|D_{obs}) &\propto \prod_{i=1}^{n} S(y_i|\boldsymbol{\psi})^{N_i - \nu_i} (N_i f(y_i|\boldsymbol{\psi}))^{\nu_i} \\ &\times \exp\left\{\sum_{i=1}^{n} [N_i \boldsymbol{x}_i'\boldsymbol{\beta} - \log(N_i!) - \exp(\boldsymbol{x}_i'\boldsymbol{\beta})]\right\} \\ &\times \prod_{i=1}^{n_0} \Big(\exp(\boldsymbol{x}_{0i}'\boldsymbol{\beta}) f(y_{0i}|\boldsymbol{\psi})\Big)^{a_0\nu_{0i}} S(y_{0i}|\boldsymbol{\psi})^{N_{0i}} \\ &\times \exp\left\{\sum_{i=1}^{n_0} \Big[N_{0i}(\log(a_0) + \boldsymbol{x}_{0i}'\boldsymbol{\beta}) - \log(N_{0i}!) - a_0 \exp(\boldsymbol{x}_{0i}'\boldsymbol{\beta})\Big]\right\} \\ &\times \pi_0(\boldsymbol{\beta}, \boldsymbol{\psi}) a_0^{\gamma_0 - 1}(1-a_0)^{\lambda_0 - 1}.\end{aligned}$$

We further assume that $\pi_0(\boldsymbol{\beta}, \boldsymbol{\psi})$ is log-concave in each component of $(\boldsymbol{\beta}, \boldsymbol{\psi})$. To run the Gibbs sampler, we need to sample from the following complete conditional posterior distributions:

(i) $[\boldsymbol{\beta}, \boldsymbol{N}, \boldsymbol{N}_0 | \boldsymbol{\psi}, a_0, D_{obs}]$;

(ii) $[\boldsymbol{\psi} | a_0, \boldsymbol{\beta}, \boldsymbol{N}, \boldsymbol{N}_0, D_{obs}]$; and

(iii) $[a_0 | \boldsymbol{\beta}, \boldsymbol{\psi}, \boldsymbol{N}_0, D_{0,obs}]$.

Now we briefly explain how to sample from each of the above conditional posterior distributions. For (i), we apply the collapsed Gibbs procedure of Liu (1994). It is easy to observe that

$$[\boldsymbol{\beta}, \boldsymbol{N}, \boldsymbol{N}_0 | \boldsymbol{\psi}, a_0, D_{obs}] = [\boldsymbol{\beta} | \boldsymbol{\psi}, a_0, D_{obs}][\boldsymbol{N}, \boldsymbol{N}_0 | \boldsymbol{\beta}, \boldsymbol{\psi}, a_0, D_{obs}]. \tag{5.2.18}$$

In (5.2.18), we draw $\boldsymbol{\beta}$ by collapsing $\boldsymbol{N}$ and $\boldsymbol{N}_0$. As discussed in Chen, Shao, and Ibrahim (2000), the collapsed Gibbs sampler is a useful tool for improving convergence and mixing of the Gibbs sampler. To sample $\boldsymbol{N}$ and $\boldsymbol{N}_0$, we need to draw

(a) $[N_i | \boldsymbol{\beta}, \boldsymbol{\psi}, a_0, D_{obs}]$ for $i = 1, 2, \ldots, n$; and

(b) $[N_{0i} | \boldsymbol{\beta}, \boldsymbol{\psi}, a_0, D_{0,obs}]$ for $i = 1, 2, \ldots, n_0$.

It is easy to observe that

$$[N_i | \boldsymbol{\beta}, \boldsymbol{\psi}, a_0, D_{obs}] \sim \mathcal{P}(S(y_i | \boldsymbol{\psi}) \exp(\boldsymbol{x}_i' \boldsymbol{\beta})) + \nu_i$$

for $i = 1, 2, \ldots, n$, and

$$[N_{0i} | \boldsymbol{\beta}, \boldsymbol{\psi}, a_0, D_{0,obs}] \sim \mathcal{P}(a_0 S(y_{0i} | \boldsymbol{\psi}) \exp(\boldsymbol{x}_{0i}' \boldsymbol{\beta}))$$

for $i = 1, 2, \ldots, n_0$. Thus, sampling $\boldsymbol{N}$ or $\boldsymbol{N}_0$ from $[\boldsymbol{N}, \boldsymbol{N}_0 | \boldsymbol{\beta}, \boldsymbol{\psi}, a_0, D_{obs}]$ is straightforward. The conditional posterior density for $[\boldsymbol{\beta} | \boldsymbol{\psi}, a_0, D_{obs}]$ has the form

$$\begin{aligned}\pi(\boldsymbol{\beta} | \boldsymbol{\psi}, a_0, D_{obs}) \propto \exp \Bigg\{ & \sum_{i=1}^{n} [\nu_i \boldsymbol{x}_i' \boldsymbol{\beta} - \exp(\boldsymbol{x}_i' \boldsymbol{\beta})(1 - S(y_i | \boldsymbol{\psi}))] \\ & + \sum_{i=1}^{n_0} [(a_0 \nu_{0i} \boldsymbol{x}_{0i}' \boldsymbol{\beta} - a_0 \exp(\boldsymbol{x}_{0i}' \boldsymbol{\beta})(1 - S(y_{0i} | \boldsymbol{\psi}))] \Bigg\} \\ & \times \pi_0(\boldsymbol{\beta}, \boldsymbol{\psi}).\end{aligned} \tag{5.2.19}$$

It can be shown that $\pi(\boldsymbol{\beta} | \boldsymbol{\psi}, a_0, D_{obs})$ is log-concave in each component of $\boldsymbol{\beta}$.

Further, the conditional posterior density for $[\psi|a_0, \boldsymbol{N}, \boldsymbol{N}_0, D_{obs}]$ is given by

$$\pi(\psi|a_0, \boldsymbol{\beta}, \boldsymbol{N}, \boldsymbol{N}_0, D_{obs}) \propto \prod_{i=1}^{n} S(y_i|\psi)^{N_i - \nu_i}\, f(y_i|\psi)^{\nu_i} \times \prod_{i=1}^{n_0} f(y_{0i}|\psi)^{a_0\nu_{0i}} S(y_{0i}|\psi)^{N_{0i}} \times \pi_0(\boldsymbol{\beta}, \psi). \quad (5.2.20)$$

Using a similar proof given by Berger and Sun (1993), we can show that $\pi(\psi|a_0, \boldsymbol{\beta}, \boldsymbol{N}, \boldsymbol{N}_0, D_{obs})$ is log-concave in each component of ψ. Therefore, we can use the adaptive rejection algorithm of Gilks and Wild (1992) to sample from (5.2.19) and (5.2.20). Finally, the conditional posterior density for $[a_0|\boldsymbol{\beta}, \psi, \boldsymbol{N}_0, D_{0,obs}]$ is of the form

$$\pi(a_0|\boldsymbol{\beta}, \psi, \boldsymbol{N}_0, D_{0,obs}) \propto \exp\left\{ \sum_{i=1}^{n_0} \Big[a_0\nu_{0i}(\log(f(y_{0i}|\psi)) + \boldsymbol{x}'_{0i}\boldsymbol{\beta}) + N_{0i}\log(a_0) - a_0 \exp(\boldsymbol{x}'_{0i}\boldsymbol{\beta}) \Big] \right\} \times a_0^{\gamma_0 - 1}(1 - a_0)^{\lambda_0 - 1}. \quad (5.2.21)$$

The conditional posterior density $\pi(a_0|\boldsymbol{\beta}, \psi, \boldsymbol{N}_0, D_{0,obs})$ is log-concave when $\gamma_0 \geq 1$ and $\lambda_0 \geq 1$, and therefore, we can use the adaptive rejection algorithm again to sample a_0 from $\pi(a_0|\boldsymbol{\beta}, \psi, \boldsymbol{N}_0, D_{0,obs})$.

We note that the introduction of the latent variables $\boldsymbol{N}$ and $\boldsymbol{N}_0$ facilitate a straightforward MCMC algorithm for sampling from the joint posterior distribution of $[\boldsymbol{\beta}, \psi, a_0|D_{obs}]$. We also note that by taking $a_0 = 0$ with probability 1, we can apply the above MCMC sampling algorithm to sample from the posterior distribution $\pi(\boldsymbol{\beta}, \psi|D_{obs})$ for $(\boldsymbol{\beta}, \psi)$ with a noninformative prior given by (5.2.9).

Example 5.1. Melanoma data. We consider the E1684 data discussed in Example 1.2. The response variable is overall survival, which is defined as the time from randomization until death. One of our main goals in this example is to compare inferences between the standard cure rate model and the alternative model. First, we compare the maximum likelihood estimates of the cure rates between the two models (5.2.1) and (5.1.1). Second, using the cure rate model in (5.2.1), we carry out a Bayesian analysis with covariates using the proposed priors in (5.2.13), and compare the results to the estimates based on the standard cure rate model in (5.1.1). Three covariates and an intercept are included in the analyses. The covariates are age (x_1), gender (x_2) (male, female), and performance status (x_3) (fully active, other). Performance status is abbreviated by PS in the tables below. After deleting missing observations, a total of $n = 284$ observations are used in the analysis. Table 5.1 provides a summary of the E1684 data. For the survival time summary in Table 5.1, the Kaplan-Meier estimate of the median survival and its corresponding 95% confidence interval (CI) are given. In all of the analyses, we standardized the age covariate to stabilize the posterior computations.

TABLE 5.1. Summary of E1684 Data.

Survival Time (y) (years)	Median	3.15
	95% CI	(2.34, 4.33)
Status (frequency))	censored	110
	death	174
Age (x_1) (years)	Mean	47.0
	Std Dev	13.0
Gender (x_2) (frequency)	Male	171
	Female	113
PS (x_3) (frequency)	Fully Active	253
	Other	31

Source: Chen, Ibrahim, and Sinha (1999).

Figure 5.1 shows three superimposed plots of the survival curve based on the Kaplan-Meier method (dashed line), the standard cure rate model (5.1.1) (dotted line), and the parametric cure rate model (5.2.1) (solid line). We see that the three plots are nearly identical, giving essentially the same results.

We now consider several analyses with the covariates included. Figure 5.2 shows a box-plot of the maximum likelihood estimates (MLE's) of the cure rates for all patients for the two models, where the label 1 denotes the standard cure rate model and the label 2 denotes the parametric cure rate model. We see that the two box-plots are very similar. The first, second, and third quartiles for the two box-plots are 0.32, 0.36, and 0.40 for the standard cure rate model and 0.33, 0.36, and 0.39 for the parametric cure rate model. We see from Figure 5.2 that the variation in the cure rate estimates from the standard model is greater than that of the parametric cure rate model. In fact, the standard deviations of the cure rate estimates are 0.06 for the standard cure rate model and 0.05 for the parametric cure rate model. The maximum likelihood estimates, their standard deviations and P-values for the parametric cure rate model are also computed and the results are reported in Table 5.2.

Several years earlier, a similar melanoma study with the same patient population was conducted by ECOG. This study, denoted by E1673, serves

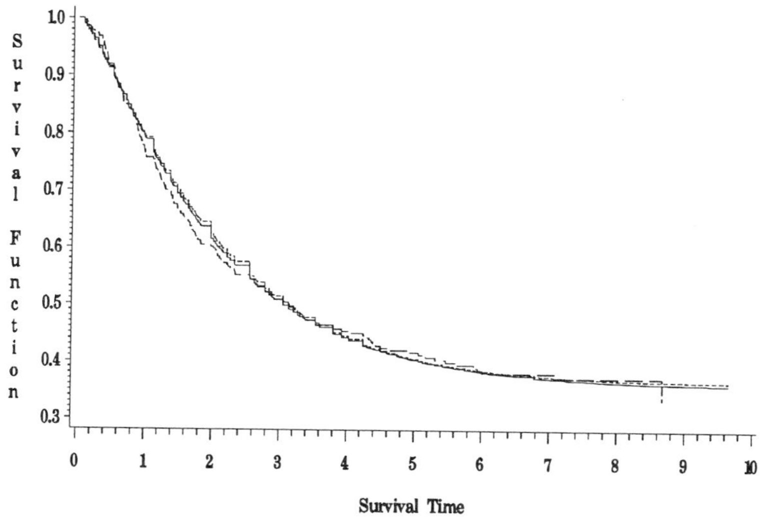

FIGURE 5.1. Superimposed survival curves for the E1684 data.

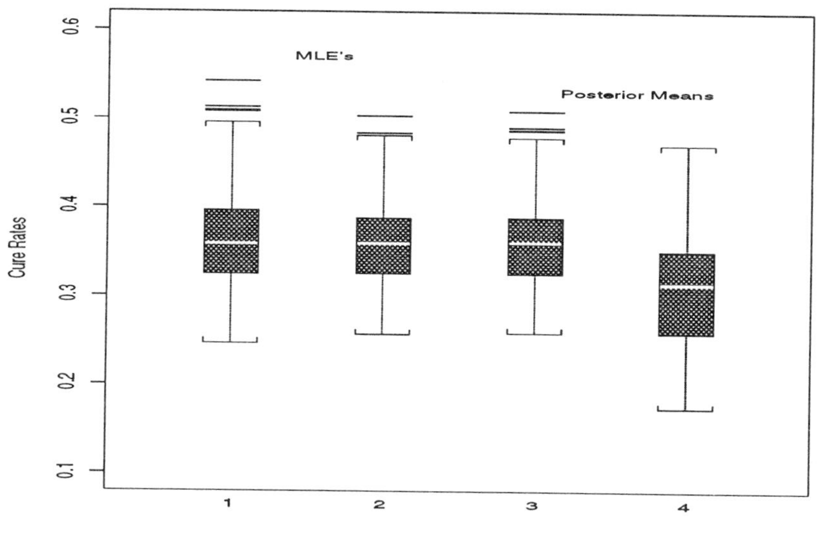

FIGURE 5.2. Box-plots of the cure rates for all patients.

TABLE 5.2. Maximum Likelihood Estimates (MLE) of the Model Parameters for the E1684 Data.

Variable	MLE	Std Dev	P-value
intercept	0.09	0.11	0.38
age	0.09	0.07	0.21
gender	−0.12	0.16	0.44
ps	−0.20	0.26	0.44
α	1.32	0.09	0.00
λ	−1.34	0.12	0.00

Source: Chen, Ibrahim, and Sinha (1999).

as the historical data for our Bayesian analysis of E1684. A total of $n_0 = 650$ patients are used in the historical data. Table 5.3 summarizes the historical data E1673. Using the E1673 study as historical data, we consider an analysis with the power prior in (5.2.13). For the initial prior for $\boldsymbol{\beta}$, we take an improper uniform prior, and for $\pi_0(\alpha|\nu_0, \tau_0)$, we take $\nu_0 = 1$ and $\tau_0 = 0.01$ to ensure a proper prior. We note that this choice for $\pi_0(\alpha|\nu_0, \tau_0)$ also guarantees log-concavity. The parameter λ is taken to have a normal distribution with mean 0 and variance 10,000.

TABLE 5.3. Summary of E1673 Data.

Survival Time (y_0) (years)	Median	8.80
	95% CI	(5.44, 11.62)
Status (frequency))	censored	257
	death	393
Age (x_{01}) (years)	Mean	48.0
	Std Dev	14.0
Gender (x_{02}) (frequency)	Male	375
	Female	275
PS (x_{03}) (frequency)	Fully Active	561
	Other	89

Source: Chen, Ibrahim, and Sinha (1999).

Table 5.4 gives posterior estimates of $\boldsymbol{\beta}$ based on several values of (γ_0, λ_0) using the model (5.2.1). In Table 5.4 we obtain, for example, $E(a_0|D_{obs}) = 0.14$ by taking $(\gamma_0, \lambda_0) = (200, 1)$. The case $a_0 = 0$ with probability 1 (w.p.1) essentially yields the MLE's of $\boldsymbol{\beta}$, α, and λ given in Table 5.2. Table 5.4 indicates a fairly robust pattern of behavior. The estimates of the posterior mean, standard deviation, or highest posterior density (HPD) intervals of $\boldsymbol{\beta}$ do not change a great deal if a low or moderate weight is given to the historical data. However, if a higher than moderate weight is given to the historical data, these posterior summaries can change a lot. For example, when the posterior mean of a_0 is less than 0.14, we see that all of the HPD intervals for $\boldsymbol{\beta}$ include 0, and when the posterior mean of a_0 is greater than or equal to 0.14, some HPD intervals for $\boldsymbol{\beta}$ do not include 0. The HPD interval for age does not include 0 when the posterior mean of a_0 is 0.29, and it includes 0 when less weight is given to the historical data. This finding is interesting, since it indicates that age is a potentially important prognostic factor for predicting survival in melanoma. Such a conclusion is not possible based on a frequentist or Bayesian analysis of the current data alone.

TABLE 5.4. Posterior Estimates of the Model Parameters with $\alpha \sim \mathcal{G}(1, 0.01)$ and $\lambda \sim N(0, 10,000)$.

$E(a_0 \mid D_{obs})$	Variable	Posterior Mean	Posterior Std Dev	95% HPD Interval
0	intercept	0.09	0.11	(−0.12, 0.30)
(w.p.1)	age	0.09	0.07	(−0.05, 0.23)
	gender	−0.12	0.16	(−0.44, 0.19)
	ps	−0.23	0.26	(−0.73, 0.28)
	α	1.31	0.09	(1.15, 1.48)
	λ	−1.36	0.12	(−1.60, −1.11)
0.14	intercept	0.25	0.10	(0.05, 0.45)
	age	0.12	0.06	(−0.00, 0.24)
	gender	−0.20	0.14	(−0.47, 0.07)
	ps	−0.09	0.22	(−0.53, 0.31)
	α	1.06	0.06	(0.95, 1.17)
	λ	−1.62	0.12	(−1.85, −1.39)
0.29	intercept	0.26	0.09	(0.08, 0.43)
	age	0.13	0.06	(0.02, 0.24)
	gender	−0.24	0.12	(−0.48, 0.00)
	ps	−0.01	0.19	(−0.38, 0.35)
	α	1.03	0.05	(0.93, 1.13)
	λ	−1.70	0.11	(−1.91, −1.50)
1	intercept	0.22	0.06	(0.11, 0.35)
(w.p.1)	age	0.16	0.04	(0.08, 0.24)
	gender	−0.32	0.09	(−0.50, −0.15)
	ps	0.14	0.13	(−0.11, 0.39)
	α	1.00	0.04	(0.93, 1.07)
	λ	−1.82	0.08	(−1.97, −1.67)

Source: Chen, Ibrahim, and Sinha (1999).

In addition, when the historical data and current data are equally weighted (i.e., $a_0 = 1$ with probability 1), the HPD intervals for age and gender both do not include 0, thus demonstrating the importance of gender in predicting overall survival. Another feature of Table 5.4 is that the posterior standard deviations of the β_j's become smaller and the HPD intervals become narrower as the posterior mean of a_0 increases. This demonstrates that incorporation of historical data can yield more precise posterior estimates of $\boldsymbol{\beta}$. For example, we see that when $a_0 = 1$, the posterior mean, standard deviation, and HPD interval for the age coefficient are 0.16, 0.04, and $(0.08, 0.24)$, respectively, whereas when we do not incorporate any historical data (i.e., $a_0 = 0$), these values are 0.09, 0.07, and $(-0.05, 0.23)$. We see that there is a large difference in these estimates, especially in the standard deviations and the HPD intervals. A partial explanation of these results is that the E1673 study has had nearly 20 years of follow-up on 650 patients, and thus the potential impact of age and gender on overall survival is much more apparent in these data than the current data E1684, which has had less than 10 years of follow-up on 284 patients, and has about 39% censoring.

Incorporation of historical data can also affect the posterior estimates of the cure rates. Figure 5.2 shows a box-plot of the posterior means of the cure rates based on no incorporation of historical data (coded 3) and a box-plot corresponding to $E(a_0|D_{obs}) = 0.29$ (coded 4). The posterior estimates in the cure rates are quite different in the model with $E(a_0|D_{obs}) = 0.29$ compared to the one with no incorporation of historical data. The mean and standard deviations are 0.36 and 0.05 for box-plot 3 ($a_0 = 0$) and 0.31 and 0.06 for box-plot 4 ($E(a_0|D_{obs}) = 0.29$). Thus we see that the mean cure rate drops from 0.36 to 0.31 when the historical data are incorporated. Again, a partial explanation of this result is due to the fact that the historical data are much more mature than the current data, with nearly 20 years of follow-up and a smaller fraction of censored cases. Thus, the results in Figure 5.2 are not surprising, and in fact appealing, since they give us a better estimate of the cure rate compared to an estimate based on the current data alone. Such a conclusion cannot be reached by a frequentist or Bayesian analysis using only the E1684 data. It is also interesting to mention that the box-plot of the posterior means of the cure rates based on $a_0 = 0$ (labeled 3) is almost identical to the box-plot of the MLE's of the cure rates (labeled 2) in Figure 5.2. This is clear from the fact that when $a_0 = 0$ and vague proper priors are chosen for $\pi_0(.)$, the posterior distribution of $\boldsymbol{\beta}$, (and hence $\boldsymbol{\theta}$) is essentially the same as the likelihood function in (5.2.7), and thus the estimates are quite similar. This again is a desirable feature of this model since it implies that we can obtain MLE's via Gibbs sampling, without doing any analytic maximizations. That is, if we take $a_0 = 0$ and choose vague proper priors for $\pi_0(.)$, the posterior means of the parameters are very close to the maximum likelihood estimates.

A detailed sensitivity analysis for the regression coefficients was conducted by varying the hyperparameters for a_0 (i.e., (γ_0, λ_0)) and varying the hyperparameters for $\psi = (\alpha, \lambda)$. The posterior estimates of the parameters are fairly robust as the hyperparameters (γ_0, λ_0) are varied. When we vary the hyperparameters for ψ, the posterior estimates of β are also robust for a wide range of hyperparameter values. For example, when fixing the hyperparameters for a_0 so that $E(a_0|D_{obs}) = 0.29$ and taking $\alpha \sim \mathcal{G}(1,1)$ and $\lambda \sim N(0, 10)$, we obtain the posterior estimates shown in Table 5.5. We see that these priors for (α, λ) are fairly informative relative to those of Table 5.4. Other moderate to informative choices of hyperparameters for (α, λ) also led to fairly robust posterior estimates of β.

TABLE 5.5. Posterior Estimates of the Model Parameters with $E(a_0|D_{obs}) = 0.29$, $\alpha \sim \mathcal{G}(1,1)$ and $\lambda \sim N(0, 10)$.

Variable	Posterior Mean	Posterior Std Dev	95% HPD Interval
intercept	0.26	0.09	(0.07, 0.42)
age	0.13	0.06	(0.02, 0.24)
gender	−0.24	0.12	(−0.48, 0.00)
ps	−0.01	0.19	(−0.38, 0.35)
α	1.02	0.05	(0.93, 1.12)
λ	−1.69	0.11	(−1.90, −1.48)

Source: Chen, Ibrahim, and Sinha (1999).

Finally, we mention that the Gibbs sampler was used to sample from the posterior distribution. A burn-in of 1000 samples was used, with autocorrelations disappearing after lag 5 for nearly all parameters, and 50,000 Gibbs iterates were used after the burn-in for all of the posterior computations. Further, all HPD intervals were computed by using an efficient Monte Carlo method of Chen and Shao (1999). In summary, we see the powerful advantages of the cure rate model (5.2.1) and the desirable features of incorporating historical data into a Bayesian analysis. The model is also computationally attractive, requiring only a straightforward adaptive rejection algorithm of Gilks and Wild (1992) for Gibbs sampling.

5.3 Semiparametric Cure Rate Model

In this section, we consider a semiparametric version of the parametric cure rate model in (5.2.1). Following Chen, Harrington, and Ibrahim (1999), we construct a finite partition of the time axis, $0 < s_1 < \ldots < s_J$, with $s_J > y_i$ for all $i = 1, 2, \ldots, n$. Thus, we have the J intervals $(0, s_1]$, $(s_1, s_2]$, $\ldots$, $(s_{J-1}, s_J]$. We thus assume that the hazard for $F(y)$ is equal to λ_j for the

j^{th} interval, $j = 1, 2, \ldots, J$, leading to

$$F(y) = 1 - \exp\left\{-\lambda_j(y - s_{j-1}) - \sum_{g=1}^{j-1} \lambda_g(s_g - s_{g-1})\right\}. \tag{5.3.1}$$

We note that when $J = 1$, $F(y)$ reduces to the parametric exponential model. With this assumption, the complete data likelihood can be written as

$$\begin{aligned} L(\boldsymbol{\beta}, \boldsymbol{\lambda}|D) \\ &= \prod_{i=1}^{n}\prod_{j=1}^{J} \exp\left\{ - (N_i - \nu_i)\nu_{ij}\left[\lambda_j(y_i - s_{j-1}) + \sum_{g=1}^{j-1} \lambda_g(s_g - s_{g-1})\right]\right\} \\ &\times \prod_{i=1}^{n}\prod_{j=1}^{J} (N_i\lambda_j)^{\nu_{ij}\nu_i} \exp\left\{-\nu_i\nu_{ij}\left[\lambda_j(y_i - s_{j-1}) + \sum_{g=1}^{j-1} \lambda_g(s_g - s_{g-1})\right]\right\} \\ &\times \exp\left\{\sum_{i=1}^{n} [N_i\boldsymbol{x}_i'\boldsymbol{\beta} - \log(N_i!) - \exp(\boldsymbol{x}_i'\boldsymbol{\beta})]\right\}, \end{aligned} \tag{5.3.2}$$

where $\boldsymbol{\lambda} = (\lambda_1, \lambda_2, \ldots, \lambda_J)'$ and $\nu_{ij} = 1$ if the ith subject failed or was censored in the jth interval, and 0 otherwise. The model in (5.3.2) is a semiparametric version of (5.2.7). There are several attractive features of the model in (5.3.2). First, we note the degree of the nonparametricity is controlled by J. The larger the J, the more nonparametric the model is. However, by picking a small to moderate J, we get more of a parametric shape for $F(y)$. This is an important aspect for the cure rate model, since the estimation of the cure rate parameter θ could be highly affected by the nonparametric nature of $F(y)$. For this reason, it may be desirable to choose small to moderate values of J for cure rate modeling. In practice, we recommend doing analyses for several values of J to see the sensitivity of the posterior estimates of the regression coefficients. The semiparametric cure rate model (5.3.2) is quite flexible, as it allows us to model general shapes of the hazard function, as well as choose the degree of parametricity in $F(y)$ through suitable choices of J. Again, since $\boldsymbol{N}$ is not observed, the observed data likelihood, $L(\boldsymbol{\beta}, \boldsymbol{\lambda}|D_{obs})$ is obtained by summing out $\boldsymbol{N}$ from (5.3.2) as in (5.2.7).

The joint power prior for $(\boldsymbol{\beta}, \boldsymbol{\lambda}, a_0)$ for (5.3.2) is given by

$$\pi(\boldsymbol{\beta}, \boldsymbol{\lambda}, a_0|D_{0,obs}) \propto \left(\sum_{\boldsymbol{N}_0} L(\boldsymbol{\beta}, \boldsymbol{\lambda}|D_0)\right)^{a_0} \pi_0(\boldsymbol{\beta}, \boldsymbol{\lambda}) a_0^{\gamma_0 - 1}(1 - a_0)^{\lambda_0 - 1}, \tag{5.3.3}$$

where (γ_0, λ_0) are specified hyperparameters for the prior distribution of a_0, $\pi_0(\boldsymbol{\beta}, \boldsymbol{\lambda})$ is the initial prior distribution for $(\boldsymbol{\beta}, \boldsymbol{\lambda})$, and $L(\boldsymbol{\beta}, \boldsymbol{\lambda}|D_0)$ is the likelihood function in (5.3.2) with D_0 in place of D. For (5.3.2), the initial prior $\pi_0(\boldsymbol{\beta}, \boldsymbol{\lambda})$ is taken to be

$$\pi_0(\boldsymbol{\beta}, \boldsymbol{\lambda}) \propto \prod_{j=1}^{J} \lambda_j^{\zeta_0 - 1} \exp(-\tau_0 \lambda_j), \tag{5.3.4}$$

so that $\boldsymbol{\beta}$ has an improper uniform initial prior and the λ_j's have i.i.d. $\mathcal{G}(\zeta_0, \tau_0)$ distributions. Taking $a_0 = 0$ with probability 1 in (5.3.2), the posterior distribution of $(\boldsymbol{\beta}, \boldsymbol{\lambda})$ based on the observed data is thus given by

$$\pi(\boldsymbol{\beta}, \boldsymbol{\lambda} | D_{obs}) \propto \left(\sum_{\boldsymbol{N}} L(\boldsymbol{\beta}, \boldsymbol{\lambda} | D) \right) \pi_0(\boldsymbol{\beta}, \boldsymbol{\lambda}), \tag{5.3.5}$$

where $\pi_0(\boldsymbol{\beta}, \boldsymbol{\lambda})$ is defined by (5.3.4). The next theorem characterizes the propriety of the posterior distribution (5.3.5) using a joint improper prior for $\pi_0(\boldsymbol{\beta}, \boldsymbol{\lambda})$.

Theorem 5.3.1 *Consider model* (5.3.2), *the initial prior* (5.3.4), *and take* $a_0 = 0$ *with probability 1 in* (5.3.3). *Assume that (i) when* $\nu_i = 1$, $y_i > 0$, *(ii) there exists* $i_1, i_2, \ldots, i_J$ *such that* $\nu_{i_j} = 1$, *and* $s_{j-1} < y_{i_j} \leq s_j$, $j = 1, 2, \ldots, J$, *and (iii)* $\zeta_0 \geq 1$ *and* $\tau_0 \geq 0$. *Further, let* $\nu_i^* = 1$ *if* $\nu_i = 1$ *and* $i \notin \{i_1, i_2, \ldots, i_J\}$, *0 otherwise, and let* X^* *be an* $n \times p$ *matrix with* i^{th} *row equal to* $\nu_i^* x_i'$, *where* X^* *has full rank* p. *Then* $\pi(\boldsymbol{\beta}, \boldsymbol{\lambda} | D_{obs})$ *in* (5.3.5) *is proper.*

A proof of Theorem 5.3.1 is left an an exercise. Theorem 5.3.1 states that, under the improper joint initial prior in (5.3.4), the joint posterior distribution of $(\boldsymbol{\beta}, \boldsymbol{\lambda})$ based on model (5.3.2) is proper under some very general conditions. Theorem 5.3.1 is useful since it gives us some very general conditions for carrying out Bayesian analyses with improper priors for cure rate models, and it also characterizes the identifiability of (5.3.2). The conditions (i)–(iii) are similar and slightly weaker to those of Theorem 5.2.1, with the weaker condition being that the covariate matrix need not be of full rank over any one interval, but of full rank over all of the aggregate intervals. Theorem 5.3.1 implies that under these mild conditions, the parameters λ_j's in the semiparametric cure rate model in (5.3.2) are identifiable. Thus, Theorem 5.3.1 gives us easy-to-check conditions for determining whether the model (5.3.2) is identifiable.

The next theorem characterizes the propriety of the power prior given in (5.3.3) for the model (5.3.2).

Theorem 5.3.2 *Consider the power prior given by* (5.3.3) *based on the model in* (5.3.2), *where the initial prior is given in* (5.3.4). *Assume that (i) when* $\nu_{0i} = 1$, $y_{i0} > 0$, *(ii) there exist distinct indices* $i_1, i_2, \ldots, i_J$, i_1^*, i_2^*, $\ldots$, i_p^* *such that* $\nu_{0i_j} = 1$, $j = 1, 2, \ldots, J$, $\nu_{0i_\ell^*} = 1$, $\ell = 1, \ldots, p$, $s_{j-1} < y_{0i_j} \leq s_j$, $j = 1, \ldots, J$, *(iii)* $\zeta_0 \geq 1$, $\tau_0 > 0$, $\gamma_0 > p$, $\lambda_0 > 0$, *or* $\zeta_0 \geq 1$, $\tau_0 = 0$, $\kappa_0 > p + J\zeta_0, \xi_0 > 0$, *and (iv)* $X_{0p}^* = (x_{0i_1^*}, \ldots, x_{0i_p^*})'$ *has full rank* p. *Then* $\pi(\boldsymbol{\beta}, \boldsymbol{\lambda}, a_0 | D_{0,obs})$ *in* (5.3.3) *is proper.*

A proof of Theorem 5.3.2 is left as an exercise.

Example 5.2. Melanoma data. We consider the E1684 and E1690 melanoma datasets discussed in Example 1.2, and carry out a Bayesian analysis of E1690 using E1684 as historical data. The response variable is taken to be relapse-free survival (RFS). We conduct all analyses here with the treatment covariate alone and use $J = 5$ for (5.3.2). In addition, we consider several choices of a_0, including $a_0 = 0$ with probability 1, $a_0 = 1$ with probability 1, $E(a_0|D_{obs}) = 0.05$, and $E(a_0|D_{obs}) = 0.33$.

TABLE 5.6. Partial Maximum Likelihood Estimates of the Hazard Ratios for E1684 and E1690.

Study	HR	Std Dev	95% Confidence Interval
E1684	1.43	0.14	(1.08, 1.89)
E1690	1.28	0.13	(1.00, 1.65)

Source: Chen, Harrington, and Ibrahim (1999).

TABLE 5.7. Posterior Estimates of the Hazard Ratios Using Noninformative Priors for E1684 and E1690.

J	Study	HR	Std Dev	95% HPD Interval	CR (OBS, IFN)
1	E1684	1.44	0.21	(1.05, 1.86)	0.23, 0.35
	E1690	1.29	0.17	(0.97, 1.62)	0.33, 0.42
5	E1684	1.44	0.21	(1.05, 1.86)	0.23, 0.36
	E1690	1.29	0.17	(0.98, 1.63)	0.32, 0.41

Source: Chen, Harrington, and Ibrahim (1999).

Table 5.6 summarizes partial likelihood estimates of the hazard ratios for both studies. Table 5.7 shows posterior estimates for E1684 and E1690 using noninformative priors for each study with the initial prior of (5.3.3), and using the treatment covariate alone. Figures 5.3 and 5.4 show the Kaplan-Meier survival curves by treatment arm for E1684 and E1690. Figure 5.5 shows the posterior hazard ratios for E1684 and E1690 under noninformative priors for (5.3.2) using $J = 1$ and $J = 5$, respectively.

Table 5.8 shows posterior estimates using the initial prior (5.3.4) for (5.3.2) based on $a_0 = 0$ with probability 1, $E(a_0|D_{obs}) = 0.05$, $E(a_0|D_{obs}) = 0.33$ and $a_0 = 1$ with probability 1, respectively. We see from Table 5.8 that when $a_0 = 0$ with probability 1, the results for $J = 1$ and $J = 5$ are quite similar, yielding similar posterior hazard ratios, posterior standard deviations, 95% HPD intervals, and posterior cure rate estimates. In addition, we observe that the posterior hazard ratios are close to the frequentist estimate of the hazard ratio (1.28) based on partial likelihood.

From Table 5.8, we also see how the estimates change as more weight is given to the historical data. The prior giving $E(a_0|D_{obs}) = 0.05$ is relatively noninformative and yields results similar to those based on $a_0 = 0$ with probability 1. When more weight is given to the historical data, the posterior hazard ratios increase and the HPD intervals become narrower and do not include 1. This is reasonable since the posterior hazard ratios based for E1684 alone were much larger than E1690 alone, and therefore as more weight is given to E1684, the greater the posterior hazard ratios and the narrower the HPD intervals. Similarly, the posterior estimates of the cure rates base on $J = 1$ and $J = 5$ in Table 5.8 are smaller when more weight is given to the historical data. This again can be explained by the fact that the posterior cure rates for each arm in E1684 alone were much smaller than those based on E1690 alone (see Table 5.7). Figure 5.6 shows the posterior distributions of the hazard ratios for (5.3.2) with $J = 1$, and $J = 5$. For comparison, a piecewise constant exponential model using $J = 5$ is also plotted. Three different values of a_0 are chosen: $E(a_0|D, D_0) = 0.05$ (dash line), $E(a_0|D_{obs}) = 0.33$ (dot line), and $a_0 = 1$ with probability 1 (solid line). From these plots, we can see how the posterior distributions change as more weight is given to the historical data.

TABLE 5.8. Posterior Estimates of E1690.

$E(a_0\|D_{obs})$	J	HR	Std Dev	95% HPD Interval	CR (OBS, IFN)
0	1	1.29	0.17	(0.97, 1.62)	0.33, 0.42
(with probability 1)	5	1.29	0.17	(0.98, 1.63)	0.32, 0.41
0.05	1	1.29	0.16	(0.98, 1.61)	0.32, 0.41
	5	1.30	0.16	(0.99, 1.63)	0.32, 0.41
0.33	1	1.31	0.15	(1.02, 1.62)	0.31, 0.40
	5	1.31	0.15	(1.03, 1.61)	0.30, 0.40
1	1	1.34	0.13	(1.09, 1.60)	0.28, 0.39
(with probability 1)	5	1.34	0.13	(1.09, 1.60)	0.28, 0.39

Source: Chen, Harrington, and Ibrahim (1999).

The intervals $(s_{j-1}, s_j]$, $j = 1, 2, \ldots, J$ were chosen so that with the combined datasets from the historical and current data, at least one failure falls in each interval. This technique for choosing J is quite reasonable and results in a stable Gibbs sampler. We also conducted sensitivity analyses on the construction of the intervals, $(s_{j-1}, s_j]$, $j = 1, 2, \ldots, J$. Three different constructions of $(s_{j-1}, s_j]$ were considered. We chose the subintervals $(s_{j-1}, s_j]$ with (i) equal numbers of failures or censored observations; (ii) approximately equal lengths subject to the restriction that at least one failure occurs in each interval; (iii) decreasing numbers of failures or

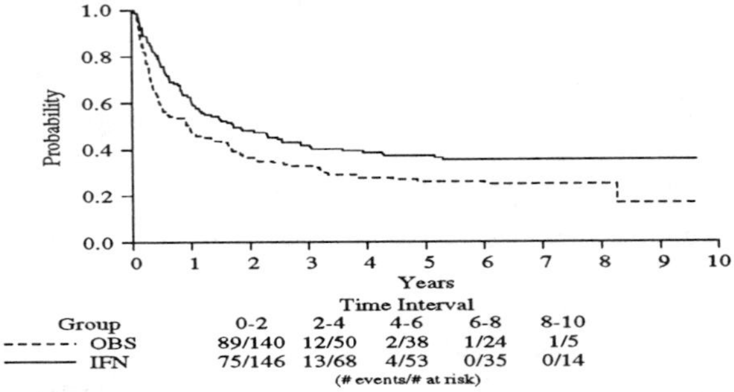

FIGURE 5.3. Kaplan-Meier RFS plots for E1684.

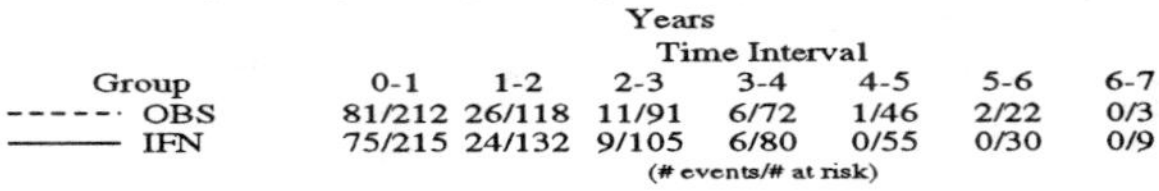

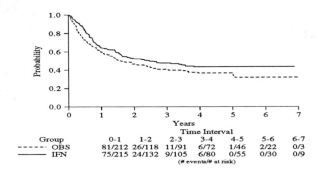

FIGURE 5.4. Kaplan-Meier RFS plots for E1690.

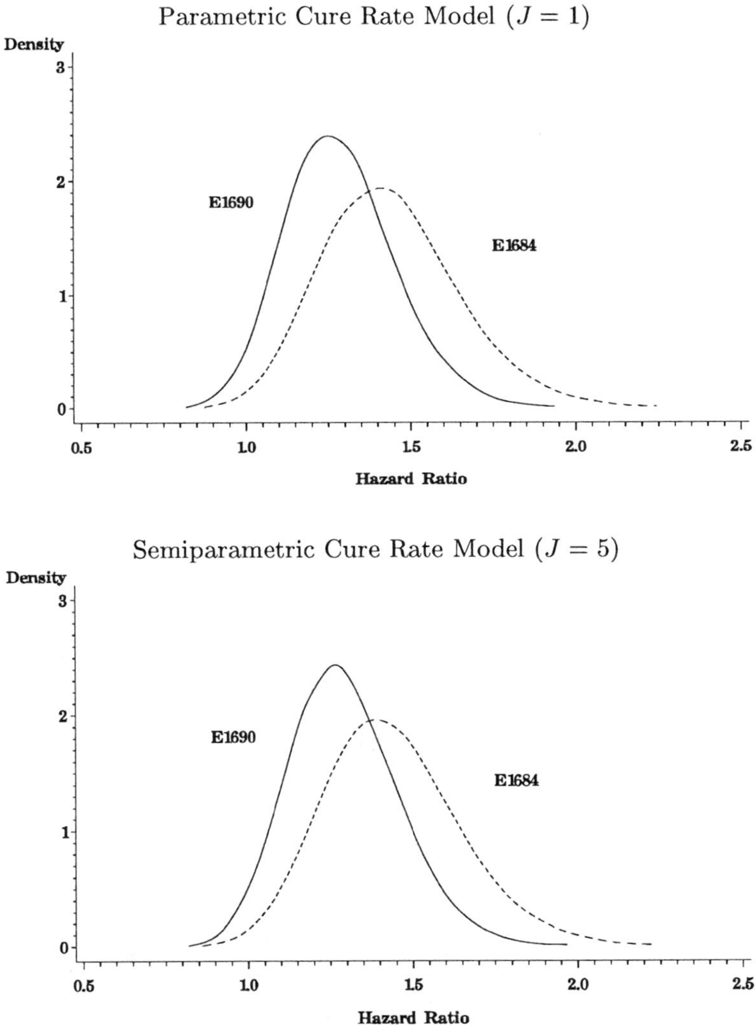

FIGURE 5.5. Plots of posterior densities of hazard ratios for E1684 and E1690.

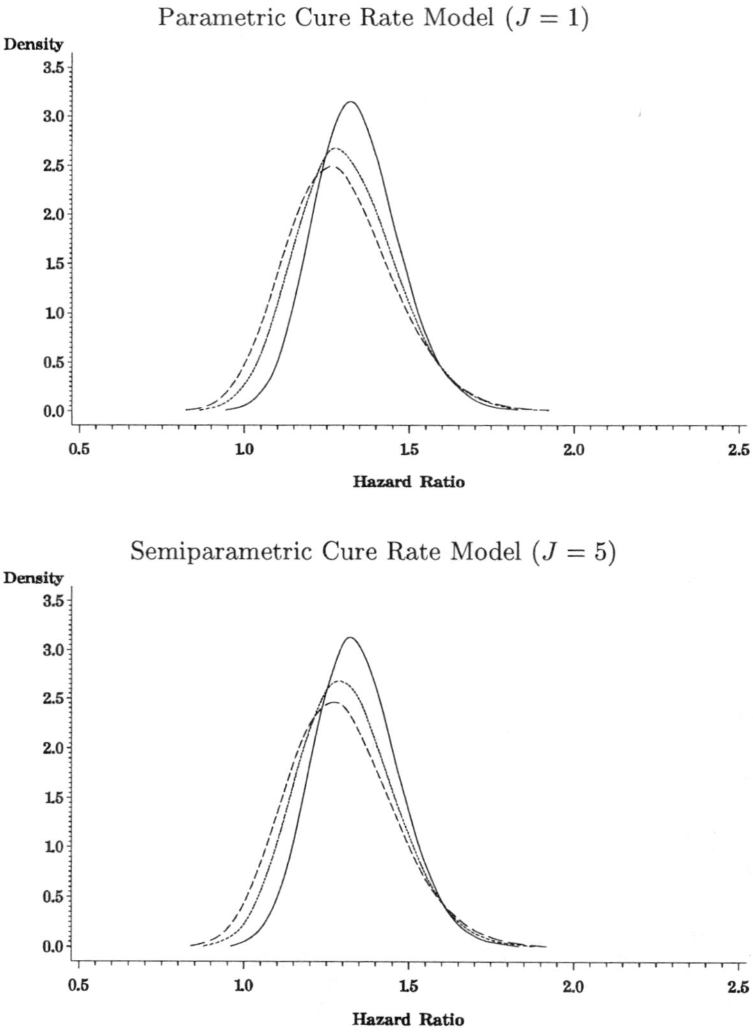

FIGURE 5.6. Plots of posterior densities of hazard ratios for full Bayesian analysis.

censored observations. More specifically, in case (iii) we took s_j to be the $((1-e^{(-j/J)})/(1-e^{-1}))^{th}$ quantile of the y_j's. The posterior estimates were quite robust with respect to these constructions.

5.4 An Alternative Semiparametric Cure Rate Model

A crucial issue with cure rate modeling, and semiparametric survival models in general, is the behavior of the model in the right tail of the survival distribution. In these models, there are typically few subjects at risk in the tail of the survival curve after sufficient follow-up, and therefore estimation of the cure rate can be quite sensitive to the choice of the semiparametric model. Thus there is a need to carefully model the right tail of the survival curve, and allow the model to be more parametric in the tail, while also allowing the model to be nonparametric in other parts of the curve. Following Ibrahim, Chen, and Sinha (2001a), we construct such a model by defining a smoothing parameter κ, $0 < \kappa < 1$ that controls the degree of parametricity in the right tail of the survival curve and it does not depend on the data. Specifically, the prior for λ_j depends on κ, such that the model converges to a parametric model in the right tail of the survival curve as $j \to \infty$. By an appropriate choice of κ, we can choose a fully nonparametric model or a fully parametric model for the right tail of the survival distribution. Also, κ will allow us some control over the degree of parametricity in the beginning and middle part of the survival distribution. A more parametric shape of the model in the right tail facilitates more stable and precise estimates of the parameters. This approach is fundamentally very different from previous approaches for semiparametric Bayesian survival analysis, which primarily focus on specifying a prior process with a mean function and possibly a smoothing parameter, in which posterior properties of both of them depend on the data.

Let $F_0(t|\psi_0)$ denote the parametric survival model we wish to choose for the right tail of the survival curve, and let $H_0(t)$ denote the corresponding cumulative baseline hazard function. Now we take the λ_j's to be independent a priori, each having a gamma prior distribution with mean

$$\mu_j = E(\lambda_j|\psi_0) = \frac{H_0(s_j) - H_0(s_{j-1})}{s_j - s_{j-1}}, \tag{5.4.1}$$

and variance

$$\sigma_j^2 = \text{Var}(\lambda_j|\psi_0, \kappa) = \mu_j \kappa^j, \tag{5.4.2}$$

where $0 < \kappa < 1$ is the smoothing parameter. We see that as $\kappa \to 0$, $\sigma_j^2 \to 0$, so that small values of κ imply a more parametric model in the right tail. In addition, we observe that as $j \to \infty$, $\sigma_j^2 \to 0$, implying that the degree of parametricity is increased at a rate governed by κ as the number of intervals increases. This property also implies that as $j \to \infty$, the survival distribution in the right tail becomes more parametric regardless of any fixed value of κ. The properties of this model are attractive. For example, if $F_0(.|\psi_0)$ is an exponential distribution, then $F_0(y|\psi_0) = 1 - \exp(-\psi_0 y)$, so that $\mu_j = \psi_0$ and $\sigma_j^2 = \psi_0 \kappa^j$. If $F_0(.|\psi_0)$ is a Weibull distribution, then $F_0(y|\psi_0) = 1 - \exp(-\gamma_0 y^{\alpha_0})$, $\psi_0 = (\alpha_0, \gamma_0)'$, so that $\mu_j = \gamma_0 \frac{(s_j^{\alpha_0} - s_{j-1}^{\alpha_0})}{s_j - s_{j-1}}$ and $\sigma_j^2 = \gamma_0 \frac{(s_j^{\alpha_0} - s_{j-1}^{\alpha_0})}{s_j - s_{j-1}} \kappa^j$.

We now formally state several properties for this model. The proofs are given in the Appendix.

Property 5.4.1 *Assume that* $\frac{s_j + s_{j-1}}{2} \to t$ *as* $s_j - s_{j-1} \to 0$. *Then for any* j, *according to this prior process,* $E(\lambda_j|\psi_0) \to h_0(t)$ *as* $s_j - s_{j-1} \to 0$, *where* $h_0(t) = \frac{d}{dt} H_0(t)$.

For example, when $F_0(y|\psi_0) = 1 - \exp(-\psi_0 y)$, then $E(\lambda_j|\psi_0) = \psi_0$ regardless of our choice of $s_1, s_2, \ldots, s_J$. When $F_0(y|\psi_0) = 1 - \exp(-\gamma_0 y^{\alpha_0})$, then $E(\lambda_j|\psi_0) \to \gamma_0 \alpha_0 t^{\alpha_0 - 1}$ as $s_j - s_{j-1} \to 0$. This assures that as j becomes large and $s_j - s_{j-1} \to 0$, then this prior process approximates any prior process with prior mean $h_0(t)$ defined on the promotion time hazard $h^*(t|\boldsymbol{\lambda})$ corresponding to (5.3.1).

Property 5.4.2 *Let* $S_p^*(y|\boldsymbol{\lambda}) = \exp(-\theta F^*(y|\boldsymbol{\lambda}))$, *where* $F^*(y|\boldsymbol{\lambda})$ *is given by* (5.3.1). *Then,* $S_p^*(y|\boldsymbol{\lambda}) \to S_p(y|\psi_0)$ *as* $\kappa \to 0$, *where* $S_p(y|\psi_0) = \exp(-\theta F_0(y|\psi_0))$.

Property 5.4.3 *Let* $f^*(y|\boldsymbol{\lambda}) = \frac{d}{dy} F^*(y|\boldsymbol{\lambda})$, *and* $h_p^*(y|\boldsymbol{\lambda}) = \theta f^*(y|\boldsymbol{\lambda})$ *denote the corresponding hazard function. Then* $h_p^*(y|\boldsymbol{\lambda}) \to \theta f_0(y|\psi_0)$ *as* $\kappa \to 0$, *where* $f_0(y|\psi_0) = \frac{d}{dy} F_0(y|\psi_0)$.

In practice, we recommend doing analyses for several values of κ, J, and $F_0(.|\psi_0)$ to examine the sensitivity of the posterior estimates to various choices of these parameters.

5.4.1 Prior Distributions

We now give joint prior specifications for the semiparametric model in (5.4.1) and (5.4.2). We specify a hierarchical model and first consider a joint (improper) noninformative prior distribution for $(\boldsymbol{\beta}, \boldsymbol{\lambda}, \psi_0)$. We specify the joint prior of these parameters as

$$\pi(\boldsymbol{\beta}, \boldsymbol{\lambda}, \psi_0) = \pi(\boldsymbol{\beta})\pi(\boldsymbol{\lambda}|\psi_0)\pi(\psi_0) \propto \pi(\boldsymbol{\beta}) \left[\prod_{j=1}^{J} \pi(\lambda_j|\psi_0)\right] \pi(\psi_0). \quad (5.4.3)$$

As noted earlier, we take each $\pi(\lambda_j|\psi_0)$ to be independent gamma densities with mean μ_j and variance σ_j^2. If $F_0(.|\psi_0)$ is an exponential distribution, then ψ_0 is a scalar, and we specify a gamma prior for it, i.e., $\pi(\psi_0) \propto \psi_0^{\zeta_0 - 1} \exp(-\tau_0 \psi_0)$, where ζ_0 and τ_0 are specified hyperparameters. If $F_0(.|\psi_0)$ is a Weibull distribution, then $\boldsymbol{\psi}_0 = (\alpha_0, \gamma_0)'$. In this case, we take a prior of the form

$$\pi(\boldsymbol{\psi}_0) = \pi(\alpha_0, \gamma_0) \propto \alpha_0^{\zeta_{\alpha_0} - 1} \exp(-\tau_{\alpha_0} \alpha_0) \gamma_0^{\zeta_{\gamma_0} - 1} \exp(-\tau_{\gamma_0} \gamma_0), \qquad (5.4.4)$$

where ζ_{α_0}, τ_{α_0}, ζ_{γ_0} and τ_{γ_0} are specified hyperparameters. For $\boldsymbol{\beta}$, we consider a uniform improper prior. The next theorem establishes the propriety of the joint posterior distribution of $(\boldsymbol{\beta}, \boldsymbol{\lambda}, \boldsymbol{\psi}_0)$, when using an exponential distribution or a Weibull distribution for $F_0(.|\boldsymbol{\psi}_0)$.

Theorem 5.4.1 *Suppose (i) when $\nu_i = 1$, $y_i > 0$, (ii) there exists $i_1, i_2, \ldots, i_J$ such that $\nu_{i_j} = 1$, and $s_{j-1} < y_{i_j} \le s_j$, $j = 1, 2, \ldots, J$, (iii) the design matrix X^* with i^{th} row equal to $\nu_i \boldsymbol{x}_i'$ is of full rank, iv) if $F_0(.|\psi_0)$ is an exponential distribution, $\zeta_0 > 0$ and $\tau_0 > \sum_{j=1}^J \frac{1}{\kappa^j} \log[(1/\kappa^j)/((y_{i_j} - s_{j-1})/2 + 1/\kappa^j)]$, and if $F_0(.|\boldsymbol{\psi}_0)$ is a Weibull distribution, $\zeta_{\gamma_0} > 0$, $\tau_{\gamma_0} \ge 0$, $\zeta_{\alpha_0} > 0$, and $\tau_{\alpha_0} > -\zeta_{\gamma_0} \log(s_J)$. Then the posterior distribution of $(\boldsymbol{\beta}, \boldsymbol{\lambda}, \boldsymbol{\psi}_0)$ is proper, i.e., $\int L(\boldsymbol{\beta}, \boldsymbol{\lambda}|D) \pi(\boldsymbol{\beta}, \boldsymbol{\lambda}, \boldsymbol{\psi}_0)\, d\boldsymbol{\beta} d\boldsymbol{\lambda} d\boldsymbol{\psi}_0 < \infty$, where $L(\boldsymbol{\beta}, \boldsymbol{\lambda}|D)$ is the likelihood function based on the observed data D.*

A proof of Theorem 5.4.1 is left as an exercise. Theorem 5.4.1 provides a very general class of improper noninformative priors for $(\boldsymbol{\beta}, \boldsymbol{\lambda}, \boldsymbol{\psi}_0)$. First we mention that in condition (iv) of Theorem 5.4.1, τ_0 can be negative, thus resulting in an improper prior for ψ_0 when $F_0(.|\psi_0)$ is exponential. Second, τ_{α_0} is also allowed to be negative, resulting in a joint improper prior for (α_0, γ_0) when $F_0(.|\boldsymbol{\psi}_0)$ is Weibull.

The power prior for this model takes the form

$$\pi(\boldsymbol{\beta}, \boldsymbol{\lambda}, \boldsymbol{\psi}_0, a_0|D_0) \propto L(\boldsymbol{\beta}, \boldsymbol{\lambda}|D_0)^{a_0} \pi_0(\boldsymbol{\beta}, \boldsymbol{\lambda}|\boldsymbol{\psi}_0) \pi_0(\boldsymbol{\psi}_0) a_0^{\xi_0 - 1} (1 - a_0)^{\lambda_0 - 1}, \qquad (5.4.5)$$

where $L(\boldsymbol{\beta}, \boldsymbol{\lambda}|D_0)$ is the likelihood function based on the observed historical data, and ξ_0 and λ_0 are prespecified hyperparameters. The initial prior for $(\boldsymbol{\beta}, \boldsymbol{\lambda}, \boldsymbol{\psi}_0)$, is given by (5.4.3) with $\pi_0(\psi_0)$ taking the form given by $\pi_0(\psi_0) \propto \psi_0^{\zeta_0 - 1} \exp(-\tau_0 \psi_0)$ or by (5.4.4) depending on the form of $F_0(.|\boldsymbol{\psi}_0)$. Following the proofs of Theorem 5.4.1 and Theorem 3 of Chen, Ibrahim, and Sinha (1999), it can be shown that the prior distribution $\pi(\boldsymbol{\beta}, \boldsymbol{\lambda}, \boldsymbol{\psi}_0, a_0|D_0)$ given by (5.4.5) is proper under some very general conditions.

Example 5.3. Melanoma data. We revisit the E1684 and E1690 trials discussed in the previous section. Our main purpose in this example is to examine the tail behavior of the model as κ, a_0, F_0, and J are varied. Of particular interest is the sensitivity of the posterior estimates of $\boldsymbol{\beta}$, $\boldsymbol{\lambda}$, and $S^*(t|\boldsymbol{\lambda}) = 1 - F^*(t|\boldsymbol{\lambda})$, as these parameters are varied, where $F^*(t|\boldsymbol{\lambda})$ is defined in (5.3.1). The E1690 study is quite suitable for our purposes here since the median follow-up for E1690 (4.33 years) is considerably smaller than E1684 (6.9 years). Thus, cure rate estimation based on the E1690 study alone, i.e., $a_0 = 0$, may be more sensitive than that of an analysis which incorporates the historical data E1684. In our example, three covariates were used, and an intercept was included in the model. The three covariates are treatment (IFN, OBS), age, which is continuous, and gender (male, female). Let $\boldsymbol{\beta} = (\beta_1, \beta_2, \beta_3, \beta_4)'$ be the regression coefficient vector corresponding to an intercept and the three covariates, respectively.

TABLE 5.9. Posterior Estimates of $\boldsymbol{\beta}$.

F_0	κ	Variable	Mean	Std Dev	95% HPD Interval
Exponential	0.05	intercept	0.183	0.096	(−0.009, 0.367)
		treatment	−0.242	0.115	(−0.469, −0.018)
		age	0.099	0.058	(−0.011, 0.214)
		gender	−0.118	0.120	(−0.361, 0.110)
	0.60	intercept	0.164	0.094	(−0.019, 0.348)
		treatment	−0.245	0.115	(−0.470, −0.020)
		age	0.098	0.058	(−0.016, 0.211)
		gender	−0.115	0.120	(−0.352, 0.119)
	0.95	intercept	0.157	0.098	(−0.038, 0.345)
		treatment	−0.242	0.116	(−0.472, −0.017)
		age	0.097	0.058	(−0.015, 0.211)
		gender	−0.113	0.121	(−0.348, 0.128)
Weibull	0.05	intercept	0.202	0.102	(0.002, 0.404)
		treatment	−0.242	0.115	(−0.471, −0.019)
		age	0.100	0.057	(−0.012, 0.213)
		gender	−0.118	0.121	(−0.358, 0.114)
	0.60	intercept	0.180	0.098	(−0.011, 0.371)
		treatment	−0.244	0.115	(−0.472, −0.021)
		age	0.098	0.058	(−0.018, 0.209)
		gender	−0.113	0.120	(−0.355, 0.117)
	0.95	intercept	0.160	0.097	(−0.034, 0.345)
		treatment	−0.244	0.116	(−0.473, −0.022)
		age	0.097	0.058	(−0.013, 0.213)
		gender	−0.115	0.120	(−0.351, 0.119)

Source: Ibrahim, Chen, and Sinha (2001a).

Table 5.9 gives posterior means, standard deviations, and 95% HPD intervals of $\boldsymbol{\beta}$ for several values of κ using the exponential and Weibull models for F_0 with $J = 10$ intervals, and when $E(a_0|D) = 0.33$. As κ is varied for a given a_0 using an exponential or Weibull F_0, we see small to moderate changes in the posterior estimates of β. As a_0 is varied, more substantial changes occur in the posterior estimates of β across values of a_0. For example, using an exponential F_0 and $\kappa = 0.05$, the posterior means, standard deviations, and 95% HPD intervals for the treatment coefficient, i.e., β_2, are -0.209, 0.130, and $(-0.461, 0.050)$ when $a_0 = 0$ with probability 1; -0.242, 0.115, and $(-0.469, -0.018)$ when $E(a_0|D) = 0.33$; and -0.277, 0.079, and $(-0.462, -0.087)$ when $a_0 = 1$ with probability 1. In general, the posterior standard deviations for $E(a_0|D) > 0$ are smaller than those for $a_0 = 0$, therefore resulting in narrower 95% HPD intervals. A partial explanation of this is that, by incorporating the historical data, more precise estimates of the regression coefficients and right tail of the survival curve are obtained. Overall, for given a_0, we conclude that the estimates of β are reasonably robust as κ is varied, but change substantially as a_0 is varied.

TABLE 5.10. Posterior Summaries of Cure Rates and $S^*(t|\lambda)$.

		Cure Rate		
F_0	κ	Mean	Std Dev	$S^*(3.5\|\lambda)$
Exponential	0.05	0.361	0.065	0.115
	0.60	0.368	0.065	0.106
	0.95	0.370	0.064	0.107
Weibull	0.05	0.354	0.065	0.136
	0.60	0.362	0.064	0.129
	0.95	0.369	0.065	0.108

Source: Ibrahim, Chen, and Sinha (2001a).

Table 5.10 shows posterior summaries of the cure rates and survival function $S^*(t|\lambda)$ for varying κ and F_0 when $E(a_0|D) = 0.33$. For a given F_0, we see moderate changes in the cure rates as κ is varied. When a_0 and κ remain fixed and F_0 is changed, we see that the estimates are quite robust. Also from Table 5.10, we can see that a monotonic increase in the mean of the cure rate estimates occurs as κ is increased. A similar phenomenon occurs with other values of a_0. In summary, Table 5.10 shows that small to moderate changes can occur in the cure rates as the degree of parametricity in the right tail of the survival curve, κ, is changed. Estimates of λ were also computed for several values of κ and a_0, assuming that F_0 is exponential. We observe that for a given a_0, the posterior estimates of λ can change moderately to considerably as κ varies. For example, with $a_0 = 0$, the posterior mean of λ_{10} is 0.617 for $\kappa = 0.05$, and 0.788 when $\kappa = 0.95$. A similar phenomenon occurs when $a_0 = 1$. These changes in λ can be summarized better by examining the estimated survival function for the non-cured patients, denoted by $S^*(t|\lambda)$. Figure 5.7 shows the posterior estimates of $S^*(t|\boldsymbol{\lambda})$ for $E(a_0|D) = 0.33$ using several values of κ. In Figure 5.7, (a) $F_0(.|\psi_0)$ is an exponential distribution; (b) $F_0(.|\psi_0)$ is a Weibull distribution; and the solid, dotted, dashed, and dot-dashed curves correspond to $\kappa = 0.05, 0.30, 0.60, 0.95$, respectively. We see from Figure 5.7 that small to moderate changes in the survival estimates occur as κ is varied. The biggest changes occur in the interval $1 \leq t \leq 5$. Table 5.10 summarizes $S^*(t|\boldsymbol{\lambda})$ at $t = 3.5$ years for several values of κ and F_0. We see from Table 5.10 that for fixed F_0 and a_0, moderate changes in $S^*(3.5|\boldsymbol{\lambda})$ occur as κ is varied. When F_0 is Weibull, bigger differences are seen. Thus, $S^*(t|\boldsymbol{\lambda})$ can be moderately sensitive to the choice of κ. We also observe that as more weight is given to the historical data (i.e., a_0 is increased), $S^*(3.5|\boldsymbol{\lambda})$ increases. For example, for a Weibull F_0 with $a_0 = 0$, $S^*(3.5|\boldsymbol{\lambda}) = 0.124$ for $\kappa = 0.05$ and 0.119 for $\kappa = 0.6$, while from Table 5.10 with $E(a_0|D) = 0.33$, we have $S^*(3.5|\boldsymbol{\lambda}) = 0.136$ for $\kappa = 0.05$ and 0.129 for $\kappa = 0.6$. This phenomenon is consistent with the notion that the cure rate decreases as more weight is given to the historical data. Finally, we note that sensitivity analyses were also carried out using several different values of J, and similar results were obtained.

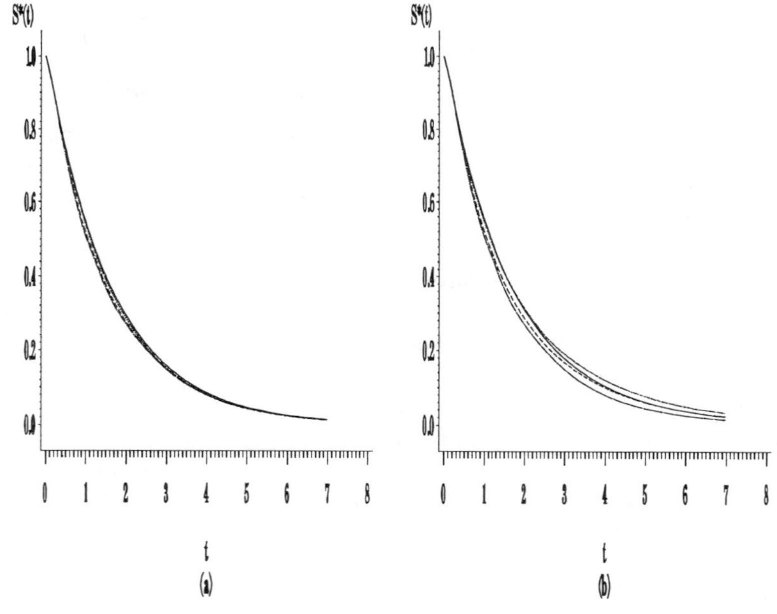

FIGURE 5.7. Plots of survival function $S^*(t|\boldsymbol{\lambda})$ for non-cured patients with $J = 10$ and $E(a_0|D) = 0.33$.

5.5 Multivariate Cure Rate Models

5.5.1 *Models*

As discussed in Section 4.3, it is often of interest to model jointly several types of failure time random variables in survival analysis, such as time to cancer relapse at two different organs, times to cancer relapse and death, times to first and second infection, and so forth. In addition, these random variables typically have joint and marginal survival curves that "plateau" beyond a certain period of follow-up, and therefore it is of great importance in these situations to develop a joint cure rate model for inference.

There does not appear to be a natural multivariate extension of the standard cure rate model in (5.1.1). Even if such an extension were available, it appears that a multivariate mixture model would be extremely cumbersome to work with from a theoretical and computational perspective. As an alternative to a direct multivariate extension of (5.1.1), we examine the model discussed in Chen, Ibrahim, and Sinha (2001), called the *multivariate cure rate model.* This model proves to be quite useful for modeling multivariate data in which the joint failure time random variables have a surviving fraction and each marginal failure time random variable also has a surviving fraction. The model is related to the univariate cure rate model discussed by Yakovlev et al. (1993) and Asselain et al. (1996). To induce the correlation structure between the failure times, we introduce a frailty term (Clayton, 1978; Hougaard, 1986b; Oakes, 1989), which is assumed to have a positive stable distribution. A positive frailty assumes that we have Cox's (Cox, 1972) proportional hazards structure conditionally (i.e., given the unobserved frailty). Thus the marginal and conditional hazards of each component have a proportional hazards structure, and thus remain in the same class of univariate cure rate models.

For clarity and ease of exposition, we will focus our discussion on the bivariate cure rate model, as extensions to the general multivariate case are quite straightforward. The bivariate cure rate model of Chen, Ibrahim, and Sinha (2001) can be derived as follows. Let $\boldsymbol{Y} = (Y_1, Y_2)'$ be a bivariate failure time, such as Y_1 = time to cancer relapse and Y_2 = time to death, or Y_1 = time to first infection, and Y_2 = time to second infection, and so forth. We assume that (Y_1, Y_2) are not ordered and have support on the upper orthant of the plane. For an arbitrary patient in the population, let $\boldsymbol{N} = (N_1, N_2)'$ denote latent (unobserved) variables for (Y_1, Y_2), respectively. We assume throughout that N_k has a Poisson distribution with mean $\theta_k w$, $k = 1, 2$, and (N_1, N_2) are independent. The quantity w is a frailty component in the model which induces a correlation between the latent variables (N_1, N_2). Here we take w to have a positive stable distribution indexed by the parameter α, denoted by $w \sim S_\alpha(1, 1, 0)$, where $0 < \alpha < 1$. Although several choices can be made for the distribution of w, the positive stable distribution is quite attractive, common, and flexible in the multivariate survival setting. In addition, it will yield several desirable properties. See Section 4.1.4 for other properties of the positive stable distribution.

Let $\boldsymbol{Z}_i = (Z_{1i}, Z_{2i})'$ denote the bivariate promotion time for the ith metastasis-competent tumor cell. The random vectors $\boldsymbol{Z}_i$, $i = 1, 2, \ldots$ are assumed to be independent and identically distributed. The cumulative distribution function of Z_{ki} is denoted by $F_k(t) = 1 - S_k(t)$, $k = 1, 2$, and F_k is independent of (N_1, N_2). The observed survival time can be defined by the random variable $Y_k = \min\{Z_{ki}, 0 \le i \le N_k\}$, where $P(Z_{k0} = \infty) = 1$ and N_k is independent of the sequence $Z_{k1}, Z_{k2}, \ldots$, for $k = 1, 2$. The survival function for $\boldsymbol{Y} = (Y_1, Y_2)'$ given w, and hence the survival function for the population given w, is given by

$$\begin{aligned}
S_{pop}(y_1, y_2 | w) &= \prod_{k=1}^{2} [P(N_k = 0) + P(Z_{k1} > t_k, \ldots, Z_{kN} > t_k, N_k \ge 1)] \\
&= \prod_{k=1}^{2} \left[\exp(-w\theta_k) + \left(\sum_{r=1}^{\infty} S_k(y_k)^r \frac{(w\theta_k)^r}{r!} \exp(-w\theta_k) \right) \right] \\
&= \prod_{k=1}^{2} \exp\left\{-w\theta_k + \theta_k w S_k(y_k)\right\} \\
&= \exp\left\{-w\left[\theta_1 F_1(y_1) + \theta_2 F_2(y_2)\right]\right\}, \qquad (5.5.1)
\end{aligned}$$

where $P(N_k = 0) = P(Y_k = \infty) = \exp(-\theta_k)$, $k = 1, 2$. We emphasize here that the primary roles of $\boldsymbol{N}$ and $\boldsymbol{Z}_i$ are that they only facilitate the construction of the model and need not have any physical or biological interpretation at all for the model to be valid. They are quite useful for the computational implementation of the model via the Gibbs sampler as discussed below and thus are defined primarily for this purpose. The model in (5.5.1) is valid for *any* time-to-event data with a cure rate structure as implied by (5.5.1) and the subsequent development. Thus the model can be useful for modeling various types of failure time data, including time to relapse, time to death, time to infection, time to complication, time to rejection, and so forth. In addition, the frailty variable w serves a dual purpose in the model — it induces the correlation between Y_1 and Y_2 and at the same time relaxes the Poisson assumption of N_1 and N_2 by adding the same extra Poisson variation through their respective means $\theta_1 w$ and $\theta_2 w$.

Following Ibragimov and Chernin (1959), the $S_\alpha(1, 1, 0)$ density for w $(0 < \alpha < 1)$ can be expressed in the form

$$f_s(w|\alpha) = a w^{-(a+1)} \int_0^1 s(u) \exp\left\{-\frac{s(u)}{w^a}\right\} du, \ w > 0, \qquad (5.5.2)$$

where

$$a = \frac{\alpha}{1-\alpha} \text{ and } s(u) = \left(\frac{\sin(\alpha\pi u)}{\sin(\pi u)}\right)^a \left(\frac{\sin[(1-\alpha)\pi u]}{\sin(\pi u)}\right),$$

and the Laplace transform of w is given by $E(\exp(-sw)) = \exp(-s^\alpha)$. A useful reference on stable distributions is Samorodnitsky and Taqqu (1994). Using the Laplace transform of w, a straightforward derivation yields the unconditional survival function

$$S_{pop}(y_1, y_2) = \exp\left\{-[\theta_1 F_1(y_1) + \theta_2 F_2(y_2)]^\alpha\right\}. \tag{5.5.3}$$

It can be shown that (5.5.3) has a proportional hazards structure if the covariates enter the model through (θ_1, θ_2).

The joint cure fraction implied by (5.5.3) is $S_{pop}(\infty, \infty) = \exp(-[\theta_1 + \theta_2]^\alpha)$. From (5.5.3), the marginal survival functions are

$$S_k(y) = \exp(-\theta_k^\alpha (F_k(y))^\alpha), \quad k = 1, 2. \tag{5.5.4}$$

Equation (5.5.4) indicates that the marginal survival functions have a cure rate structure with probability of cure $\exp(-\theta_k^\alpha)$ for Y_k, $k = 1, 2$. It is important to note in (5.5.4) that each marginal survival function has a proportional hazards structure as long as the covariates, $\boldsymbol{x}$, only enter through θ_k. The marginal hazard function is given by $\alpha\theta_k^\alpha f_k(y)(F_k(y))^{\alpha-1}$, with attenuated covariate effect $(\theta_k(x))^\alpha$, and $f_k(y)$ is the survival density corresponding to $F_k(y)$. This property is similar to the earlier observations made by Oakes (1989) for the ordinary bivariate stable frailty survival model.

In addition, we can express the marginal survival functions in (5.5.4) in terms of standard cure rate models. We can write

$$\begin{aligned} S_k(y) &= \exp(-\theta_k^\alpha (F_k(y))^\alpha) \\ &= \exp(-\theta_k^\alpha) + (1 - \exp(-\theta_k^\alpha)) \left(\frac{\exp(-\theta_k^\alpha (F_k(y))^\alpha) - \exp(-\theta_k^\alpha)}{1 - \exp(-\theta_k^\alpha)}\right) \\ &= \exp(-\theta_k^\alpha) + (1 - \exp(-\theta_k^\alpha)) S_k^*(y), \end{aligned} \tag{5.5.5}$$

where

$$S_k^*(y) = \frac{\exp(-\theta_k^\alpha (F_k(y))^\alpha) - \exp(-\theta_k^\alpha)}{1 - \exp(-\theta_k^\alpha)}, \quad k = 1, 2.$$

It is easily shown that $S_k^*(y)$ defines a proper survivor function. Thus (5.5.5) is a standard cure rate model with cure rate given by $\pi_k = \exp(-\theta_k^\alpha)$ and survivor function for the non-cured population given by $S_k^*(y)$ for $k = 1, 2$.

The parameter α $(0 < \alpha < 1)$ is a scalar parameter that is a measure of association between (Y_1, Y_2). Small values of α indicate high association between (Y_1, Y_2). As $\alpha \to 1$, this implies less association between (Y_1, Y_2) which can be seen from (5.5.3). Following Clayton (1978) and Oakes (1989), we can compute a local measure of dependence, denoted, $\theta^*(y_1, y_2)$ defined by (4.4.1), as a function of α. For the multivariate cure rate model in (5.5.3), $\theta^*(t_1, t_2)$ is well defined, and is given by

$$\theta^*(y_1, y_2) = \alpha^{-1}(1-\alpha)\left(\theta_1 F_1(y_1) + \theta_2 F_2(y_2)\right)^{-\alpha} + 1. \quad (5.5.6)$$

We see that $\theta^*(y_1, y_2)$ in (5.5.6) decreases in (y_1, y_2). That is, the association between (Y_1, Y_2) is greater when (Y_1, Y_2) are small and the association decreases over time. Such a property is desirable, for example, when Y_1 denotes time to relapse and Y_2 denotes time to death. Finally, we mention that a global measure of dependence such as Kendall's τ or the Pearson correlation coefficient is not well defined for the multivariate cure rate model (5.5.3) since no moments for cure rate models exist due to the improper survival function.

The multivariate cure rate model presented here is attractive in several respects. First, the model has a proportional hazards structure for the population hazard, conditionally as well as marginally, when covariates are entered through the cure rate parameter, and thus has an appealing interpretation. Also, the model is computationally feasible. In particular, by introducing latent variables, efficient MCMC algorithms can be developed that enable us to sample from the joint posterior distribution of the parameters. Specifically, we discuss below a modified version of the collapsed Gibbs technique of Liu (1994) for efficient Gibbs sampling from the posterior distribution.

5.5.2 The Likelihood Function

The likelihood function for this model can be obtained as follows. Suppose we have n subjects, and let N_{ki} denote the number of latent risks for the ith subject, $i = 1, 2, \ldots, n$, $k = 1, 2$. Further, we assume that the N_{ki}'s are independent Poisson random variables with mean $w_i\theta_k$, $i = 1, 2, \ldots, n$ for $k = 1, 2$. We also assume the $w_i \sim S_\alpha(1, 1, 0)$, and the w_i's are *i.i.d.*. We emphasize here that the N_{ki}'s are not observed, and can be viewed as latent variables in our model formulation. Further, suppose $Z_{ki1}, Z_{ki2}, \ldots, Z_{ki,N_{ki}}$ are the independent latent times for the N_{ki} latent risks for the *ith* subject, which are unobserved, and all have cumulative distribution function $F_k(.)$, $i = 1, 2, \ldots, n$, $k = 1, 2$. In this subsection, we specify a parametric form for $F_k(.)$, such as a Weibull or gamma distribution. We denote the indexing parameter (possibly vector valued) by $\boldsymbol{\psi}_k$, and thus write $F_k(.|\boldsymbol{\psi}_k)$ and $S_k(.|\boldsymbol{\psi}_k)$. For example, if $F_k(.|\boldsymbol{\psi}_k)$ corresponds to a Weibull distribution, then $\boldsymbol{\psi}_k = (\xi_k, \lambda_k)'$, where ξ_k is the shape parameter and λ_k is the scale parameter. Let y_{ki} denote the failure time or censoring time for subject i for the kth component, and let indicator $\nu_{ki} = 1$ if y_{ki} is an observed failure time and 0 if it is a censoring time. Let $\boldsymbol{y}_k = (y_{k1}, y_{k2}, \ldots, y_{kn})$, $\boldsymbol{\nu}_k = (\nu_{k1}, \nu_{k2}, \ldots, \nu_{kn})$, $\boldsymbol{N}_k = (N_{k1}, N_{k2}, \ldots N_{kn})$, $k = 1, 2$, and $\boldsymbol{w} = (w_1, w_2, \ldots, w_n)'$. The complete data is given by $D = (n, \boldsymbol{y}_1, \boldsymbol{y}_2, \boldsymbol{\nu}_1, \boldsymbol{\nu}_2, \boldsymbol{N}_1, \boldsymbol{N}_2, \boldsymbol{w})$, where $\boldsymbol{N}_1$, $\boldsymbol{N}_2$, and $\boldsymbol{w}$ are unobserved random vectors, and the observed data is given by $D_{obs} = (n, \boldsymbol{y}_1, \boldsymbol{y}_2, \boldsymbol{\nu}_1, \boldsymbol{\nu}_2)$. Further, let $\boldsymbol{\theta} = (\theta_1, \theta_2)'$ and $\boldsymbol{\psi} = (\boldsymbol{\psi}_1', \boldsymbol{\psi}_2')'$. The likelihood function of $(\boldsymbol{\theta}, \boldsymbol{\psi})$ based on the complete data D is given by

$$L(\boldsymbol{\theta}, \boldsymbol{\psi}|D) = \left(\prod_{k=1}^{2}\prod_{i=1}^{n} S_k(y_{ki}|\boldsymbol{\psi}_k)^{N_{ki}-\nu_{ki}} \left(N_{ki} f_k(y_{ki}|\boldsymbol{\psi}_k)\right)^{\nu_{ki}}\right) \times \exp\left\{\sum_{i=1}^{n}(N_{ki}\log(w_i\theta_k) - \log(N_{ki}!) - w_i\theta_k)\right\}, \quad (5.5.7)$$

where $f_k(y_{ki}|\boldsymbol{\psi}_k)$ is the density corresponding to $F_k(y_{ki}|\boldsymbol{\psi}_k)$. We assume a Weibull density for $f_k(y_{ki}|\boldsymbol{\psi}_k)$, so that

$$f_k(y|\boldsymbol{\psi}_k) = \xi_k y^{\xi_k - 1} \exp\left\{\lambda_k - y^{\xi_k}\exp(\lambda_k)\right\}. \quad (5.5.8)$$

To construct the likelihood function of the observed data, we integrate (5.5.7) with respect to $(\boldsymbol{N}, \boldsymbol{w})$ assuming a $S_\alpha(1,1,0)$ density for each w_i, denoted by $f_s(w_i|\alpha)$. We are led to the following theorem.

Theorem 5.5.1 *The likelihood function based on the observed data, denoted $L(\boldsymbol{\theta}, \boldsymbol{\psi}, \alpha|D_{obs})$, is given by*

$$\begin{aligned}
&L(\boldsymbol{\theta}, \boldsymbol{\psi}, \alpha|D_{obs}) \\
&\equiv \int_{R^{+n}} L(\boldsymbol{\theta}, \boldsymbol{\psi}|D) \times \left[\prod_{i=1}^{n} f_s(w_i|\alpha)\right] d\boldsymbol{w} \\
&= \theta_1^{d_1}\theta_2^{d_2}\alpha^{d_1+d_2}\left[\prod_{k=1}^{2}\prod_{i=1}^{n} f_k(y_{ki}|\boldsymbol{\psi}_k)^{\nu_{ki}}\right] \\
&\quad \times \prod_{i=1}^{n}\left\{[\theta_1 F_1(y_{1i}|\boldsymbol{\psi}_1) + \theta_2 F_2(y_{2i}|\boldsymbol{\psi}_2)]^{(\alpha-1)(\nu_{1i}+\nu_{2i})}\right\} \\
&\quad \times \prod_{i=1}^{n}\left[\alpha^{-1}(1-\alpha)(\theta_1 F_1(y_{1i}|\boldsymbol{\psi}_1) + \theta_2 F_2(y_{2i}|\boldsymbol{\psi}_2))^{-\alpha} + 1\right]^{\nu_{1i}\nu_{2i}} \\
&\quad \times \prod_{i=1}^{n}\exp\left\{-(\theta_1 F_1(y_{1i}|\boldsymbol{\psi}_1) + \theta_2 F_2(y_{2i}|\boldsymbol{\psi}_2))^{\alpha}\right\}, \qquad (5.5.9)
\end{aligned}$$

where $f_s(w_i|\alpha)$ denotes the probability density function of w_i defined by (5.5.2), $d_k = \sum_{i=1}^{n}\nu_{ki}$ for $k = 1, 2$, $R^{+n} = R^+ \times R^+ \times \cdots \times R^+$, and $R^+ = (0, \infty)$.

The proof is left an exercise.

As before, we incorporate covariates for the cure rate model (5.5.3) through the cure rate parameter $\boldsymbol{\theta}$. Let $\boldsymbol{x}_i' = (x_{i1}, x_{i2}, \ldots, x_{ip})$ denote the $p \times 1$ vector of covariates for the ith subject, and let $\boldsymbol{\beta}_k =$

$(\beta_{k1}, \beta_{k2}, \ldots, \beta_{kp})'$ denote the corresponding vector of regression coefficients for failure time random variable Y_k, $k = 1, 2$. We relate $\boldsymbol{\theta}$ to the covariates by

$$\theta_{ki} \equiv \theta(\boldsymbol{x}_i'\boldsymbol{\beta}_k) = \exp(\boldsymbol{x}_i'\boldsymbol{\beta}_k),$$

so that the cure rate for subject i is

$$\exp(-\theta_{ki}) = \exp(-\exp(\boldsymbol{x}_i'\boldsymbol{\beta}_k)),$$

for $i = 1, 2, \ldots, n$ and $k = 1, 2$. Letting $\boldsymbol{\beta} = (\boldsymbol{\beta}_1', \boldsymbol{\beta}_2')'$, we can write the observed data likelihood of $(\boldsymbol{\beta}, \boldsymbol{\psi}, \alpha)$ as

$$\begin{aligned}
&L(\boldsymbol{\beta}, \boldsymbol{\psi}, \alpha | D_{obs}) \\
&= \left(\alpha^{d_1+d_2} \prod_{k=1}^{2} \prod_{i \in \mathcal{D}_k} \exp(\boldsymbol{x}_i'\boldsymbol{\beta}_k) \right) \left[\prod_{k=1}^{2} \prod_{i=1}^{n} f_k(y_{ki}|\boldsymbol{\psi}_k)^{\nu_{ki}} \right] \\
&\quad \times \prod_{i=1}^{n} \left\{ [\exp(\boldsymbol{x}_i'\boldsymbol{\beta}_1) F_1(y_{1i}|\boldsymbol{\psi}_1) + \exp(\boldsymbol{x}_i'\boldsymbol{\beta}_2) F_2(y_{2i}|\boldsymbol{\psi}_2)]^{(\alpha-1)(\nu_{1i}+\nu_{2i})} \right\} \\
&\quad \times \prod_{i=1}^{n} \left\{ \frac{1-\alpha}{\alpha} [\exp(\boldsymbol{x}_i'\boldsymbol{\beta}_1) F_1(y_{1i}|\boldsymbol{\psi}_1) + \exp(\boldsymbol{x}_i'\boldsymbol{\beta}_2) F_2(y_{2i}|\boldsymbol{\psi}_2)]^{-\alpha} + 1 \right\}^{\nu_{1i}\nu_{2i}} \\
&\quad \times \prod_{i=1}^{n} \exp\left\{ -(\exp(\boldsymbol{x}_i'\boldsymbol{\beta}_1) F_1(y_{1i}|\boldsymbol{\psi}_1) + \exp(\boldsymbol{x}_i'\boldsymbol{\beta}_2) F_2(y_{2i}|\boldsymbol{\psi}_2))^{\alpha} \right\},
\end{aligned} \tag{5.5.10}$$

where $\mathcal{D}_k$ consists of those patients who failed according to Y_k, $k = 1, 2$, $D_{obs} = (n, \boldsymbol{y}_1, \boldsymbol{y}_2, X, \boldsymbol{\nu}_1, \boldsymbol{\nu}_2)$, X is the $n \times p$ matrix of covariates, $f_k(y_{ki}|\boldsymbol{\psi}_k)$ is given by (5.5.8) and $S_k(y_{ki}|\boldsymbol{\psi}_k) = \exp(-y_{ki}^{\xi_k} \exp(\lambda_k))$.

5.5.3 *The Prior and Posterior Distributions*

We consider a joint improper prior for $(\boldsymbol{\beta}, \boldsymbol{\psi}, \alpha) = (\boldsymbol{\beta}_1, \boldsymbol{\beta}_2, \boldsymbol{\psi}_1, \boldsymbol{\psi}_2, \alpha)$ of the form

$$\begin{aligned}
\pi(\boldsymbol{\beta}, \boldsymbol{\psi}, \alpha) &= \pi(\boldsymbol{\beta}_1, \boldsymbol{\beta}_2, \boldsymbol{\psi}_1, \boldsymbol{\psi}_2, \alpha) \propto \pi(\boldsymbol{\psi}_1)\pi(\boldsymbol{\psi}_2) I(0 < \alpha < 1) \\
&= \prod_{k=1}^{2} \pi(\xi_k, \lambda_k) I(0 < \alpha < 1),
\end{aligned} \tag{5.5.11}$$

where $I(0 < \alpha < 1) = 1$ if $0 < \alpha < 1$, and 0 otherwise. Thus, (5.5.11) implies that $\boldsymbol{\beta}$, $\boldsymbol{\psi}$, and α are independent a priori, $(\boldsymbol{\beta}_1, \boldsymbol{\beta}_2)$ are independent a priori with an improper uniform prior, α has a proper uniform prior over the interval $(0, 1)$, and $(\boldsymbol{\psi}_1, \boldsymbol{\psi}_2)$ are independent and identically distributed as $\pi(\boldsymbol{\psi}_k)$ a priori. We will assume that

$$\pi(\xi_k, \lambda_k) = \pi(\xi_k | \nu_0, \tau_0) \pi(\lambda_k),$$

where

$$\pi(\xi_k|\delta_0, \tau_0) \propto \xi_k^{\delta_0 - 1} \exp\{-\tau_0 \xi_k\}, \quad \text{and} \quad \pi(\lambda_k) \propto \exp\{-c_0 \lambda_k^2\},$$

and δ_0, τ_0, and c_0 are specified hyperparameters. With these specifications, the posterior distribution of $(\boldsymbol{\beta}, \boldsymbol{\psi}, \alpha)$ based on the observed data $D_{obs} = (n, \boldsymbol{y}_1, \boldsymbol{y}_2, X, \boldsymbol{\nu}_1, \boldsymbol{\nu}_2)$ is given by

$$\pi(\boldsymbol{\beta}, \boldsymbol{\psi}, \alpha|D_{obs}) \propto L(\boldsymbol{\beta}, \boldsymbol{\psi}, \alpha|D_{obs}) \prod_{k=1}^{2} \pi(\xi_k|\delta_0, \tau_0)\pi(\lambda_k), \tag{5.5.12}$$

where $L(\boldsymbol{\beta}, \boldsymbol{\psi}, \alpha|D_{obs})$ is given by (5.5.10). We are led to the following theorem concerning the propriety of the posterior distribution in (5.5.12) using the noninformative improper prior (5.5.11).

Theorem 5.5.2 *Let X_k^* be an $n \times p$ matrix with rows $\nu_{ki} x_{ki}'$ for $k = 1, 2$. Then if (i) X_k^* is of full rank for $k = 1, 2$, (ii) $\pi(\lambda_k)$ is proper, and (iii) $\tau_0 > 0$ and $\delta_0 > -\min\{d_1, d_2\}$, then the posterior given in* (5.5.12) *is proper.*

The proof of Theorem 5.5.2 is left an an exercise. Note that the conditions given in Theorem 5.5.2 are sufficient but *not* necessary for the propriety of the posterior distribution. However, the conditions stated in the theorem are quite general and typically satisfied for most datasets. We also note that a proper prior for ξ_k is not required in order to obtain a proper posterior. This can be observed from condition (iii) because $\pi(\xi_k|\nu_0, \tau_0)$ is no longer proper when $\nu_0 < 0$. We note that Theorem 5.5.2 only requires that $\pi(\lambda_k)$ be any proper prior. Although several choices can be made, we will take independent normal densities for $\pi(\lambda_k)$, $k = 1, 2$, in the remainder of this section.

5.5.4 Computational Implementation

Following Chen, Ibrahim, and Sinha (2001), we present a modified version of the collapsed Gibbs technique of Liu (1994) to sample from the posterior distribution. This technique results in an efficient Gibbs sampling scheme which reduces the correlations between the parameters and the latent variables. We note here that MCMC methods for multivariate survival data have also been examined by Qiou, Ravishanker, and Dey (1999).

From (5.5.2), it can be shown that $f_s(w|\alpha)$ is obtained by marginalizing with respect to u from the joint density

$$f(w, u|\alpha) = aw^{-(a+1)} s(u) \exp\left\{-\frac{s(u)}{w^a}\right\}, \quad w > 0, \ 0 < u < 1.$$

This relationship plays an important role in the implementation of the Gibbs sampler.

To facilitate the Gibbs sampler, we introduce several auxiliary (latent) variables. We note here that Gibbs sampling using auxiliary variables has been used by many in the Bayesian literature, including Besag and Green (1993) and Higdon (1998). The auxiliary variables are $\boldsymbol{N} = (N_1, N_2)$, where $\boldsymbol{N}_k = (N_{k1}, N_{k2}, \ldots N_{kn})$ for $k = 1, 2$, $\boldsymbol{w} = (w_1, w_2, \ldots, w_n)$, and $\boldsymbol{u} = (u_1, u_2, \ldots, u_n)$. The joint posterior distribution of $(\boldsymbol{\beta}, \boldsymbol{\psi}, \alpha, \boldsymbol{N}, \boldsymbol{w}, \boldsymbol{u}|D_{obs})$ is given by

$$\begin{aligned}\pi(\boldsymbol{\beta}, \boldsymbol{\psi}, \alpha, \boldsymbol{N}, \boldsymbol{w}, \boldsymbol{u}|D_{obs}) \propto & \prod_{k=1}^{2}\prod_{i=1}^{n} S_k(y_{ki}|\boldsymbol{\psi}_k)^{N_{ki}-\nu_{ki}} \left(N_{ki} f_k(y_{ki}|\boldsymbol{\psi}_k)\right)^{\nu_{ki}} \\ & \times \exp\left\{\sum_{i=1}^{n}(N_{ki}\log(w_i\theta_{ki}) - \log(N_{ki}!) - w_i\theta_{ki})\right\} \\ & \times \prod_{i=1}^{n}\left[a w_i^{-(a+1)} s(u_i) \exp\left\{-\frac{s(u_i)}{w_i^a}\right\}\right] \\ & \times \prod_{k=1}^{2}\Big(\pi(\xi_k|\delta_0, \tau_0)\pi(\lambda_k)\Big), \end{aligned} \tag{5.5.13}$$

where $\theta_{ki} = \exp(\boldsymbol{x}_i'\boldsymbol{\beta}_k)$, $\delta_0 > -\min\{d_1, d_2\}$, $\tau_0 > 0$, and $c_0 > 0$. To run the Gibbs sampler, we need to sample from the following conditional distributions: $[\boldsymbol{\psi}|\boldsymbol{\beta}, \alpha, \boldsymbol{N}, \boldsymbol{w}, \boldsymbol{u}, D_{obs}]$ and $[\boldsymbol{\beta}, \alpha, \boldsymbol{N}, \boldsymbol{w}, \boldsymbol{u}|\boldsymbol{\psi}, D_{obs}]$.

The conditional posterior density for $[\boldsymbol{\psi}|\boldsymbol{\beta}, \alpha, \boldsymbol{N}, \boldsymbol{u}, D_{obs}]$ is given by

$$\begin{aligned}\pi(\boldsymbol{\psi}|\boldsymbol{\beta}, \alpha, \boldsymbol{N}, \boldsymbol{u}, D_{obs}) \propto \prod_{k=1}^{2} \xi_k^{d_k+\delta_0-1} \exp\Big\{ & d_k\lambda_k + \sum_{i=1}^{n}\nu_{ki}\xi_k\log(y_{ki}) \\ & - N_{ki}e^{\lambda_k}y_{ki}^{\xi_k} - \tau_0\xi_k - c_0\lambda_k^2\Big\}.\end{aligned}$$

Using a similar proof given by Berger and Sun (1993), we can show that $\pi(\boldsymbol{\psi}|\boldsymbol{\beta}, \alpha, \boldsymbol{N}, \boldsymbol{u}, D_{obs})$ is log-concave in ξ_k or λ_k for $k = 1, 2$. Thus, the adaptive rejection algorithm of Gilks and Wild (1992) can be used here to sample $\boldsymbol{\psi}$.

Sampling from $[\boldsymbol{\beta}, \alpha, \boldsymbol{N}, \boldsymbol{w}, \boldsymbol{u}|\boldsymbol{\psi}, D_{obs}]$ is the most challenging and expensive part of this algorithm. Sampling from the five complete conditional distributions may result in high correlations between $(\boldsymbol{\beta}, \alpha, \boldsymbol{N}, \boldsymbol{w}, \boldsymbol{u})$ due to the high dimension of the latent vectors. To remedy this potential problem, we apply the collapsed Gibbs procedure of Liu (1994). It is easy to observe that

$$[\boldsymbol{\beta}, \alpha, \boldsymbol{N}, \boldsymbol{w}, \boldsymbol{u}|\boldsymbol{\psi}, D_{obs}] = [\boldsymbol{\beta}, \alpha, \boldsymbol{w}, \boldsymbol{u}|\boldsymbol{\psi}, D_{obs}][\boldsymbol{N}|\boldsymbol{\beta}, \alpha, \boldsymbol{w}, \boldsymbol{u}, \boldsymbol{\psi}, D_{obs}]. \tag{5.5.14}$$

In (5.5.14), we draw $(\boldsymbol{\beta}, \alpha, \boldsymbol{w}, \boldsymbol{u})$ by collapsing $\boldsymbol{N}$, which is crucial for achieving convergence in the MCMC algorithm. For $[\boldsymbol{\beta}, \alpha, \boldsymbol{w}, \boldsymbol{u}|\boldsymbol{\psi}, D_{obs}]$, we draw $\boldsymbol{\beta}$ from $[\boldsymbol{\beta}|\alpha, \boldsymbol{w}, \boldsymbol{u}, \boldsymbol{\psi}, D_{obs}]$ and $(\alpha, \boldsymbol{w}, \boldsymbol{u})$ from $[\alpha, \boldsymbol{w}, \boldsymbol{u}|\boldsymbol{\beta}, \boldsymbol{\psi}, D_{obs}]$.

After some algebra, we derive the density of $[\boldsymbol{\beta}|\alpha, \boldsymbol{w}, \boldsymbol{u}, \boldsymbol{\psi}, D_{obs}]$ as

$$\pi(\boldsymbol{\beta}|\alpha, \boldsymbol{w}, \boldsymbol{u}, \boldsymbol{\psi}, D_{obs}) \propto \exp\left\{\sum_{k=1}^{2}\sum_{i=1}^{n}[\nu_{ki}\boldsymbol{x}_i'\boldsymbol{\beta}_k - w_i F_k(y_{ki}|\psi_k)\exp(\boldsymbol{x}_i'\boldsymbol{\beta}_k)]\right\}. \tag{5.5.15}$$

It is easy to see that $\pi(\boldsymbol{\beta}|\alpha, \boldsymbol{w}, \boldsymbol{u}, \boldsymbol{\psi}, D_{obs})$ is log-concave in each component of $\boldsymbol{\beta}$ and thus we can use the adaptive rejection algorithm to draw $\boldsymbol{\beta}$. To draw $[\alpha, \boldsymbol{w}, \boldsymbol{u}|\boldsymbol{\beta}, \boldsymbol{\psi}, D_{obs}]$, we use the collapsed Gibbs procedure one more time. That is, we draw α from $[\alpha|\boldsymbol{\beta}, \boldsymbol{\psi}, D_{obs}]$ by collapsing $\boldsymbol{w}$ and $\boldsymbol{u}$, and then draw $(\boldsymbol{w}, \boldsymbol{u})$ from $[\boldsymbol{w}, \boldsymbol{u}|\alpha, \boldsymbol{\beta}, \boldsymbol{\psi}, D_{obs}]$.

The conditional posterior density for $[\alpha|\boldsymbol{\beta}, \boldsymbol{\psi}, D_{obs}]$ can be written as

$$\pi(\alpha|\boldsymbol{\beta}, \boldsymbol{\psi}, D_{obs}) \propto L(\boldsymbol{\beta}, \boldsymbol{\psi}, \alpha|D_{obs}), \tag{5.5.16}$$

where $L(\boldsymbol{\beta}, \boldsymbol{\psi}, \alpha|D_{obs})$ is given by (5.5.10). Generating α from (5.5.16) is not trivial since $\pi(\alpha|\boldsymbol{\beta}, \boldsymbol{\psi}, D_{obs})$ is not log-concave. Therefore, we consider the following Metropolis-Hastings algorithm with a "de-constraint" transformation to draw α. Since $0 < \alpha < 1$, we let

$$\alpha = \frac{e^{\eta}}{1+e^{\eta}}, \quad -\infty < \eta < \infty. \tag{5.5.17}$$

Then

$$\pi(\eta|\boldsymbol{\beta}, \boldsymbol{\psi}, D_{obs}) = \pi(\alpha|\boldsymbol{\beta}, \boldsymbol{\psi}, D_{obs})\frac{e^{\eta}}{(1+e^{\eta})^2}.$$

Instead of directly sampling α, we generate η by choosing a normal proposal $N(\hat{\eta}, \hat{\sigma}_{\hat{\eta}}^2)$, where $\hat{\eta}$ is the maximizer of the logarithm of $\pi(\eta|\boldsymbol{\beta}, \boldsymbol{\psi}, D_{obs})$ and $\hat{\sigma}_{\hat{\eta}}^2$ is the minus of the inverse of the second derivative of $\log \pi(\eta|\boldsymbol{\beta}, \boldsymbol{\psi}, D_{obs})$ evaluated at $\eta = \hat{\eta}$, that is,

$$\hat{\sigma}_{\hat{\eta}}^{-2} = -\left.\frac{d^2 \log \pi(\eta|\boldsymbol{\beta}, \boldsymbol{\psi}, D_{obs})}{d\eta^2}\right|_{\eta=\hat{\eta}}.$$

The algorithm to generate η operates as follows:

Step 1. Let η be the current value.

Step 2. Generate a proposal value η^* from $N(\hat{\eta}, \hat{\sigma}_{\hat{\eta}}^2)$.

Step 3. A move from η to η^* is made with probability

$$\min\left\{\frac{\pi(\eta^*|\boldsymbol{\beta}, \boldsymbol{\psi}, D_{obs})\phi\left(\frac{\eta-\hat{\eta}}{\hat{\sigma}_{\hat{\eta}}}\right)}{\pi(\eta|\boldsymbol{\beta}, \boldsymbol{\psi}, D_{obs})\phi\left(\frac{\eta^*-\hat{\eta}}{\hat{\sigma}_{\hat{\eta}}}\right)}, 1\right\},$$

where ϕ is the standard normal probability density function.

After we obtain η, we compute α by using (5.5.17).

Similarly, we can derive the joint conditional density for $(\boldsymbol{w}, \boldsymbol{u})$, which takes the form

$$\pi(\boldsymbol{w}, \boldsymbol{u}|\alpha, \boldsymbol{\beta}, \boldsymbol{\psi}, D_{obs}) \propto \prod_{i=1}^{n} \left[w_i^{\nu_i} \exp\{-w_i \sum_{k=1}^{2} \exp(\boldsymbol{x}_i'\boldsymbol{\beta}_k)(1 - S_k(y_{ki}|\boldsymbol{\psi}_k))\} \times w_i^{-(a+1)} s(u_i) \exp\left\{ -\frac{s(u_i)}{w_i^a} \right\} \right], \quad (5.5.18)$$

where $\nu_i = \nu_{1i} + \nu_{2i}$ and $a = \alpha/(1-\alpha)$. Now, we use the ratio-of-uniforms (ROU) method and a rejection algorithm (for example, see Devroye, 1986, pp. 40–65, 194–205) to draw (w_i, u_i) for $i = 1, 2, \ldots, n$. More specifically, the ROU algorithm for drawing w_i requires the following steps:

Step 1. Compute

$$a^* = \sup(\pi^*(w_i|u_i, \alpha, \boldsymbol{\beta}, \boldsymbol{\psi}, D_{obs}))^{1/2}$$

and

$$b^* = \sup w_i(\pi^*(w_i|u_i, \alpha, \boldsymbol{\beta}, \boldsymbol{\psi}, D_{obs}))^{1/2},$$

where

$$\pi^*(w_i|u_i, \alpha, \boldsymbol{\beta}, \boldsymbol{\psi}, D_{obs}) = w_i^{\nu_i} \exp\{-w_i \sum_{k=1}^{2} \exp(\boldsymbol{x}_i'\boldsymbol{\beta}_k)(1 - S_k(y_{ki}|\boldsymbol{\psi}_k))\} \times w_i^{-(a+1)} \exp\left\{ -\frac{s(u_i)}{w_i^a} \right\}. \quad (5.5.19)$$

Step 2. Draw ζ from $U(0, a^*)$ and ω from $U(0, b^*)$.

Step 3. Return $w_i = \zeta/\omega$ if $\zeta^2 \leq \pi^*(\zeta/\omega|u_i, \alpha, \boldsymbol{\beta}, \boldsymbol{\psi}, D_{obs})$; otherwise, go to Step 2.

The rejection algorithm for sampling u_i operates as follows:

Step 1. Independently generate u_i and v from $U(0, 1)$.

Step 2. Return u_i if $v \leq \frac{s(u_i)}{w_i^a} \exp\left\{ -\frac{s(u_i)}{w_i^a} \right\}$; otherwise, go to Step 1.

Finally, we draw $\boldsymbol{N}$ from $[\boldsymbol{N}|\boldsymbol{\beta}, \alpha, \boldsymbol{w}, \boldsymbol{u}, \boldsymbol{\psi}, D_{obs}]$. Since

$$N_{ki}|\boldsymbol{\beta}, \alpha, \boldsymbol{w}, \boldsymbol{u}, \boldsymbol{\psi}, D_{obs} \sim \mathcal{P}(w_i S_k(y_{ki}|\boldsymbol{\psi}_k) \exp(\boldsymbol{x}_i'\boldsymbol{\beta}_k)) + \nu_{ki}$$

for $k = 1, 2$ and $i = 1, 2, \ldots, n$, sampling $\boldsymbol{N}$ from its conditional posterior distribution is straightforward.

The introduction of latent variables indeed converts an intractable and nearly impossible computational problem (which involves direct sampling from the posterior with the likelihood based on the observed data given in (5.5.10)), into an attractive one, in which the parameters are sampled from a posterior based on the complete-data likelihood. Throughout the entire

MCMC implementational scheme, we use the collapsed Gibbs technique of Liu (1994). Thus, instead of sampling α directly from its conditional distribution $\pi(\alpha|\boldsymbol{w}, \boldsymbol{u}, \boldsymbol{\beta}, \boldsymbol{\psi}, D_{obs})$ as in Buckle (1995), we sample α from its marginal posterior distribution $\pi(\alpha|\boldsymbol{\beta}, \boldsymbol{\psi}, D_{obs})$. Similarly, we draw from $\pi(\boldsymbol{\beta}|\boldsymbol{w}, \boldsymbol{u}, \boldsymbol{\psi}, \alpha, D_{obs})$ instead of $\pi(\boldsymbol{\beta}|\boldsymbol{N}, \boldsymbol{w}, \boldsymbol{u}, \boldsymbol{\psi}, \alpha, D_{obs})$. By doing these two steps, we reduce the intra-correlations between α and $(\boldsymbol{w}, \boldsymbol{u})$, and $\boldsymbol{\beta}$ and $\boldsymbol{N}$, respectively. Therefore the convergence of the induced Markov chain is improved.

Example 5.4. Melanoma data. To illustrate the methodology, we consider the E1684 melanoma clinical trial discussed in Example 1.7. Our purpose in this example is to illustrate the multivariate cure rate model in (5.5.3) and demonstrate several of its properties. We emphasize here that the multivariate cure rate model discussed in this section is valid only when sufficient follow-up is available on all of the time-to-event endpoints and the calendar date of entry is assumed to be noninformative on the outcome variables.

TABLE 5.11. Summary of Covariates for a Subset of the E1684 Data.

Age (x_1) (years)		Gender (x_2) (frequency)		PS (x_3) (frequency)	
mean	46.663	Male	165	Fully Active	243
Std Dev	12.818	Female	109	Other	31

Source: Chen, Ibrahim, and Sinha (2001).

We consider the two failure time random variables, Y_1 = time to relapse from randomization and Y_2 = relapse to death. We note that all of the patients who died in this study had also relapsed. In this example, we use a subset of the E1684 data discussed in Example 1.2 by removing 10 cases, in which the time to relapse is the same as the time to death. Three covariates and an intercept are included in the model. The covariates are age (x_1), gender (x_2) (male, female), and performance status (x_3) (fully active, other). Performance status is abbreviated by PS in the tables below. A summary of survival and relapse times is given in Table 1.9. Table 5.11 gives statistical summaries for the covariates for this subset. A total of $n = 274$ observations are used in the analysis. In all of the computations, we standardized all the covariates to have mean 0 and standard deviation 1 in order to improve the convergence of the MCMC algorithm. Specifically, standardizing the covariates greatly reduces the correlation between the intercept term and the other regression coefficients. We use the noninformative improper prior in (5.5.11), with $\pi(\boldsymbol{\beta}) \propto 1$, $\lambda_k \sim N(0, 10{,}000)$, $\xi_k \sim \mathcal{G}(1, 0.01)$, and independent for each $k = 1, 2$. Also, we take a uniform prior for α on the interval $(0, 1)$. In this example, 50,000 MCMC iterations were used in all of the computations after a burn-in of 1000 iterations. Convergence was checked using the methods discussed in Cowles and Carlin (1996). Specifically, trace plots, autocorrelations, and Gelman-Rubin statistics (Gelman and Rubin, 1992) were computed, and convergence was observed to occur before 500 iterations.

TABLE 5.12. Posterior Estimates for A Subset of E1684 Data.

Parameter	Mean	Std Dev	95% HPD interval
β_{10}	0.234	0.116	(0.004, 0.459)
β_{11}	0.072	0.101	(−0.122, 0.274)
β_{12}	0.008	0.104	(−0.202, 0.206)
β_{13}	0.020	0.105	(−0.191, 0.222)
β_{20}	0.922	0.170	(0.585, 1.256)
β_{21}	0.147	0.116	(−0.077, 0.383)
β_{22}	−0.195	0.122	(−0.434, 0.041)
β_{23}	−0.199	0.121	(−0.435, 0.039)
α	0.709	0.066	(0.585, 0.840)
ξ_1	1.258	0.101	(1.064, 1.457)
ξ_2	1.496	0.124	(1.253, 1.737)
λ_1	−0.852	0.178	(−1.209, −0.516)
λ_2	−1.421	0.240	(−1.895, −0.966)

Source: Chen, Ibrahim, and Sinha (2001).

Table 5.12 gives posterior estimates of $\boldsymbol{\beta} = (\boldsymbol{\beta}_1', \boldsymbol{\beta}_2')$, $\boldsymbol{\xi} = (\xi_1, \xi_2)'$, $\boldsymbol{\lambda} = (\lambda_1, \lambda_2)'$, and α, where $\boldsymbol{\beta}_k = (\beta_{k0}, \beta_{k1}, \beta_{k2}, \beta_{k3})'$, $k = 1, 2$. We see from Table 5.12 that all of the highest posterior density (HPD) intervals for the regression coefficients of the covariates contain 0. Also, from Table 5.12, we see that the posterior mean of α is 0.709, with a 95% HPD interval of (0.585, 0.840). As discussed in Section 5.5.1, this indicates a moderate association between time to relapse and relapse to death for these data, as was expected. A plot of the marginal posterior distribution of α is given in Figure 5.8. We see in this figure that the posterior distribution of α appears quite symmetric with a mode at 0.699. Figure 5.9 shows a box-plot of the posterior means of the cure rates for each failure time variable. We note that when covariates are included in the model, each subject has an individual cure rate. From Figure 5.9, we see that the median cure rate for time to relapse (0.285) is much higher than the median cure rate for relapse to death (0.103). In general, there is much more variability in the estimated cure rates for the relapse to death variable. Table 5.13 gives numerical summaries for both box-plots of Figure 5.9. In Table 5.13, IQR denotes interquartile range.

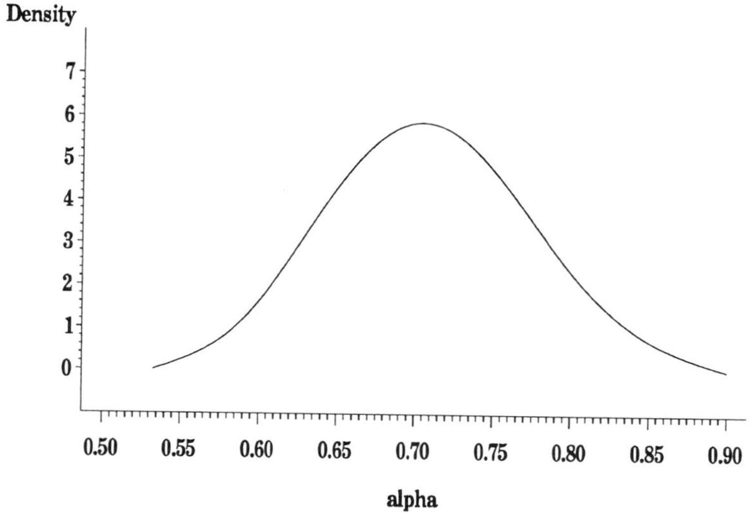

FIGURE 5.8. The marginal posterior density of α.

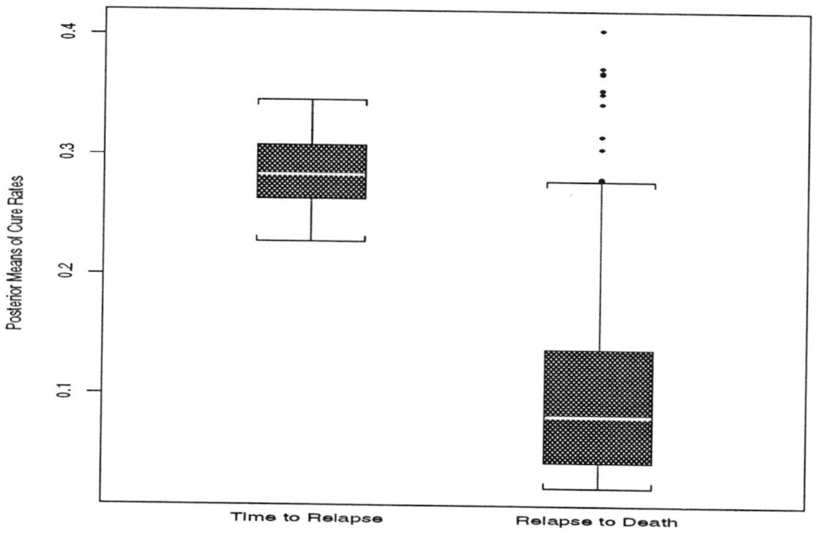

FIGURE 5.9. Box-plots of the posterior means of the cure rates for all patients.

TABLE 5.13. Summary of Box-Plots for A Subset of E1684 Data.

Failure time	mean	Std Dev	median	IQR	min	max
time to relapse	0.285	0.026	0.283	0.045	0.228	0.346
relapse to death	0.103	0.078	0.082	0.094	0.023	0.405

Source: Chen, Ibrahim, and Sinha (2001).

Figure 5.10 shows two superimposed plots, where plot (a) represents time to relapse and plot (b) represents relapse to death. The covariates are not used in constructing plots (a) and (b). In plot (a), the two superimposed plots correspond to the Kaplan-Meier estimate of survival, and the maximum likelihood estimate of the marginal survival function based on the multivariate cure rate model. We see that the two curves in plot (a) are nearly identical, and appear to plateau after approximately six years of follow-up. In plot (b), the relapse to death variable appears to plateau after approximately four years of follow-up. Figure 5.11 shows a three-dimensional plot of the posterior mean survival surface based on average age for males with fully active performance status. We see in this plot how the survival curve plateaus for each failure time variable. The joint survival function approaches a joint cure fraction, and the marginal survival functions each approach a cure fraction. From this figure, it is clear that the estimated cure rate for the time to relapse variable is larger than the estimated cure rate for the relapse to death variable.

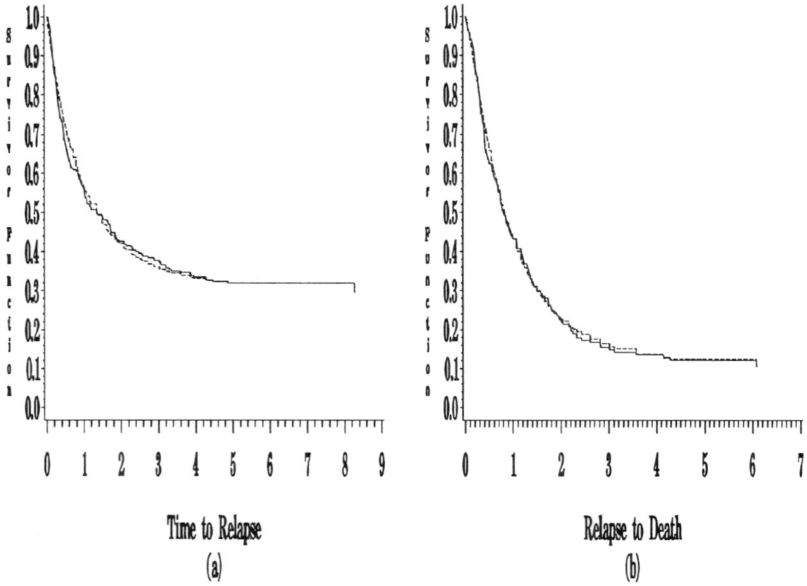

FIGURE 5.10. Superimposed survival curves for a subset of E1684.

Finally, we compared the individual fits of the univariate models for Y_1 and Y_2 to the marginal models induced by the bivariate cure rate model. In general, the 95% HPD intervals for the univariate model were all narrower than those based on the multivariate model. For example, for Y_1, the 95% HPD interval for β_1 was $(-0.083, 0.199)$, which is narrower than the 95% HPD interval for β_1 given in Table 5.12. In general, there is no general trend to these HPD intervals. The width of the HPD intervals for the multivariate model depend on the frailty distribution and the data, and therefore these intervals can be narrower or wider than intervals based on the corresponding univariate model. Therefore, in general, it becomes difficult to assess the efficiency in the multivariate model, since the width of the HPD intervals and posterior standard deviations heavily depend on the frailty distribution and the dataset at hand.

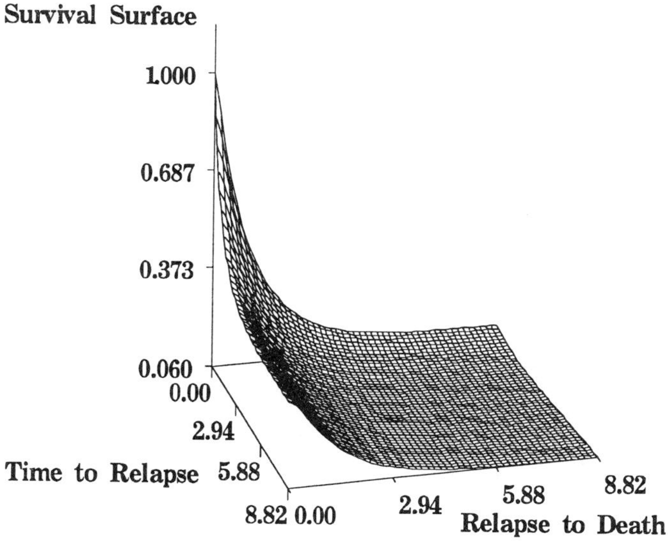

FIGURE 5.11. The bivariate posterior survival surface.

Appendix

Proof of Theorem 5.2.1. By summing out the unobserved latent variable $\boldsymbol{N}$, the complete-data likelihood given in (5.2.8) reduces to

$$\sum_{\boldsymbol{N}} L(\beta, \psi|D) = \prod_{i=1}^{n} \Big(\theta_i f(y_i|\psi)\Big)^{\nu_i} \exp\{-\theta_i(1 - S(y_i|\psi))\}. \qquad (5.A.1)$$

To prove Theorem 5.2.1, we first show that there exists a constant $M > 1$ such that

$$\Big(\theta_i f(y_i|\psi)\Big)^{\nu_i} \exp\{-\theta_i(1 - S(y_i|\psi))\} \le \alpha^{\nu_i} M. \tag{5.A.2}$$

When $\nu_i = 0$, (5.A.2) is obviously true since $\exp\{-\theta_i(1 - S(y_i|\psi))\} \le 1$. For $\nu_i = 1$, the left-hand side of (5.A.2) can be rewritten as

$$y_i^{-1} \frac{\alpha {y_i}^{\alpha} e^{\lambda} \exp(-e^{\lambda} y_i^{\alpha})}{1 - \exp(-e^{\lambda} y_i^{\alpha})} \Big[(1 - S(y_i|\psi))\theta_i \exp\Big\{ -\theta_i(1 - S(y_i|\psi))\Big\}\Big]. \tag{5.A.3}$$

Let

$$g_1(z) = \frac{ze^{-z}}{1 - e^{-z}}, \quad g_2(z) = ze^{-z} \text{ for } z > 0.$$

Then, it can be shown that there exists a common constant $g_0 > 0$ such that

$$g_1(z) \le g_0, \text{ and } g_2(z) \le g_0 \text{ for all } z > 0. \tag{5.A.4}$$

Using (5.A.4), (5.A.3) is less than or equal to $y_i^{-1}\alpha g_0^2$. Thus, taking $M^* = g_0^2 \max_{i:\ \nu_i=1} \{y_i^{-1}\}$ and $M = \max\{1, M^*\}$, we obtain (5.A.2).

Since X^* is of full rank, there must exist k linearly independent row vectors $\boldsymbol{x}'_{i_1}, \boldsymbol{x}'_{i_2}, \ldots, \boldsymbol{x}'_{i_k}$ such that $\nu_{i_1} = \nu_{i_2} = \cdots = \nu_{i_k} = 1$. Using (5.A.1) and (5.A.2),

$$\begin{aligned} &\int_{-\infty}^{\infty}\int_0^{\infty}\int_{R^k} \sum_{\boldsymbol{N}} L(\boldsymbol{\beta}, \psi|D)\pi(\alpha|\nu_0, \tau_0)\pi(\lambda)\, d\boldsymbol{\beta}\, d\alpha\, d\lambda \\ &\le \int_{-\infty}^{\infty}\int_0^{\infty}\int_{R^k} (\alpha M)^{d-k} \prod_{j=1}^{k} \Bigg[f(y_{i_j}|\psi) \exp\Big\{\boldsymbol{x}'_{i_j}\boldsymbol{\beta} \\ &\quad - (1 - S(y_{i_j}|\psi)) \exp(\boldsymbol{x}'_{i_j}\boldsymbol{\beta})\Big\}\Bigg] \pi(\alpha|\nu_0, \tau_0)\pi(\lambda)\, d\boldsymbol{\beta}\, d\alpha\, d\lambda, \end{aligned} \tag{5.A.5}$$

where R^k denotes k dimensional Euclidean space. Now we make the transformation $u_j = \boldsymbol{x}'_{i_j}\boldsymbol{\beta}$ for $j = 1, 2, \ldots, k$. This is a one-to-one linear transformation from $\boldsymbol{\beta}$ to $\boldsymbol{u} = (u_1, u_2, \ldots, u_k)'$. Thus, (5.A.5) is proportional to

$$\int_{-\infty}^{\infty}\int_{0}^{\infty}\int_{R^k}\alpha^{d-k}\prod_{j=1}^{k} f(y_{i_j}|\psi)\exp\left\{u_j-(1-S(y_{i_j}|\psi))\exp(u_j)\right\}$$
$$\times\,\pi(\alpha|\nu_0,\tau_0)\pi(\lambda)\;du\;d\alpha\;d\lambda$$
$$=\int_{-\infty}^{\infty}\int_{0}^{\infty}\alpha^{d-k}\prod_{j=1}^{k}\left[f(y_{i_j}|\psi)\int_{-\infty}^{\infty}\exp\left\{u_j-(1-S(y_{i_j}|\psi))\exp(u_j)\right\}du_j\right]$$
$$\times\,\pi(\alpha|\nu_0,\tau_0)\pi(\lambda)\;d\alpha\;d\lambda. \qquad (5.A.6)$$

Integrating out u, (5.A.6) reduces to

$$\int_{-\infty}^{\infty}\int_{0}^{\infty}\alpha^{d-k}\left[\prod_{j=1}^{k}\frac{f(y_{i_j}|\psi)}{(1-S(y_{i_j}|\psi))}\right]\pi(\alpha|\nu_0,\tau_0)\pi(\lambda)\;d\alpha\;d\lambda. \qquad (5.A.7)$$

In (5.A.7), using (5.A.4), we have

$$\frac{f(y_{i_j}|\psi)}{(1-S(y_{i_j}|\psi))}=\frac{\alpha(y_{i_j})^{\alpha-1}e^{\lambda}\exp(-e^{\lambda}y_{i_j}{}^{\alpha})}{1-\exp(-e^{\lambda}y_{i_j}{}^{\alpha})}\le K_0\alpha,$$

where $K_0=g_0\max_{1\le j\le k}\left\{y_{i_j}^{-1}\right\}$. Thus, (5.A.7) is less than or equal to

$$\int_{-\infty}^{\infty}\int_{0}^{\infty}(\alpha)^{d-k}\prod_{j=1}^{k}\left[K_0\alpha\right]\pi(\alpha|\nu_0,\tau_0)\pi(\lambda)\;d\alpha\;d\lambda$$
$$=K_0^k\int_{-\infty}^{\infty}\int_{0}^{\infty}\alpha^{d}\pi(\alpha|\nu_0,\tau_0)\pi(\lambda)\;d\alpha\;d\lambda<\infty$$

by conditions (ii) and (iii). This completes the proof. □

Proof of Theorem 5.2.2. To prove Theorem 5.2.2, it suffices to show that

$$\int_{R^k} L_1(\boldsymbol{\beta},\boldsymbol{\psi}^*|D_{obs})d\boldsymbol{\beta}=\infty. \qquad (5.A.8)$$

Using (5.1.1) and (5.2.10), $L_1(\boldsymbol{\beta},\boldsymbol{\psi}^*|D_{obs})$ can be written as

$$L_1(\boldsymbol{\beta},\boldsymbol{\psi}^*|D_{obs})=\prod_{i=1}^{n}\Bigg\{\left[(1-G(\boldsymbol{x}_i'\boldsymbol{\beta}))f^*(y_i|\boldsymbol{\psi}^*)\right]^{\nu_i}$$
$$\times\left[G(\boldsymbol{x}_i'\boldsymbol{\beta})+(1-G(\boldsymbol{x}_i'\boldsymbol{\beta}))S^*(y_i|\boldsymbol{\psi}^*)\right]^{1-\nu_i}, \qquad (5.A.9)$$

where $f^*(t|\psi^*)$ is the density function corresponding to $S^*(t|\psi^*)$. From (5.A.9), it can be shown that

$$L_1(\boldsymbol{\beta}, \psi^*|D_{obs}) \geq \prod_{i=1}^{n}((1 - G(\boldsymbol{x}_i'\boldsymbol{\beta}))\,[f^*(y_i|\psi^*)]^{\nu_i}\,[S^*(y_i|\psi^*)]^{1-\nu_i}\,. \tag{5.A.10}$$

Thus, it is obvious that the integration of (5.A.10) over $\boldsymbol{\beta}$ equals ∞, since $L_1(\boldsymbol{\beta}, \psi^*|D_{obs})$ is equivalent to a binomial regression likelihood which has all the responses "failures." This proves the theorem. □

<u>**Proof of Theorem 5.2.3.**</u> Similar to (5.A.2), we have

$$\left(\theta_{0i} f(y_{0i}|\psi)\right)^{\nu_{0i}} \exp\{-\theta_{0i}(1 - S(y_{0i}|\psi))\} \leq \alpha^{\nu_{0i}} M_0,$$

where $M_0 > 1$ is a constant. Since X_0^* is of full rank, there must exist k linearly independent row vectors $\boldsymbol{x}_{0i_1}', \boldsymbol{x}_{0i_2}', \ldots, \boldsymbol{x}_{0i_k}'$ such that $\nu_{0i_1} = \nu_{0i_2} = \cdots = \nu_{0i_k} = 1$. Following the proof of Theorem 5.2.1, we have

$$\begin{aligned}
&\int_0^1 \int_{-\infty}^{\infty} \int_0^{\infty} \int_{R^k} \left[\sum_{\boldsymbol{N}_0} L(\boldsymbol{\beta}, \psi|D_0)\right]^{a_0} \pi_0(\alpha|\nu_0, \tau_0) \\
&\qquad \times \pi_0(\lambda) a_0^{\gamma_0 - 1}(1 - a_0)^{\lambda_0 - 1}\, d\boldsymbol{\beta}\, d\alpha\, d\lambda\, da_0 \\
&\leq \int_0^1 \int_{-\infty}^{\infty} \int_0^{\infty} \int_{R^k} (\alpha M_0)^{a_0(d_0 - k)} \prod_{j=1}^{k} \Bigg[(f(y_{0i_j}|\psi))^{a_0} \exp\Big\{a_0 \boldsymbol{x}_{0i_j}'\boldsymbol{\beta} \\
&\qquad - a_0(1 - S(y_{0i_j}|\psi)) \exp(\boldsymbol{x}_{0i_j}'\boldsymbol{\beta})\Big\}\Bigg] \pi_0(\alpha|\nu_0, \tau_0)\pi_0(\lambda) \\
&\qquad \times a_0^{\gamma_0 - 1}(1 - a_0)^{\lambda_0 - 1}\, d\boldsymbol{\beta}\, d\alpha\, d\lambda\, da_0.
\end{aligned} \tag{5.A.11}$$

Since $M_0 \geq 1$ and $0 < a_0 < 1$, $M_0^{a_0} \leq M_0$. Taking the transformation $u_{0j} = \boldsymbol{x}_{0i_j}'\boldsymbol{\beta}$ for $j = 1, 2, \ldots, k$ and ignoring the constant, (5.A.11) is less than or equal to

$$\int_0^1 \int_{-\infty}^{\infty} \int_0^{\infty} \int_{R^k} \alpha^{a_0(d_0-k)} \prod_{j=1}^{k} \Bigg[\left(f(y_{0i_j}|\psi)\right)^{a_0} \exp\Big\{ a_0 u_{0j} - a_0(1 - S(y_{0i_j}|\psi)) \exp(u_{0j}) \pi_0(\alpha|\nu_0, \tau_0) \Big\} \Bigg] \pi_0(\alpha|\nu_0, \tau_0)$$
$$\times \pi_0(\lambda) a_0^{\gamma_0 - 1}(1 - a_0)^{\lambda_0 - 1}\, d\boldsymbol{u}_0\, d\alpha\, d\lambda\, da_0$$
$$= \int_0^1 \int_{-\infty}^{\infty} \int_0^{\infty} \alpha^{a_0(d_0-k)} \prod_{j=1}^{k} \Bigg[\left(f(y_{i_j}|\psi)\right)^{a_0} \int_{-\infty}^{\infty} \exp\Big\{ a_0 u_{0j} - a_0(1 - S(y_{0i_j}|\psi)) \exp(u_{0j}) \Big\} du_{0j} \Bigg] \pi_0(\alpha|\nu_0, \tau_0)$$
$$\times \pi_0(\lambda) a_0^{\gamma_0 - 1}(1 - a_0)^{\lambda_0 - 1}\, d\alpha\, d\lambda\, da_0, \qquad (5.A.12)$$

where $\boldsymbol{u}_0 = (u_{01}, u_{02}, \ldots, u_{0k})'$. Integrating out $\boldsymbol{u}_0$, (5.A.12) reduces to

$$\int_0^1 \int_{-\infty}^{\infty} \int_0^{\infty} \alpha^{a_0(d_0-k)} \prod_{j=1}^{k} \left[\frac{f(y_{0i_j}|\psi)}{(1 - S(y_{0i_j}|\psi))} \right]^{a_0} \frac{\Gamma(a_0)}{a_0^{a_0}}$$
$$\times \pi_0(\alpha|\nu_0, \tau_0) \pi_0(\lambda) a_0^{\gamma_0 - 1}(1 - a_0)^{\lambda_0 - 1}\, d\alpha\, d\lambda\, da_0, \qquad (5.A.13)$$

where $\Gamma(\cdot)$ denotes the gamma function. Using (5.A.4), it can be shown that

$$\left[\frac{f(y_{0i_j}|\psi)}{(1 - S(y_{0i_j}|\psi))} \right]^{a_0} \leq K_1 \alpha^{a_0},$$

where K_1 is a positive constant. Since $0 < a_0 < 1$,

$$\frac{\Gamma(a_0)}{a_0^{a_0}} = \frac{a_0^{-1}\Gamma(a_0 + 1)}{a_0^{a_0}} \leq K_2 a_0^{-1},$$

where K_2 is a positive constant. Using the above two inequalities, (5.A.13) is less than or equal to

$$\int_0^1 \int_{-\infty}^{\infty} \int_0^{\infty} \alpha^{a_0(d_0-k)} \prod_{j=1}^{k} \left[K_1 \alpha^{a_0} K_2 a_0^{-1} \right] \pi_0(\alpha|\nu_0, \tau_0)$$
$$\times \pi_0(\lambda) a_0^{\gamma_0 - 1}(1 - a_0)^{\lambda_0 - 1}\, d\alpha\, d\lambda\, da_0$$

$$\leq (K_1 K_2)^k \int_0^1 \int_{-\infty}^{\infty} \int_0^{\infty} a_0^{-k}(1 + \alpha^{d_0}) \pi_0(\alpha|\nu_0, \tau_0) \pi_0(\lambda)$$
$$\times a_0^{\gamma_0 - 1}(1 - a_0)^{\lambda_0 - 1}\, d\alpha\, d\lambda\, da_0 < \infty$$

by conditions (ii)–(iv). This completes the proof. $\square$

Proof of Property 5.4.1. From the definition of $E(\lambda_j|\psi_0)$ in (5.4.1), we have

$$\begin{aligned}E(\lambda_j|\psi_0) =& \frac{H_0(s_j) - H_0(s_{j-1})}{s_j - s_{j-1}}\\ =& \frac{H_0\left(\frac{s_j+s_{j-1}}{2} + \frac{s_j-s_{j-1}}{2}\right) - H_0\left(\frac{s_j+s_{j-1}}{2} - \frac{s_j-s_{j-1}}{2}\right)}{s_j - s_{j-1}}.\end{aligned}$$

Now using the definition of a derivative of the function $H_0(.)$ and the assumption that $\frac{s_j+s_{j-1}}{2} \to t$, we have

$$\lim_{s_j - s_{j-1} \to 0} E(\lambda_j|\psi_0) = \frac{d}{dt}H_0(t) = h_0(t).$$

□

Proof of Property 5.4.2. Note that for $t \in (s_j, s_{j-1})$,

$$\begin{aligned}E[H^*(t)|\psi_0] =& \mu_j(t - s_{j-1}) + \sum_{g=1}^{j-1} \mu_g(s_g - s_{g-1})\\ =& \left(\frac{t - s_{j-1}}{s_j - s_{j-1}}\right)[H_0(s_j) - H_0(s_{j-1})] + H_0(s_{j-1}). \qquad (5.A.14)\end{aligned}$$

Note that the right-hand side of (5.A.14) equals $H_0(t)$ when $t \in \{s_1, s_2, \cdots, s_J\}$ and for other values of t it is only approximately equal in the sense that as $s_j - s_{j-1} \to 0$, the right-hand side of (5.A.14) approaches $H_0(t)$. Again,

$$\text{Var}[H^*(t|\boldsymbol{\lambda})|\psi_0] = (t - s_{j-1})^2 \mu_j \kappa^j + \sum_{g=1}^{j-1} \mu_g \kappa^g (s_g - s_{g-1}) \to 0$$

as $\kappa \to 0$. This shows that $H^*(t|\boldsymbol{\lambda}) \to H_0(t)$ in probability for $t \in \{s_1, \cdots, s_J\}$ and that implies

$$F^*(t|\boldsymbol{\lambda}) = 1 - \exp(-H^*(t|\boldsymbol{\lambda})) \to F_0(t|\psi_0)$$

in probability. So, given θ, the random survival function $S_p^*(t|\boldsymbol{\lambda}) = \exp[-\theta F^*(t|\boldsymbol{\lambda})]$ converges to $\exp[-\theta F_0(t|\psi_0)]$ in probability as $\kappa \to 0$. This convergence is exact for $t \in \{s_1, s_2, \cdots, s_J\}$ and approximate for other values of t.

□

Proof of Property 5.4.3. From Chen, Ibrahim, and Sinha (1999), it is known that $h_p^*(t|\boldsymbol{\lambda}) = \theta f^*(t|\boldsymbol{\lambda}) = \theta\lambda_j \exp(-\lambda_j t)$, where

$$h_p^*(t|\boldsymbol{\lambda}) = \lim_{\delta \downarrow 0} \frac{P(T \in (t, t+\delta)|T > t, \theta, F^*)}{\delta},$$

with the probability evaluated under the model in (5.3.2). For $t \in (s_{j-1}, s_j)$, the moment generating function $m_j(t)$ of λ_j, under the prior process on F^* given (ψ_0, κ), is given by

$$m_j(t) = E[\exp(-\lambda_j t)|\mu_j, \kappa] \;=\; \left(\frac{\mu_j}{\mu_j + \mu_j \kappa^j t}\right)^{\frac{\mu_j}{\kappa^j}}. \tag{5.A.15}$$

Using (5.A.15), we can show that $E[\lambda_j \exp(-\lambda_j t)|\mu_j, \kappa] = -m_j'(t)$ approaches $\mu_j \exp(-\mu_j t)$ as $\kappa \to 0$ and $\text{Var}[\lambda_j \exp(-\lambda_j t)|\mu_j, \kappa] = 0$. Then it follows that $h_p^*(t|\boldsymbol{\lambda}) = \theta f^*(t|\boldsymbol{\lambda}) = \theta\lambda_j \exp(-\lambda_j t)$ approaches $\theta\mu_j \exp(-\mu_j t) \simeq \theta f_0(t|\psi_0)$ in probability as $\kappa \to 0$. □

Exercises

5.1 Derive (5.2.5) and (5.2.6).

5.2 Derive (5.2.7).

5.3 Consider the E1684 and E1673 melanoma datasets discussed in Example 5.1. Write a Gibbs sampling program to reproduce the results of Table 5.4.

5.4 Derive (5.3.2).

5.5 Prove Theorem 5.3.1.

5.6 Prove Theorem 5.3.2.

5.7 Prove Theorem 5.4.1.

5.8 Use (4.4.1) to derive (5.5.6) for the multivariate cure rate model in (5.5.3).

5.9 Prove Theorem 5.5.1.

5.10 Prove Theorem 5.5.2.

5.11 For the multivariate cure rate model,

(a) Use (5.5.13) to show that the conditional posterior density $\pi(\boldsymbol{\beta}|\alpha, \boldsymbol{w}, \boldsymbol{u}, \boldsymbol{\gamma}, D_{obs})$ for $\boldsymbol{\beta}$ is given by (5.5.15).

(b) Show that $\pi(\boldsymbol{\beta}|\alpha, \boldsymbol{w}, \boldsymbol{u}, \boldsymbol{\gamma}, D_{obs})$ is log-concave in each component of $\boldsymbol{\beta}$.

5.12 Derive the conditional posterior density $\pi(\boldsymbol{w}, \boldsymbol{u}|\alpha, \boldsymbol{\beta}, \boldsymbol{\gamma}, D_{obs})$ given by (5.5.18).

5.13 Verify (5.A.1).

5.14 Show (5.A.4).

5.15 ACCELERATED FAILURE TIME MODEL
Suppose y follows an accelerated failure time model if

$$\log(y) = \boldsymbol{x}'\boldsymbol{\beta} + \epsilon,$$

where ϵ has some distribution (see Chapter 10 for more details).

(a) Show that when y follows a cure rate model and y also follows an accelerated failure time model, then the survival function can be written as

$$S(y|\boldsymbol{x}) = \exp\left\{-\theta F_0(\varphi(\boldsymbol{x})y)\right\} \tag{5.E.1}$$

for some $\theta > 0$ and a proper cdf $F_0(y)$.

(b) Find $S(y|\boldsymbol{x})$ when $F_0(y) = 1 - \exp(-\lambda_0 y)$ and $\varphi(\boldsymbol{x}) = \exp(\boldsymbol{x}'\boldsymbol{\beta})$.

(c) Consider the E1690 data discussed in Example 5.2, with no covariates. Suppose the population survival function is given by

$$S_1(y) = \exp\left\{-\theta(1 - \exp(-\lambda_0\gamma t))\right\}. \tag{5.E.2}$$

Note that $S_1(y)$ here corresponds to a accelerated failure time model.

(i) Derive the likelihood function of the parameters corresponding to (5.E.2).

(ii) Derive the conditional posteriors for all of the parameters.

(iii) Suppose that θ, λ_0, and γ have independent gamma prior distributions. Write a computer program to carry out Gibbs sampling from the joint posterior assuming noninformative gamma priors for all of the parameters.

(d) Repeat part (c) using

$$S_2(y) = \exp\left\{-\theta\gamma(1 - \exp(-\lambda_0 t))\right\}. \tag{5.E.3}$$

Note that $S_2(y)$ here corresponds to a cure rate model.

(e) Plot $E(S_1(y)|D)$ versus $\hat{S}_{1,KM}(y)$, where $\hat{S}_{1,KM}(y)$ denotes the Kaplan-Meier estimate of $S_1(y)$. Also, plot $E(S_2(y)|D)$ versus $\hat{S}_{2,KM}(y)$, where $\hat{S}_{2,KM}(y)$ denotes the Kaplan-Meier estimate of $S_2(y)$. For both models, comments on the plots, and on which model appears to fit better.

5.16 Consider the setup of Exercise 5.15 with the inclusion of the treatment covariate into the analysis, so that

$$S_1(y) = \exp\left\{-\theta(1 - \exp(-\lambda_0 t \exp(\boldsymbol{x}'\boldsymbol{\beta})))\right\}$$

and

$$S_2(y) = \exp\left\{-\theta \exp(\boldsymbol{x}'\boldsymbol{\beta})(1 - \exp(-\lambda_0 t))\right\}.$$

(a) Repeat parts (a)–(e) of Exercise 5.15.

(b) Plot $-\log(\hat{S}_{KM}(y))$ versus y for both treatment groups, and comment on the fit of the two models corresponding to $S_1(y)$ and $S_2(y)$.

5.17 Repeat Exercise 5.16 using a Weibull model for $F_0(y)$, and thus $F_0(y) = 1 - \exp(-\lambda_0 y^{\alpha_0})$, where α_0 has a noninformative gamma prior.

5.18 One way to interpret the cure rate model of (5.2.1) is via a discrete frailty survival model. Suppose we consider a conditional hazard $h(y|N) = h^*(y)N$, where N is the frailty term, and $N \sim \mathcal{P}(\theta)$.

(a) Show that the unconditional survival function is given by $S(y) = \exp\{-\theta F^*(y)\}$, where $F^*(y)$ is the cdf corresponding to $h^*(y)$.

(b) Also show that for the cure rate model of (5.2.1), $h_{pop}(y) = h^*(y)E(N|Y > y)$. We note that this result is similar to $h(y) = h_0(y)E(w|Y > y)$ for the frailty model.

(c) We revisit the random M-site model for multiple event time data discussed in the exercises of Chapter 4. Suppose that there are N tumor sites within a rat and $N \sim \mathcal{P}(\theta)$. Let $U_1, U_2, \ldots, U_N$ denote the time to tumor for these sites and assume that the U_i's are i.i.d. with cdf $F(y)$. Consider $U_{(1)}$, which is the time to first tumor appearance. Show that the survival function of $U_{(1)}$ is given by $S_u(y) = \exp(-\theta F(y))$, which is the same as in (5.2.1).

6
Model Comparison

Model comparison is a crucial part of any statistical analysis. Due to recent computational advances, sophisticated techniques for Bayesian model comparison in survival analysis are becoming increasingly popular. There has been a recent surge in the statistical literature on Bayesian methods for model comparison, including articles by George and McCulloch (1993), Madigan and Raftery (1994), Ibrahim and Laud (1994), Laud and Ibrahim (1995), Kass and Raftery (1995), Chib (1995), Chib and Greenberg (1998), Raftery, Madigan, and Volinsky (1995), George, McCulloch, and Tsay (1996), Raftery, Madigan, and Hoeting (1997), Gelfand and Ghosh (1998), Clyde (1999), and Chen, Ibrahim, and Yiannoutsos (1999). Articles focusing on Bayesian approaches to model comparison in the context of survival analysis include Madigan and Raftery (1994), Raftery, Madigan, and Volinsky (1995), Sinha, Chen, and Ghosh (1999), Ibrahim, Chen, and Sinha (2001b), Ibrahim and Chen (1998), Ibrahim, Chen, and MacEachern (1999), Sahu, Dey, Aslanidou, and Sinha (1997), Aslanidou, Dey, and Sinha (1998), Chen, Harrington, and Ibrahim (1999), and Ibrahim, Chen, and Sinha (2001a).

The scope of Bayesian model comparison is quite broad, and can be investigated via Bayes factors, model diagnostics, and goodness of fit measures. We discuss model diagnostics in Chapter 10. In many situations, one may want to compare several models which are not nested. Such comparisons are common in survival analysis, since, for example, we may want to compare a fully parametric model versus a semiparametric model, or a cure rate model versus a Cox model, and so forth. In this chapter, we discuss several methods for Bayesian model comparison, including Bayes factors and posterior model probabilities, the Bayesian Information Criterion (BIC), the Conditional Predictive Ordinate (CPO), and the L measure.

6.1 Posterior Model Probabilities

Perhaps the most common method of Bayesian model assessment is the computation of posterior model probabilities. The Bayesian approach to model selection is straightforward in principle. One quantifies the prior uncertainties via probabilities for each model under consideration, specifies a prior distribution for each of the parameters in each model, and then uses Bayes theorem to calculate posterior model probabilities. Let m denote a specific model in the model space $\mathcal{M}$, and let $\boldsymbol{\theta}^{(m)}$ denote the parameter vector associated with model m. Then, by Bayes theorem the posterior probability of model m is given by

$$p(m|D) = \frac{p(D|m)p(m)}{\sum_{m \in \mathcal{M}} p(D|m)p(m)}, \tag{6.1.1}$$

where D denotes the data,

$$p(D|m) = \int L(\boldsymbol{\theta}^{(m)}|D)\pi(\boldsymbol{\theta}^{(m)})\, d\boldsymbol{\theta}^{(m)}, \tag{6.1.2}$$

$L(\boldsymbol{\theta}^{(m)}|D)$ is the likelihood, and $p(m)$ denotes the prior probability of model m.

In Bayesian model selection, specifying meaningful prior distributions for the parameters in each model is a difficult task requiring contextual interpretations of a large number of parameters. A need arises then to look for some useful, automated specifications. Reference priors can be used in many situations to address this. In some cases, however, they lead to ambiguous posterior probabilities, and require problem-specific modifications such as those in Smith and Spiegelhalter (1980). Berger and Pericchi (1996) have proposed the Intrinsic Bayes Factor, which provides a generic solution to the ambiguity problem. However, reference priors exclude the use of any real prior information one may have. Even if one overcomes the problem of specifying priors for the parameters in the various models, there remains the question of choosing prior probabilities $p(m)$ for the models themselves. A uniform prior on the model space $\mathcal{M}$ may not be desirable in situations where the investigator has prior information on each subset model. To overcome difficulties in prior specification, power priors can be used to specify priors for $\boldsymbol{\theta}^{(m)}$ as well as in specifying $p(m)$ for all $m \in \mathcal{M}$. We now describe this in the context of Bayesian variable selection.

6.1.1 *Variable Selection in the Cox Model*

Variable selection is one of the most frequently encountered problems in statistical data analysis. In cancer or AIDS clinical trials, for example, one often wishes to assess the importance of certain prognostic factors such as treatment, age, gender, or race in predicting survival outcome. Most of the existing literature addresses variable selection using criterion-based methods such as the Akaike Information Criterion (AIC) (Akaike, 1973) or Bayesian Information Criterion (BIC) (Schwarz, 1978). As is well known, Bayesian variable selection is often difficult to carry out because of the challenge in

(i) specifying prior distributions for the regression parameters for all possible models in $\mathcal{M}$;

(ii) specifying a prior distribution on the model space; and

(iii) computations.

In this subsection, we demonstrate how to make the variable selection problem easier in survival analysis through semiautomatic specifications of the prior distributions and the development of efficient computational algorithms. Let p denote the number of covariates for the full model and let $\mathcal{M}$ denote the model space. We enumerate the models in $\mathcal{M}$ by $m = 1, 2, \ldots, \mathcal{K}$, where $\mathcal{K}$ is the dimension of $\mathcal{M}$ and model $\mathcal{K}$ denotes the full model. Also, let $\boldsymbol{\beta}^{(\mathcal{K})} = (\beta_0, \beta_1, \ldots, \beta_{p-1})'$ denote the regression coefficients for the full model including an intercept, and let $\boldsymbol{\beta}^{(m)}$ denote a $p_m \times 1$ vector of regression coefficients for model m with an intercept, and a specific choice of $p_m - 1$ covariates. We write $\boldsymbol{\beta}^{(\mathcal{K})} = (\boldsymbol{\beta}^{(m)'}, \boldsymbol{\beta}^{(-m)'})'$, where $\boldsymbol{\beta}^{(-m)}$ is $\boldsymbol{\beta}^{(\mathcal{K})}$ with $\boldsymbol{\beta}^{(m)}$ deleted. We now consider Bayesian variable selection for the Cox model described in Section 3.2.4 using the model in (3.2.14). Recall that this model is based on a discretized gamma process on the baseline hazard function with independent increments. Under model m, the likelihood can be written as

$$L(\boldsymbol{\beta}^{(m)}, \boldsymbol{\delta} | D^{(m)}) = \prod_{j=1}^{J} \left\{ \exp\left\{-\delta_j (a_j + b_j)\right\} \prod_{k \in \mathcal{D}_j} \left[1 - \exp\{-\eta_k^{(m)} T_j\}\right] \right\}, \tag{6.1.3}$$

where $\eta_k^{(m)} = \exp(\boldsymbol{x}_k^{(m)'} \boldsymbol{\beta}^{(m)})$, $\boldsymbol{x}_k^{(m)}$ is a $p_m \times 1$ vector of covariates for the ith individual under model m, $X^{(m)}$ denotes the $n \times p_m$ covariate matrix of rank p_m, and $D^{(m)} = (n, \boldsymbol{y}, X^{(m)}, \boldsymbol{\nu})$ denotes the data under model m. The rest of the terms in (6.1.3) are defined in (3.2.14), (3.2.15), and (3.2.16). We have written the model here assuming that $\boldsymbol{\delta} = (\delta_1, \delta_2, \ldots, \delta_J)'$ does not depend on m. This is reasonable here, since our primary goal is variable selection, that is, to determine the dimension of $\boldsymbol{\beta}^{(m)}$. In this light, $\boldsymbol{\delta}$ can be viewed as a nuisance parameter in the variable selection problem. A more general version of the model can be constructed by letting $\boldsymbol{\delta}$ depend on m.

Following (3.3.4), the power prior under model m can be written as

$$\pi(\boldsymbol{\beta}^{(m)}, \boldsymbol{\delta}, a_0 | D_0^{(m)}) \propto L(\boldsymbol{\beta}^{(m)}, \boldsymbol{\delta} | D_0^{(m)})^{a_0} \pi_0(\boldsymbol{\beta}^{(m)} | c_0) \times \pi_0(\boldsymbol{\delta} | \boldsymbol{\theta}_0) \pi(a_0 | \alpha_0, \lambda_0), \tag{6.1.4}$$

where $D_0^{(m)} = (n_0, \boldsymbol{y}_0, X_0^{(m)}, \boldsymbol{\nu}_0)$ is the historical data under model m,

$$\pi_0(\boldsymbol{\delta} | \boldsymbol{\theta}_0) \propto \prod_{j=1}^{J} \delta_j^{f_{0j}-1} \exp\left\{-\delta_j g_{0j}\right\},$$

$$\pi(a_0 | \alpha_0, \lambda_0) \propto a_0^{\alpha_0 - 1}(1 - a_0)^{\lambda_0 - 1},$$

and $\boldsymbol{\theta}_0 = (f_{01}, g_{01}, \ldots, f_{0M}, g_{0M})'$ and (α_0, λ_0) are prespecified hyperparameters. For the purposes of prior elicitation, it is easier to work with $\mu_0 = \alpha_0/(\alpha_0 + \lambda_0)$ and $\sigma_0^2 = \mu_0(1 - \mu_0)(\alpha_0 + \lambda_0 + 1)^{-1}$.

An attractive feature of the power prior for $\boldsymbol{\beta}^{(m)}$ in variable selection problems is that it is semiautomatic in the sense that one only needs a one time input of $(D_0^{(m)}, c_0, \boldsymbol{\theta}_0, \alpha_0, \lambda_0)$ to generate the prior distributions for all $m \in \mathcal{M}$.

In addition to the strategies mentioned in Section 3.3.2, choices of prior parameters for $\boldsymbol{\delta}$ can be made in several ways. One may take vague choices of prior parameters for the δ_j's such as $f_{0j} \propto g_{0j}(s_j - s_{j-1})$ and take g_{0j} small. This choice may be suitable if there is little prior information available on the baseline hazard rate. More informative choices for $\boldsymbol{\theta}_0$ can be made by incorporating the historical data $D_0^{(m)}$ into the elicitation process. A suitable choice of f_{0i} would be an increasing estimate of the baseline hazard rate. To construct such an estimate under model m, we can fit a Weibull model via maximum likelihood using $D_0^{(m)} = (n_0, \boldsymbol{y}_0, X_0^{(m)}, \boldsymbol{\nu}_0)$ as the data. Often, the fit will result in a strictly increasing hazard. We denote such a hazard by $h^*(s|D_0^{(m)})$. Thus, we can take $f_{0j} = b_{0j} h^*(s_j|D_0^{(m)})$, where $b_{0j} = s_j - s_{j-1}$. In the event that the fitted Weibull model results in a constant or decreasing hazard, doubt is cast on the appropriateness of the gamma process as a model for the hazard, and we do not recommend this elicitation method. There are numerous other approaches to selecting this baseline hazard. Alternative classes of parametric models may be fit to $D_0^{(m)}$ or a nonparametric method such as that of Padgett and Wei (1980) may be used to construct an increasing hazard.

6.1.2 *Prior Distribution on the Model Space*

Let the initial prior for the model space be denoted by $p_0(m)$. Given the historical data $D_0^{(m)}$, the prior probability of model m for the current study based on an update of y_0 via Bayes theorem is given by

$$p(m) \equiv p(m|D_0^{(m)}) = \frac{p(D_0^{(m)}|m)p_0(m)}{\sum_{m\in\mathcal{M}} p(D_0^{(m)}|m)p_0(m)}, \tag{6.1.5}$$

where

$$p(D_0|m) = \int L(\boldsymbol{\beta}^{(m)}, \boldsymbol{\delta}|D_0^{(m)})\pi_0(\boldsymbol{\beta}^{(m)}|d_0)\pi_0(\boldsymbol{\delta}|\boldsymbol{\kappa}_0)\; d\boldsymbol{\beta}^{(m)}\; d\boldsymbol{\delta}, \tag{6.1.6}$$

$L(\boldsymbol{\delta}, \boldsymbol{\beta}^{(m)}|D_0^{(m)})$ is the likelihood function of the parameters based on $D_0^{(m)}$, $\pi_0(\boldsymbol{\beta}^{(m)}|d_0)$ is the initial prior for $\boldsymbol{\beta}^{(m)}$ given in (6.1.4) with d_0 replacing c_0, and and $\pi_0(\boldsymbol{\delta}|\boldsymbol{\kappa}_0)$ is the initial prior for $\boldsymbol{\delta}$ in (3.3.2) with $\boldsymbol{\kappa}_0$ replacing $\boldsymbol{\theta}_0$. Following Section 3.3, we take $\pi_0(\boldsymbol{\beta}^{(m)}|d_0)$ to be a $N_{p_m}(0, d_0 W_0^{(m)})$, where $W_0^{(m)}$ is the submatrix of the diagonal matrix $W_0^{(\mathcal{K})}$ corresponding to model m. Large values of d_0 will tend to increase the prior probability for model m. Thus, the prior probability of model m for the current study is precisely the posterior probability of m given the historical data $D_0^{(m)}$, that is $p(m) \equiv p(m|D_0^{(m)})$. This choice for $p(m)$ has several additional nice interpretations. First, $p(m)$ corresponds to the usual Bayesian update of $p_0(m)$ using $D_0^{(m)}$ as the data. Second, as $d_0 \to 0$, $p(m)$ reduces to $p_0(m)$. Therefore, as $d_0 \to 0$, the historical data $D_0^{(m)}$ have a minimal impact in determining $p(m)$. On the other hand, as $d_0 \to \infty$, $\pi_0(\boldsymbol{\beta}^{(m)}|d_0)$ plays a minimal role in determining $p(m)$, and in this case, the historical data play a larger role in determining $p(m)$. The parameter d_0 thus serves as a tuning parameter to control the impact of $D_0^{(m)}$ on the prior model probability $p(m)$. It is important to note that we use a scalar parameter c_0 in constructing the power prior $\pi(\boldsymbol{\beta}^{(m)}, \boldsymbol{\delta}, a_0|D_0^{(m)})$ in (6.1.4), while we use a *different* scalar parameter d_0 in determining $p(m)$. This development provides us with great flexibility in specifying the prior distribution for $\boldsymbol{\beta}^{(m)}$ as well as the prior model probabilities $p(m)$. Finally, we note that when there is little information about the relative plausibility of the models at the initial stage, taking $p_0(m) = \frac{1}{\mathcal{K}}$, $m = 1, 2, \ldots, \mathcal{K}$, *a priori* is a reasonable "neutral" choice.

6.1.3 Computing Prior and Posterior Model Probabilities

To compute $p(m)$ in (6.1.5), we follow the Monte Carlo approach of Ibrahim and Chen (1998) to estimate all of the prior model probabilities using a single Gibbs sample from the full model. In the context of Bayesian variable selection for logistic regression, Chen, Shao, and Ibrahim (2000) use a similar idea to compute the prior model probabilities. This method involves computing the marginal distribution of the data via ratios of normalizing constants and it requires posterior samples *only* from the *full model* for computing the prior probabilities for all possible models. The method is thus very efficient for variable selection. The technical details of this method are given as follows.

The marginal density $p(D_0^{(m)}|m)$ corresponds precisely to the normalizing constant of the joint posterior density of $(\boldsymbol{\beta}^{(m)}, \boldsymbol{\delta})$. That is,

$$p(D_0^{(m)}|m) = \int L(\boldsymbol{\beta}^{(m)}, \boldsymbol{\delta}|D_0^{(m)})\pi_0(\boldsymbol{\beta}^{(m)}|d_0)\pi_0(\boldsymbol{\delta}|\boldsymbol{\kappa}_0)\, d\boldsymbol{\beta}^{(m)}\, d\boldsymbol{\delta}. \tag{6.1.7}$$

Suppose that under the full model, we have a sample $\{(\boldsymbol{\beta}_{0,l}^{(\mathcal{K})}, \boldsymbol{\delta}_{0,l}),\ l = 1, 2, \ldots, L\}$ from

$$\pi_0(\boldsymbol{\beta}^{(\mathcal{K})}, \boldsymbol{\delta}|D_0^{(\mathcal{K})}) \propto \pi_0^*(\boldsymbol{\beta}^{(\mathcal{K})}, \boldsymbol{\delta}|D_0^{(\mathcal{K})}), \tag{6.1.8}$$

where

$$\pi_0^*(\boldsymbol{\beta}^{(\mathcal{K})}, \boldsymbol{\delta}|D_0^{(\mathcal{K})}) = L(\boldsymbol{\beta}^{(m)}, \boldsymbol{\delta}|D_0^{(m)})\pi_0(\boldsymbol{\beta}^{(m)}|d_0)\pi_0(\boldsymbol{\delta}|\boldsymbol{\kappa}_0). \tag{6.1.9}$$

Following the Monte Carlo method of Chen and Shao (1997a), the prior probability of model m can be estimated by

$$\begin{aligned}\hat{p}(m) =& \hat{p}(m|D_0^{(m)}) \\ =& \frac{\frac{1}{L}\sum_{l=1}^{L} \frac{\pi_0^*(\boldsymbol{\beta}_{0,l}^{(m)}, \boldsymbol{\delta}_{0,l}|D_0^{(m)}) w_0(\boldsymbol{\beta}_{0,l}^{(-m)}|\boldsymbol{\beta}_{0,l}^{(m)})}{\pi_0^*(\boldsymbol{\beta}_{0,l}^{(\mathcal{K})}, \boldsymbol{\delta}_{0,l}|D_0^{(\mathcal{K})})} p_0(m)}{\frac{1}{L}\sum_{j=1}^{\mathcal{K}}\sum_{l=1}^{L} \frac{\pi_0^*(\boldsymbol{\beta}_{0,l}^{(j)}, \boldsymbol{\delta}_{0,l}|D_0^{(j)}) w_0(\boldsymbol{\beta}_{0,l}^{(-j)}|\boldsymbol{\beta}_{0,l}^{(j)})}{\pi_0^*(\boldsymbol{\beta}_{0,l}^{(\mathcal{K})}, \boldsymbol{\delta}_{0,l}|D_0^{(\mathcal{K})})} p_0(j)},\end{aligned} \tag{6.1.10}$$

where $\pi_0^*(\boldsymbol{\beta}^{(m)}, \boldsymbol{\delta}|D_0^{(m)})$ is given by (6.1.9) with m replacing $\mathcal{K}$, and $w_0(\boldsymbol{\beta}^{(-m)} \mid \boldsymbol{\beta}^{(m)})$ is a *completely* known conditional density whose support is contained in, or equal to, the support of the conditional density of $\boldsymbol{\beta}^{(-m)}$ given $\boldsymbol{\beta}^{(m)}$ with respect to the full model joint prior distribution (6.1.8). Note that the choice of the weight function $w_0(\boldsymbol{\beta}^{(-m)}|\boldsymbol{\beta}^{(m)})$ is somewhat arbitrary. However, in Chen and Shao (1997b), it is shown that the best choice of $w_0(\boldsymbol{\beta}^{(-m)}|\boldsymbol{\beta}^{(m)})$ is the conditional density of $\boldsymbol{\beta}^{(-m)}$ given $\boldsymbol{\beta}^{(m)}$ with respect to the joint prior density $\pi_0(\boldsymbol{\beta}^{(\mathcal{K})}, \boldsymbol{\delta}|D_0^{(\mathcal{K})})$ given by (6.1.8). Since a closed form expression of this conditional density is not available, we follow an empirical procedure given in Section 9.1 of Chen, Shao, and Ibrahim (2000) to select $w_0(\boldsymbol{\beta}^{(-m)}|\boldsymbol{\beta}^{(m)})$. Specifically, using the sample $\{(\boldsymbol{\beta}_{0,l}^{(\mathcal{K})}, \boldsymbol{\delta}_{0,l}),\ l = 1, 2, \ldots, L\}$, we construct the mean and covariance matrix of $\boldsymbol{\beta}^{(\mathcal{K})}$, denoted by $(\tilde{\boldsymbol{\beta}}_0, \tilde{\Sigma}_0)$, and then we choose $w_0(\boldsymbol{\beta}^{(-m)}|\boldsymbol{\beta}^{(m)})$ to be the conditional density of the p-dimensional normal distribution, $N_p(\tilde{\boldsymbol{\beta}}_0, \tilde{\Sigma}_0)$, for $\boldsymbol{\beta}^{(-m)}$ given $\boldsymbol{\beta}^{(m)}$. Thus, $w_0(\boldsymbol{\beta}^{(-m)}|\boldsymbol{\beta}^{(m)})$ is given by

$$w_0(\boldsymbol{\beta}^{(-m)}|\boldsymbol{\beta}^{(m)}) = (2\pi)^{-(p-p_m)/2}|\tilde{\Sigma}_{11.2m}|^{-1/2} \times \exp\{-\tfrac{1}{2}(\boldsymbol{\beta}^{(-m)} - \tilde{\boldsymbol{\mu}}_{11.2m})'\tilde{\Sigma}_{11.2m}^{-1}(\boldsymbol{\beta}^{(-m)} - \tilde{\boldsymbol{\mu}}_{11.2m})\}, \quad (6.1.11)$$

where

$$\tilde{\Sigma}_{11.2m} = \tilde{\Sigma}_{11m} - \tilde{\Sigma}_{12m}\tilde{\Sigma}_{22m}^{-1}\tilde{\Sigma}_{12m}',$$

$\tilde{\Sigma}_{11m}$ is the covariance matrix from the marginal distribution of $\boldsymbol{\beta}^{(-m)}$, $\tilde{\Sigma}_{12m}$ consists of the covariances between $\boldsymbol{\beta}^{(-m)}$ and $\boldsymbol{\beta}^{(m)}$, and $\tilde{\Sigma}_{22m}$ is the covariance matrix of the marginal distribution of $\boldsymbol{\beta}^{(m)}$ with respect to the joint normal distribution $N_p(\tilde{\boldsymbol{\beta}}_0, \tilde{\Sigma}_0)$ for $\boldsymbol{\beta}^{(\mathcal{K})}$. Also in (6.1.11),

$$\tilde{\boldsymbol{\mu}}_{11.2m} = \tilde{\boldsymbol{\mu}}^{(-m)} + \tilde{\Sigma}_{12m}\tilde{\Sigma}_{22m}^{-1}(\boldsymbol{\beta}^{(m)} - \tilde{\boldsymbol{\mu}}^{(m)}),$$

where $\tilde{\boldsymbol{\mu}}^{(-m)}$ is the mean of the normal marginal distribution of $\boldsymbol{\beta}^{(-m)}$ and $\tilde{\boldsymbol{\mu}}^{(m)}$ is the mean of the normal marginal distribution of $\boldsymbol{\beta}^{(m)}$. A nice feature of this procedure is that $w_0(\boldsymbol{\beta}^{(-m)}|\boldsymbol{\beta}^{(m)})$ is calculated in an automatic fashion.

There are several advantages of the above Monte Carlo procedure:

(i) we need only one random draw from $\pi_0(\boldsymbol{\beta}^{(\mathcal{K})}, \boldsymbol{\delta}|D_0^{(\mathcal{K})})$, which greatly eases the computational burden;

(ii) it is more numerically stable since we calculate ratios of the densities in (6.1.10); and

(iii) in (6.1.10), $\pi_0(\boldsymbol{\beta}^{(\mathcal{K})}, \boldsymbol{\delta}|D_0^{(\mathcal{K})})$ plays the role of a ratio importance sampling density (see Chen and Shao, 1997a) which needs to be known only up to a normalizing constant since this common constant cancels out in the calculation.

Now, we discuss how to compute the posterior model probabilities. Using Bayes' theorem, the posterior probability of model k can be written as

$$p(m|D^{(m)}) = \frac{p(D^{(m)}|m)p(m)}{\sum_{j\in\mathcal{M}} p(D^{(j)}|j)p(j)}, \quad (6.1.12)$$

where $p(m)$ is estimated by (6.1.10),

$$p(D^{(m)}|m) = \int L(\boldsymbol{\beta}^{(m)}, \boldsymbol{\delta}|D^{(m)})\pi(\boldsymbol{\beta}^{(m)}, \boldsymbol{\delta}, a_0|D_0^{(m)})\, d\boldsymbol{\beta}^{(m)}\, d\boldsymbol{\delta}\, da_0 \quad (6.1.13)$$

denotes the marginal distribution of the data $D^{(m)}$ for the current study under model m, $L(\boldsymbol{\beta}^{(m)}, \boldsymbol{\delta}|D^{(m)})$ is the likelihood function given by (6.1.3), and $\pi(\boldsymbol{\beta}^{(m)}, \boldsymbol{\delta}, a_0|D_0^{(m)})$ is the joint prior density defined in (6.1.4). The marginal density $p(D^{(m)}|m)$ is precisely the normalizing constant of the joint posterior density of $(\boldsymbol{\beta}^{(m)}, \boldsymbol{\delta}, a_0)$.

Computing the posterior model probability $p(m|D^{(m)})$ given by (6.1.12) requires a different Monte Carlo method other than the one for computing the prior model probability $p(m)$ given by (6.1.5). We give a brief explanation as follows. From (6.1.13), it can be seen that the calculation of posterior probabilities requires evaluating

$$\begin{aligned} p(D^{(m)}|m) &= \int L(\boldsymbol{\beta}^{(m)}, \boldsymbol{\delta}|D^{(m)})\pi(\boldsymbol{\beta}^{(m)}, \boldsymbol{\delta}, a_0|D_0^{(m)})\, d\boldsymbol{\beta}^{(m)}\, d\boldsymbol{\delta}\, da_0 \\ &= \int L(\boldsymbol{\beta}^{(m)}, \boldsymbol{\delta}|D^{(m)})\frac{\pi^*(\boldsymbol{\beta}^{(m)}, \boldsymbol{\delta}, a_0|D_0^{(m)})}{c_m}\, d\boldsymbol{\beta}^{(m)}\, d\boldsymbol{\delta}\, da_0, \end{aligned} \tag{6.1.14}$$

where the unnormalized joint prior density

$$\begin{aligned} \pi^*(\boldsymbol{\beta}^{(m)}, \boldsymbol{\delta}, a_0|D_0^{(m)}) &= L(\boldsymbol{\beta}^{(m)}, \boldsymbol{\delta}|D_0^{(m)})^{a_0}\pi_0(\boldsymbol{\beta}^{(m)}|c_0) \\ &\quad \times \pi_0(\boldsymbol{\delta}|\boldsymbol{\theta}_0)\pi(a_0|\alpha_0, \lambda_0), \end{aligned} \tag{6.1.15}$$

and the normalizing constant for the joint prior density is given by

$$c_m = \int \pi^*(\boldsymbol{\beta}^{(m)}, \boldsymbol{\delta}, a_0|D_0^{(m)})\, d\boldsymbol{\beta}^{(m)}\, d\boldsymbol{\delta}\, da_0. \tag{6.1.16}$$

Due to the complexity of (6.1.15), a closed form of c_m is not available. Therefore, computing $p(D^{(m)}|m)$ requires evaluating the ratio of two analytically intractable integrals, which is essentially a ratio of two normalizing constants. However, to compute similar quantities for the prior model probability $p(m)$ given by (6.1.5), we only need to evaluate

$$\int L(\boldsymbol{\beta}^{(m)}, \boldsymbol{\delta}|D_0^{(m)})\pi_0(\boldsymbol{\beta}^{(m)}|d_0)\pi_0(\boldsymbol{\delta}|\boldsymbol{\kappa}_0)\, d\boldsymbol{\beta}^{(m)}\, d\boldsymbol{\delta},$$

since closed forms of $\pi_0(\boldsymbol{\beta}^{(m)}|d_0)$ and $\pi_0(\boldsymbol{\delta}|\boldsymbol{\kappa}_0)$ are available.

To develop a more efficient Monte Carlo method to calculate posterior model probabilities, we first present a useful theoretical result. Let $\pi(\boldsymbol{\beta}^{(-m)}|D_0^{(\mathcal{K})})$ and $\pi(\boldsymbol{\beta}^{(-m)}|D^{(\mathcal{K})})$ denote the respective marginal prior and posterior densities of $\boldsymbol{\beta}^{(-m)}$ obtained from the full model. Then it can be shown that the posterior probability of model m is given by

$$p(m|D^{(m)}) = \frac{\frac{\pi(\boldsymbol{\beta}^{(-m)}=0|D^{(\mathcal{K})})}{\pi(\boldsymbol{\beta}^{(-m)}=0|D_0^{(\mathcal{K})})}p(m)}{\sum_{j=1}^{\mathcal{K}} \frac{\pi(\boldsymbol{\beta}^{(-j)}=0|D^{(\mathcal{K})})}{\pi(\boldsymbol{\beta}^{(-j)}=0|D_0^{(\mathcal{K})})}p(j)}, \tag{6.1.17}$$

where $\pi(\boldsymbol{\beta}^{(-m)} = 0|D_0^{(\mathcal{K})})$ and $\pi(\boldsymbol{\beta}^{(-m)} = 0|D^{(\mathcal{K})})$ are the marginal prior and posterior densities of $\boldsymbol{\beta}^{(-m)}$ evaluated at $\boldsymbol{\beta}^{(-m)} = 0$ for $m = 1, 2, \ldots, \mathcal{K}$. The proof of (6.1.17) directly follows from Theorem 9.1.1 of Chen, Shao, and Ibrahim (2000), and thus the details are omitted here for brevity. The result given in (6.1.17) is attractive since it shows that the posterior probability $p(m|D^{(m)})$ is simply a function of the prior model probabilities $p(m)$ and the marginal prior and posterior density functions of $\beta^{(-m)}$ for the full model evaluated at $\beta^{(-m)} = 0$. In (6.1.17), we use (6.1.10) to compute the prior model probability $p(m)$, and for notational convenience we let $\pi(\beta^{(-\mathcal{K})} = 0|D^{(\mathcal{K})}) = \pi(\beta^{(-\mathcal{K})} = 0|D_0^{(\mathcal{K})}) = 1$. Due to the complexity of the prior and posterior distributions, the analytical forms of $\pi(\beta^{(-m)}|D_0^{(\mathcal{K})})$ and $\pi(\beta^{(-m)}|D^{(\mathcal{K})})$ are not available. However, we can adopt the importance-weighted marginal posterior density estimation (IWMDE) method of Chen (1994) to estimate these marginal prior and posterior densities. The IWMDE method requires using only two respective Markov chain Monte Carlo (MCMC) samples from the prior and posterior distributions for the full model, making the computation of complicated posterior model probabilities feasible. It directly follows from the IWMDE method that a simulation consistent estimator of $\pi(\beta^{(-m)} = 0|D^{(\mathcal{K})})$ is given by

$$\begin{aligned}
&\hat{\pi}(\beta^{(-m)} = 0|D^{(\mathcal{K})}) \\
&\quad = \frac{1}{L}\sum_{l=1}^{L} w(\beta_l^{(-m)}|\beta_l^{(m)}) \frac{\pi(\beta_l^{(m)}, \beta^{(-m)} = 0, \delta_l, a_{0,l}|D^{(\mathcal{K})})}{\pi(\beta_l^{(\mathcal{K})}, \delta_l, a_{0,l}|D^{(\mathcal{K})})},
\end{aligned}$$

where $w(\beta^{(-m)}|\beta^{(m)})$ is a completely known conditional density of $\beta^{(-m)}$ given $\beta^{(m)}$, whose support is contained in, or equal to, the support of the conditional density of $\beta^{(-m)}$ given $\beta^{(m)}$ with respect to the full model joint posterior distribution, $\{(\beta_l^{(\mathcal{K})}, \delta_l, a_{0,l}), l = 1, 2, \ldots, L\}$ is a sample from the joint posterior distribution $\pi(\beta^{(\mathcal{K})}, \delta, a_0|D^{(\mathcal{K})})$. To construct a good $w(\beta^{(-m)}|\beta^{(m)})$, we can use a procedure similar to the one used to construct $w_0(\beta^{(-m)}|\beta^{(m)})$ in (6.1.10) for calculating the prior model probabilities. Similarly, we can obtain $\hat{\pi}(\beta^{(-m)} = 0|D_0^{(\mathcal{K})})$, an estimate of $\pi(\beta^{(-m)} = 0|D_0^{(\mathcal{K})})$, using a sample from the joint prior distribution $\pi(\beta^{(\mathcal{K})}, \delta, a_0|D_0^{(\mathcal{K})})$. We use an example given in Ibrahim and Chen (1998) to illustrate the above Bayesian variable selection method.

Example 6.1. Multiple myeloma data. We examine in more detail the multiple myeloma study E2479 discussed in Example 1.1. Our main goal in this example is to illustrate the prior elicitation and variable selection techniques discussed in this section. We also examine the sensitivity of the posterior probabilities to the choices of (μ_0, σ_0^2), c_0, and d_0. We have two similar studies, Study 1, and the current study, Study 2. Initially, there were nine covariates for both studies. However, one covariate, proteinuria, had nearly 50% missing values and therefore was deleted from the analysis. Thus, a total of $n = 339$ observations were available from Study 2, with 8 observations being right censored. Our analysis used $p = 8$ covariates.

These are blood urea nitrogen (x_1), hemoglobin (x_2), platelet count (x_3) (1 if normal, 0 if abnormal), age (x_4), white blood cell count (x_5), bone fractures (x_6), percentage of the plasma cells in bone marrow (x_7), and serum calcium (x_8). To ease the computational burden, we standardized all of the variables. In fact, the standardization helped the numerical stability in the implementation of the adaptive rejection algorithm (Gilks and Wild, 1992) for sampling the regression coefficients from the posterior distribution. Study 1 consisted of $n_0 = 65$ observations of which 17 were right censored. We use the information in Study 1 as historical data for Study 2. The dataset for Study 1 can be found in Krall, Uthoff, and Harley (1975). A plot of the estimated baseline hazards for the two studies shows that the increasing baseline hazard rate assumption is plausible.

We conduct sensitivity analyses with respect to (i) c_0, (ii) d_0, and (iii) (μ_0, σ_0^2). These are shown in Tables 6.1, 6.2, and 6.3, respectively. To compute the prior and posterior model probabilities, 50,000 Gibbs iterations were used to get convergence. We use $J = 28$, with the intervals chosen so that with the combined datasets from the historical and current data, at least one failure or censored observation falls in each interval. This technique for choosing J is reasonable and preserves the consistency in the interpretation of $\boldsymbol{\delta}$ for the two studies. We take $W_0^{(\mathcal{K})}$ to be the diagonal elements of the inverse of the Fisher information matrix T_0 in (3.3.9) of $\boldsymbol{\beta}^{(\mathcal{K})}$ based on the Cox's partial likelihood, where $\mathcal{K} = 2^8$ in this example. In addition, we take a uniform initial prior on the model space, that is, $p_0(m) = \frac{1}{\mathcal{K}}$ for $m = 1, 2, \ldots, \mathcal{K}$, $\boldsymbol{\theta}_0 = \boldsymbol{\kappa}_0$ and use $f_{0j} = s_j - s_{j-1}$ if $s_j - s_{j-1} \geq 1$ and $f_{0j} = 1.1$ if $s_j - s_{j-1} < 1$, and $g_{0j} = 0.001$. For the last interval, we take $g_{0j} = 10$ for $j = J$ since very little information in the data is available for this last interval. The above choices of f_{0j} and g_{0j} ensure the log-concavity of $\pi_0(\boldsymbol{\delta}|\boldsymbol{\theta}_0)$, as this is required in sampling $\boldsymbol{\delta}$ from its conditional prior and posterior distributions. A stepwise variable selection procedure in SAS for the current study yields $(x_2, x_3, x_4, x_7, x_8)$ as the top model.

TABLE 6.1. The Posterior Model Probabilities for $(\mu_0, \sigma_0^2) = (0.5, 0.004)$, $d_0 = 3$ and Various Choices of c_0.

c_0	m	$p(m)$	$p(D\|m)$	$p(m\|D)$
3	$(x_1, x_2, x_3, x_4, x_5, x_7, x_8)$	0.015	0.436	0.769
10	$(x_1, x_2, x_3, x_4, x_5, x_7, x_8)$	0.015	0.310	0.679
30	$(x_1, x_2, x_3, x_4, x_5, x_7, x_8)$	0.015	0.275	0.657

Source: Ibrahim and Chen (1998).

Table 6.1 gives the model with the largest posterior probability using $(\mu_0, \sigma_0^2) = (0.5, 0.004)$, (i.e., $\alpha_0 = \lambda_0 = 30$) for several values of c_0. For each value of c_0 in Table 6.1, the model $(x_1, x_2, x_3, x_4, x_5, x_7, x_8)$ obtains the largest posterior probability, and thus model choice is not sensitive to these values. In addition, for $d_0 = 3$ and for any $c_0 \geq 3$, the $(x_1, x_2, x_3, x_4, x_5, x_7, x_8)$ model obtains the largest posterior probability. Although not shown in Table 6.1, values of $c_0 < 3$ do not yield $(x_1, x_2, x_3, x_4, x_5, x_7, x_8)$ as the top model. Thus, model choice may become sensitive to the choice of c_0 when $c_0 < 3$.

From Table 6.2, we see how the prior model probability is affected as d_0 is changed. In each case, the true model obtains the largest posterior probability. Under the settings of Table 6.2, the $(x_1, x_2, x_3, x_4, x_5, x_7, x_8)$ model obtains the largest prior probability when $d_0 \geq 3$. With values of $d_0 < 3$, however, model choice may be sensitive to the choice of d_0. For example, when $d_0 = 0.0001$ and $c_0 = 10$, the top model is $(x_1, x_2, x_4, x_5, x_7, x_8)$ with posterior probability of 0.42 and the second-best model is $(x_1, x_2, x_3, x_4, x_5, x_7, x_8)$ with posterior probability of 0.31. Finally, we mention that as both c_0 and d_0 become large, the $(x_1, x_2, x_3, x_4, x_5, x_7, x_8)$ model obtains the largest posterior model probability.

TABLE 6.2. The Posterior Model Probabilities for $(\mu_0, \sigma_0^2) = (0.5, 0.004)$, $c_0 = 3$ and Various Choices of d_0.

d_0	m	$p(m)$	$p(D\|m)$	$p(m\|D)$
5	$(x_1, x_2, x_3, x_4, x_5, x_7, x_8)$	0.011	0.436	0.750
10	$(x_1, x_2, x_3, x_4, x_5, x_7, x_8)$	0.005	0.436	0.694
30	$(x_1, x_2, x_3, x_4, x_5, x_7, x_8)$	0.001	0.436	0.540

Source: Ibrahim and Chen (1998).

In addition, Tables 6.1 and 6.2 indicate a monotonic decrease in the posterior probability of model $(x_1, x_2, x_3, x_4, x_5, x_7, x_8)$ as c_0 and d_0 are increased. This indicates that there is a moderate impact of the historical data on model choice.

Table 6.3 shows a sensitivity analysis with respect to (μ_0, σ_0^2). Under these settings, model choice is not sensitive to the choice of (μ_0, σ_0^2). We see that in each case, $(x_1, x_2, x_3, x_4, x_5, x_7, x_8)$ obtains the largest posterior probability. In addition, there is a monotonic increase in the posterior model probability as more weight is given to the historical data.

Although not shown here, several different partitioning schemes for the choices of the intervals $(s_{j-1}, s_j]$ were considered following a procedure similar to Ibrahim, Chen, and MacEachern (1999). The prior and posterior

TABLE 6.3. The Posterior Model Probabilities for $c_0 = 10$, $d_0 = 10$ and Various Choices of (μ_0, σ_0^2).

(μ_0, σ_0^2)	m	$p(m)$	$p(D\|m)$	$p(m\|D)$
(0.5, 0.008)	$(x_1, x_2, x_3, x_4, x_5, x_7, x_8)$	0.005	0.274	0.504
(0.5, 0.004)	$(x_1, x_2, x_3, x_4, x_5, x_7, x_8)$	0.005	0.310	0.558
$(0.98, 3.7 \times 10^{-4})$	$(x_1, x_2, x_3, x_4, x_5, x_7, x_8)$	0.005	0.321	0.572

Source: Ibrahim and Chen (1998).

model probabilities do not appear to be too sensitive to the choices of the intervals $(s_{j-1}, s_j]$. This is a comforting feature of this approach since it allows the investigator some flexibility in choosing these intervals. As indicated by Ibrahim, Chen, and MacEachern (1999), sensitivity analyses on (f_{0j}, g_{0j}) show that as long as the shape parameter f_{0j} remains fixed, the results are not too sensitive to changes in the scale parameter g_{0j}. However, when we fix the scale parameter and vary the shape parameter, the results can be sensitive.

6.2 Criterion-Based Methods

Bayesian methods for model comparison usually rely on posterior model probabilities or Bayes factors, and it is well known that to use these methods, proper prior distributions are needed when the number of parameters in the two competing models are different. In addition, posterior model probabilities are generally sensitive to the choices of prior parameters, and thus one cannot simply select vague proper priors to get around the elicitation issue. Alternatively, criterion-based methods can be attractive in the sense that they do not require proper prior distributions in general, and thus have an advantage over posterior model probabilities in this sense. However, posterior model probabilities are intrinsically well calibrated since probabilities are relatively easy to interpret, whereas criterion-based methods are generally not easy to calibrate or interpret. Thus, one potential criticism of criterion-based methods for model comparison is that they generally do not have well-defined calibrations.

Recently, Ibrahim, Chen, and Sinha (2001b) propose a Bayesian criterion called the *L measure*, for model assessment and model comparison, and propose a calibration for it. The L measure can be used as a general model assessment tool for comparing models and assessing goodness of fit for a particular model, and thus in this sense, the criterion is potentially quite versatile. To facilitate the formal comparison of several models, Ibrahim, Chen, and Sinha (2001b) also propose a calibration for the L measure by deriving the marginal prior predictive density of the difference between the L measures of the candidate model and the true model. This calibrating marginal density is called the *calibration distribution.* Since, in practice, the true model will not be known, we use the criterion minimizing model in place of the true model, and derive the calibration distribution based on the criterion minimizing model. Thus an L measure statistic and its corresponding calibration distribution are computed for each candidate model.

In the next two subsections, we present the general formulation of the L measure criterion and its calibration, discuss MCMC strategies for computing it, and illustrate the method using an example involving interval censored data.

6.2.1 The L Measure

Consider an experiment that yields the data $\boldsymbol{y} = (y_1, y_2, \ldots, y_n)'$. Denote the joint sampling density of the y_i's by $f(\boldsymbol{y}|\boldsymbol{\theta})$, where $\boldsymbol{\theta}$ is a vector of indexing parameters. We allow the y_i's to be fully observed, right censored, or interval censored. In the right censored case, y_i may be a failure time or a censoring time. In the interval censored case, we only observe the interval $[a_{l_i}, a_{r_i}]$ in which y_i occurred. Let $\boldsymbol{z} = (z_1, z_2, \ldots, z_n)'$ denote future values of a replicate experiment. That is, $\boldsymbol{z}$ is a future response vector with the same sampling density as $\boldsymbol{y}|\boldsymbol{\theta}$. The idea of using a future response vector $\boldsymbol{z}$ in developing a criterion for assessing a model or comparing several models has been well motivated in the literature by Geisser (1993) and the many references therein, Ibrahim and Laud (1994), Laud and Ibrahim (1995), and Gelfand and Ghosh (1998).

Let $\eta(\cdot)$ be a known function, and let $y_i^* = \eta(y_i)$, $z_i^* = \eta(z_i)$, $\boldsymbol{y}^* = (y_1^*, y_2^*, \ldots, y_n^*)'$, and $\boldsymbol{z}^* = (z_1^*, z_2^*, \ldots, z_n^*)'$. For example, in survival analysis, it is common to take the logarithms of the survival times, and thus in this case $\eta(y_i) = \log(y_i) = y_i^*$. Also, $\eta(y_i) = \log(y_i)$ is a common transformation in Poisson regression. It is also common to take $\eta(\cdot)$ to be the identity function (i.e., $\eta(y_i) = y_i$), as in normal linear regression or logistic regression, so that in this case, $y_i^* = y_i$ and $z_i^* = z_i$.

We modify the general formulation of Gelfand and Ghosh (1998) to develop the L measure. For a given model, we first define the statistic

$$L_1(\boldsymbol{y}^*, \boldsymbol{b}) = E\left[(\boldsymbol{z}^* - \boldsymbol{b})'(\boldsymbol{z}^* - \boldsymbol{b})\right] + \delta(\boldsymbol{y}^* - \boldsymbol{b})'(\boldsymbol{y}^* - \boldsymbol{b}), \quad (6.2.1)$$

where the expectation is taken with respect to the posterior predictive distribution of $\boldsymbol{z}^*|\boldsymbol{y}^*$. The posterior predictive density of $\boldsymbol{z}^*|\boldsymbol{y}^*$ is given by

$$\pi(\boldsymbol{z}^*|\boldsymbol{y}^*) = \int f(\boldsymbol{z}^*|\boldsymbol{\theta})\pi(\boldsymbol{\theta}|\boldsymbol{y}^*)\, d\boldsymbol{\theta}, \quad (6.2.2)$$

where $\boldsymbol{\theta}$ denotes the vector of indexing parameters, $f(\boldsymbol{z}^*|\boldsymbol{\theta})$ is the sampling distribution of the future vector $\boldsymbol{z}^*$, and $\pi(\boldsymbol{\theta}|\boldsymbol{y}^*)$ denotes the posterior distribution of $\boldsymbol{\theta}$. The statistic in (6.2.1) takes the form of a weighted discrepancy measure. The vector $\boldsymbol{b} = (b_1, b_2, \ldots, b_n)'$ is an arbitrary location vector to be chosen and δ is a nonnegative scalar that weights the discrepancy based on the future values relative to the observed data. The general criterion in (6.2.1) is a special case of a class considered by Gelfand and Ghosh (1998), which are motivated from a Bayesian decision theoretic viewpoint. We refer the reader to their paper for a more general motivation and discussion. Setting $\boldsymbol{b} = \boldsymbol{y}^*$ in (6.2.1) yields the criterion of Ibrahim and Laud (1994).

In scalar notation, (6.2.1) can be written as

$$L_1(\boldsymbol{y}^*, \boldsymbol{b}) = \sum_{i=1}^{n} \{\text{Var}(z_i^*|\boldsymbol{y}^*) + (\mu_i - b_i)^2 + \delta(y_i^* - b_i)^2\}, \tag{6.2.3}$$

where $\mu_i = E(z_i^*|\boldsymbol{y}^*)$. Thus we see that (6.2.3) has the appealing decomposition as a sum involving the predictive variances plus two squared "bias" terms, $(\mu_i - b_i)^2$ and $\delta(y_i^* - b_i)^2$, where δ is a weight for the second bias component.

We follow Gelfand and Ghosh (1998) by selecting $\boldsymbol{b}$ as the minimizer of (6.2.3). Gelfand and Ghosh (1998) show that the $\boldsymbol{b}$ which minimizes (6.2.3) is

$$\hat{\boldsymbol{b}} = (1 - \nu)\boldsymbol{\mu} + \nu\, \boldsymbol{y}^*, \tag{6.2.4}$$

where $\boldsymbol{\mu} = (\mu_1, \mu_2, \ldots, \mu_n)'$, $\nu = \delta/(\delta + 1)$, which upon substitution in (6.2.3) leads to the criterion

$$L_2(y^*) = \sum_{i=1}^{n} \text{Var}(z_i^*|y_i^*) + \nu \sum_{i=1}^{n} (\mu_i - y_i^*)^2. \tag{6.2.5}$$

Clearly, $0 \le \nu < 1$, where $\nu = 0$ if $\delta = 0$, and $\nu \to 1$ as $\delta \to \infty$. The quantity ν plays a major role in (6.2.5). It can be interpreted as a weight term in the squared bias component of (6.2.5), and appears to have a lot of potential impact on the ordering of the models, as well as characterizing the properties of the L measure and calibration distribution. Ibrahim and Laud (1994) use $\nu = 1$, and thus give equal weight to the squared bias and variance components. However, there is no theoretical justification for such a weight, and indeed, using $\nu = 1$ may not be desirable in certain situations. Allowing ν to vary between zero and one gives the user a great deal of flexibility in the tradeoff between bias and variance, and therefore results in values of ν that are more desirable than others. This raises the question of whether certain values of ν are "optimal" in some sense for model selection purposes. Ibrahim, Chen, and Sinha (2001b) address this optimality issue for the linear model, and theoretically show that certain values of ν yield highly desirable properties of the L measure and the calibration distribution compared to other values of ν. They demonstrate that the choice of ν has much potential influence on the properties of the L measure, calibration distribution, and model choice in general. Based on their theoretical exploration, $\nu = \frac{1}{2}$ is a desirable and justifiable choice for model selection

If $\boldsymbol{y}^*$ is fully observed, then (6.2.5) is straightforward to compute. However, if $\boldsymbol{y}^*$ contains right censored or interval censored observations, then (6.2.5) is computed by taking the expectation of these censored observations with respect to the posterior predictive distribution of the censored observations. Let $\boldsymbol{y}^* = (\boldsymbol{y}^*_{\text{obs}}, \boldsymbol{y}^*_{\text{cens}})$, where $\boldsymbol{y}^*_{\text{obs}}$ denotes the completely observed components of $\boldsymbol{y}^*$, and $\boldsymbol{y}^*_{\text{cens}}$ denotes the censored components. Here, we assume that $\boldsymbol{y}^*_{\text{cens}}$ is a random quantity and $\boldsymbol{a}_l < \boldsymbol{y}^*_{\text{cens}} < \boldsymbol{a}_r$, where $\boldsymbol{a}_l$ and $\boldsymbol{a}_r$ are known. For ease of exposition, we let $D = (n, \boldsymbol{y}^*_{\text{obs}}, \boldsymbol{a}_l, \boldsymbol{a}_r)$ denote the observed data. Then (6.2.5) is modified as

$$L(\boldsymbol{y}^*_{\text{obs}}) = E_{\boldsymbol{y}^*_{\text{cens}}|D}[1\{\boldsymbol{a}_l < \boldsymbol{y}^*_{\text{cens}} < \boldsymbol{a}_r\} L_2(\boldsymbol{y}^*)], \tag{6.2.6}$$

where $1\{\boldsymbol{a}_l < \boldsymbol{y}^*_{\text{cens}} < \boldsymbol{a}_r\}$ is a generic indicator function taking the value 1 if $\boldsymbol{a}_l < \boldsymbol{y}^*_{\text{cens}} < \boldsymbol{a}_r$ and 0 otherwise, and the expectation $E_{\boldsymbol{y}^*_{\text{cens}}|D}$ is taken with respect to the posterior predictive distribution $f(\boldsymbol{y}^*_{\text{cens}}|\boldsymbol{\theta})\pi(\boldsymbol{\theta}|D)$. Note that $\boldsymbol{a}_l < \boldsymbol{y}^*_{\text{cens}} < \boldsymbol{a}_r$ means that the double inequalities hold for each component of these vectors. If, for example, all n observations are censored, then the above notation means $a_{l_i} < y^*_{\text{cens},i} < a_{r_i}$, $i = 1, \ldots, n$, where $\boldsymbol{a}_l = (a_{l_1}, \ldots, a_{l_n})'$, $\boldsymbol{a}_r = (a_{r_1}, \ldots, a_{r_n})'$, and $\boldsymbol{y}^*_{\text{cens}} = (y^*_{\text{cens},1}, \ldots, y^*_{\text{cens},n})'$. We shall call (6.2.6) the L *measure*. Small values of the L measure imply a good model. Specifically, we can write (6.2.6) as

$$L(\boldsymbol{y}^*_{\text{obs}}) = \int \int_{\boldsymbol{a}_l}^{\boldsymbol{a}_r} L_2(\boldsymbol{y}^*) f(\boldsymbol{y}^*_{\text{cens}}|\boldsymbol{\theta})\pi(\boldsymbol{\theta}|D)\, d\boldsymbol{y}^*_{\text{cens}}\, d\boldsymbol{\theta}, \tag{6.2.7}$$

where $f(\boldsymbol{y}^*_{\text{cens}}|\boldsymbol{\theta})$ is the sampling density of $\boldsymbol{y}^*_{\text{cens}}$ and $\pi(\boldsymbol{\theta}|D)$ is the posterior density of $\boldsymbol{\theta}$ given the observed data D. If $\boldsymbol{y}^*$ has right censored observations, then $\boldsymbol{a}_r = \infty$, and $\boldsymbol{a}_l$ is a vector of censoring times. If $\boldsymbol{y}^*$ has interval censored observations, then $(\boldsymbol{a}_l, \boldsymbol{a}_r)$ is a sequence of finite interval censoring times. If $\boldsymbol{y}^*$ is fully observed, that is, $\boldsymbol{y}^*_{\text{obs}} = \boldsymbol{y}^*$, then (6.2.6) reduces to (6.2.5), and therefore $L(\boldsymbol{y}^*_{\text{obs}}) \equiv L_2(\boldsymbol{y}^*)$ in this case.

It can be shown that (6.2.5) can be expressed as a posterior expectation, so that

$$L_2(\boldsymbol{y}^*) = \sum_{i=1}^{n} \{E_{\boldsymbol{\theta}|D}(E[(z_i^*)^2|\boldsymbol{\theta}]) - \mu_i^2\} + \nu \sum_{i=1}^{n} (\mu_i - y_i^*)^2, \tag{6.2.8}$$

where $\mu_i = E_{\boldsymbol{\theta}|D}[E(z_i^*|\boldsymbol{\theta})]$, and the expectation $E_{\boldsymbol{\theta}|D}$ is taken with respect to the posterior distribution $\pi(\boldsymbol{\theta}|D)$. Thus (6.2.5) and (6.2.6) can be computed by sampling from the posterior distribution of $\boldsymbol{\theta}$ via MCMC methods. Once the posterior samples of $\boldsymbol{\theta}$ are obtained, (6.2.8) and (6.2.6) can be evaluated. More specifically, suppose that $\{\boldsymbol{\theta}_q,\ q = 1, 2, \ldots, Q\}$ is an MCMC sample from $\pi(\boldsymbol{\theta}|D)$ and $\{\boldsymbol{y}^*_{\text{cens},q},\ q = 1, 2, \ldots, Q\}$ is an MCMC sample from the truncated posterior predictive distribution $1\{\boldsymbol{a}_l < \boldsymbol{y}^*_{\text{cens}} < \boldsymbol{a}_r\} f(\boldsymbol{y}^*_{\text{cens}}|\boldsymbol{\theta})\pi(\boldsymbol{\theta}|D)$. Then an MC estimate of $L(\boldsymbol{y}^*_{\text{obs}})$ is given by

$$\hat{L}(\boldsymbol{y}^*_{obs}) = \sum_{i=1}^{n} \left\{ \frac{1}{Q} \sum_{q=1}^{Q} \left(E\left[(z_i^*)^2 | \boldsymbol{\theta}_q \right] \right) - \hat{\mu}_i^2 \right\} + \nu \left\{ \sum_{\{i:\ y_i^* \text{ observed}\}} (\hat{\mu}_i - y_i^*)^2 \right.$$
$$\left. + \frac{1}{Q} \sum_{q=1}^{Q} \left[\sum_{\{i:\ y_i^* \text{ censored}\}} (\hat{\mu}_i - y^*_{\text{cens},iq})^2 \right] \right\}, \tag{6.2.9}$$

where $\hat{\mu}_i = (1/Q) \sum_{q=1}^{Q} E(z_i^*|\boldsymbol{\theta}_q)$, and $y^*_{\text{cens},iq}$ is the ith component of $\boldsymbol{y}^*_{\text{cens},q}$. In the cases where $E[(z_i^*)^2|\boldsymbol{\theta}]$ and $E(z_i^*|\boldsymbol{\theta})$ are not analytically available, we need an MCMC sample $\{(\boldsymbol{z}_q^*, \boldsymbol{\theta}_q),\ q = 1, 2, \ldots, Q\}$ from the joint distribution $f(\boldsymbol{z}^*|\boldsymbol{\theta})\pi(\boldsymbol{\theta}|D)$. Then, in (6.2.9), we replace

$$\frac{1}{Q} \sum_{q=1}^{Q} (E[(z_i^*)^2 | \boldsymbol{\theta}_q]) \quad \text{and} \quad \frac{1}{Q} \sum_{q=1}^{Q} E(z_i^* | \boldsymbol{\theta}_q)$$

by

$$\frac{1}{Q} \sum_{r=1}^{Q} \left(z_{i,q}^* \right)^2 \quad \text{and} \quad \frac{1}{Q} \sum_{q=1}^{Q} z_{i,q}^*,$$

where $z^*_{i,q}$ is the ith component of $\boldsymbol{z}_q^*$. Thus, computing $L(\boldsymbol{y}^*_{\text{obs}})$ is relatively straightforward.

6.2.2 The Calibration Distribution

Ibrahim, Chen, and Sinha (2001b) propose a calibration for the L measure $L(\boldsymbol{y}^*_{\text{obs}})$. To motivate the calibration distribution, let c denote the candidate model under consideration, and let t denote the true model. Further, let $L_c(\boldsymbol{y}^*_{\text{obs}})$ denote the L measure for the candidate model c, and let $L_t(\boldsymbol{y}^*_{\text{obs}})$ denote the L measure for the true model t. Now consider the difference in L measures

$$D(\boldsymbol{y}^*_{\text{obs}}, \nu) \equiv L_c(\boldsymbol{y}^*_{\text{obs}}) - L_t(\boldsymbol{y}^*_{\text{obs}}). \tag{6.2.10}$$

The quantity in (6.2.10) is a random variable in $\boldsymbol{y}^*_{\text{obs}}$, and depends on ν. It measures the discrepancy in the L measure values between the candidate model and the true model. To calibrate the L measure, we construct the marginal distribution of $D(\boldsymbol{y}^*_{\text{obs}}, \nu)$, computed with respect to the prior predictive distribution of $\boldsymbol{y}^*_{\text{obs}}$ under the true model t, denoted by

$$p_t(\boldsymbol{y}^*_{\text{obs}}) = \int f_t(\boldsymbol{y}^*_{\text{obs}}|\boldsymbol{\theta})\pi_t(\boldsymbol{\theta})\ d\boldsymbol{\theta}. \tag{6.2.11}$$

Thus, the calibration distribution is defined as

$$p_{L_c} \equiv p(D(\boldsymbol{y}^*_{\text{obs}}, \nu)). \tag{6.2.12}$$

Thus p_{L_c} is the marginal distribution of $D(\boldsymbol{y}^*_{\text{obs}}, \nu)$, computed with respect to $p_t(\boldsymbol{y}^*_{\text{obs}})$. We refer to p_{L_c} as the *calibration distribution* for the candidate model c throughout. We see that p_{L_c} is a univariate distribution, and therefore easily tabulated and plotted. If the candidate model is "close" to the true model, then p_{L_c} should have a mean (or mode) that is close to zero, and much of its mass should be centered around this point. On the other hand, if the candidate model and true model are far apart, then p_{L_c} will have a mean (or mode) that is far from zero. One obvious advantage in having an entire distribution as the calibration is that one is able to make plots of it, and derive various summary statistics from it, such as its mean, mode, and HPD intervals. We see that p_{L_c} is computed for every candidate model c, where $c \neq t$, and therefore changes with every c. We also see from (6.2.12) that for p_{L_c} to be well defined, we need a proper prior distribution for $\boldsymbol{\theta}$. This definition of the calibration distribution in (6.2.12) is appealing since it avoids the potential problem of a double use of the data as discussed by Bayarri and Berger (1999).

The definition of p_{L_c} depends on the data only through $\boldsymbol{y}^*_{\text{obs}}$. When we have right censored data, $\boldsymbol{y}^*_{\text{obs}}$ consists of the observed failure times, and thus p_{L_c} in this case is a function only of the observed failure times. Therefore, its computation does not depend on any of the censoring times. In situations where all of the observations are censored, $\boldsymbol{y}^*_{\text{obs}}$ consists of the empty set. In this case, the definition of p_{L_c} in (6.2.12) must be slightly modified. For interval censored data, we "impute" each interval censored observation by sampling from the truncated prior predictive distribution, where the truncation is taken to be the endpoints of the interval censored observation. Thus, if y_i^* is interval censored in the interval $[a_{l_i}, a_{r_i}]$, then we impute y_i^* by replacing it with a sample of size 1 from

$$p_t(y_i^*) \propto \int f_t(y_i^*|\boldsymbol{\theta})\pi_t(\boldsymbol{\theta})\, d\boldsymbol{\theta}, \quad a_{l_i} < y_i^* < a_{r_i},$$

$i = 1, \ldots, n$. We denote the sampled value by $\tilde{y}_i^*$ for each interval censored observation, thus obtaining $\tilde{\boldsymbol{y}}^* = (\tilde{y}_1^*, \ldots, \tilde{y}_n^*)'$ for the n interval censored observations. We then treat $\tilde{\boldsymbol{y}}^*$ as $\boldsymbol{y}^*_{\text{obs}}$. That is, we set $\tilde{\boldsymbol{y}}^* = \boldsymbol{y}_{\text{obs}}$, and compute p_{L_c} using $\tilde{\boldsymbol{y}}^* = \boldsymbol{y}^*_{\text{obs}}$. Thus, in the interval censored case, $\tilde{\boldsymbol{y}}^*$ can be viewed as *pseudo-observations* needed to form $\boldsymbol{y}^*_{\text{obs}}$ in order to facilitate the computation of p_{L_c}. This is a reasonable technique for obtaining the calibration distribution p_{L_c} when all of the observations are interval censored, and produces good results as demonstrated in Example 6.2.

Once p_{L_c} is computed, several statistical summaries can be obtained from it to summarize the calibration. These include various HPD intervals and the mean of $D(\boldsymbol{y}^*_{\text{obs}}, \nu)$. The mean of the calibration distribution is denoted by

$$\mu_c(\nu) = E_t(D(\boldsymbol{y}^*_{\text{obs}}, \nu)), \tag{6.2.13}$$

where $E_t(\cdot)$ denotes the expectation with respect to the prior predictive distribution of the true model. This summary, $\mu_c(\nu)$, is attractive since it measures, on average, how close the centers are of the candidate and true models. If the candidate model is a good model, then $\mu_c(\nu)$ should be close to 0, whereas if the candidate model is far from the true model, then $\mu_c(\nu)$ should be far from 0. We note that $\mu_c(\nu)$ depends on the candidate model and therefore changes with every c. If $c = t$, then $\mu_c(\nu) = 0$ for all ν.

Since the true model t will not be known in practice, we use the criterion minimizing model $t_{\min}$ in place of t for computing (6.2.10). Thus, in practice, we compute

$$\hat{D}(\boldsymbol{y}^*_{\text{obs}}, \nu) = L_c(\boldsymbol{y}^*_{\text{obs}}) - L_{t_{\min}}(\boldsymbol{y}^*_{\text{obs}}), \tag{6.2.14}$$

and

$$\hat{p}_{L_c} = p(\hat{D}(\boldsymbol{y}^*_{\text{obs}}, \nu)), \tag{6.2.15}$$

where $\hat{p}_{L_c}$ is computed with respect to the prior predictive distribution of the criterion minimizing model. Also, $\mu_c(\nu)$ is estimated by $\hat{\mu}(\nu)$, where $\hat{\mu}(\nu) = E_{t_{\min}}[\hat{D}(\boldsymbol{y}^*_{\text{obs}}, \nu)]$.

Finally, we briefly describe how to compute the calibration distribution p_{L_c} via MCMC sampling. For illustrative purposes, we consider only the case where $\boldsymbol{y}^*_{\text{obs}}$ is *not empty* as the computation is even much simpler when $\boldsymbol{y}^*_{\text{obs}}$ is empty. From (6.2.12), it can be seen that computing the calibration distribution requires the following two steps: for a candidate model c,

(i) generate a pseudo-observation $\tilde{\boldsymbol{y}}^*$ from the prior predictive distribution $f_t(\boldsymbol{y}^*|\boldsymbol{\theta})\pi_t(\boldsymbol{\theta})$; and

(ii) set $\boldsymbol{y}^*_{\text{obs}} = \tilde{\boldsymbol{y}}^*$ and use the method described in Subsection 6.2.1 to obtain an MC estimate of $L(\boldsymbol{y}^*_{\text{obs}})$.

We repeat (i) and (ii) Q times to obtain an MCMC sample of $L_c(\boldsymbol{y}^*_{\text{obs}})$. Then we repeat (i) and (ii) Q times using the criterion minimizing model to obtain an MCMC sample of $L_{t_{\min}}(\boldsymbol{y}^*_{\text{obs}})$. Using these MCMC samples, we can compute the entire calibration distribution p_{L_c}, for example, by using the kernel method (see Silverman, 1986). We note that step (ii) may be computationally intensive. However, the entire computational procedure is quite straightforward.

Example 6.2. Breast cancer data. We consider the breast cancer data, given in Table 1.4, from Finkelstein and Wolfe (1985), which consists of a dataset of (case-2) interval censored data. In this dataset, 46 early breast cancer patients receiving only radiotherapy (covariate value $x = 0$) and 48 patients receiving radio chemotherapy ($x = 1$) were monitored for cosmetic changes through weekly clinic visits.

Sinha, Chen, and Ghosh (1999) consider a semiparametric Bayesian analysis of these data using three models based on a discretized version of the Cox model (Cox, 1972). Specifically, the hazard, $\lambda(y|x)$, is taken to be a piecewise constant function with $\lambda(y|x) = \lambda_k \theta_k^x$ for $y \in I_k$, where $\theta_k = e^{\beta_k}$, $I_k = (a_{k-1}, a_k]$ for $k = 1, 2, \ldots, J$, $0 = a_0 < a_1 < \cdots < a_J = \infty$, and J is the total number of grid intervals. The length of each grid can be taken to be sufficiently small to approximate any hazard function for all practical purposes.

In this example, we consider the following three models:

$\mathcal{M}_1$: (i) $\lambda_k \overset{\text{indep}}{\sim} \mathcal{G}(\eta_k, \gamma_k)$ for $k = 1, 2, \ldots, J$; and
(ii) $\beta_k = \beta$ for $k = 1, 2, \ldots, J$; $\beta \sim N(\beta_0, w_0^2)$.

$\mathcal{M}_2$: (i) λ_k's have the same prior as in model $\mathcal{M}_1$; and
(ii) $\beta_{k+1}|\beta_1, \beta_2, \ldots, \beta_k \sim N(\beta_k, w_k^2)$ for $k = 0, 1, \ldots, J-1$.

$\mathcal{M}_3$: (i) $\alpha_{k+1}|\alpha_1, \alpha_2, \ldots, \alpha_k \sim N(\alpha_k, v_k^2)$, where $\alpha_k = \log(\lambda_k)$ for $k = 0, 1, \ldots, J-1$;
(ii) Same as in $\mathcal{M}_2$.

A detailed description of the motivation for $\mathcal{M}_1$, $\mathcal{M}_2$, and $\mathcal{M}_3$ can be found in Sinha, Chen, and Ghosh (1999). Our purpose here is to compute the L measure and the calibration distribution for the three models using noninformative priors. The prior parameters for $\mathcal{M}_1$ are given by $\eta_k = 0.2$, $\gamma_k = 0.4$ for $k = 1, 2, \ldots, J$, $\beta_0 = 0$, and $w_0 = 2.0$. The prior parameters for $\mathcal{M}_2$ and $\mathcal{M}_3$ are given in Section 3.4 and Exercise 3.10, respectively.

TABLE 6.4. L Measure and Calibration Summaries for Breast Cancer Data.

Model	L Measure	$\mu_c(\frac{1}{2})$	50% HPD	95% HPD
1*	80.45	–	–	–
2	87.24	5.36	(5.23, 5.83)	(4.24, 6.34)
3	113.54	28.91	(28.71, 29.61)	(27.23, 30.23)

* Criterion minimizing model.
Source: Ibrahim, Chen, and Sinha (2001b).

The complete data likelihoods and the implementational details of the Gibbs sampler to sample from the resulting posterior distributions can be found in Section 3.4 and Exercise 3.10. For the breast cancer data, we implement the Gibbs sampling algorithms and use 50,000 MCMC iterations for computing the L measures and calibration distributions. We also choose $y^* = \eta(y) = \log(y)$ in (6.2.7). Table 6.4 shows the results using $\nu = \frac{1}{2}$, and reveals that model $\mathcal{M}_1$ is the criterion minimizing model with an L measure value of 80.45. Models $\mathcal{M}_2$ and $\mathcal{M}_3$ have $\mu_c(\frac{1}{2})$ values of $\mu_2(\frac{1}{2}) = 5.36$ and $\mu_3(\frac{1}{2}) = 28.91$, respectively, and therefore, model $\mathcal{M}_2$ is much closer to the criterion minimizing model than model $\mathcal{M}_3$. This is also clearly displayed in Figure 6.1, which gives the calibration distributions for models $\mathcal{M}_2$ and $\mathcal{M}_3$. We see from Figure 6.1 that there is a wide separation between p_{L_2} and p_{L_3}, and p_{L_2} has smaller dispersion than p_{L_3}. The HPD intervals for models $\mathcal{M}_2$ and $\mathcal{M}_3$ do not contain 0. We conclude here that both models $\mathcal{M}_2$ and $\mathcal{M}_3$ are sufficiently different from one another as well as being sufficiently different from the criterion minimizing model. We note that other choices of prior parameters yield similar L measure values and calibration distributions, and thus the results are not very sensitive to the choice of the prior distributions.

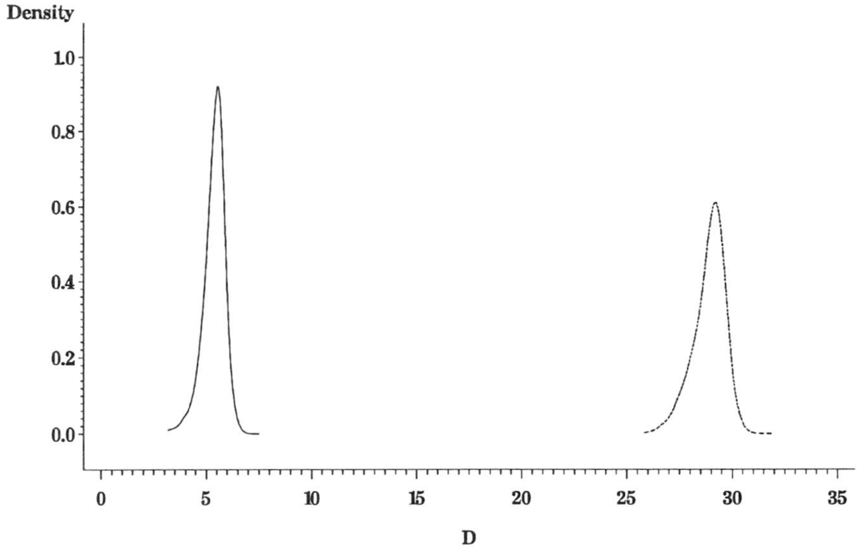

FIGURE 6.1. Calibration distributions for breast cancer data; solid curve: model $\mathcal{M}_2$, and dashed curve: model $\mathcal{M}_3$.

6.3 Conditional Predictive Ordinate

The Conditional Predictive Ordinate (CPO) statistic is a very useful model assessment tool which has been widely used in the statistical literature under various contexts. For a detailed discussion of the CPO statistic and its applications to model assessment, see Geisser (1993), Gelfand, Dey, and Chang (1992), Dey, Chen, and Chang (1997), and Sinha and Dey (1997).

For the ith observation, the CPO statistic is defined as

$$\text{CPO}_i = f(y_i|D^{(-i)}) = \int f(y_i|\boldsymbol{\beta}, \boldsymbol{\lambda}, \boldsymbol{x}_i)\pi(\boldsymbol{\beta}, \boldsymbol{\lambda}|D^{(-i)})\, d\boldsymbol{\beta}\, d\boldsymbol{\lambda}, \tag{6.3.1}$$

where y_i denotes the response variable and $\boldsymbol{x}_i$ is the vector of covariates for case i, $D^{(-i)}$ denotes the data with the ith case deleted, and $\pi(\boldsymbol{\beta}, \boldsymbol{\lambda}|D^{(-i)})$ is the posterior density of $(\boldsymbol{\beta}, \boldsymbol{\lambda})$ based on the data $D^{(-i)}$. From (6.3.1), we see that CPO_i is the marginal posterior predictive density of y_i given $D^{(-i)}$, and can be interpreted as the height of this marginal density at y_i. Thus, large values of CPO_i imply a better fit of the model.

For most models for survival data, a closed form of CPO_i is not available. However, a Monte Carlo estimator of CPO_i can be obtained using a single MCMC sample from the posterior distribution $\pi(\boldsymbol{\beta}, \boldsymbol{\lambda}|D)$, where D denotes the data including all cases. The implementational details for computing CPO_i can be found in Chapter 10 of Chen, Shao, and Ibrahim (2000).

For comparing two competing models, we examine the CPO_i's under both models. The observation with a larger CPO value under one model will support that model over the other. Therefore, a plot of CPO_i's under both models against observation number should reveal that the better model has the majority of its CPO_i's above those of the poorer fitting model. In comparing several competing models, the CPO_i values under all models can be plotted against the observation number in a single graph.

An alternative to CPO plots is the summary statistic called the logarithm of the Pseudomarginal likelihood (LPML) (see Geisser and Eddy, 1979), defined as

$$\text{LPML} = \sum_{i=1}^{n} \log(\text{CPO}_i). \tag{6.3.2}$$

In the context of survival data, the statistic LPML has been discussed by Gelfand and Mallick (1995) and Sinha and Dey (1997). To compare LPML's from two different studies for a given model, we propose to use a modification of (6.3.2), which is the average LPML, given by

$$\text{ALPML} = \frac{\text{LPML}}{n}, \tag{6.3.3}$$

where n is the sample size. The statistic ALPML can be interpreted as the relative pseudomarginal likelihood.

We see from (6.3.1) that LPML is always well defined as long the posterior predictive density is proper. Thus, LPML is well defined under improper priors, and in addition, it is very computationally stable. Therefore, LPML has a clear advantage over the Bayes factor as a model assessment tool, since it is well known that the Bayes factor is not well defined with improper priors, and is generally quite sensitive to vague proper priors. Thus, the Bayes factor is not applicable for many of our models here, since we consider several models involving improper priors. In addition, the LPML statistic also has clear advantages over other model selection criteria, such as the L *measure* discussed in Section 6.2. The L measure is a Bayesian criterion requiring finite second moments of the sampling distribution of y_i, whereas the LPML statistic does not require existence of any moments. Since the cure rate models in (5.2.8) and (5.3.2) have improper survival functions, no moments of the sampling distribution exist, and therefore the L measure is not well defined for these models. Thus, for the models considered here, the LPML statistic is well motivated. Next, we use three examples to demonstrate the CPO statistic.

Example 6.3. Melanoma data. In this example, we use the E1684 and E1690 melanoma datasets discussed in Example 1.2 to illustrate the CPO statistic.

First, we consider the piecewise exponential (PE) model from Chapter 3 given by (3.1.1), and the semiparametric cure rate (SPCR) model given by (5.3.2) from Chapter 5. For the piecewise exponential model in (3.1.1), we consider a fully parametric analysis (i.e., $J = 1$) and a semiparametric analysis using $J = 5$. For the semiparametric cure rate model in (5.3.2), we use $J = 5$. We note that $J = 1$ in (5.3.2) corresponds to a fully parametric cure rate model. In addition, we consider several choices of a_0, including $a_0 = 0$ and $a_0 = 1$ with probability 1, $E(a_0|D) = 0.05$, and $E(a_0|D) = 0.30$.

TABLE 6.5. CPO Statistics for E1684 and E1690.

Study	Model	ALPML
E1684	PE ($J = 5$)	-1.3775
E1690	PE ($J = 5$)	-1.2232
E1684	SPCR ($J = 1$)	-1.3407
E1690	SPCR ($J = 1$)	-1.2172
E1684	SPCR ($J = 5$)	-1.3439
E1690	SPCR ($J = 5$)	-1.2184

Source: Chen, Harrington, and Ibrahim (1999).

Table 6.5 shows results of ALPML for the E1684 and E1690 studies separately, based on $a_0 = 0$ with probability 1 and the initial prior (3.1.5) for model (3.1.1) and initial prior (5.3.4) with $\zeta_0 = 1$ and $\tau_0 = 0$ for model (5.3.2). We see from Table 6.5 that the results for PE and SPCR are quite similar, yielding similar ALPML statistics. In addition, the PE model with $J = 5$ gives comparable results to the cure rate models. However, the exponential model (i.e., the PE model with $J = 1$) yields a smaller CPO statistic relative to the other models, indicating a poorer fit. These results suggest that the SPCR models appear to provide a more adequate fit to the E1690 data compared to the exponential model and are comparable to, but slightly better than, the PE model with $J = 5$.

Table 6.6 shows the results of the LPML's when $a_0 = 0$ with probability 1, $E(a_0|D) \approx 0.05$, 0.20, 0.30, and 0.60, and $a_0 = 1$ with probability 1. Table 6.6 is quite informative. First, we see that $J = 5$ is better than $J = 1$ or $J = 10$. However, for the SPCR model, $J = 1$ and $J = 5$ are fairly close. Second, for both $J = 1$ or $J = 5$, the cure rate model yields a better fit than the PE model. Third, the incorporation of the E1684 data into the analysis improves the model fit. Fourth, for all the cases, LPML is a concave function of $E(a_0|D)$ (see Figure 6.2). This is an interesting feature in LPML in that it demonstrates that there is an "optimal" weight for the historical data with respect to the statistic LPML, and thus this property is potentially very useful in selecting a model.

TABLE 6.6. LPML Statistics for PE and SPCR Models.

Model	$E(a_0\|D)$	$J = 1$	$J = 5$	$J = 10$
PE	0	−575.60	−522.30	−523.62
	0.05	−575.45	−522.05	−523.20
	0.20	−575.23	−521.67	−522.39
	0.30	−575.13	−521.59	−522.12
	0.60	−574.95	−521.61	−522.02
	1	−574.64	−522.24	−522.71
SPCR	0	−519.75	−520.24	−524.42
	0.05	−519.61	−519.89	−523.82
	0.20	−519.39	−519.43	−522.83
	0.30	−519.34	−519.31	−522.53
	0.60	−519.40	−519.67	−522.56
	1	−519.67	−520.16	−522.97

Source: Chen and Ibrahim (2001).

Now we consider the LPML statistic using the alternative semiparametric cure rate (ASPCR) model of Ibrahim, Chen, and Sinha (2001a) described in Section 5.4. Our main purpose in this example is to use the LPML statistic to help us assess the goodness of fit of the models for different choices of κ, a_0, $F_0(.|\psi_0)$, and J. The values for the other hyperparameters involved in the ASPCR model are given in Example 5.3.

Table 6.7 gives the LPML statistics for several values of κ using the exponential and Weibull models for F_0 with $J = 10$ intervals, and when $E(a_0|D) = 0.33$. As κ is varied for a given a_0 using an exponential or Weibull F_0, we see that, for given a_0, the LPML statistic is always largest for the smallest value of κ. This indicates that when the model is more parametric in the right tail of the survival curve, the better the fit. We also observe that for fixed κ, LPML is a concave function of a_0. As an example, for $\kappa = 0.05$ and an exponential F_0, the LPML statistics are -518.87 for $a_0 = 0$ with probability 1, -518.08 when $E(a_0|D) = 0.33$, and -518.21 for $a_0 = 1$ with probability 1. Compared to the LPML Statistics given in Table

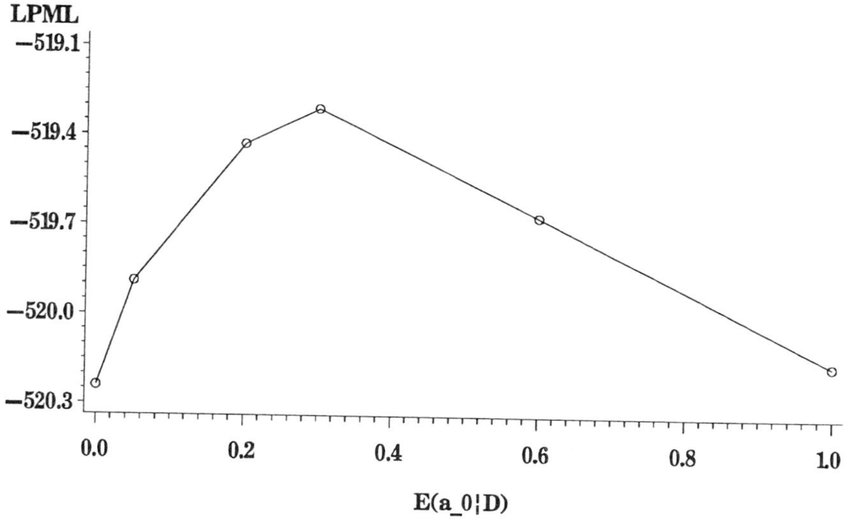

FIGURE 6.2. Plot of LPML's for SPCR with $J = 5$.

6.6, we can see that the ASPCR model with $\kappa = 0.05$ and an exponential F_0 is better than the PE and SPCR models.

TABLE 6.7. LPML Statistics for ASPCR Models with Several Values of a_0, F_0, and κ.

F_0	κ	$a_0 = 0$	$E(a_0\|D) = 0.33$	$a_0 = 1$
Exponential	0.05	−518.87	−518.08	−518.21
	0.30	−519.64	−518.74	−518.98
	0.60	−521.44	−520.39	−521.03
	0.95	−524.22	−522.14	−522.30
Weibull	0.05	−520.55	−519.27	−519.04
	0.30	−520.40	−519.24	−519.26
	0.60	−521.49	−520.46	−521.05
	0.95	−524.26	−522.11	−522.36

Source: Ibrahim, Chen, and Sinha (2001a).

Example 6.4. Kidney infection data (Example 4.3 continued). As another example, we compare two frailty models (Model I and Model II) discussed in Example 4.3.

Figure 6.3 plots the log CPO ratios for Model I versus Model II against observation numbers. Thus, a point bigger than 0 supports Model I. In Figure 6.3, □ represents a censored observation and ∘ represents a failure time. It shows that approximately 50% of the observations support Model I. The difference in LPML's between Model I and Model II is approximately

0.92. Thus, Model I is slightly better than Model II based on the CPO statistic.

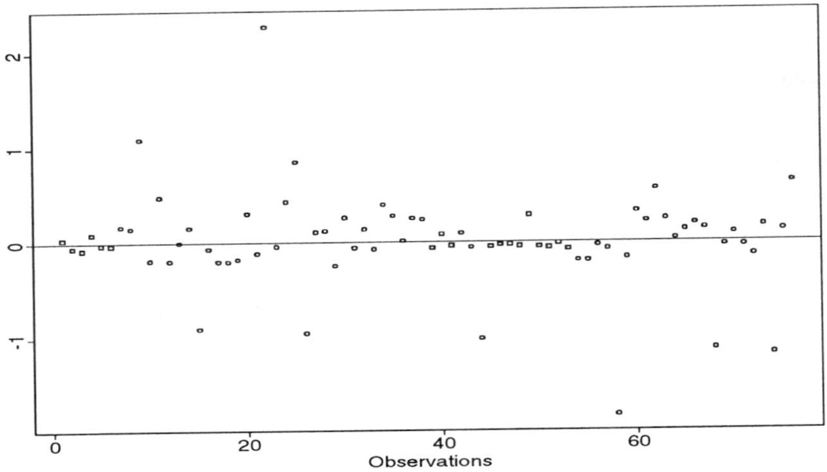

FIGURE 6.3. Plot of log CPO ratios for Model I versus Model II for kidney infection data.

Sahu, Dey, Aslanidou, and Sinha (1997) also consider the L measure to compare Model I and Model II. Using (6.2.5) with $\nu = 0.5$, they obtained L measure values of 1.63 and 103.31 for Model I and Model II, respectively. Thus, Model II is much worse than Model I based on the L measure criterion.

Example 6.5. Breast cancer data (Example 6.2 continued). For interval censored data, the CPO statistic for the ith observation is defined as

$$\text{CPO}_i = P(Y_i \in (a_{l_i}, a_{r_i}]|x_i, D^{(-i)}),$$

where $D^{(-i)}$ denotes the interval censored data with the ith patient removed. CPO_i is the posterior predictive probability of the observed data for the ith patient given the modified data $D^{(-i)}$. Let $\boldsymbol{\theta}$ denote the vector of model parameters. Following Sinha, Chen, and Ghosh (1999), CPO_i can be computed as

$$\text{CPO}_i = \left(\text{E}\left[\frac{1}{P(Y_i \in (a_{l_i}, a_{r_i}]|\boldsymbol{\theta}, x_i)}\right]\right)^{-1}, \tag{6.3.4}$$

where the expectation is taken with respect to the joint posterior $\pi(\boldsymbol{\theta}|D)$. Note that

$$P(Y_i \in (a_{l_i}, a_{r_i}]|\boldsymbol{\theta}, z_i) = \exp\left\{-\sum_{k=1}^{l_i} \lambda_k \theta_k^{x_i} \tilde{\Delta}_k\right\} - \exp\left\{-\sum_{k=1}^{r_i} \lambda_k \theta_k^{x_i} \tilde{\Delta}_k\right\},$$

where $\tilde{\Delta}_k = a_k - a_{k-1}$ and $\theta_k = \exp(\beta_k)$. Thus, CPO_i can be easily computed by using (6.3.4). A Monte Carlo estimate of CPO_i is given by

$$\begin{aligned}\widehat{\text{CPO}}_i =& \text{E}\left[\frac{1}{P(Y_i \in (a_{l_i}, a_{r_i}]|\boldsymbol{\theta}, x_i)}\right]\\ =& \frac{1}{L}\sum_{l=1}^{L}\left[\exp\left\{-\sum_{k=1}^{l_i} \lambda_{kl} \theta_{kl}^{x_i} \tilde{\Delta}_k\right\} - \exp\left\{-\sum_{k=1}^{r_i} \lambda_{kl} \theta_{kl}^{x_i} \tilde{\Delta}_k\right\}\right]^{-1},\end{aligned}$$

where $\{\theta_{kl},\ l = 1, 2, \cdots, L\}$ (L is large) is an MCMC sample from the posterior distribution $\pi(\boldsymbol{\theta}|D)$.

Using the values of the hyperparameters given in Example 6.2 and $L = 50,000$, we compute the LPML statistics. The values of the LPML's are -157.61 and -188.33 for $\mathcal{M}_1$ and $\mathcal{M}_2$, respectively, and the CPO_i's are displayed in Figure 6.4. Based on the LPML statistics, it is clear that $\mathcal{M}_1$ is more preferable than $\mathcal{M}_2$. The plots of the pairwise log CPO ratios are consistent with the single summary measure LPML. In Figure 6.4, 84% of the log CPO ratios for $\mathcal{M}_1$ versus $\mathcal{M}_2$ are positive. Therefore, the data support $\mathcal{M}_1$ instead of $\mathcal{M}_2$, which is also consistent with the L measure criterion.

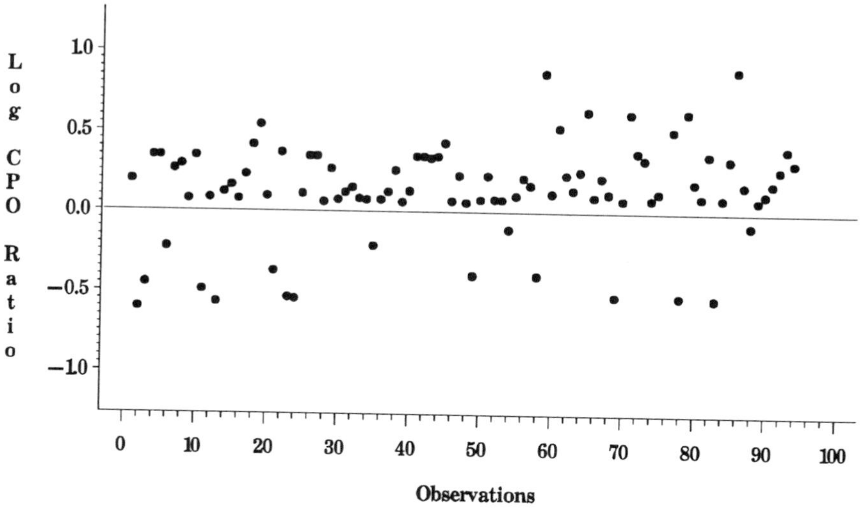

FIGURE 6.4. Plot of log CPO ratios for $\mathcal{M}_1$ versus $\mathcal{M}_2$ for breast cancer data.

6.4 Bayesian Model Averaging

A popular approach to model selection is Bayesian Model Averaging (BMA). In this approach, one bases inference on an average of all possible models in the model space $\mathcal{M}$, instead of a single "best" model. Suppose $\mathcal{M} = \{\mathcal{M}_1, \mathcal{M}_2, \ldots, \mathcal{M}_{\mathcal{K}}\}$, and let Δ denote the quantity of interest such as a future observation, a set of regression coefficients, or the utility of a course of action. Then, the posterior distribution of Δ is given by

$$\pi(\Delta|D) = \sum_{k=1}^{\mathcal{K}} \pi(\Delta|D, \mathcal{M}_k)p(\mathcal{M}_k|D), \tag{6.4.1}$$

where D denotes the data, $\pi(\Delta|D, \mathcal{M}_k)$ is the posterior distribution of Δ under model $\mathcal{M}_k$, and $p(\mathcal{M}_k|D)$ is the posterior model probability. Equation (6.4.1), called BMA, consists of an average of the posterior distributions under each model weighted by the corresponding posterior model probabilities. The motivation behind BMA is based on the notion that a single "best" model ignores uncertainty about the model itself, which can result in underestimated uncertainties about quantities of interest, whereas BMA in (6.4.1) incorporates model uncertainty.

Averaging over all possible models as in (6.4.1) provides better predictive ability, as measured by a logarithmic scoring rule, than using any single model $\mathcal{M}_k$:

$$-E\left[\log\left\{\sum_{j=1}^{\mathcal{K}} \pi(\Delta|D, \mathcal{M}_j)p(\mathcal{M}_j|D)\right\}\right] \leq -E[\log\{\pi(\Delta|D, \mathcal{M}_k)\}] \tag{6.4.2}$$

for $k = 1, 2, \ldots, \mathcal{K}$, where Δ is the observable to be predicted and the expectation is taken with respect to $\sum_{j=1}^{\mathcal{K}} \pi(\Delta|D, \mathcal{M}_j)p(\mathcal{M}_j|D)$. This result follows from the nonnegativity of the Kullback–Leibler information divergence.

The implementation of BMA is difficult for two reasons. First, $p(\mathcal{M}_k|D)$ can be difficult to compute. Second, the number of terms in (6.4.1) can be enormous. One solution to reduce the number of possible models in (6.4.1) involves applying the Occam's window algorithm of Madigan and Raftery (1994). Two basic principles underly this ad hoc approach. First, if a model predicts the data far less well than the model that provides the best predictions, then it has effectively been discredited and should no longer be considered. Thus models not belonging to

$$\mathcal{A}' = \left\{\mathcal{M}_k : \frac{\max_l\{p(\mathcal{M}_l|D)\}}{p(\mathcal{M}_k|D)} \leq C\right\} \tag{6.4.3}$$

are excluded from (6.4.1), where C is chosen by the data analyst and $\max_l\{p(\mathcal{M}_l|D)\}$ denotes the model with the highest posterior probability. A common choice of C is $C = 20$. The number of models in Occam's window increases as C decreases. Second, appealing to Occam's razor, models that receive less support from the data than any other simpler models are excluded. That is, models from (6.4.1) are excluded if they belong to

$$\mathcal{B} = \left\{ \mathcal{M}_k : \exists\, \mathcal{M}_l \in \mathcal{M}, \mathcal{M}_l \subset \mathcal{M}_k, \frac{p(\mathcal{M}_l|D)}{p(\mathcal{M}_k|D)} > 1 \right\}. \tag{6.4.4}$$

Thus (6.4.1) is replaced by

$$\pi(\Delta|D) = \frac{\sum_{\mathcal{M}_k \in \mathcal{A}} \pi(\Delta|D, \mathcal{M}_k) p(D|\mathcal{M}_k) p(\mathcal{M}_k)}{\sum_{\mathcal{M}_k \in \mathcal{A}} p(D|\mathcal{M}_k) p(\mathcal{M}_k)}, \tag{6.4.5}$$

where $\mathcal{A} = \mathcal{A}' \backslash \mathcal{B} \in \mathcal{M}$, $p(D|\mathcal{M}_k)$ is the marginal likelihood of the data D under model $\mathcal{M}_k$, and $p(\mathcal{M}_k)$ denotes the prior model probability.

This strategy greatly reduces the number of possible models in (6.4.1), and now all that is required is a search strategy to identify the models in $\mathcal{A}$. Two further principles underly the search strategy. The first principle–Occam's window–concerns interpreting the ratio of posterior model probabilities $p(\mathcal{M}_1|D)/p(\mathcal{M}_0|D)$, where $\mathcal{M}_0$ is a model with one less predictor than $\mathcal{M}_1$. If there is evidence for $\mathcal{M}_0$, then $\mathcal{M}_1$ is rejected, but to reject $\mathcal{M}_0$, stronger evidence for the larger model $\mathcal{M}_1$ is required. These principles fully define the strategy. Madigan and Raftery (1994) provide a detailed description of the algorithm and mention that the number of terms in (6.4.1) is often reduced to fewer than 25.

The second approach for reducing the number of terms in (6.4.1) is to approximate (6.4.1) using an MCMC approach. Madigan and York (1995) propose the MCMC model composition (MC^3) methodology, which generates a stochastic process that moves through the model space. A Markov chain $\{\mathcal{M}(l), l = 1, 2, \ldots\}$ is constructed with state space $\mathcal{M}$ and equilibrium distribution $p(\mathcal{M}_k|D)$. If this Markov chain is simulated for $l = 1, 2, \ldots, L$, then under certain regularity conditions, for any function $g(\mathcal{M}_k)$ defined on $\mathcal{M}$, the average

$$\hat{G} = \frac{1}{L} \sum_{l=1}^{L} g(\mathcal{M}(l))$$

converges almost surely to $E(g(\mathcal{M}_k))$ as $L \to \infty$. To compute (6.4.1) in this fashion, set $g(\mathcal{M}_k) = \pi(\Delta|D, \mathcal{M}_k)$. To construct the Markov chain, define a neighborhood $\text{nbd}(\mathcal{M}_*)$ for each $\mathcal{M}_* \in \mathcal{M}$ that consists of the model $\mathcal{M}_*$ itself and the set of models with either one variable more or one variable fewer than $\mathcal{M}_*$. Define a transition matrix q by setting $q(\mathcal{M}_* \to \mathcal{M}'_*) = 0$ for all $\mathcal{M}'_* \notin \text{nbd}(\mathcal{M}_*)$ and $q(\mathcal{M}_* \to \mathcal{M}'_*)$ constant for all $\mathcal{M}'_* \in \text{nbd}(\mathcal{M}_*)$. If the chain is currently in state $\mathcal{M}_*$, then we proceed by drawing $\mathcal{M}'_*$ from $q(\mathcal{M}_* \to \mathcal{M}'_*)$. It is then accepted with probability

$$\min\left\{1, \frac{p(\mathcal{M}'_*|D)}{p(\mathcal{M}_*|D)}\right\}.$$

Otherwise, the chain stays in state $\mathcal{M}_*$.

To compute $p(D|\mathcal{M}_k)$, Raftery (1996) suggests the use of the Laplace approximation, leading to

$$\log(p(D|\mathcal{M}_k)) = \log(L(\hat{\boldsymbol{\theta}}_k|D, \mathcal{M}_k)) - p_k \log(n) + O(1), \tag{6.4.6}$$

where n is the sample size, $L(\hat{\boldsymbol{\theta}}_k|D, \mathcal{M}_k)$ is the likelihood function, $\hat{\boldsymbol{\theta}}_k$ is the maximum likelihood estimate (MLE) of $\boldsymbol{\theta}_k$ under model $\mathcal{M}_k$, and p_k is the number of parameters in model $\mathcal{M}_k$. This is the Bayesian information criterion (BIC) approximation derived by Schwarz (1978). In fact, (6.4.6) is much more accurate for many practical purposes than its $O(1)$ error term suggests. Kass and Wasserman (1995) show that when $\mathcal{M}_j$ and $\mathcal{M}_k$ are nested and the amount of information in the prior distribution is equal to that in one observation, then the error in (6.4.6) is $O(n^{-1/2})$, under certain assumptions, rather than $O(1)$. Raftery (1996) gives further empirical evidence for the accuracy of this approximation.

6.4.1 BMA for Variable Selection in the Cox Model

Volinsky, Madigan, Raftery, and Kronmal (1997) discuss how to carry out variable selection in the Cox model using BMA. Equation (6.4.1) has three components, each posing its own computational difficulties. The predictive distribution $\pi(\Delta|D, \mathcal{M}_k)$ requires integrating out the model parameter $\boldsymbol{\theta}_k$. The posterior model probabilities $p(\mathcal{M}_k|D)$ similarly involve the calculation of an integrated likelihood. Finally, the models that fall into $\mathcal{A}$ must be located and evaluated efficiently.

In (6.4.1) the predictive distribution of Δ given a particular model $\mathcal{M}_k$ is found by integrating out the model parameter $\boldsymbol{\theta}_k$:

$$\pi(\Delta|D, \mathcal{M}_k) = \int \pi(\Delta|\boldsymbol{\theta}_k, D, \mathcal{M}_k)\pi(\boldsymbol{\theta}_k|D, \mathcal{M}_k)\, d\boldsymbol{\theta}_k. \tag{6.4.7}$$

This integral does not have a closed form solution for Cox models. Volinsky, Madigan, Raftery, and Kronmal (1997) use the MLE approximation:

$$\pi(\Delta|D, \mathcal{M}_k) \approx \pi(\Delta|\hat{\boldsymbol{\theta}}_k, D, \mathcal{M}_k). \tag{6.4.8}$$

In the context of model uncertainty, this approximation was used by Taplin (1993) and found it to give an excellent approximation in his time series regression problem; it was subsequently used by Taplin and Raftery (1994) and Draper (1995).

In regression models for survival analysis, analytic evaluation of $p(D|\mathcal{M}_k)$ is not possible in general, and an analytic or computational approximation is needed. In regular statistical models (roughly speaking, those in which the MLE is consistent and asymptotically normal), $p(D|\mathcal{M}_k)$ can be approximated by (6.4.6) via the Laplace method (Raftery, 1996).

Equation (6.4.1) requires the specification of model priors. When there is little prior information about the relative plausibility of the models considered, taking them all to be equally likely *a priori* is a reasonable "neutral" choice. Volinsky et al. (1997) adopt this choice in their paper. With very large model spaces (up to 10^{12} models) involving several kinds of models and about 20 datasets, Volinsky et al. (1997) found no perverse effects from putting a uniform prior over the models (see Raftery, Madigan, and Volinsky, 1997; Madigan and Raftery, 1994; Madigan, Andersson, Perlman, and Volinsky, 1996). When prior information about the importance of a variable is available, a prior probability on model $\mathcal{M}_k$ can be specified as

$$p(\mathcal{M}_k) = \prod_{j=1}^{p} {\pi_j}^{\delta_{kj}} (1 - \pi_j)^{1-\delta_{kj}}, \tag{6.4.9}$$

where $\pi_j \in [0, 1]$ is the prior probability that $\boldsymbol{\theta}_j \neq 0$, δ_{kj} is an indicator of whether or not variable j is included in model $\mathcal{M}_k$. Assigning $\pi_j = 0.5$ for all j corresponds to a uniform prior across the model space, while $\pi_j < 0.5$ for all j imposes a penalty for large models. Using $\pi_j = 1$ ensures that variable j is included in all models. Using this framework, elicitation of prior probabilities for models is straightforward and avoids the need to elicit priors for a large number of models. For an alternative approach when expert information is available, see Madigan and York (1995).

6.4.2 Identifying the Models in $\mathcal{A}'$

Volinsky et al. (1997) propose a procedure which requires identifying the best model and averaging over only those models with posterior probability of at least $1/C$ of the posterior probability of the best model (Equation (6.4.3)). They define the best model as the model with the largest BIC, corresponding to the model with the highest posterior model probability. Therefore a quick way of screening the models, without fitting them all, and of targeting those that are close in posterior probability to the best one, is needed. Then, the summation in (6.4.3) is over this reduced set of models.

Such a procedure exists for linear regression. Regression by leaps and bounds (Furnival and Wilson, 1974) is an efficient algorithm which provides the top q models of each model size, where q is designated by the user, plus the MLE $\hat{\boldsymbol{\theta}}_k$, Var$(\hat{\boldsymbol{\theta}}_k)$, and the coefficient of multiple determination R^2 for each model $\mathcal{M}_k$ returned. For two models A and B, where A and B are each subsets of the full parameter set, if $A \subset B$ then RSS(A) $>$ RSS(B), where RSS is the residual sum of squares. Using this fact, the method eliminates large portions of the model space by sweep operations on the matrix:

$$\begin{pmatrix} X'X & X'\boldsymbol{y} \\ \boldsymbol{y}'X & \boldsymbol{y}'\boldsymbol{y} \end{pmatrix}, \tag{6.4.10}$$

where X is the matrix of covariates and $\boldsymbol{y}$ is the vector of responses.

Lawless and Singhal (1978) developed a modification of the leaps and bounds algorithm for nonlinear regression models, which provides an approximate likelihood ratio test statistic (and therefore an approximate BIC value). The method proceeds as follows. Let $\boldsymbol{\theta}$ be the parameter vector of the full model and let $\boldsymbol{\theta}_k$ be the vector for a given submodel $\mathcal{M}_k$. Rewrite $\boldsymbol{\theta}_k$ as $(\boldsymbol{\theta}_1', \boldsymbol{\theta}_2')'$ so that model $\mathcal{M}_k$ corresponds to the submodel $\boldsymbol{\theta}_2 = 0$. Also, let

$$V = \mathcal{I}^{-1} = \begin{pmatrix} V_{11} & V_{12} \\ V_{12}' & V_{22} \end{pmatrix}$$

denote the inverse observed information matrix. If $L(\hat{\boldsymbol{\theta}}|D)$ is the maximized likelihood under the full (unrestricted) model, and $L(\tilde{\boldsymbol{\theta}}|D)$ is the maximized likelihood under $\boldsymbol{\theta}_2 = 0$, then

$$\Lambda = -2[\log L(\tilde{\boldsymbol{\theta}}|D) - \log L(\hat{\boldsymbol{\theta}}|D)]$$

is the usual likelihood ratio statistic for the test of the submodel versus the full model while

$$\Lambda^* = \hat{\boldsymbol{\theta}}_2' {V_{22}}^{-1} \hat{\boldsymbol{\theta}}_2$$

is an approximation to Λ based on the Wald statistic. Finally, replace the matrix in (6.4.10) with

$$\begin{pmatrix} \mathcal{I} & \mathcal{I}\hat{\boldsymbol{\theta}} \\ \hat{\boldsymbol{\theta}}'\mathcal{I} & \hat{\boldsymbol{\theta}}'\mathcal{I}\hat{\boldsymbol{\theta}} \end{pmatrix}$$

and perform the same matrix sweep operators from the leaps and bounds algorithm on this matrix. As a result the function provides

(i) an estimate of the best q proportional hazards models for each model size;

(ii) the likelihood ratio test (LRT) approximation Λ^* for each model;

(iii) an approximation to $\tilde{\theta}$, the MLE for the parameters of the submodel; and

(iv) the asymptotic covariance matrix ${V_{11}}^{-1}$.

As long as q is large enough, this procedure returns the models in $\mathcal{A}'$ plus many models not in $\mathcal{A}'$. We can use the approximate LRT to reduce the remaining subset of models to those most likely to be in $\mathcal{A}'$. This reduction step keeps only the models whose posterior probabilities are at least $1/C^*$ of the posterior model probability (PMP) of the best model, where C^* is greater than C. Volinsky et al. (1997) suggest using $C^* = C^2$, which is found to be large enough to virtually guarantee that no models in $\mathcal{A}'$ will be lost.

Kuk (1984) first applies this algorithm to Cox models in order to find the single best model. Volinsky et al. (1997) use it to help locate the models in $\mathcal{A}'$ that are to be averaged over. The returned models are fitted by a standard survival analysis program, the exact BIC value for each one, which corresponds to a posterior model probability given by (6.4.6), is calculated and those models not in $\mathcal{A}'$ are eliminated. The posterior model probabilities are then normalized over the model set. BMA parameter estimates and standard errors are calculated by taking weighted averages of the estimates and errors from the individual models, using the posterior model probabilities as weights.

Finally, the posterior probability that the regression coefficient for a variable is nonzero is computed, by adding the posterior probabilities of the models which contain that variable. Following Kass and Raftery (1995), standard rules of thumb for interpreting this posterior probability are given in Table 6.8.

TABLE 6.8. Rules of Thumb for Interpreting Posterior Probability (PP).

PP	Interpretation
$< 50\%$	evidence against the effect
$50 - 75\%$	weak evidence for the effect
$75 - 95\%$	positive evidence
$95 - 99\%$	strong evidence
$> 99\%$	very strong evidence

Source: Volinsky, Madigan, Raftery, and Kronmal (1997).

6.4.3 Assessment of Predictive Performance

As discussed in Volinsky, Madigan, Raftery, and Kronmal (1997), we assess the relative values of BMA and that of competing methods on the basis of their predictive performance. To assess performance, we randomly split the data into two halves and use an analogue of the logarithmic scoring rule of Good (1952). First, we apply each model selection method to the first half of the data, called the *build data* (D^B). The corresponding coefficient estimates define a predictive density for each person in the second half of the data (*test data*, or D^T). Then, a log score for any given model $\mathcal{M}_k$ is based on the observed ordinate of the predictive density for the subjects in D^T:

$$\sum_{d \in D^T} \log \pi(d|D^B, \mathcal{M}_k). \tag{6.4.11}$$

Similarly, the predictive log score for BMA is

$$\sum_{d \in D^T} \log \left\{ \sum_{\mathcal{M}_k \in \mathcal{M}^*} \pi(d|D^B, \mathcal{M}_k) p(\mathcal{M}_k|D^B) \right\}, \tag{6.4.12}$$

where $\mathcal{M}^*$ is the set of BMA-selected models.

However, the Cox model does not directly provide a predictive density. Rather it provides an estimated predictive cumulative distribution function (cdf) which is a step function (Breslow, 1975) and therefore does not lead to differentiation into a density. In the spirit of Cox's partial likelihood, we use an alternative to the predictive density:

$$\pi(d|D^B, \mathcal{M}_k) = \left(\frac{\exp(\boldsymbol{x}_i' \hat{\boldsymbol{\beta}}_k)}{\sum_{\ell \in \mathcal{R}_i} \exp(\boldsymbol{x}_\ell' \hat{\boldsymbol{\beta}}_k)} \right)^{\nu_i},$$

where $\boldsymbol{\theta}_k = \boldsymbol{\beta}_k$ denotes the vector of regression coefficients under model $\mathcal{M}_k$, $\mathcal{R}_i$ is the risk set at time y_i, and ν_i is the censoring indicator. By substituting this into (6.4.11) and (6.4.12) above, we now have an analogue to a log score called the *partial predictive score* (PPS). Using the PPS, we can compare BMA to any single model selected. The partial predictive score is greater for the method which gives higher probability to the events that occur in the test set.

To compare methods based on their *predictive discrimination*, namely, how well they sort the subjects in the test set into discrete risk categories (high, medium, low risk), Volinsky et al. (1997) proposed the following procedure to assess predictive discrimination of a single model in the context of a cardiovascular health study:

Step 1. Fit the model to the build data to get estimated coefficients $\hat{\boldsymbol{\beta}}$.

Step 2. Calculate risk scores $(\boldsymbol{x}_i' \hat{\boldsymbol{\beta}})$ for each subject in the build data.

Step 3. Define low, medium, and high risk groups for the model by the empirical 33rd and 66th percentiles of the risk scores.

Step 4. Calculate risk scores for the test data and assign each subject to a risk group.

Step 5. Extract the subjects who are assessed as being in a higher risk group by one method than by another, and tabulate what happened to those subjects over the study period.

To assess predictive discrimination for BMA, we must take account of the multiple models that we average over. We replace the first two steps above with

Step 1′. Fit each model $\mathcal{M}_k \in \mathcal{A}'$ to get estimated coefficients $\hat{\boldsymbol{\beta}}_k$.

Step 2′. Calculate risk scores $(\boldsymbol{x}_i' \hat{\boldsymbol{\beta}}_k)$ under model $\mathcal{M}_k$ for each person in the build data. A person's risk score under BMA is the weighted average of these: $\sum_{\mathcal{M}_k \in \mathcal{A}'} (\boldsymbol{x}_i' \hat{\boldsymbol{\beta}}_k) p(\mathcal{M}_k | D^B)$.

Example 6.6. The cardiovascular health study. The cardiovascular health study (CHS), with over 95% censoring, provides an opportunity to compare BMA with model selection methods in the presence of heavy censoring. Fried et al. (1991) described the complete sample design and study methods as well as specific protocols for the classification of some of independent variables (e.g., congestive heart failure, abnormal ejection fraction). The CHS is a longitudinal, observational study, funded by the National Institutes of Health (NIH), and started in June 1989 to study cardiovascular disease in patients aged 65 and over. The CHS studies 23 possible risk factors for stroke on 4504 subjects, 172 of whom did suffer a stroke during the study. One of the goals of the CHS researchers is to design a risk assessment model from these variables. To this end, all 23 variables can be collected via a patient's medical history or a noninvasive exam. A summary of the cardiovascular health data is given in Volinsky et al. (1997). The 23 variables are age (years) (x_1), diuretic use (x_2), regular aspirin use (x_3), systolic blood pressure (mmHg) (x_4), creatinine (mg/dl) (x_5), diabetes (x_6), EGG atrial fibrillation (x_7), abnormal ejection fraction (x_8), stenosis of common carotid artery (x_9), timed walk (x_{10}), LV hypertrophy (x_{11}), self-reported atrial fibrillation (x_{12}), diastolic blood pressure (mmHg) (x_{13}), congestive heart failure (x_{14}), fibrinogen (mg/dl) (x_{15}), sex (x_{16}), fasting glucose (mg/dl) (x_{17}), HDL (mg/dl) (x_{18}), LDL (mg/dl) (x_{19}), antihypertension medication (x_{20}), insulin (uU/dl) (x_{21}), history of cardiovascular disease (x_{22}), and pack years smoked (x_{23}).

The model chosen by a stepwise (backward elimination) procedure, starting with all 23 variables, included the following 10 variables: age, diuretic, aspirin, systolic blood pressure, creatinine, diabetes, atrial fibrillation (by ecg), stenosis, timed walk, and left ventricular (LV) hypertrophy. The model with the highest approximate posterior probability was the same as the stepwise model except that LV hypertrophy was not included. In the list of models provided by BMA, the stepwise model appears fourth. Table 6.9 contains the variables included in any of the top five models. Inference about independent variables is expressed in terms of the posterior probability that the parameter does not equal zero. Table 6.10 contains the posterior means, standard deviations, and posterior probabilities of the variables.

Figure 6.5 shows the posterior probability that each regression coefficient is nonzero, plotted against the corresponding P-value from stepwise variable selection. Overall, the posterior probabilities imply weaker evidence for effects than do the P-values. This is partly due to the fact that P-values overstate confidence because they ignore model uncertainty. However, even when there is no model uncertainty, P-values arguably overstate the evidence for an effect (see Edwards, Lindman, and Savage, 1963; Berger and Delampady, 1987; Berger and Sellke, 1987).

TABLE 6.9. Top Five Models for the Cardiovascular Health Data.

k	$\mathcal{M}_k$	PMP (in %)
1	$(x_1, x_2, x_3, x_4, x_5, x_6, x_7, x_9, x_{10})$	1.7
2	$(x_1, x_3, x_4, x_5, x_6, x_7, x_9, x_{10}, x_{11})$	1.6
3	$(x_1, x_3, x_4, x_5, x_6, x_7, x_9, x_{10})$	1.4
4	$(x_1, x_2, x_3, x_4, x_5, x_6, x_7, x_9, x_{10}, x_{11})$	1.4
5	$(x_1, x_3, x_4, x_5, x_6, x_7, x_8, x_9, x_{10})$	1.1

Source: Volinsky, Madigan, Raftery, and Kronmal (1997).

TABLE 6.10. Posterior Estimates of Variables for the Cardiovascular Health Data.

Var	Mean	SD	$P(\beta_j \neq 0)$	Var	Mean	SD	$P(\beta_j \neq 0)$
x_1	0.04	0.01	0.89	x_{13}	0.01	0.01	0.00
x_2	0.40	0.17	0.59	x_{14}	0.79	0.37	0.24
x_3	0.52	0.16	0.99	x_{15}	0.00	0.00	0.00
x_4	0.02	0.00	1.00	x_{16}	0.37	0.18	0.21
x_5	0.43	0.14	0.64	x_{17}	0.00	0.00	0.00
x_6	0.42	0.10	1.00	x_{18}	-0.01	0.01	0.07
x_7	0.83	0.30	0.66	x_{19}	0.00	0.00	0.24
x_8	0.46	0.22	0.26	x_{20}	0.05	0.23	0.00
x_9	0.34	0.12	0.90	x_{21}	0.00	0.00	0.00
x_{10}	0.47	0.10	1.00	x_{22}	0.24	0.17	0.05
x_{11}	0.60	0.25	0.50	x_{23}	0.00	0.00	0.00
x_{12}	0.62	0.31	0.18				

Source: Volinsky, Madigan, Raftery, and Kronmal (1997).

TABLE 6.11. P-values and Posterior Probabilities for Eight Variables.

Variable	P-value	$P(\beta_j \neq 0)$
Age (x_1)	0.007	0.89
Diuretic use (x_2)	0.026	0.59
ECG At. fib. (x_7)	0.005	0.66
Stenosis (x_9)	0.004	0.90
LV hyper. (x_{11})	0.022	0.50
Creatinine (x_{14})	0.002	0.64
Sex (x_{16})	0.053	0.21
LDL (x_{19})	0.058	0.24

Source: Volinsky, Madigan, Raftery, and Kronmal (1997).

For the four variables, systolic blood pressure, timed walk, diabetes and daily aspirin, the posterior probabilities and the P-values agree that there is very strong evidence for an effect (P-value < 0.001 and $P(\beta_j \neq 0) > 0.99$). For the eight variables in Table 6.11, however, the two approaches lead to qualitatively different conclusions. Each P-value overstates the evidence for an effect. For the first four of the variables, the P-value would lead to the effect being called "highly significant" (P-value < 0.01), while the posterior probability indicates the evidence to be positive but not strong. For the next two variables (LV hyper. and Diuretic use), the P-value is "significant" (P-value < 0.05), but the posterior probabilities indicate the evidence for an effect to be weak. For the last two variables (sex and LDL), the P-values are "marginally significant" (P-value < 0.06), but the posterior probabilities actually indicate (weak) evidence *against* an effect.

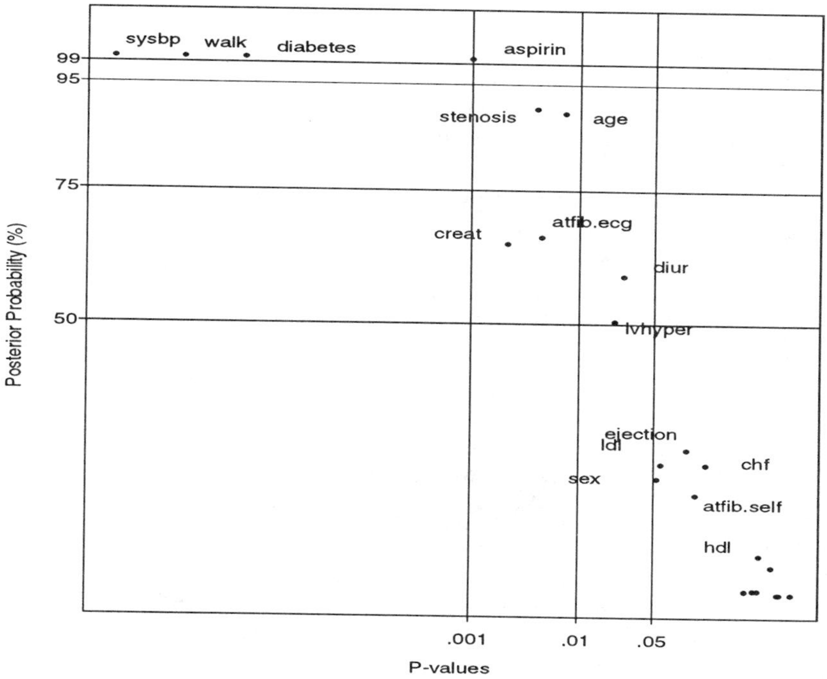

FIGURE 6.5. Plot of the posterior probability that each regression coefficient is nonzero, against the corresponding P-value.

For the remaining 11 variables, P-values and posterior probabilities agree in saying that there is little or no evidence for an effect. However, posterior probabilities enable one to make one distinction that P-values cannot. One may fail to reject the null hypothesis of "no effect" because either (a) there are not enough data to detect an effect, or (b) the data provide evidence *for* the null hypothesis. P-values cannot distinguish between these two situations, but posterior probabilities can. Thus, for example, for the ejection fraction, $P(\beta_j \neq 0) = 0.26$, so that the data are indecisive, while for diastolic blood pressure, $P(\beta_j \neq 0) = 0.00$, indicating strong evidence *for* the null hypothesis of no effect. The posterior probability of "no effect" can be viewed as an approximation to the posterior probability of the effect being "small," namely, $P(|\beta_j| < \varepsilon)$, provided that ε is at most about one-half of a standard error (see Berger and Delampady, 1987).

Six of the variables are known to be risk factors for the general population: age, systolic blood pressure, diabetes, atrial fibrillation, antihypertensive medication, and smoking (Kannel, McGee, and Gordon, 1976). An analysis was performed by increasing the prior probability for these variables to 0.75. The results showed no change in the conclusion for diabetes and blood pressure (both stayed at 1.00), or for smoking and antihypertensive medication (both remained under 0.10). However, for atrial fibrillation as measured by ECG, the posterior probability is 0.92 (in the original analysis it was 0.66). This sensitivity is probably because only about 3% of the subjects had this condition. The posterior probability for age jumps from 0.90 in the original analysis to 0.99.

For assessing predictive performance, the data were randomly split into two parts such that an equal number of events (86 strokes) occurred in each part. The result for BMA is compared with those for stepwise model selection and for the single model with the highest posterior probability. The predictive scores (PPS) are -641.6 for the top PMP model, -632.8 for the stepwise model, and -632.8 for BMA. A higher score (less negative) indicates better predictive performance. The top model and stepwise model may be different than those in the previous section since they are built using only half the data. The difference in PPS of 11.7 can be viewed as an increase in predictive performance *per event* by a factor of $\exp(11.7/86) = 1.15$ or by about 15%. This means that BMA predicts who is at risk for stroke 15% more effectively than a method which picks the model with the best posterior model probability (or 3.5% more effectively than a stepwise method).

Predictive discrimination shows the benefit of using Bayesian model averaging in another way. Table 6.12 shows the classification of the 2252 patients in the test data, and whether those patients had a stroke in the study period. The patients assigned to the high-risk group by BMA had more strokes than did those assigned high risk by other methods; similarly those assigned to the low-risk group by BMA had fewer strokes. Table 6.13 summarizes the outcome for the patients who were given higher risk by one method over another. The patients assigned to a higher risk group by BMA had more strokes and were stroke-free for a shorter period of time. Both tables show that Bayesian model averaging is better at indicating which patients are at risk of strokes.

TABLE 6.12. Classification of 2252 Patients in the Test Data.

		Assigned Risk Group		
		Low	Medium	High
BMA	Stroke-free	751	770	645
	Stroke	7	24	55
Stepwise	Stroke-free	750	799	617
	Stroke	8	27	51
Top PMP	Stroke-free	724	801	641
	Stroke	10	28	48

Source: Volinsky, Madigan, Raftery, and Kronmal (1997).

TABLE 6.13. Summary of the outcomes for the Patients Who Were Given Higher Risk by One Method over Another.

Estimated Risk	No. of Strokes	No. of Subjects	Mean Surv Time
BMA > top PMP	11	147	1169
BMA < top PMP	1	160	1243
BMA > stepwise	10	165	1191
BMA < stepwise	5	133	1243

Source: Volinsky, Madigan, Raftery, and Kronmal (1997).

Although there is much literature regarding the dangers of model selection and of stepwise methods in general, a glance at popular survival analysis texts (e.g., Fleming and Harrington, 1991; Kalbfleisch and Prentice, 1980) indicates that stepwise is still the status quo for variable selection in Cox models. Similarly, P-values from the final model are the standard reported evidence for a variable's importance. By applying BMA to Cox models, we can account for the model uncertainty ignored by stepwise methods. BMA shifts the focus away from selecting a single model. While stepwise methods lead to hard inclusion or exclusion of each variable, the posterior probability that the parameter is nonzero supplied by BMA can be more informative.

Also, there may be some evidence for the predictive value of a variable, even if it is not in the "best" selected model. For example, the variable ejection fraction is not in either the stepwise selected model or the model with the highest posterior probability, but the posterior probability that the parameter is nonzero is an indecisive 0.26, indicating that it should not necessarily be discarded in future work. In fact it is included in any prediction with weight proportional to its posterior probability. This indecisive result may be due to a lack of power in the data rather than to the variable not having appreciable predictive value.

The partial predictive score allows a direct comparison of predictive performance of one Cox model to another, and for the CHS it shows increased predictive performance for BMA over model selection techniques. Altman and Anderson (1989) claimed that the choice of model used for prediction "appears of limited importance" in their example. On the contrary, for the CHS the predictive discrimination analysis appears to show that a substantive difference can be made in evaluating who is at high risk for stroke. Thus, it may ultimately allow better targeting of patients susceptible to stroke so that preventive resources can be allocated more efficiently.

To implement BMA for proportional hazard models, Volinsky et al. (1997) have written a series of S-PLUS functions. The function `bic.surv` performs the Bayesian model averaging and outputs all information about the models selected, while `summary.bs` condenses this information to output only posterior model probabilities, posterior variable probabilities, and the variables included in the selected models. The functions allow user specification of parameters to limit the number of models returned and of priors for independent variables. All the software can be obtained by sending the message "`send bic.surv from S`" to `statlib@stat.cmu.edu` or on the website "`http://lib.stat.cmu.edu/S/bic.surv`".

6.5 Bayesian Information Criterion

Volinsky and Raftery (2000) discuss the use of the Bayesian Information Criterion (BIC) for model selection in models for censored survival data. They show that posterior model probabilities can be approximated using BIC, a simple expression that involves only the maximized likelihood, the sample size, and the number of risk factors in the model. Following Volinsky and Raftery (2000), we show that when there is censoring, an improved approximation is obtained by replacing the total sample size by the number of uncensored cases. This revised approximation preserves, and may improve, the asymptotic properties of the BIC approximation, and also provides better results in practice for some standard epidemiological datasets.

The Bayesian framework for hypothesis testing uses Bayes factors to quantify the evidence for one hypothesized model against another (see Kass and Raftery, 1995). Schwarz (1978) derives the BIC as a large sample approximation to twice the logarithm of the Bayes factor. For a model $\mathcal{M}_k$ parameterized by an p_k-dimensional vector $\boldsymbol{\theta}_k$,

$$\text{BIC} = -2\{\ell_k(\hat{\boldsymbol{\theta}}_k) - \ell_0(\hat{\boldsymbol{\theta}}_0)\} + (p_k - p_0)\log(n), \tag{6.5.1}$$

where $\ell_k(\hat{\boldsymbol{\theta}}_k)$ and $\ell_0(\hat{\boldsymbol{\theta}}_0)$ are the log maximized likelihoods under $\mathcal{M}_k$ and a reference model $\mathcal{M}_0$, whose parameter has dimension p_0, where n is the sample size. With nested models, BIC equals the standard likelihood ratio test statistic minus a complexity penalty which depends on the degrees of freedom of the test of $\mathcal{M}_0$ against $\mathcal{M}_k$. BIC provides an approximation to the Bayes factor which can readily be computed from the output of standard statistical software packages (see Kass and Raftery, 1995; Raftery, 1995).

The derivation of BIC involves a Laplace approximation to the Bayes factor, and ignores terms of constant order, so that the difference between BIC and twice the log Bayes factor does not vanish asymptotically in general, although it becomes inconsequential in large samples. However, Kass and Wasserman (1995) show that under certain nonrestrictive regularity conditions, the difference between BIC and twice the log Bayes factor does tend to zero for a specific choice of prior on the parameters. They argue that this implicit prior is a reasonable one.

Kass and Wasserman (1995) note that the "sample size" n which appears in the penalty term of (6.5.1) must be carefully chosen. In censored data models such as the proportional hazards model of Cox (1972), subjects contribute different amounts of information to the likelihood, depending on whether or not they are censored. Volinsky and Raftery (2000) have found that in the penalty term for BIC, substituting d, the number of uncensored events, for n, the total number of individuals, results in an improved criterion without sacrificing the asymptotic properties shown by Kass and Wasserman (1995).

Standard Bayesian testing procedures use the Bayes factor (BF), which is the ratio of integrated likelihoods for two competing models. Kass and Raftery (1995) derive BIC as an approximation to twice the difference in log integrated likelihoods, so that the difference in BIC between two models approximates twice the logarithm of the Bayes factor. Hence,

$$\frac{2\log(\mathrm{BF}) - \mathrm{BIC}}{2\log(\mathrm{BF})} \to 0, \quad \text{as } n \to \infty.$$

However,

$$2\log(\mathrm{BF}) - \mathrm{BIC} \not\to 0, \quad \text{as } n \to \infty. \tag{6.5.2}$$

Equation (6.5.2) implies that, for general priors on the parameters, $2\log(\mathrm{BF}) - \mathrm{BIC}$ has a nonvanishing asymptotic error of constant order, i.e., of order $O(1)$. Since the absolute value of BIC increases with n, the error tends to zero as a proportion of BIC. Therefore, BIC has the undesirable property that for any constant c, $\mathrm{BIC} + c$ also approximates twice the log Bayes factor to the same order of approximation as BIC itself. This, $O(1)$ error suggests that the BIC approximation is somewhat crude, and may perform poorly for small samples.

Kass and Wasserman (1995) show that with nested models, under a particular prior on the parameters, the constant order asymptotic error disappears, and they argue that this prior can reasonably be used for inference purposes. Following the notation of their paper, let $\boldsymbol{y} = (y_1, y_2, \ldots, y_n)'$ be i.i.d. observations from a family parameterized by $(\boldsymbol{\theta}, \boldsymbol{\psi})$, with $\dim(\boldsymbol{\theta}, \boldsymbol{\psi}) = p$ and $\dim(\boldsymbol{\theta})=p_0$. Our goal is to test H_0: $\boldsymbol{\psi} = \boldsymbol{\psi}_0$ against H_1: $\boldsymbol{\psi} \in R^{p-p_0}$

using the Bayes factor,

$$\text{BF} = \frac{p(\boldsymbol{y}|H_0)}{p(\boldsymbol{y}|H_1)} = \frac{\int_{R^{p_0}} L(\boldsymbol{\theta}, \boldsymbol{\psi}_0|\boldsymbol{y})\pi_0(\boldsymbol{\theta})\, d\boldsymbol{\theta}}{\int_{R^p} L(\boldsymbol{\theta}, \boldsymbol{\psi}|\boldsymbol{y})\pi(\boldsymbol{\theta}, \boldsymbol{\psi})\, d\boldsymbol{\theta}\, d\boldsymbol{\psi}}, \tag{6.5.3}$$

where $L(\boldsymbol{\theta}, \boldsymbol{\psi}|\boldsymbol{y})$ is the likelihood function, and $\pi_0(\boldsymbol{\theta})$ and $\pi(\boldsymbol{\theta}, \boldsymbol{\psi})$ are the prior distributions under H_0 and H_1, respectively. The BIC for testing H_0 versus H_1 is:

$$\text{BIC} = -2\{\ell_1(\hat{\boldsymbol{\theta}}, \hat{\boldsymbol{\psi}}) - \ell_0(\hat{\boldsymbol{\theta}}_0)\} + (p - p_0)\log(n), \tag{6.5.4}$$

where $\ell_0(\hat{\boldsymbol{\theta}}_0)$ and $\ell_1(\hat{\boldsymbol{\theta}}, \hat{\boldsymbol{\psi}})$ are the log maximized likelihood functions under H_0 and H_1, respectively.

Let $I(\boldsymbol{\theta}, \boldsymbol{\psi})$ be the $p \times p$ Fisher information matrix of $(\boldsymbol{\theta}, \boldsymbol{\psi})$ under H_1, let $I_{\boldsymbol{\theta}\boldsymbol{\psi}}(\boldsymbol{\theta}, \boldsymbol{\psi}_0)$ denote the information matrix $(-E(\frac{\partial^2 \ell(\boldsymbol{\theta}, \boldsymbol{\psi})}{\partial \boldsymbol{\theta} \partial \boldsymbol{\psi}}))$ evaluated at $(\boldsymbol{\theta}, \boldsymbol{\psi}_0)$, and let $\pi_{\boldsymbol{\psi}}(\boldsymbol{\psi})$ be the marginal prior density of $\boldsymbol{\psi}$ under H_1. The main result of Kass and Wasserman (1995) is as follows. If the following conditions hold:

(i) the parameters are *null orthogonal*, that is, $I_{\boldsymbol{\theta}\boldsymbol{\psi}}(\boldsymbol{\theta}, \boldsymbol{\psi}_0) = 0$ for all $\boldsymbol{\theta}$,

(ii) the MLE $\hat{\boldsymbol{\psi}}$ satisfies $\hat{\boldsymbol{\psi}} - \boldsymbol{\psi}_0 = O_p(n^{-1/2})$, and

(iii) $-\frac{1}{n}D^2 l(\hat{\theta}, \hat{\psi}) - I(\theta, \psi) = O_p(n^{-1/2})$,

then

$$2\log \text{BF} = \text{BIC} - 2\log\{(2\pi)^{(p_0-p)/2}|I_{\boldsymbol{\psi}\boldsymbol{\psi}}(\hat{\boldsymbol{\theta}}, \boldsymbol{\psi}_0)|^{-1/2}\pi_{\boldsymbol{\psi}}(\hat{\boldsymbol{\psi}})\} + O_p(n^{-1/2}). \tag{6.5.5}$$

In addition, if $\pi_{\boldsymbol{\psi}}(\boldsymbol{\psi})$ is a standard multivariate normal density with location $\boldsymbol{\psi}_0$ and covariance matrix $\left[I_{\boldsymbol{\psi}\boldsymbol{\psi}}(\boldsymbol{\theta}, \boldsymbol{\psi}_0)\right]^{-1}$, the asymptotic error of constant order (the second term on the right in (6.5.5)) will vanish, leaving

$$2\log(\text{BF}) = \text{BIC} + O_p(n^{-1/2}). \tag{6.5.6}$$

If the prior on $\boldsymbol{\psi}$ is not of this form, then this error term in (6.5.5) gives the constant order asymptotic error in BIC as an approximation to twice the log Bayes factor. This result has an important implication. BIC is a Bayesian procedure which does not require the specification of a prior, but it approximates a Bayes factor which is based on a particular prior for the parameter of interest. Therefore, when using BIC to compare models, the Kass-Wasserman result defines an implicit prior which BIC uses. This prior, which we call the *overall unit information prior*, is appealing, because it is a normal distribution centered around $\boldsymbol{\psi}_0$ with the amount of information in the prior equal to the average amount of information in one observation. Since the prior is based on only one observation, it is vague yet proper.

6.5.1 Model Selection Using BIC

Model selection criteria such as BIC are often used to select variables in regression problems. Here, we use BIC to determine the best models (where models are variable subsets) in a class of censored survival models.

When censoring is present it is unclear whether the penalty in BIC should use n, the number of observations, or d, the number of events. When using the partial likelihood, there are only as many terms in the partial likelihood as there are events d. Kass and Wasserman (1995) indicate that the term used in the penalty should be the rate at which the Hessian matrix of the log-likelihood function grows, which suggests that d is the correct quantity to use. However, if we are to use a revised version of BIC, it is important that the new criterion continue to have the asymptotic properties that Kass and Wasserman derived. In fact, our revised BIC does have these properties, with a slightly modified outcome. Let us alter condition (iii) to be

$$-\frac{1}{d}D^2 l(\hat{\boldsymbol{\theta}}, \hat{\boldsymbol{\psi}}) - I_u(\boldsymbol{\theta}, \boldsymbol{\psi}) = O_p(n^{-1/2}), \tag{6.5.7}$$

where $I_u(\boldsymbol{\theta}, \boldsymbol{\psi})$ is the expected Fisher information for one uncensored observation (the *uncensored unit information*). If this holds, then (6.5.6) is true, and the new BIC (with d in the penalty) is an $O_p(n^{-1/2})$ approximation to twice the Bayes factor where the prior variance on $\boldsymbol{\theta}$ is now equal to the inverse of the uncensored unit information. By using d in the penalty instead of n, it can be shown that this asymptotic result holds, the only difference being in the implicit prior on the parameter.

To investigate which of these criteria is better, we first consider a simple exponential censored survival model. We use a conjugate unit information prior for this model, and find that BIC provides a closer approximation to twice the log Bayes factor when n is replaced by d in equation (6.5.1). Then, we consider the Cox proportional hazards model. The theorems stated in the subsequent subsections are proved in Volinsky and Raftery (2000).

6.5.2 Exponential Survival Model

Consider n subjects, where subject i has observed time y_i, and let $\boldsymbol{y} = (y_1, y_2, \ldots, y_n)'$. The censoring indicator ν_i describes whether y_i is a survival time ($\nu_i = 1$) or a censoring time ($\nu_i = 0$). Also let $D = (\boldsymbol{y}, \boldsymbol{\nu})$, where $\boldsymbol{\nu} = (\nu_1, \nu_2, \ldots, \nu_n)'$. We consider testing H_0: $y_i \sim \mathcal{E}(\lambda_0)$ against H_1: $y_i \sim \mathcal{E}(\lambda)$; $\lambda \sim \mathcal{G}(a, b)$, where y_i is the actual time of death for the ith subject (which will be unobserved if the observed time is a censoring time). BIC corresponds to an implicit normal prior, but the conjugate gamma prior in H_1 seems more appropriate here because it puts mass only on positive values of λ, while the normal prior puts some mass on negative values of λ, which should be excluded. We are interested in how well BIC approximates twice the log Bayes factor (6.5.3), where

$$p(y|H_i) = \int L(\lambda|D, H_i)\pi(\lambda|H_i)d\,\lambda,$$

$L(\lambda|D, H_i)$ is the likelihood function and $\pi(\lambda|H_i)$ denotes the prior distribution of λ under H_i.

Since the prior on λ is not normal, the BIC approximation to twice the log Bayes factor will have asymptotic error of constant order. We consider two unit-information gamma priors, one corresponding to using n in the penalty, the other to using d, where $d = \sum_{i=1}^{n} \nu_i$. These priors have mean equal to the null hypothetical value (λ_0), and variance equal to the inverse expected information in one observation (for n) or in one uncensored observation (for d). In the n case, this yields a $\mathcal{G}(q, q/\lambda_0)$ distribution, where q is the proportion of the data that is censored. In the d case, this yields a $\mathcal{G}(1, 1/\lambda_0)$ distribution. We call these the *overall unit information prior* and the *uncensored unit information prior*, respectively. Using these priors, we can calculate the asymptotic difference between $2 \log \text{BF}$ and BIC for both n and d. We denote this asymptotic error by AE_n or AE_d, depending on which penalty term is used in BIC. The following theorem shows that, asymptotically, using d in the penalty term gives a closer approximation to twice the log Bayes factor.

Theorem 6.5.1 *Consider comparing the exponential survival models H_0 and H_1 using BIC under independent censoring. Then $|AE_d| < |AE_n|$.*

In the context of epidemiological studies, this means that using BIC with d in the penalty will provide a better approximation to the Bayes factor for comparing two competing models.

6.5.3 The Cox Proportional Hazards Model

In the Cox proportional hazards model, estimation of $\boldsymbol{\beta}$ is commonly based on the Cox's partial likelihood, namely,

$$\text{PL}(\boldsymbol{\beta}) = \prod_{i=1}^{n} \left(\frac{\exp(\boldsymbol{x}_i'\boldsymbol{\beta})}{\sum_{\ell \in \mathcal{R}_i} \exp(\boldsymbol{x}_\ell'\boldsymbol{\beta})} \right)^{\nu_i}, \tag{6.5.8}$$

where $\mathcal{R}_i$ is the set of individuals at risk at time y_i, $\boldsymbol{x}_i$ is the vector of covariates for the ith individual, and ν_i is an censoring indicator (uncensored $= 1$).

Suppose that we have survival data on n individuals with independent survival times $y_1, y_2, \cdots, y_n$, and independent censoring indicators $\nu_1, \nu_2, \ldots, \nu_n$. Consider the models H_1: $h_i(t) = \lambda_0(t)\exp(\boldsymbol{x}_i'\boldsymbol{\beta})$, and H_0: $h_i(t) = \lambda_0^{(0)}(t)\exp\{(\boldsymbol{x}_i^{(0)})'\boldsymbol{\beta}^{(0)}\}$, where $\boldsymbol{\beta}^{(0)}$ is a subset of $\boldsymbol{\beta}$ and $\boldsymbol{x}_i^{(0)}$ is the corresponding subset of $\boldsymbol{x}_i$. Then, H_0 and H_1 are nested models, with $\boldsymbol{\theta} = \boldsymbol{\beta}^{(0)}$, $(\boldsymbol{\theta}, \boldsymbol{\psi}) = \boldsymbol{\beta}$, $p_0 = \dim(\boldsymbol{\beta}^{(0)})$ and $p = \dim(\boldsymbol{\beta})$ in the previous notation. The following theorems show that the conclusions from Kass and Wasserman (1995) hold for Cox proportional hazards models, whether n or d is used in the BIC penalty.

Theorem 6.5.2 *Under null orthogonality and independent censoring, equation* (6.5.5) *holds for Cox models when n is used in the penalty for BIC, and the normal overall unit information prior is used.*

Theorem 6.5.3 *Under null orthogonality and independent censoring, equation* (6.5.6) *holds for Cox models when d is used in the penalty for BIC, and the normal uncensored unit information prior is used.*

It follows that (6.5.6) holds for the Cox model, and that the $O(n^{-1/2})$ result also holds for both the overall and the uncensored unit information priors with the appropriate penalty.

Example 6.7. Evaluation of the unit information priors. The choice of d or n in BIC determines the implicit prior variance. A desirable reference prior has most of its mass concentrated in that region of the parameter space in which parameter estimates tend to fall, yet is rather flat over that area. Volinsky and Raftery (2000) considered three commonly cited survival datasets, including the cardiovascular health study (CHS) (see Example 6.6), and two others analyzing lung cancer (PBC) (Prentice, 1973; Kalbfleisch and Prentice, 1980) and liver disease (V.A.) (Dickson, Fleming, and Weisner, 1985; Fleming and Harrington, 1991). They compared the priors for each choice of penalty to the parameter estimates eventually calculated.

TABLE 6.14. Comparison of Mean Implicit Prior Standard Deviation for the Two Different Penalties in BIC.

Data	p	n	% Censored	Prior SD (d)	Prior SD (n)	Range of $\hat{\beta}_i$
CHS	23	4504	96	1.0	5.0	$(-0.09, 0.34)$
PBC	12	312	60	1.0	1.8	$(-0.29, 0.34)$
V.A.	8	137	7	1.0	1.0	$(-0.65, 0.43)$

Source: Volinsky and Raftery (2000).

For each of the studies, both the overall and the uncensored unit information priors were computed. These three datasets include a total of 43 candidate independent variables for predicting the risk of onset of a disease. Table 6.14 compares the unit information priors for the three datasets. For clarity, the results have been standardized by the uncensored unit information prior standard deviation. In the cardiovascular health study, the overall unit-information prior standard deviations are on average five times greater than the average of the uncensored unit-information prior standard deviations, reflecting the large amount of censoring in these data. The parameter estimates (measured in units of one uncensored unit information prior standard deviation) range from -0.09 to 0.34, indicating that a prior standard deviation of 5.0 is much more spread out than necessary. For the three datasets, the range of all the estimates is $(-0.65, 0.43)$, indicating that the uncensored unit information prior (with a prior standard deviation of 1 on this scale) is more realistic because it covers the distribution of estimates very well without being much more spread out.

Figure 6.6 is a histogram of the estimated coefficients in the three datasets, with overall and uncensored unit-information priors also shown. The plot shows that the uncensored unit-information prior puts more prior mass on what is more likely to occur, yet is still a rather conservative prior which allows for outlying estimates. It thus seems more satisfactory in practice for these datasets.

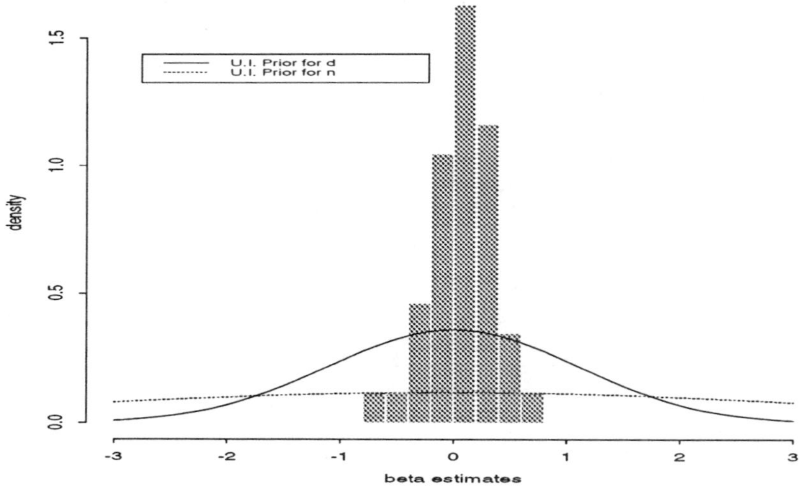

FIGURE 6.6. Histogram of standardized estimated coefficients from the three datasets.

Example 6.8. The cardiovascular health study (Example 6.6 continued). The purpose of this example is to show the effect of changing the penalty term on the analysis as well as on predictive performance. For these data, d is approximately 4% of n, so the choice of the penalty term will have a substantial impact on the analysis. In what follows we refer to the analyses that use n and d in the penalty term as the *n-analysis* and the *d-analysis*, respectively.

For the n-analysis, the best model according to BIC has seven variables (regular aspirin use, systolic blood pressure, creatinine level, diabetes, presence of atrial fibrillation, stenosis of the carotid artery, and a timed walk). When d is used, the penalty for additional parameters is smaller, and the best model now contains nine variables — the same as above plus age and diuretic use. The question of whether or not to include age in the final model is an interesting one. Although stroke incidence is known to increase with age, it is possible that age has no inherent effect on stroke beyond its effect on the other covariates. The d-analysis implies that age is indeed an important variable independent of the others. It may be acting as a surrogate for unidentified variables, or perhaps age itself is inherently a risk factor for stroke. Articles written in the medical literature on this dataset (Fried et al., 1991; Manolio et al., 1996) support the conclusion of the d-analysis that both age and diuretic use are important independent risk factors.

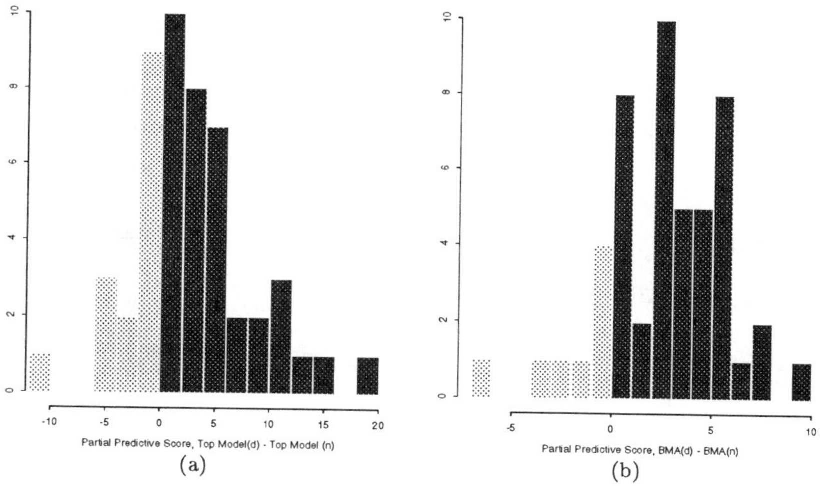

FIGURE 6.7. Histograms showing the difference in partial predictive score $(d - n)$ on 50 splits of the data.

Another way of assessing the models is through predictive performance. A model with improved prediction will be able to provide a better risk evaluation of future patients. The data were split in half to create a model building set and a validation set. The models are assessed using an analogue of Good's log score called the partial predictive score (PPS), which is discussed in detail in Section 6.4.3. For 50 different splits of the data, Volinsky and Raftery (2000) calculated the difference in the PPS. Figure 6.7 shows the histograms of these 50 differences. In Figure 6.7, a positive result (shown in black) indicates a "win" for d. On the left are the results of picking the single best model, and the results from model averaging are shown on the right. In 35 of the 50 splits, the d-best model performed better than the n-best model, with an average of a 3.4% improvement in predictive performance per event. Similar improvements in predictive performance have been shown to make a substantive difference in risk classification models discussed in Volinsky et al. (1997).

From this example, we get two main benefits: a better evaluation of the individual risk factor's association with the outcome, and a better way of predicting who is most susceptible to disease in the future. Since the risk factor data are noninvasive and easily collected, the d-best model can help to provide an improved risk assessment model for cardiovascular disease.

Exercises

6.1 Suppose we have a sample $\{(\boldsymbol{\beta}_{0,l}^{(\mathcal{K})}, \boldsymbol{\delta}_{0,l}),\ l = 1, 2, \ldots, L\}$ from $\pi_0(\boldsymbol{\beta}^{(\mathcal{K})}, \boldsymbol{\delta}|D_0^{(\mathcal{K})})$ in (6.1.8). Show that under some regularity conditions such as ergodicity, the estimated prior model probability $\hat{p}(m)$ given by (6.1.10) is a simulation consistent estimator of $p(m)$. That is,

$$\lim_{L\to\infty} \hat{p}(m) = \lim_{L\to\infty} \frac{\frac{1}{L}\sum_{l=1}^{L} \frac{\pi_0^*(\boldsymbol{\beta}_{0,l}^{(m)}, \boldsymbol{\delta}_{0,l}|D_0^{(m)}) w_0(\boldsymbol{\beta}_{0,l}^{(-m)}|\boldsymbol{\beta}_{0,l}^{(m)})}{\pi_0^*(\boldsymbol{\beta}_{0,l}^{(\mathcal{K})}, \boldsymbol{\delta}_{0,l}|D_0^{(\mathcal{K})})} p_0(m)}{\frac{1}{L}\sum_{j=1}^{\mathcal{K}}\sum_{l=1}^{L} \frac{\pi_0^*(\boldsymbol{\beta}_{0,l}^{(j)}, \boldsymbol{\delta}_{0,l}|D_0^{(j)}) w_0(\boldsymbol{\beta}_{0,l}^{(-j)}|\boldsymbol{\beta}_{0,l}^{(j)})}{\pi_0^*(\boldsymbol{\beta}_{0,l}^{(\mathcal{K})}, \boldsymbol{\delta}_{0,l}|D_0^{(\mathcal{K})})} p_0(j)}$$

$$= p(m), \text{ almost surely,}$$

where $\pi_0^*(\boldsymbol{\beta}^{(\mathcal{K})}, \boldsymbol{\delta}|D_0^{(\mathcal{K})})$ is defined in (6.1.9).

6.2 Let $\pi_l(\boldsymbol{\theta})$, $l = 1, 2$, be two densities, each of which is known up to a normalizing constant:

$$\pi_l(\boldsymbol{\theta}) = \frac{q_l(\boldsymbol{\theta})}{c_l}, \quad \boldsymbol{\theta} \in \Theta_l,$$

where Θ_l is the support of π_l, and the unnormalized density $q_l(\boldsymbol{\theta})$ can be evaluated at any $\boldsymbol{\theta} \in \Theta_l$ for $l = 1, 2$. Then, the ratio of two normalizing constants is defined as $r = c_1/c_2$. Also let $\pi(\boldsymbol{\theta})$ be an arbitrary density with the support $\Theta = \Theta_1 \cup \Theta_2$. Given a random or MCMC sample $\{\boldsymbol{\theta}_1, \boldsymbol{\theta}_2, \ldots, \boldsymbol{\theta}_n\}$ from π, a ratio importance sampling (RIS) estimator of r, proposed by Chen and Shao (1997a), is given by

$$\hat{r} = \frac{\sum_{i=1}^{n} q_1(\boldsymbol{\theta}_i)/\pi(\boldsymbol{\theta}_i)}{\sum_{i=1}^{n} q_2(\boldsymbol{\theta}_i)/\pi(\boldsymbol{\theta}_i)}. \tag{6.E.1}$$

Chen and Shao (1997a) also show that the relative mean-square error $\mathrm{RE}^2(\hat{r}) = E(\hat{r} - r)^2/r^2$ has the asymptotic form

$$\mathrm{RE}^2(\hat{r}_{\mathrm{RIS}}) = \frac{1}{n} E_\pi \left[\frac{\{\pi_1(\boldsymbol{\theta}) - \pi_2(\boldsymbol{\theta})\}^2}{\pi^2(\boldsymbol{\theta})} \right] + o\Big(\frac{1}{n}\Big), \tag{6.E.2}$$

where the expectation E_π is taken with respect to π.

(a) Show that $\hat{p}(m)$ in (6.1.10) is a RIS estimator by appropriately choosing q_1, q_2, and π in (6.E.1).

(b) The simulation standard error of $\hat{p}(m)$ is defined as the square root of the estimated variance of $\hat{p}(m)$. Using (6.E.2), show that a first-order approximation of the simulation standard error (s.e.) of $\hat{p}(m)$ is given by

$$\begin{aligned} &s.e.(\hat{p}(m)) \\ &= \frac{1}{\hat{d} L^{1/2}} \left[\frac{1}{L} \sum_{l=1}^{L} \left\{ \frac{\pi_0^*(\boldsymbol{\beta}_{0,l}^{(m)}, \boldsymbol{\delta}_{0,l} | D_0^{(m)}) w_0(\boldsymbol{\beta}_{0,l}^{(-m)} | \boldsymbol{\beta}_{0,l}^{(m)})}{\pi_0^*(\boldsymbol{\beta}_{0,l}^{(\mathcal{K})}, \boldsymbol{\delta}_{0,l} | D_0^{(\mathcal{K})})} p_0(m) \right. \right. \\ &\quad \left. \left. - \ \hat{p}(m) \sum_{j=1}^{\mathcal{K}} \frac{\pi_0^*(\boldsymbol{\beta}_{0,l}^{(j)}, \boldsymbol{\delta}_{0,l} | D_0^{(j)}) w_0(\boldsymbol{\beta}_{0,l}^{(-j)} | \boldsymbol{\beta}_{0,l}^{(j)})}{\pi_0^*(\boldsymbol{\beta}_{0,l}^{(\mathcal{K})}, \boldsymbol{\delta}_{0,l} | D_0^{(\mathcal{K})})} p_0(j) \right\}^2 \right]^{1/2}, \end{aligned} \tag{6.E.3}$$

where

$$\hat{d} = \frac{1}{L} \sum_{l=1}^{L} \sum_{j=1}^{\mathcal{K}} \left\{ \frac{\pi_0^*(\boldsymbol{\beta}_{0,l}^{(j)}, \boldsymbol{\delta}_{0,l} | D_0^{(j)}) w_0(\boldsymbol{\beta}_{0,l}^{(-j)} | \boldsymbol{\beta}_{0,l}^{(j)})}{\pi_0^*(\boldsymbol{\beta}_{0,l}^{(\mathcal{K})}, \boldsymbol{\delta}_{0,l} | D_0^{(\mathcal{K})})} p_0(j) \right\},$$

and $\{(\boldsymbol{\beta}_{0,l}^{(\mathcal{K})}, \boldsymbol{\delta}_{0,l}),\ l = 1, 2, \ldots, L\}$ is an MCMC sample from $\pi_0(\boldsymbol{\beta}^{(\mathcal{K})}, \boldsymbol{\delta} | D_0^{(\mathcal{K})})$ in (6.1.8).

6.3 Show that the posterior model probability $p(m|D^{(m)})$ given in (6.1.17) is invariant to scale changes in the covariates.

6.4 As an alternative to the prior distribution for $(\boldsymbol{\beta}^{(m)}, \boldsymbol{\delta})$ given in (6.1.4), Ibrahim, Chen, and MacEachern (1999) propose the joint prior

$$\pi(\boldsymbol{\beta}^{(m)}, \boldsymbol{\delta} | D_0^{(m)}) = \pi(\boldsymbol{\beta}^{(m)} | D_0^{(m)}) \pi(\boldsymbol{\delta} | \boldsymbol{\theta}_0), \tag{6.E.4}$$

where $\pi(\boldsymbol{\delta}|\boldsymbol{\theta}_0)$ is the same as the one given in (6.1.4),

$$\boldsymbol{\beta}^{(m)} | D_0^{(m)} \sim N_{p_m}(\boldsymbol{\mu}_0^{(m)}, c_0 (T_0^{(m)})^{-1}), \tag{6.E.5}$$

and $\boldsymbol{\mu}_0^{(m)}$ and $T_0^{(m)}$ are obtained by (3.3.8) and (3.3.9) with $D_0^{(m)}$ replacing D_0. For the historical data from Study 1 described in Example 6.1 and using (6.E.4) with the same prior $\pi(\boldsymbol{\delta}|\boldsymbol{\theta}_0)$ given in Example 6.1, compute the prior means and prior standard deviations of the regression coefficients for models $(x_2, x_3, x_4, x_7, x_8)$, $(x_1, x_2, x_3, x_4, x_5, x_7, x_8)$, and $(x_1, x_2, x_3, x_4, x_5, x_6, x_7, x_8)$ for $c_0 = 1$, 3, 10, and 30.

6.5 In (6.1.6), Ibrahim, Chen, and MacEachern (1999) let d_0 depend on m, and propose the following algorithm for determining $d_0^{(m)}$. They note that small values of $d_0^{(m)}$ will tend to increase the prior probability for model m. They let $d_0^{(m)}$ depend on p_m as follows. Define $v_0^{(m)} = \frac{1}{1+d_0^{(m)}}$ and take $v_0^{(m)}$ to be of the form $v_0^{(m)} = ba^{\frac{1}{p_m}}$ where b and a are specified prior parameters in $[0,1]$. The decomposition of $v_0^{(m)}$ into this form provides flexibility in assigning a prior for $\mathcal{M}$. The parameters a and b play an important role in this formulation, especially in determining prior probabilities for models of different size. A proposed method of specifying a and b is as follows. Let p denote the number of covariates for the full model, and suppose, for example, the investigator wishes to assign prior weights in the range $r_1 = 0.10$ for the single covariate model ($i.e., p_m = 1$) to $r_2 = 0.15$ for the full model ($i.e., p_m = p$). This leads to the equations $r_1 = ba = 0.10$ and $r_2 = ba^{\frac{1}{p}} = 0.15$ for which a closed form solution for a and b can be obtained.

Consider the historical data with $p = 8$ covariates from Study 1 described in Example 6.1 and $(r_1, r_2) = (0.10, 0.15)$.

(a) Find a and b.

(b) Using (6.1.10) and (6.1.5) with the above-proposed prior $\pi_0(\boldsymbol{\beta}^{(m)}|d_0^{(m)})$ and the same prior $\pi(\boldsymbol{\delta}|\boldsymbol{\theta}_0)$ given in Example 6.1, compute the prior model probabilities $(p(m))$ for models $(x_2, x_3, x_4, x_7, x_8)$, $(x_1, x_2, x_3, x_4, x_5, x_7, x_8)$, and $(x_1, x_2, x_3, x_4, x_5, x_6, x_7, x_8)$.

(c) Compute the simulation standard error (s.e.) of $\hat{p}(m)$ for each of the models in part (b) using (6.E.3).

6.6 Consider the multiple myeloma E2479 data and the historical data used in Example 6.1. Using the prior distribution for $(\boldsymbol{\beta}^{(m)}, \boldsymbol{\delta})$ given in Exercise 6.4 and the prior model probabilities discussed in Exercise 6.5, compute the posterior model probabilities $p(m|D^{(m)})$ for models $(x_2, x_3, x_4, x_7, x_8)$, $(x_1, x_2, x_3, x_4, x_5, x_7, x_8)$, and $(x_1, x_2, x_3, x_4, x_5, x_6, x_7, x_8)$.

6.7 Consider an analysis of the multiple myeloma data given in Example 6.1 using a piecewise constant hazard model as given by (3.1.1) in Chapter 3.

(a) Find $L(\boldsymbol{\beta}^{(m)}, \boldsymbol{\lambda}|D^{(m)})$.

(b) Now consider only two models in the model space, with $m_1 = (x_2, x_3, x_4, x_7, x_8)$ and $m_2 = (x_1, x_2, x_3, x_4, x_7, x_8)$, and $p(m_1) = p(m_2) = 1/2$. Consider the prior given by (3.3.6) in Chapter 3. Compute $\boldsymbol{\mu}_0^{(m)}$ and $(T_0^{(m)})^{-1}$, for $m = 1, 2$, using the historical data.

(c) Show that

$$\left[\phi(\boldsymbol{\beta}^{(m)}|\boldsymbol{\mu}_0^{(m)}, (T_0^{(m)})^{-1})\right]^{a_0} \propto \phi(\boldsymbol{\beta}^{(m)}|\boldsymbol{\mu}_0^{(m)}, a_0^{-1}(T_0^{(m)})^{-1}),$$

where $\phi(.|\mu, \Sigma)$ denotes the multivariate normal density with mean $\boldsymbol{\mu}$ and covariance matrix Σ. This establishes the connection between the prior in (3.3.6) and the power prior.

(d) Compute $p(m|D^{(m)})$ using $\pi(\boldsymbol{\beta}^{(m)}, \boldsymbol{\lambda}) = \pi(\boldsymbol{\beta}^{(m)})\pi(\boldsymbol{\lambda})$, where $\pi(\boldsymbol{\beta}^{(m)})$ is given in part (b), $\pi(\boldsymbol{\lambda})$ is a product of J independent gamma densities with vague hyperparameters, and $a_0 = 0.5$.

(e) Repeat part (d) using $a_0 = 0.01$, 0.1, 0.7, and 1, and assess the sensitivity of the posterior probabilities to the choice of a_0.

6.8 Consider an analysis of the E1684 and E1690 melanoma data given in Example 1.2 using a piecewise constant hazard model as given by (3.1.1) in Chapter 3. Consider the two models, $m_1 = (x_1, x_2, x_3)$ and $m_2 = (x_2, x_3)$, with $p(m_1) = p(m_2) = 1/2$. Further assume that

$$\pi(\boldsymbol{\beta}^{(m)}, \boldsymbol{\lambda}) = \pi_1(\boldsymbol{\lambda})\pi_2(\beta_1^{(m)})\pi_3(\beta_2, \beta_3)$$

where $\pi_2(\beta_1^{(2)})$ has a probability mass of 1 at $\beta_1^{(2)} = 0$. Assume a noninformative prior for $\pi(\boldsymbol{\lambda})$ as in Exercise 6.7.

(a) A reasonable prior for $\pi_2(\beta_1^{(1)})$ is a $N(\mu_0, \sigma_0^2)$ prior. We can elicit μ_0 and σ_0^2 as follows: γ_1 (< 0) is the value which represents the clinically significant IFN ($x_1 = 2$) effect, that is, a clinically significant IFN effect should reduce the hazard by a factor $\exp(\gamma_1)$. Then given γ_1, we take

$$P\left(Z > \frac{\gamma_1 - \mu_0}{\sigma_0}\right) = 0.01, \tag{6.E.6}$$

a priori, where $Z = \frac{\beta_1^{(1)} - \mu_0}{\sigma_0}$ has a $N(0, 1)$ distribution, a priori. Also, we specify γ_2, which is, say, the lower 10th percentile of the prior of $\beta_1^{(1)}$, given by

$$P\left(Z < \frac{\gamma_2 - \mu_0}{\sigma_0}\right) = 0.10. \tag{6.E.7}$$

Thus γ_2 is the quantity that, a priori, reduces the IFN effect on the hazard by a factor smaller than $\exp(\gamma_2)$ with a 10% chance. Suppose that $\gamma_1 = -0.20$ and $\gamma_2 = -0.40$. Use (6.E.6) and (6.E.7) to find μ_0 and σ_0.

(b) Now suppose $\pi(\beta_2, \beta_3) = N_2\left(\begin{pmatrix}\mu_{20}\\ \mu_{30}\end{pmatrix}, \Sigma_{20}\right)$. Find μ_{20}, μ_{30}, and Σ_{20} using the prior in (3.3.6) with historical data covariates (x_{20}, x_{30}).

(c) Compute $p(m|D^{(m)})$ for $m = 1, 2$.

6.9 (a) Repeat the analysis of Exercise 6.8 using $\pi_2(\beta_1^{(1)}) = N(0, 10^6)$ (i.e., a noninformative prior).

(b) Compute $p(m|D^{(m)})$ for part (a) and compare your results with those of Exercise 6.8. What are your conclusions?

(c) Based on your results in part (b), which prior specification for $\pi_2(\beta_1^{(1)})$ is more reasonable and why?

6.10 A Simulation Study for Variable Selection

Suppose we generate survival times y_i, $i = 1, 2 \ldots, n$, from a mixture of Weibull densities taking the form

$$f(y_i) = \psi f_1(y_i) + (1 - \psi) f_2(y_i), \tag{6.E.8}$$

where

$$f_1(y_i) = 0.8\lambda_i(\lambda_i y_i)^{(0.8-1)} \exp\left\{-(\lambda_i y_i)^{0.8}\right\},$$

$$f_2(y_i) = 2\lambda_i^2 y_i \exp\{-(\lambda_i y_i)^2\},$$

$0 \le \psi \le 1$, $n = 500$, $\lambda_i = \exp(2x_{i1} + 1.5x_{i2})$ and the $x_i = (x_{i1}, x_{i2})'$ are independent bivariate normal random vectors each with mean $\mu = (1, 1)'$ and covariance matrix $\Sigma = \begin{pmatrix}0.2 & 0.07\\ 0.07 & 0.1\end{pmatrix}$. In addition, 10% of the observations are randomly right censored. Two additional covariates (x_{i3}, x_{i4}) are independently randomly generated, each having a normal distribution with mean 1 and variance 0.25. Further, x_{i3} and x_{i4} are independent. Notice that $f_1(y_i)$ has a decreasing hazard function while $f_2(y_i)$ has an increasing hazard function. The full model ($p = 4$) thus contains four covariates (x_1, x_2, x_3, x_4), and the true model contains the covariates (x_1, x_2). To obtain the prior prediction $\boldsymbol{y}_0$ of $\boldsymbol{y}$, we generate a new set of covariates that have the same distribution as the set corresponding to the y_i's. Then, using this new set of covariates, we generate another independent set of 500 survival times, denoted w_i, from model (6.E.8). We take the prior prediction $\boldsymbol{y}_0 = (y_{01}, y_{02}, \ldots, y_{0n})'$ as a perturbation of $\boldsymbol{w} = (w_1, w_2, \ldots, w_n)'$, taking the form $y_{0i} = w_i + l_0\, \hat{\sigma}\, z_i$, $i = 1, 2, \ldots, n$, where l_0 is a scalar multiple, $\hat{\sigma}$ is the standard deviation of the w_i's, and the z_i are i.i.d.

truncated standard normal random variates. Simulate the datasets with $l_0 = 0$, 0.5, and 1, and $\psi = 0$, 0.5, and 1.
For $\pi(\delta|\boldsymbol{\theta}_0)$ and $\pi_0(\boldsymbol{\delta})$, we take $J = 60$, with each $\delta_j \sim G(s_j - s_{j-1}, 0.1)$, $j = 1, 2, \ldots, J$, with the intervals chosen so that with the combined datasets at least one failure or censored observation falls in each interval.

(a) For each simulated dataset, compute the posterior model probabilities for all 16 possible subset models using (6.1.5) and (6.1.12) for $(\alpha_0, \lambda_0) = (5,5)$, $(30,30)$, and $(100,100)$, $c_0 = 1$, 3, 5, 10, and 100, and $d_0 = 5$, 10, and 30.
(b) For each simulated dataset, compute the posterior model probabilities for all 16 possible subset models using the prior for the model parameters given in Exercise 6.4 with $c_0 = 1$, 3, 5, 10, and 100, and the prior on model space described in Exercise 6.5 with $(r_1, r_2) = (0.10, 0.15)$.
(c) Comment on your findings from this simulation study.

6.11 A SIMULATION STUDY FOR L MEASURE
Generate data y_i from the Weibull model

$$f(y_i|\alpha, \lambda_i) = \alpha^{y_i - 1} \exp\{\lambda_i - y_i^\alpha \exp(\lambda_i)\},$$

where $\lambda_i = \beta_0 + \beta_1 x_i$, $i = 1, 2, \ldots, n$, $n = 200$, $\alpha = 2$, $\boldsymbol{\beta} = (\beta_0, \beta_1) = (1, -1)$, and 10% of the observations are randomly right censored with censoring times $t_i = 0.75 * y_i$. Also generate the covariate x_i as i.i.d. $N(0,1)$ variates. Suppose that the prior distribution for $\boldsymbol{\beta}$ is $N_2(0, 4I_2)$, where I_2 is a 2×2 identity matrix, and the prior distribution for α is a gamma density with shape and scale parameters $(1, 0.5)$. Consider three models:

(1) the true Weibull model from which the data are generated;
(2) the Weibull model with a random α parameter; and
(3) the exponential model ($\alpha = 1$).

Take $y_i^* = \eta(y_i) = \log(y_i)$. Compute the L measures and calibration distributions for these three models.

6.12 Suppose $D = (\boldsymbol{y}, X, \boldsymbol{\nu})$, where $\boldsymbol{y} = (y_1, y_2, \ldots, y_n)'$ denotes the vector of n independent survival times, X is the matrix of covariates with the ith row $\boldsymbol{x}_i'$, and $\boldsymbol{\nu} = (\nu_1, \nu_2, \ldots, \nu_n)'$ is the vector of censoring indicators with $\nu_i = 1$ if y_i is a failure time and $\nu_i = 0$ if y_i is right censored. Let $f(y_i|\boldsymbol{\beta}, \boldsymbol{\lambda}, \boldsymbol{x}_i)$ denote the density of y_i and $S(y_i|\boldsymbol{\beta}, \boldsymbol{\lambda}, \boldsymbol{x}_i)$ the survival function. Also let $\pi(\boldsymbol{\beta}, \boldsymbol{\lambda})$ denote the prior distribution.

(a) Write out the likelihood functions $L(\boldsymbol{\beta}, \boldsymbol{\lambda}|D)$ and $L(\boldsymbol{\beta}, \boldsymbol{\lambda}|D^{(-i)})$, where $D^{(-i)}$ denotes the data with the ith case removed.

(b) Let $\pi(\boldsymbol{\beta}, \boldsymbol{\lambda}|D^{(-i)})$ denote the posterior distribution based on the data $D^{(-i)}$. Show that the CPO can be written as follows:

$$\begin{aligned}\mathrm{CPO}_i =& f(y_i|D^{(-i)}) = \int f(y_i|\boldsymbol{\beta}, \boldsymbol{\lambda}, \boldsymbol{x}_i)\pi(\boldsymbol{\beta}, \boldsymbol{\lambda}|D^{(-i)})\ d\boldsymbol{\beta}\ d\boldsymbol{\lambda} \\ =& \left(\int \frac{1}{f(y_i|\boldsymbol{\beta}, \boldsymbol{\lambda}, \boldsymbol{x}_i)}\pi(\boldsymbol{\beta}, \boldsymbol{\lambda}|D)\ d\boldsymbol{\beta}\ d\boldsymbol{\lambda}\right)^{-1}, \qquad (6.E.9)\end{aligned}$$

when $\nu_i = 1$, and

$$\begin{aligned}\mathrm{CPO}_i =& P(Y_i \geq y_i|D^{(-i)}) = \int S(y_i|\boldsymbol{\beta}, \boldsymbol{\lambda}, \boldsymbol{x}_i)\pi(\boldsymbol{\beta}, \boldsymbol{\lambda}|D^{(-i)})\ d\boldsymbol{\beta}\ d\boldsymbol{\lambda} \\ =& \left(\int \frac{1}{S(y_i|\boldsymbol{\beta}, \boldsymbol{\lambda}, \boldsymbol{x}_i)}\pi(\boldsymbol{\beta}, \boldsymbol{\lambda}|D)\ d\boldsymbol{\beta}\ d\boldsymbol{\lambda}\right)^{-1}, \qquad (6.E.10)\end{aligned}$$

when $\nu_i = 0$.

(c) Using (6.E.9), (6.E.10), and an MCMC sample $\{(\boldsymbol{\beta}_l, \boldsymbol{\lambda}_l),\ l = 1, 2, \ldots, L\}$ from $\pi(\boldsymbol{\beta}, \boldsymbol{\lambda}|D)$, derive a simulation consistent estimator of CPO_i.

6.13 Consider the simulation study described in Exercise 6.11. Compute the LPML statistics defined by (6.3.2) for the three models defined in Exercise 6.11.

6.14 Consider the datasets discussed in Example 6.1 and the prior distribution given in (6.1.4).

(a) Using the same values of prior hyperparameters given in Example 6.1 for $\pi(\boldsymbol{\delta}|\boldsymbol{\theta}_0)$, compute the LPML statistics for models $(x_2, x_3, x_4, x_7, x_8)$, $(x_1, x_2, x_3, x_4, x_5, x_7, x_8)$, and $(x_1, x_2, x_3, x_4, x_5, x_6, x_7, x_8)$ for $c_0 = 1$, 3, 10, and 30, and $(\alpha_0, \lambda_0) = (10, 10)$, $(30, 30)$, $(50, 50)$, and $(100, 100)$.

(b) Are the LPML statistics sensitive to the prior hyperparameters c_0 and (α_0, λ_0)?

(c) Does the top posterior probability model match the top LPML model?

6.15 Write a computer program to verify all of the entries in Table 6.4.

6.16 Write a computer program to verify all of the entries in Tables 6.6 and 6.7.

6.17 Prove (6.4.2).

6.18 Using the E1684 data with the E1673 data as historical data, write a computer program to produce a table similar to that of Table 6.10. For the priors, use those of Table 5.4 with $E(a_0|D_{obs}) = 0.29$.

6.19 Prove (6.5.6).

6.20 Prove Theorem 6.5.1.

6.21 Prove Theorem 6.5.2.

6.22 Using the E1684 data with all of the the covariates given in Table 5.1, carry out variable subset selection with the BIC criterion defined in (6.5.1) using the cure rate model defined by (5.2.1) for the data.

6.23 Prove Theorem 6.5.3.

7 Joint Models for Longitudinal and Survival Data

7.1 Introduction

Joint models for survival and longitudinal data have recently become quite popular in cancer and AIDS clinical trials, where a longitudinal biologic marker such as CD4 count or immune response can be an important predictor of survival. Often in clinical trials where the primary endpoint is time to an event, patients are also monitored longitudinally with respect to one or more biologic endpoints throughout the follow-up period. This may be done by taking immunologic or virologic measures in the case of infectious diseases or perhaps with a questionnaire assessing the quality of life after receiving a particular treatment. Often these longitudinal measures are incomplete or may be prone to measurement error. These measurements are also important because they may be predictive of survival. Therefore methods which can model both the longitudinal and the survival components jointly are becoming increasingly essential in most cancer and AIDS clinical trials.

To motivate these models, we first discuss three different settings where such methodology is advantageous and often used. These are AIDS clinical trials, cancer vaccine trials and quality of life studies. In this chapter we will discuss several methodologies for joint modeling, focusing on Bayesian approaches.

7.1.1 Joint Modeling in AIDS Studies

In clinical trials of therapies for diseases associated with human immunodeficiency virus (HIV), immunologic and virologic markers are measured repeatedly over time on each patient. The interval lengths vary between data collection times and missing data is quite common. These markers are prone to measurement error and high within patient variability due to biological fluctuations (see Lange, Carlin, and Gelfand, 1992; Taylor et al., 1989; Hoover et al., 1992). Modeling these covariates over time is preferable to using the raw data as noted by Tsiatis, DeGruttola, and Wulfsohn (1995), DeGruttola and Tu (1994), DeGruttola et al. (1993), LaValley and DeGruttola (1996), and Tsiatis et al. (1992). In addition, models provide estimates for time points where data are not available. Many HIV clinical trials focus on the opportunistic infections (OI) associated with HIV disease where the survival endpoint is the time to development of the OI. In these trials, immunologic and virologic markers might be utilized as time-varying predictor variables.

The most common measure used to assess immunological health of an HIV patient is the CD4+ lymphocyte count, or CD4 count for short. Higher CD4 counts indicate a stronger immune system that is more prepared to resist infection. Lower CD4 counts indicate a higher risk of an OI. Viral load is a measure of the amount of virus in the blood plasma. A lower viral load is preferable and may indicate successful treatment of the disease. A patient's success on treatment is often evaluated by these two markers. When a patient begins a successful treatment regimen, the viral load may drop drastically and fall below a detectable level. The CD4 count may take longer to respond or may not respond at all. As viral load decreases, we may expect the CD4 count to increase as the immune system has time to recover. However, CD4 count is slower to respond than viral load. Because of this complex relationship between the immunologic and virologic markers, we may want a multivariate model for the longitudinal covariates. These trajectories are generally difficult to model parametrically; therefore, we may want to allow for more flexibility in the curve by considering nonparametric models, as considered by Brown et al. (2000) and Taylor, Cumberland, and Sy (1994).

7.1.2 Joint Modeling in Cancer Vaccine Trials

In cancer vaccine (immunotherapy) trials, vaccinations are given to patients to raise the patient's antibody levels against the tumor cells. In these studies, the time-to-event endpoint is often time to disease progression or time to death. A successful vaccine activates the patient's immune system against future tumor growth. In this case, a patient's antibody production increases, indicating an increase in the bodies' immune strength. Therefore, measuring these antibodies helps the clinician to evaluate the immunity level. Ibrahim, Chen, and Sinha (2000) consider joint modeling for a cancer vaccine study in malignant melanoma (See Example 1.7) which is also discussed in further detail in Section 7.3. They perform a survival analysis adjusting for longitudinal immunological measures. The primary measures of antibody response are the IgG and IgM antibody titres. The levels of these markers are conjectured to be associated with the clinical outcome and are therefore monitored during follow-up. These markers are prone to measurement error; therefore, the raw data should not be used as covariates in a survival analysis. A method which jointly models the longitudinal marker as well as the survival outcome is therefore necessary.

7.1.3 Joint Modeling in Health-Related Quality of Life Studies

The collection of quality of life (QOL) data in clinical trials has become increasingly common, particularly when the survival benefit of a treatment is anticipated to be small or modest. In fact, one might argue that for a patient, quality of life is at times an even more important factor in treatment decisions than any modest survival benefit. Although this type of data provides much useful information for the decision-making process of both patient and physician, the challenges encountered in the collection and analysis of QOL data make it hard to provide meaningful statements about QOL differences by treatment. A QOL survey instrument is typically administered to study participants at a number of prespecified time points during treatment and follow-up. Complete QOL data for patients at all of the specified collection times is frequently unavailable due to adverse events such as treatment toxicities or disease progression. Patients who are very ill when they report to the clinic may be less likely to complete the QOL instrument, and clinic personnel may feel that it is unethical to ask a patient to complete such a form when the patient feels so poorly. Therefore it is quite plausible that the missingness of QOL data is related to the patient's QOL at the assessment time, and strong evidence that QOL data is generally not missing at random has become accepted in the literature (see Bernhard et al., 1998). It is well known that such nonignorable missingness often leads to serious biases and must be taken into account at the time of analysis. These considerations lead to the development of joint models for longitudinal and survival data, where the longitudinal measure is QOL, and the survival component of the model acts as a type of nonignorable missing data mechanism. Troxel at al. (1998) summarize missing data issues that arise in the analysis of quality of life data in cancer clinical trials and discuss software that can be used to fit such models. One such issue is that missingness of QOL data is often not monotone, yielding observations with nonignorable intermittent missingness. Another important issue is that a subject's QOL data is frequently subject to informative censoring by a terminal event such as death or disease progression. Herring and

Ibrahim (2001) discuss frequentist-based approaches for dealing with these types of situations.

7.2 Methods for Joint Modeling of Longitudinal and Survival Data

Often in time-to-event studies, patients are monitored throughout the study period with biologic measurements taken to evaluate health status. Statistical packages are widely available to perform survival analyses with time-dependent covariates. However, if the covariates are measured with error, the analysis becomes more complex. Simply including the raw measurements in the survival analysis leads to bias as pointed out by Prentice (1982). In this section, to motivate the Bayesian paradigm, we will first review several frequentist methods discussed in the statistical literature for jointly modeling survival and longitudinal data. First, we will look at a partial likelihood method developed by Tsiatis, DeGruttola, and Wulfsohn (1995). Next, we will review frequentist methods presented by DeGruttola and Tu (1994), Wulfsohn and Tsiatis (1997), and Hogan and Laird (1997). Then we will discuss the Bayesian joint modeling approaches taken by Faucett and Thomas (1996), Ibrahim, Chen, and Sinha (2000), and Wang and Taylor (2000).

7.2.1 Partial Likelihood Models

The hazard function of survival time with time-dependent covariates $\mathcal{X}^*(t)$ is generally expressed as

$$h(t|\mathcal{X}^*(t)) = \lim_{\delta \to 0} \frac{1}{\delta} P(t \le T \le t + \delta | T \ge t, \mathcal{X}^*(t)),$$

where $\mathcal{X}^*(t)$ is the covariate history up to time t and T is the true survival time.

In the presence of right censoring we only observe $Y = \min(T, C)$, where C is a potential censoring time and the failure indicator ν, which is equal to 1 if the individual is observed to fail ($T \le C$), and 0 otherwise. Therefore, we can write the hazard for those who fail as

$$\lim_{\delta \to 0} \frac{1}{\delta} P(t \le Y \le t + \delta, \nu = 1 | Y \ge t, \mathcal{X}^*(t)).$$

The proportional hazards model relates the hazard to time-dependent covariates,

$$h(t|\mathcal{X}^*(t)) = h_0(t)\varphi(\mathcal{X}^*(t), \boldsymbol{\beta}),$$

where $\varphi(\mathcal{X}^*(t), \boldsymbol{\beta})$ is a function of the covariate history specified up to an unknown parameter or vector of parameters $\boldsymbol{\beta}$. Leaving the underlying baseline hazard $h_0(t)$ unspecified, one approach for estimating $\boldsymbol{\beta}$ is to maximize the Cox's partial likelihood

$$\prod_{i=1}^{n}\left[\varphi(\mathcal{X}_i^*(y_i),\boldsymbol{\beta})/\sum_{j=1}^{n}\varphi(\mathcal{X}_j^*(y_i),\boldsymbol{\beta})I(y_j\geq y_i)\right]^{\nu_i}, \tag{7.2.1}$$

where $I(y_j \geq y_i)$ is the indicator function so that $I(y_j \geq y_i) = 1$ if $y_j \geq y_i$ and 0 otherwise, $\mathcal{X}_i^*(t)$ is the covariate history for the ith case, y_i is the observed survival time, and ν_i is the censoring indicator, where $\nu_i = 1$ if y_i is a failure time and $\nu_i = 0$ otherwise, for $i = 1, 2, \ldots, n$.

In our present setting, the true covariate history, $\mathcal{X}^*(t)$, is not available. However, we may have observations, $X(t)$, representing some function of the true covariate, $X^*(t)$, which we refer to here as the *trajectory function.*

Tsiatis, DeGruttola, and Wulfsohn (1995) present a computationally straightforward and easy-to-implement approach which reduces the bias in a model with time-varying covariates measured with error. They use asymptotic approximations to show consistency of estimates for modeling the longitudinal data separately, then plugging the estimates into a Cox proportional hazards model. Estimation and inference for the survival model are carried out using the partial likelihood theory developed by Cox (1972, 1975). In the case where the trajectory function $X^*(t)$ and $X(t)$ have the same dimension, Tsiatis, DeGruttola, and Wulfsohn (1995) specify the longitudinal model as

$$X(t) = X^*(t) + \epsilon(t),$$

where $\epsilon(t)$ is measurement error with $E(\epsilon(t)) = 0$, $\text{Var}(\epsilon(t)) = \sigma^2$ and $\text{Cov}(\epsilon(t_1), \epsilon(t_2)) = 0$, $t_1 \neq t_2$, and $X^*(t)$ is the trajectory function. Letting $\mathcal{X}(t) = \{X(t_1), X(t_2), \ldots, X(t_j);\ t_j \leq t\}$ denote the history of the observed covariate up to time t leads to the hazard

$$h(t|\mathcal{X}(t)) = \int h(t|\mathcal{X}^*(t), \mathcal{X}(t))dP(\mathcal{X}^*(t)|\mathcal{X}(t), Y \geq t).$$

Further assumptions that neither the measurement error nor the timing of the visits prior to time t are prognostic yield

$$h(t|\mathcal{X}^*(t), \mathcal{X}(t)) = h(t|\mathcal{X}^*(t)) = h_0(t)\varphi(\mathcal{X}^*(t), \boldsymbol{\beta}).$$

Combining the previous two expressions results in

$$h(t|\mathcal{X}(t)) = h_0(t)E[\varphi(\mathcal{X}^*(t), \boldsymbol{\beta})|X(t_1), ..., X(t_j), t_j \leq t, Y \geq t]. \tag{7.2.2}$$

Denote the conditional expectation in (7.2.2) by $E(t, \boldsymbol{\beta})$. If $E(t, \boldsymbol{\beta})$ were known, we could estimate $\boldsymbol{\beta}$ by maximizing Cox's partial likelihood,

$$\prod_{i=1}^{n}\left[E_i(y_i,\boldsymbol{\beta})/\sum_{j=1}^{n}E_j(y_i,\boldsymbol{\beta})I(y_j\geq y_i)\right]^{\nu_i}, \tag{7.2.3}$$

where $E_i(t, \boldsymbol{\beta}) = E[\varphi(\mathcal{X}_i^*(t), \boldsymbol{\beta})|\mathcal{X}_i(t), Y_i \geq t]$, $\mathcal{X}_i^*(t)$ and $\mathcal{X}_i(t)$ denote the histories of true and observed covariates, and Y_i is the observed survival time for $i = 1, 2, \ldots, n$.

Consider the case when the hazard is only a function of the univariate current value, $X(t)$. For the relative risk formulation of the original Cox model, $\varphi(x, \beta) = e^{x\beta}$. Thus, to compute $E(t, \beta)$ in (7.2.2), we must estimate

$$E[\exp\{\beta \mathcal{X}^*(t)\}|\mathcal{X}(t), Y \geq t],$$

which is the moment generating function of the conditional distribution $[\mathcal{X}^*(t)|\mathcal{X}(t), Y \geq t]$. Assuming a normal approximation to this conditional distribution, we are led to the moment generating function

$$\exp\{\beta\mu(t|\mathcal{X}(t)) + \beta^2\sigma^2(t|\mathcal{X}(t))/2\},$$

where

$$\mu(t|\mathcal{X}(t)) = E\{X^*(t)|\mathcal{X}(t), Y \geq t\}$$

and

$$\sigma^2(t|\mathcal{X}(t)) = \text{Var}\{X^*(t)|\mathcal{X}(t), Y \geq t\}.$$

At each event time, a new model for the covariate is fit given all the covariate data up to that event time. The fitted value for that time is then plugged into the model. For models that are more complicated than the Cox model or the additive relative risk model, Tsiatis, DeGruttola, and Wulfsohn (1995) recommend using the first-order approximation:

$$\begin{aligned} E(t, \boldsymbol{\beta}) &= E[\varphi(X^*(t), \boldsymbol{\beta})|\mathcal{X}(t), Y \geq t)] \\ &\approx \varphi(E(X^*(t)|\mathcal{X}(t), Y \geq t), \boldsymbol{\beta}). \end{aligned}$$

This approximation allows any type of model to be used to fit the longitudinal covariates. The conditional expected values of the covariates $X^*(t)$ are simply plugged into (7.2.3) to get the maximum likelihood estimates of the regression parameter $\boldsymbol{\beta}$ in the proportional hazards model. It is not clear how appropriate this approximation is or how it can be validated. Use of this approximation gives the investigator a method which can be easily implemented using existing software. The analyst simply fits the longitudinal data using an appropriate model, then includes the fitted values from this model in a Cox proportional hazards model. Variance estimation follows from calculating the observed information from the partial likelihood function. However, alternative formulations would be more desirable because it would be easier to validate model assumptions and would make more efficient use of the data.

7.2.2 Joint Likelihood Models

Although the two-stage procedure discussed in the previous section allows for an easy analysis of the data with existing software packages and reduces bias over using the raw covariate data $X(t)$ directly in a Cox model, a modeling approach that makes more efficient use of the data by modeling the outcomes jointly may be more desirable. Several approaches have been taken to model survival and longitudinal outcomes through a joint likelihood model. These approaches have the advantage of making more efficient use of the data since information about survival also goes into modeling the covariate process. They produce unbiased estimates and do not rely on approximations for incorporation of more complex covariate trajectories.

One approach is based on a model for survival conditional on the observed longitudinal covariate with the joint likelihood equal to $f_{Y|X} f_X$. This model is used when time to event is the primary outcome of interest and longitudinal measurements may help predict the outcome. The second approach sets the model up as $f_{X|Y} f_Y$. This second approach is more often used in longitudinal studies where one might want to account for time to loss of follow-up. DeGruttola and Tu (1994) propose such an approach and extend the general random effects model to the analysis of longitudinal data with informative censoring. A similar model for informatively censored longitudinal data was proposed by Schluchter (1992) and Schluchter, Greene, and Beck (2000).

DeGruttola and Tu (1994) jointly model survival times and disease progression using normally distributed random effects. Assuming that these two outcomes are independent given the random effects, the joint likelihood is easily specified. The maximum likelihood estimates of the unknown parameters are obtained using the EM algorithm. Their model can be described as follows. Consider a sample of n subjects, indexed by i, each of whom has m_i observations of a marker of disease progression. Let X_i be an $m_i \times 1$ vector, whose elements X_{ij} are the observed values of the marker for the ith person on the jth occasion of measurement for $i = 1, 2, \ldots, n$; $j = 1, 2, \ldots, m_i$. Let $y_i = \min(t_i, c_i)$, where t_i and c_i denote the survival and censoring times for the ith subject, respectively. Then, the mixed effects model of the disease progression marker is

$$X_i = X_i^* + \epsilon_i$$

and

$$X_i^* = T_i \boldsymbol{\alpha} + Z_i \boldsymbol{b}_i,$$

where $\boldsymbol{\alpha}$ is a $p \times 1$ vector of unknown parameters, T_i is a known full-rank $m_i \times p$ design matrix, $\boldsymbol{b}_i \sim N(0, \Psi)$ denotes a $k \times 1$ vector of unknown individual effects, the $\boldsymbol{b}_i$'s are i.i.d., Z_i is a known full-rank $m_i \times k$ design matrix, $\boldsymbol{\epsilon}_i \sim N_{m_i}(0, \sigma^2 I_i)$ is a vector of residuals, and I_i is an $m_i \times m_i$ identity matrix. The following normal mixed effects model is used for the survival times (or a monotone transformation of survival, as appropriate)

$$y_i = \boldsymbol{w}_i' \boldsymbol{\zeta} + \boldsymbol{\lambda}' \boldsymbol{b}_i + r_i,$$

where $\boldsymbol{\zeta}$ is $q \times 1$ vector of unknown parameters, $\boldsymbol{w}_i = (w_{i1}, w_{i2}, \ldots, w_{iq})'$ is a $q \times 1$ known design matrix, $\boldsymbol{\lambda}$ is a $k \times 1$ vector of unknown parameters, $r_i \sim N(0, \omega^2)$, $\omega > 0$, and y_i is the survival time or some monotonic transformation of survival time such as the log of survival time. The longitudinal marker and survival times are independent conditional on the random effects; therefore, the complete data log-likelihood, i.e., the likelihood that would apply if $\boldsymbol{b}_i$ and y_i were observed, is written as

$$l_c = \sum_{i=1}^{n} \log[\phi(X_i|\boldsymbol{b}_i, \boldsymbol{\alpha}, \sigma^2)\phi(\boldsymbol{b}_i|\Psi)\phi(y_i|\boldsymbol{b}_i, \boldsymbol{\zeta}, \omega^2)], \tag{7.2.4}$$

where $\phi(.|.)$ denotes the appropriate normal probability density function. Estimation of the unknown parameters is accomplished using the EM algorithm, a technique which iterates between solving for the expected values of functions of the unobserved data (random effects and errors in this case) given the observed data and the maximum likelihood estimates of the parameters until convergence. The sufficient statistics for the M-step are

$$\sum_{i=1}^{n} \boldsymbol{b}_i\boldsymbol{b}_i', \ \sum_{i=1}^{n} \boldsymbol{\epsilon}_i'\boldsymbol{\epsilon}_i, \ \sum_{i=1}^{n} r_i^2, \ \sum_{i=1}^{n} Z_i\boldsymbol{b}_i, \tag{7.2.5}$$

$$\sum_{i=1}^{n} y_i\boldsymbol{b}_i, \text{ and } \sum_{i=1}^{n} w_{ji}\boldsymbol{b}_i \ (1 \le j \le q). \tag{7.2.6}$$

Let $D_{obs} = \{(y_i, X_i), \quad i = 1, 2, \ldots, n\}$ denote the observed data. The parameters to be estimated are $\boldsymbol{\alpha}$, $\boldsymbol{\zeta}$, $\boldsymbol{\lambda}$, σ^2, ω^2, and Ψ, which are collectively denoted by $\boldsymbol{\theta}$. The parameter estimates in the M-step are given by

$$\hat{\sigma}^2 = \left(\sum_{i=1}^{n} m_i\right)^{-1} E\left(\sum_{i=1}^{n} \boldsymbol{\epsilon}_i'\boldsymbol{\epsilon}_i|\hat{\boldsymbol{\theta}}, D_{obs}\right), \quad \hat{\omega}^2 = n^{-1}E\left(\sum_{i=1}^{n} r_i^2|\hat{\boldsymbol{\theta}}, D_{obs}\right),$$

and

$$\hat{\Psi} = n^{-1}E\left(\sum_{i=1}^{n} \boldsymbol{b}_i\boldsymbol{b}_i'|\hat{\boldsymbol{\theta}}, D_{obs}\right),$$

where $\hat{\boldsymbol{\theta}}$ denotes the estimate of $\boldsymbol{\theta}$ at the current EM iteration. The parameters $\boldsymbol{\lambda}$, $\boldsymbol{\alpha}$ and $\boldsymbol{\zeta}$ are estimated by

$$\hat{\boldsymbol{\lambda}} = \left(\sum_{i=1}^{n} \boldsymbol{b}_i\boldsymbol{b}_i'\right)^{-1} \left(\sum_{i=1}^{n} (y_i - \boldsymbol{w}_i\hat{\boldsymbol{\zeta}})\boldsymbol{b}_i\right), \quad \hat{\boldsymbol{\alpha}} = \left(\sum_{i=1}^{n} \mathbf{T}_i'\mathbf{T}_i\right)^{-1} \sum_{i=1}^{n} T_i'(X_i - Z_i\boldsymbol{b}_i), \tag{7.2.7}$$

and

$$\hat{\zeta} = \left(\sum_{i=1}^{n} \boldsymbol{w}_i \boldsymbol{w}_i' \right)^{-1} \sum_{i=1}^{n} \boldsymbol{w}_i (y_i - \hat{\boldsymbol{\lambda}}' \boldsymbol{b}_i). \tag{7.2.8}$$

The covariance matrix of the estimates of the parameters of interest $(\boldsymbol{\alpha}, \boldsymbol{\zeta})$ at convergence of the EM algorithm can be obtained by using Louis's formula (Louis, 1982), and the detailed derivation is thus left as an exercise.

It is not clear how robust this model is to departures from parametric assumptions, especially in the assumption of normality for monotonic transformations of survival times. The link between failure times and longitudinal outcomes is not obvious or easily interpreted since it is made through the random effects alone. There is no parameter linking failure time to the longitudinal covariates directly.

Wulfsohn and Tsiatis (1997) also propose a joint likelihood model. They assume a proportional hazards model for survival conditional on the longitudinal marker. A random effects model is used to model the covariate and measurement error. Maximum likelihood estimates of all parameters in the joint model are obtained using the EM algorithm with numeric quadrature and a Newton-Raphson approximation performed at each iteration. The model is then applied to data from an HIV clinical trial. Denote by t_i the true survival time for individual i. Suppose we observe $y_i = \min(t_i, c_i)$, where c_i corresponds to a potential censoring time, and the censoring indicator ν_i, which takes 1 if the failure is observed and 0 otherwise. Assume the covariate for individual i is measured at time $\boldsymbol{t}_i = (t_{ij} : t_{ij} \leq y_i)$, where t_{ij} is the time from randomization for $j = 1, 2, \ldots, m_i$. The random effects model for the longitudinal covariate is

$$X_{ij} = X_{ij}^* + \epsilon_{ij} \tag{7.2.9}$$

and

$$X_{ij}^* = b_{0i} + b_{1i} t_{ij}, \tag{7.2.10}$$

for $j = 1, 2, \ldots, m_i$, where

$$\boldsymbol{b}_i = (b_{0i}, b_{1i})' \sim N_2(\boldsymbol{b}, \Psi), \quad \epsilon_i \sim N_{m_i}(0, \sigma^2 I_i),$$

$\boldsymbol{b} = (b_0, b_1)'$, $\Psi = (\Psi_{jj^*})$ is a 2×2 unknown covariance matrix, and I_i is an $m_i \times m_i$ identity matrix, for $i = 1, 2, \ldots, n$. The hazard function for the proportional hazards model is given by

$$h(t|\boldsymbol{b}_i, X_i, \boldsymbol{t}_i) = h(t|\boldsymbol{b}_i) = h_0(t) \exp\{\beta(b_{0i} + b_{1i} t)\}.$$

The complete data log-likelihood is given by

$$\sum_{i=1}^{n} \left[\sum_{i=1}^{m_i} \log f(X_{ij}|\boldsymbol{b}_i, \sigma^2) + \log f(\boldsymbol{b}_i|\boldsymbol{b}, \Psi) + \log f(y_i, \nu_i|\boldsymbol{b}_i, h_0, \beta) \right],$$

where $f(X_{ij}|\boldsymbol{b}_i, \sigma^2)$ and $f(\boldsymbol{b}_i|\boldsymbol{b}, \Psi)$ are the densities of $N(b_{0i} + b_{1i}t_{ij}, \sigma^2)$ and $N(\boldsymbol{b}, \Psi)$ distributions, respectively, and

$$\begin{aligned} f(y_i, \nu_i|\boldsymbol{b}_i, h_0, \beta) =& [h_0(y_i)\exp\{\beta(b_{0i} + b_{1i}y_i)\}]^{\nu_i} \\ & \times \exp\left[-\int_0^{y_i} h_0(u)\exp\{\beta(b_{0i} + b_{1i}u)\}du\right]. \end{aligned}$$

The quantities estimated in the M-step of the EM algorithm are

$$\hat{\boldsymbol{b}} = \sum_{i=1}^{n} E_i(\boldsymbol{b}_i)/n,$$

$$\hat{\Psi} = \sum_{i=1}^{n} E_i[(\boldsymbol{b}_i - \hat{\boldsymbol{b}})(\boldsymbol{b}_i - \hat{\boldsymbol{b}})']/n,$$

$$\hat{\sigma}^2 = \sum_{i=1}^{n}\sum_{i=1}^{m_i} E_i(y_{ij} - b_{0i} - b_{1i}t_{ij})^2 / \sum_{i=1}^{n} m_i,$$

and

$$\hat{h}_0(u) = \sum_{i=1}^{n} \frac{\nu_i I(y_i = u)}{\sum_{j=1}^{n} E_j[\exp\{\beta(b_{0i} + b_{1i}u)\}]I(y_j \geq u)}.$$

Here $E_i(\boldsymbol{b}_i)$ denotes the conditonal expectation of $\boldsymbol{b}_i$ given the observed data and the current parameter estimates. The MLE of the parameter of interest, β, is updated at each iteration using a one-step Newton-Raphson algorithm giving an estimate at the kth iteration of

$$\hat{\beta}_k = \hat{\beta}_{k-1} + I^{-1}_{\hat{\beta}_{k-1}} S_{\hat{\beta}_{k-1}},$$

where $S_{\hat{\beta}_{k-1}}$ is the score for β at the $(k-1)^{th}$ iteration and $I_{\hat{\beta}_{k-1}}$ is the information for β at the $(k-1)^{th}$ iteration.

The E-step involves calculating the expected values in the equations for the M-step. The conditional expectation of any function φ of the random effects is denoted by $E_i \equiv E\{\varphi(\boldsymbol{b}_i)|y_i, \nu_i, X_i, \boldsymbol{t}_i, \hat{\boldsymbol{\theta}}\}$, where $\hat{\boldsymbol{\theta}}$ denotes the estimate of $\boldsymbol{\theta}$ at the current EM iteration. Thus

$$E_i = \frac{\int_{-\infty}^{\infty} \varphi(\boldsymbol{b}_i) f(y_i, \nu_i|\boldsymbol{b}_i, \hat{h}_0, \hat{\beta}) f(\boldsymbol{b}_i|X_i, \boldsymbol{t}_i, \hat{\boldsymbol{b}}, \hat{\Psi}, (\hat{\sigma}^2)) d\boldsymbol{b}_i}{\int_{-\infty}^{\infty} f(y_i, \nu_i|\boldsymbol{b}_i, \hat{h}_0, \hat{\beta}) f(\boldsymbol{b}_i|X_i, \boldsymbol{t}_i, \hat{\boldsymbol{b}}, \hat{\Psi}, \hat{\sigma}^2) d\boldsymbol{b}_i},$$

which was evaluated using a two-point Gauss-Hermite quadrature formula.

To estimate the variance of $\hat{\beta}$ at EM convergence, Wulfsohn and Tsiatis (1997) define the profile score $S_\beta(\hat{\boldsymbol{\theta}}_{-\beta}(\hat{\beta}))$ to be the derivative of the log-likelihood with respect to β evaluated at $\hat{\beta}$. The remaining parameters, $\boldsymbol{\theta}_{-\beta}$, are estimated using restricted maximum likelihood estimates which are calculated by using a separate EM algorithm applied to the likelihood and keeping $\hat{\beta}$ fixed. They estimate the variance by calculating this score over several values of β, then they fit a line to these estimates and take the negative inverse of the slope of this line as the estimate of the variance. This involves implementing the EM algorithm several times to get estimates for the other parameters for each value of β to estimate this line. In their model and application, this linear approximation appeared valid.

Some advantages of this approach are that it makes more efficient use of the data than the two-stage approach suggested by Tsiatis, DeGruttola, and Wulfsohn (1995), it uses the full likelihood in estimation, and makes a direct link between the survival and longitudinal covariate. However, the parametric form of the trajectory function, X_{ij}^*, may be inappropriate in some settings. Also, the EM algorithm is slow to converge; therefore, it may not be feasible to extend the model of the trajectory function to the multivariate case. It took two hours to obtain estimates for a dataset with 137 individuals with 24 failures and approximately six covariate measurements after using reasonable starting values for the parameters. There is no mention of the amount of time required to get an estimate for the variance of $\hat{\beta}$. Although more complex models can be fit, it is not clear how computationally intensive they would be, or whether the assumption of linearity of $S_\beta(\hat{\boldsymbol{\theta}}_{-\beta}(\hat{\beta}))$ would still be reasonable.

Example 7.1. AIDS data. Wulfsohn and Tsiatis (1997) applied their method to a dataset from a double-blind placebo controlled trial of patients with advanced HIV. Of the total of 281 patients, 144 were randomized to received zidovudine (ZDV), and the remaining 137 patients were given a placebo. Measurements of CD4 count were taken prior to treatment and approximately every four weeks while on therapy. The study lasted for 18 weeks. The goal of the study was to understand the CD4 trajectories of the patients and to evaluate the strength of the relationship between the CD4 trajectory and survival. The unobserved trajectory here is assumed to be a true "biological marker" according to the definition of Prentice (1989) in that the treatment effect on survival is expressed through its effect on the marker, which then affects survival. Thus, given the marker trajectory, there is no treatment effect on survival. The data was fit to both this model and the two-stage model proposed by Tsiatis, DeGruttola, and Wulfsohn (1995) in which the unobservable marker trajectory estimates are re-evaluated for each risk time. Table 7.1 compares the estimates of the longitudinal covariate from the two models for placebo group results. In Table 7.1, TSM and JM denote the two-stage model and the joint model, respectively. Estimates for the 8th and 16th event times are shown for the two-stage model.

The parameter which describes the strength of the relationship between CD4 and survival, β, was estimated as -0.3029 compared to the two-stage model which estimated it as -0.284. The slope of the fitted line to the profile score for β is estimated to be -41.980, giving an estimate of the

TABLE 7.1. Parameter Estimates for AIDS Data.

Model	Event	b_0	b_1	σ^2	Ψ_{11}	Ψ_{12}	Ψ_{22}
TSM	8th	4.23	−0.0050	0.301	1.18	0.0016	0.000032
	16th	4.27	−0.0045	0.305	1.15	0.0024	0.000029
JM		4.17	−0.0047	0.396	1.11	0.0027	0.000014

Source: Wulfsohn and Tsiatis (1997).

standard error of 0.154 in contrast to 0.144 in the two-stage model. They explain that the increase in the variance estimate is due to the random effects being influenced by the uncertainty in the estimated growth curve parameters, therefore incorporating more variability. Also, they explain the increase in the estimate of standard error for pure measurement error in joint estimation as due to overfitting in the two-stage model.

7.2.3 Mixture Models

Hogan and Laird (1997) use a different approach for constructing the joint likelihood model. They are interested in explaining how the time to some event, such as study dropout, affects a longitudinal outcome. They build the joint likelihood using a conditional-marginal approach by specifying the joint distribution of (X, Y) as $f(X, Y) = f(X|Y)f(Y)$. Parameters in this model have different interpretations than the models discussed in the previous sebsections. Let $\boldsymbol{y} = (y_1, y_2, \ldots, y_n)'$ denote the vector of survival times, and let $\boldsymbol{\nu} = (\nu_1, \nu_2, \ldots, \nu_i)'$ be the vector of censoring indicators. The longitudinal model is specified as

$$(X_i|y_i) = W_i(y_i)\boldsymbol{\alpha} + U_i(y_i)\boldsymbol{b}_i + \boldsymbol{\epsilon}_i,$$

where $X_i = (X_{i1}, X_{i2}, \ldots, X_{i,m_i})'$ is a $m_i \times 1$ vector of measurements recorded at times t_{i1}, t_{i2}, $\ldots$, t_{i,m_i}, $W_i(y_i)$ is the $m_i \times p$ design matrix, which may depend on survival time y_i, $\boldsymbol{\alpha}$ is a $p \times 1$ vector of parameters, $U_i(y_i)$ is the $m_i \times k$ matrix, which may also depend on y_i, $\boldsymbol{b}_i \sim N_k(0, \Psi)$ is independent of $\epsilon_i \sim N_{m_i}(0, \sigma^2 I_i)$, and I_i is the $m_i \times m_i$ identity matrix, for $i = 1, 2, \ldots, n$.

Let $\mathcal{S} = \{s_1, s_2, ..., s_L\}$ be the set of ordered event times. Let $\pi_l = P(Y = s_l)$ with $\sum_{l=1}^{L} \pi_l = 1$. In addition to $\boldsymbol{y}$ and $\boldsymbol{\nu}$, the event time data for the ith subject also consists of a multinomial vector of indicators, $\boldsymbol{\delta}_i = (\delta_{i1}, \delta_{i2}, ..., \delta_{iL})'$, where $\delta_{il} = I\{y_i = s_l\}$, $P(\delta_{il} = 1) = \pi_l$. The missing observations in $\boldsymbol{\delta}_i$ are induced by independent right censoring of the event times. When y_i is a censored event time, then $\nu_i = 0$ and $\boldsymbol{\delta}_i = (0, 0, \ldots, 0, *, \ldots, *)'$ is incomplete ($*$ denotes a missing observation). The final zero occurs at element $\max\{l : s_l \leq y_i\}$. For example, if $\mathcal{S} = \{1, 3, 8\}$ and $(y_i, \nu_i) = (4, 0)$, then $\boldsymbol{\delta}_i = (0, 0, *)'$ in the presence of right censoring.

If the data is completely observed, maximum likelihood estimation is straightforward. Let $\boldsymbol{\theta} = (\boldsymbol{\alpha}, \Psi, \sigma^2, \boldsymbol{\pi})$, where $\boldsymbol{\pi} = (\pi_1, \pi_2, \ldots, \pi_{L-1})'$. Let $D = \{D_i = (X_i, \boldsymbol{\delta}_i, W_i, U_i) : i = 1, 2, ..., n\}$ denote the completely observed data. Also, let $D_{il} = D_i(s_l) = (X_i, \delta_{il}, W_{il}, U_{il})$ denote an individual's complete data and covariates, setting $y_i = s_l$. Then the complete data likelihood is

$$L(\boldsymbol{\theta}|D) = \left\{ \prod_{i=1}^{n} \prod_{l=1}^{L} [f_{X|\boldsymbol{\delta}}(\boldsymbol{\alpha}, \Psi, \sigma^2 | D_{il})]^{\delta_{il}} \right\} \left\{ \prod_{l=1}^{L} \pi_l^{\delta_{+l}} \right\}, \tag{7.2.11}$$

where $\delta_{+l} = \sum_{i=1}^{n} \delta_{il}$. Note that $f_{X|\boldsymbol{\delta}}(\boldsymbol{\alpha}, \Psi, \sigma^2 | D_{il})$ must be found by integrating out the random effects. Since the likelihood factors over the parameters $(\boldsymbol{\alpha}, \Psi, \sigma^2)$ and $\boldsymbol{\pi}$, it may be maximized by separately maximizing $f_{X|\boldsymbol{\delta}}$ and $f_{\boldsymbol{\delta}}$ using commercially available software packages. When event times are censored, the EM algorithm can be used to maximize the complete data log-likelihood

$$\sum_{i=1}^{n} \sum_{l=1}^{L} \delta_{il} \log[\pi_l f_{X|\boldsymbol{\delta}}(\boldsymbol{\alpha}, \Psi, \sigma^2 | D_{il})].$$

In the E-step, we evaluate the expected value of the log-likelihood conditional on the observed data, D_{obs}. The only missing quantities in this model are the δ_{il}'s after censoring. The expected value calculated in the E-step is

$$\begin{aligned} & E[\delta_{il} | \boldsymbol{\theta}^{(r)}, D_{obs}] \\ =& P(\delta_{il} = 1 | \boldsymbol{\theta}^{(r)}, D_{obs}) \\ =& \begin{cases} \dfrac{\pi_l^{(r)} f_{X|\boldsymbol{\delta}}(\boldsymbol{\alpha}^{(r)}, \Psi^{(r)}, (\sigma^2)^{(r)} | D_{il}) I\{s_l > y_i\}}{\sum_{k=1}^{L} \pi_k^{(r)} f_{X|\boldsymbol{\delta}}(\boldsymbol{\alpha}^{(r)}, \Psi^{(r)}, (\sigma^2)^{(r)} | D_{ik}) I\{s_k > y_i\}}, & \text{event censored,} \\ \delta_{il}, & \text{event observed,} \end{cases} \end{aligned}$$

where $\boldsymbol{\theta}^{(r)} = (\boldsymbol{\alpha}^{(r)}, \Psi^{(r)}, (\sigma^2)^{(r)}, \boldsymbol{\pi}^{(r)})$ is the value from the rth iteration of the EM algorithm. The likelihood factors as in the complete case; therefore $(\boldsymbol{\alpha}^{(r+1)}, \Psi^{(r+1)}, (\sigma^2)^{(r)})$ are found by maximizing the likelihood for a mixed model, and

$$\pi_l^{(r+1)} = n^{-1} \sum_{i=1}^{n} E[\delta_{il} | \boldsymbol{\theta}^{(r)}, D_{obs}].$$

Hogan and Laird (1997) recommend using the empirical Fisher information for variance estimation when the sample size is large enough, otherwise they recommend using the observed Fisher information. This approach is best suited for studies where the primary interest is in the longitudinal outcome and a time to event variable has a significant impact on this outcome. However, in settings where survival is the primary study endpoint as discussed in Section 7.1, this model may not be as appropriate.

7.3 Bayesian Methods for Joint Modeling of Longitudinal and Survival Data

The methods discussed in the previous sections for joint modeling have all been based on a frequentist approach. However, it may be advantageous to take a Bayesian approach to solving this problem. In the Bayesian paradigm, asymptotic approximations are not necessary, model assessment is more straightforward, computational implementation is typically much easier, and historical data can be easily incorporated into the inference procedure. In this section, we will review three Bayesian approaches taken by Faucett and Thomas (1996), Ibrahim, Chen, and Sinha (2000), and Wang and Taylor (2000). The Bayesian methods we will review use the same general approach to building the model as Wulfsohn and Tsiatis (1997). However, there is an important difference in specifying the hazard function that affects the survival component's likelihood contribution. A piecewise approximation is made for the underlying hazard by discretizing the time scale into $j = 1, 2, ..., J$ intervals and setting $h_0(t) = \lambda_j,\ s_{j-1} < t \le s_j$. When J is large, the step function approximates a smooth function.

Faucett and Thomas (1996) take a Bayesian approach to solving the same random effects and proportional hazards models as Wulfsohn and Tsiatis (1997), which are defined by (7.2.9) and (7.2.10). They use noninformative priors on all the parameters in order to achieve similar results to maximum likelihood approaches. The authors specified improper uniform priors for $\boldsymbol{b}$ and β, and improper priors for $(\Psi, \sigma^2, \boldsymbol{\lambda})$, where $\boldsymbol{\lambda} = (\lambda_1, \lambda_2, \ldots, \lambda_J)'$. These priors take the form $\pi(\Psi) \propto |\Psi|^{-3/2}$, $\pi(\sigma^2) \propto 1/\sigma^2$ and $\pi(\lambda_j) \propto 1/\lambda_j$. Because these priors are improper, the posterior may be improper.

The full conditional posteriors of the random effects to be used in Gibbs sampling can be found using Bayes' rule. For example, the conditional posterior of the random intercept is

$$\begin{aligned}&[b_{0i}|b_{1i}, \boldsymbol{b}, \Psi, \sigma^2, h_0(t), \beta, \{X_{ij}\}, y_i, \nu_i]\\ &\propto [\{X_{ij}\}|\boldsymbol{b}_i, \sigma^2] \times [b_{0i}|b_{1i}, \boldsymbol{b}, \Psi] \times [y_i, \nu_i|\boldsymbol{b}_i, h_0(t), \beta],\end{aligned}$$

where

$$[\{X_{ij}\}|\boldsymbol{b}_i, \sigma^2] \sim N(b_{0i} + b_{1i}t_{ij}, \sigma^2),$$

$$[b_{0i}|b_{1i}, \boldsymbol{b}, \Psi] \sim N\left(b_0 + \frac{\Psi_{12}}{\Psi_{22}}(b_{1i} - b_1),\ \Psi_{11}(1 - \frac{\Psi_{12}^2}{\Psi_{11}\Psi_{22}})\right), \qquad (7.3.1)$$

and

$$\begin{aligned}f(y_i, \nu_i|\boldsymbol{b}_i, h_0(t), \beta) &\propto [h_0(y_i)\exp\{\beta(b_{0i} + b_{1i}y_i)\}]^{\nu_i}\\ &\quad\times \exp\left[-\int_0^{y_i} h_0(u)\exp\{\beta(b_{0i} + b_{1i}u)\}du\right]. \qquad (7.3.2)\end{aligned}$$

Since the resulting posterior distributions of the random effects do not have a closed form, adaptive rejection sampling is used to generate samples from the posterior. The posteriors for the means of the random effects, b_0 and b_1, are $N(\bar{b}_0 - (\bar{b}_1 - b_1)\frac{\Psi_{12}}{\Psi_{22}}, \Psi_{11}(1 - \frac{\Psi_{12}^2}{\Psi_{11}\Psi_{22}}))$ and $N(\bar{b}_1 - (\bar{b}_0 - b_0)\frac{\Psi_{12}}{\Psi_{11}}, \Psi_{22}(1-\frac{\Psi_{12}^2}{\Psi_{11}\Psi_{22}}))$, respectively, where $\bar{b}_0$ and $\bar{b}_1$ are the means across subjects of the currently sampled values of the random effects. The full conditional posterior distribution of Ψ is a Wishart distribution, and therefore samples are easily generated from this distribution. The conditional posterior of σ^2 is $\mathcal{IG}\left(\sum_{i=1}^n m_i/2, \sum_{i=1}^n \sum_{i=1}^{m_i}[X_{ij} - (b_{0i} + b_{1i}t_{ij})]^2/2\right)$, where $\mathcal{IG}(\alpha, \tau)$ denotes the inverse gamma distribution with density $\pi(\sigma^2|\alpha, \tau) \propto (\sigma^2)^{-(\alpha+1)} \exp(-\tau/\sigma^2)$ for $\alpha > 0$ and $\tau > 0$. The conditional posterior distribution for a single step of the baseline hazard λ_k has a gamma distribution:

$$\lambda_k \sim \mathcal{G}(\alpha_{\lambda_k}, \tau_{\lambda_k}),$$

where the shape parameter

$$\alpha_{\lambda_k} = \sum_{i=1}^n \nu_i I(s_{k-1} < y_i \leq s_k),$$

and the scale parameter

$$\tau_{\lambda_k} = \sum_{i=1}^n I(y_i > s_{k-1}) \int_{s_{k-1}}^{\min(y_i, s_k)} \exp[\beta(b_{0i} + b_{1i}u)]du.$$

The conditional posterior distribution of β is proportional to the likelihood function of the survival parameters in (7.3.2). Faucett and Thomas (1996) used rejection sampling with a normal guess density with mean equal to the maximum likelihood estimate of β at the current iteration and variance equal to twice the inverse of the observed information evaluated at the MLE to generate samples from this distribution.

Ibrahim, Chen, and Sinha (2000) present a Bayesian joint model for bivariate longitudinal and survival data. For their longitudinal model, they assume that two covariates $(X_1(t), X_2(t))$ are observed which both measure the true unobservable univariate antibody measure $(X^*(t))$. The model is specified as

$$X_{i1}(t) = X_i^*(t) + \epsilon_{i1}(t) \tag{7.3.3}$$

and

$$X_{i2}(t) = \alpha_0 + \alpha_1 X_i^*(t) + \epsilon_{i2}(t), \tag{7.3.4}$$

where $\epsilon_{i1} \sim N(0, \sigma_1^2)$ and $\epsilon_{i2} \sim N(0, \sigma_2^2)$. Further motivation of the model in (7.3.3) and (7.3.4) is given in Example 7.2. $X_i^*(t)$ is modeled via a trajectory function, $g_{\gamma_i}(t)$, which may be either linear or quadratic in nature. The survival times (y_i's) are modeled using a proportional hazards model assuming the random errors of the longitudinal component are not prognostic of the survival time. The hazard function is given by

$$h(t|\mathcal{X}_i^*(t^*), \mathcal{X}_i(t^*), z_i) = h(t|\mathcal{X}_i^*(t^*), z_i) = h_0(t)\exp\{\beta_1 X_i^*(t^*) + z_i'\boldsymbol{\beta}_2\}, \tag{7.3.5}$$

where $\mathcal{X}_i^*(t^*)$ and $\mathcal{X}_i(t^*)$ denote the histories of X_i^* and X_i up to time t^*. In (7.3.5), $\boldsymbol{z}_i$ denotes a $p \times 1$ vector of baseline covariates for subject i. We assume that the $\boldsymbol{z}_i$'s are not measured with error. In addition, $\boldsymbol{\beta}_2$ is a $p \times 1$ vector of regression coefficients corresponding to $\boldsymbol{z}_i$. Setting $X_i^*(t) = g_{\boldsymbol{\gamma}_i}(t)$, the quadratic parametric trajectory function, for example, takes on the form $\gamma_{i0} + \gamma_{i1}t + \gamma_{i2}t^2$. Let $t_{i1}, t_{i2}, \ldots, t_{im_i}$ denote the times at which the measurements X_{ij} are taken, and let $g_{\boldsymbol{\gamma}_i}(t_{ij})$ denote the trajectory function evaluated at t_{ij}.

Conditional on the subject-specific trajectory parameters, $\boldsymbol{\gamma}_i = (\gamma_{i0}, \gamma_{i1}, \gamma_{i2})'$, the observed trajectories and survival times are independent. Let $\mathbf{X}_{i1} = (x_{i11}, x_{i21}, \ldots, x_{im_i1})'$ and $\mathbf{X}_{i2} = (x_{i12}, x_{i22}, \ldots, x_{im_i2})'$, and let $X_1 = (X_{11}, X_{21}, \ldots, X_{n1})$, and $X_2 = (X_{12}, X_{22}, \ldots, X_{n2})$. Also let y_i denote the event time for the ith subject, which may be right censored, and let $\boldsymbol{y} = (y_{1,2}, \ldots, y_n)'$ denote the vector of event times. Further, let $\boldsymbol{\nu} = (\nu_1, \nu_2, \ldots, \nu_n)'$ denote the vector of censoring indicators, where $\nu_i = 1$ indicates a failure and $\nu_i = 0$ indicates a right censored observation, and $\boldsymbol{z} = (z_1', z_2', \ldots, z_n')'$. We take $h_0(t)$ to be a constant λ_j over the time intervals $I_j = (s_{j-1}, s_j]$, for $j = 1, 2, \ldots, J$, where $s_0 = 0 < s_1 < \ldots < s_J < s_{J+1} = \infty$. Then, the likelihood function is written as

$$L = f_1(X_1|\boldsymbol{\gamma}, \sigma_1) f_2(X_2|\boldsymbol{\gamma}, \sigma_2, \alpha_0, \alpha_1) f_3(\boldsymbol{y}, \boldsymbol{\nu}, \boldsymbol{z}|\boldsymbol{\lambda}, \beta_1, \boldsymbol{\beta}_2, \boldsymbol{\gamma}), \tag{7.3.6}$$

where f_1 and f_2 correspond to the longitudinal contribution to the likelihood and are specified as

$$f_1(X_1|\boldsymbol{\gamma}, \sigma_1) \propto \sigma_1^{-M} \exp\Big\{-\sum_{i=1}^{n}\sum_{j=1}^{m_i}(x_{ij1} - g_{\boldsymbol{\gamma}_i}(t_{ij}))^2/(2\sigma_1^2)\Big\} \tag{7.3.7}$$

and

$$f_2(X_2|\boldsymbol{\gamma}, \sigma_2, \alpha_0, \alpha_1) \propto \sigma_2^{-M} \exp\Big\{\sum_{i=1}^{n}\sum_{j=1}^{m_i}(x_{ij2} - (\alpha_0 + \alpha_1 g_{\boldsymbol{\gamma}_i}(t_{ij})))^2/(2\sigma_2^2)\Big\}. \tag{7.3.8}$$

The contribution of the survival component to the likelihood is

$$f_3(\boldsymbol{y}, \boldsymbol{\nu}, \boldsymbol{z}|\boldsymbol{\lambda}, \beta_1, \boldsymbol{\beta}_2, \boldsymbol{\gamma}) \propto \left\{-\sum_{j=1}^{J}\sum_{i=1}^{n}\lambda_j B_{ij}\right\}\left(\prod_{j=1}^{J}\lambda_j^{d_j}\right) \times \exp\left\{\sum_{j=1}^{J}\sum_{i=1}^{n}\nu_i(\beta_1 g_{\boldsymbol{\gamma}_i}(t_{ij}^*) + \boldsymbol{\beta}_2' z_i)\right\}, \tag{7.3.9}$$

where d_j is the number of failures in the kth time interval, t^*_{ij} is the the most recent time at which a measurement was taken, and $M = \sum_{i=1}^n m_i$. In (7.3.9), the computational algorithm for B_{ij} proceeds as follows:

(i) If $y_i < s_{j-1}$, $B_{ij} = 0$.

(ii) If $y_i > s_j$, let $j_{i1} = \max\{l : t^*_{il} \le s_{j-1}\}$ and $j_{i2} = \max\{l : t^*_{il} \le s_j\}$, where t^*_{il} is the rescaled t_{il} so that t^*_{il} has the same unit as s_i. Then if $j_{i1} = j_{i2}$,

$$B_{ij} = (s_j - s_{j-1}) \exp\left\{\beta_1 g_{\gamma_i}(t_{ij_{i1}}) + \beta_2' z_i\right\},$$

and if $j_{i1} < j_{i2}$,

$$\begin{aligned} B_{ij} =& (t^*_{i,j_{i1}+1} - s_{j-1}) \exp\left\{\beta_1 g_{\gamma_i}(t_{i,j_{i1}+1}) + \beta_2' z_i\right\} \\ &+ \sum_{l=j_{i1}+1}^{j_{i2}} (t^*_{i,l+1} - t^*_{il}) \exp\left\{\beta_1 g_{\gamma_i}(t_{il}) + \beta_2' z_i\right\} \\ &+ (s_j - t^*_{ij_{i2}}) \exp\left\{\beta_1 g_{\gamma_i}(t_{ij_{i2}}) + \beta_2' z_i\right\}. \end{aligned}$$

(iii) If $s_{j-1} < y_i \le s_j$, consider j_{i1} and j_{i2} defined in (ii). Then if $j_{i1} = j_{i2}$ or $y_i \le t^*_{i,j_{i1}+1}$ when $j_{i1} < j_{i2}$, then

$$B_{ij} = (y_i - s_{j-1}) \exp\left\{\beta_1 g_{\gamma_i}(t_{ij_{i1}}) + \beta_2' z_i\right\}.$$

Otherwise, we define

$$\begin{aligned} B_{ij} =& (t^*_{i,j_{i1}+1} - s_{j-1}) \exp\left\{\beta_1 g_{\gamma_i}(t_{i,j_{i1}+1}) + \beta_2' z_i\right\} \\ &+ \sum_{l=j_{i1}+1}^{k_i} (t^*_{i,l+1} - t^*_{il}) \exp\left\{\beta_1 g_{\gamma_i}(t_{il}) + \beta_2' z_i\right\} \\ &+ (y_i - t^*_{ik_i}) \exp\left\{\beta_1 g_{\gamma_i}(t_{ik_i}) + \beta_2' z_i\right\}, \end{aligned}$$

where $j_{i1} + 1 \le k_i \le j_{i2}$ is chosen so that $t^*_{ik_i} < y_i \le t^*_{i,k_i+1}$.

We note that when j_{il} $(l = 1, 2)$ does not exist, we define $j_{il} = 1$, and the calculation of B_{ij} needs a minor adjustment. In the likelihood function (7.3.9), we have invoked an approximation, that is,

$$\int_0^{y_i} h_0(u) \exp\left\{\beta_1 g_{\gamma_i}(u) + \beta_2' z_i\right\} du \approx \sum_{i=1}^n \sum_{j=1}^J \lambda_j B_{ij}.$$

The priors are all chosen to be noninformative with $\lambda_j \sim \mathcal{G}(a_j, b_j)$, $j = 1, 2, \ldots, J$, $\beta_1 \sim N(\zeta_1, v_1^2)$, $\beta_2 \sim N(\zeta_2, v_2^2)$ (for $p = 1$), $\alpha_0 \sim N(\zeta_3, v_3^2)$, $\alpha_1 \sim N(\zeta_4, v_4^2)$, $\sigma_1^2 \sim \mathcal{IG}(a_{01}, b_{01})$, $\sigma_2^2 \sim \mathcal{IG}(a_{02}, b_{02})$, and $\gamma_i \overset{iid}{\sim} N_3(\mu_0, \Sigma_0)$, where $\mu_0 = (\mu_{01}, \mu_{02}, \mu_{03})$, and

$$\Sigma_0 = \begin{pmatrix} \sigma_{01}^2 & 0 & 0 \\ 0 & \sigma_{02}^2 & 0 \\ 0 & 0 & \sigma_{03}^2 \end{pmatrix}.$$

Additional normal and inverse Wishart priors are placed on μ_0 and Σ_0, respectively. That is, $\mu_{01} \sim N(\xi_{01}, v_{01}^2)$, $\mu_{02} \sim N(\xi_{02}, v_{02}^2)$, $\mu_{03} \sim N(\xi_{03}, v_{03}^2)$, $\sigma_{01}^2 \sim \mathcal{IG}(e_{01}, f_{01})$, $\sigma_{02}^2 \sim \mathcal{IG}(e_{02}, f_{02})$, and $\sigma_{03}^2 \sim \mathcal{IG}(e_{03}, f_{03})$. Gibbs sampling is used to sample from the posterior. The detailed derivations of the conditional distributions for the Gibbs sampler are left as an exercise.

Example 7.2. Cancer vaccine data. Ibrahim, Chen, and Sinha (2000) examine the cancer vaccine data discussed in Example 1.8. Table 1.10 shows a statistical summary of the IgG (X_1) and IgM (X_2) measures for each treatment arm and the number of missing antibody titres at each time point. Both the IgG and IgM antibody measures are related to the true unobservable univariate antibody level X^*. From a biological perspective, a change in X_1 often implies a change in X^* over a suitable window of time on average, and hence X_1 is deemed more important as a source of information on X^* compared to X_2. This is why X_1 is assumed to be on an unbiased scale for X^*. Thus the form of (7.3.3) and (7.3.3) are based on sound biological considerations. Based on fitting a Cox model, the hazard ratio is 2.61 in favor of treatment A with a P-value of 0.02, and the 95% confidence interval is (1.17, 5.82). Natural logarithms of IgG and IgM were used in all analyses. Noninformative priors were used for all of the models in the analyses below. For example, for the quadratic trajectory model with $p = 1$ using $J = 8$, we take $a_j = b_j = 0$, $j = 1, 2, \ldots, 8$, $v_j^2 = v_{0j}^2 = 100$, $j = 1, 2, 3, 4$, $\xi_1 = \xi_2 = \xi_3 = \xi_4 = \xi_{01} = \xi_{02} = \xi_{03} = 0$, $a_{01} = a_{02} = 0$, $b_{01} = b_{02} = 0.001$, $e_{0j} = f_{0j} = 0.001$, $j = 1, 2, 3$.

TABLE 7.2. Parameter Estimates for Cancer Vaccine Data.

Parameter	Mean	SD	Parameter	Mean	SD
β_1	−0.28	0.20	σ_{01}^2	1.40	0.41
β_2	1.02	0.43	σ_{02}^2	0.04	0.06
λ_1	0.05	0.03	σ_{03}^2	0.06	0.09
λ_2	0.05	0.04	α_0	2.49	0.30
λ_3	0.06	0.05	α_1	0.38	0.09
λ_4	0.08	0.09	σ_1^2	3.88	0.38
λ_5	0.17	0.20	σ_2^2	6.22	0.54
λ_6	0.03	0.04	μ_{01}	2.76	0.19
λ_7	0.04	0.06	μ_{02}	5.21	0.59
λ_8	0.26	0.48	μ_{03}	−3.88	0.60

Source: Ibrahim, Chen, and Sinha (2000).

The posterior estimates of the parameters are given in Table 7.2. The posterior mean and SD are 0.38 and 0.09 for α_1, and −3.88 and 0.60 for μ_{03}, respectively. Moreover, the 95% highest posterior density (HPD) intervals

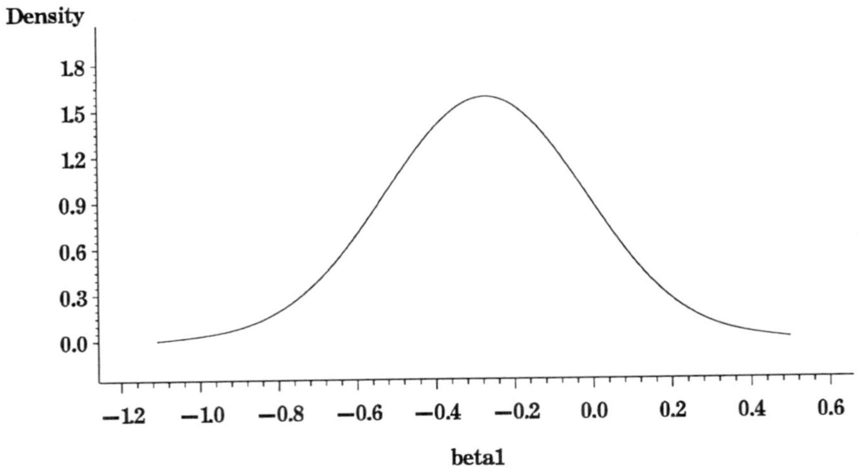

FIGURE 7.1. Marginal posterior density for β_1 for cancer vaccine data.

for α_1 and μ_{03} are $(0.20, 0.57)$ and $(-4.99, -2.68)$, which do not contain 0. This is strong evidence for a quadratic trajectory model. The posterior mean and standard deviation of β_1 is -0.28 and 0.20, respectively. A plot of the marginal posterior density of β_1 is also given in Figure 7.1. The posterior estimate of β_1 is negative and far from 0, implying that there is an association between relapse free survival and antibody titre count. That is, the magnitude of β_1 implies that increased antibody titres counts are moderately associated with longer relapse-free survival. This phenomenon is also confirmed by the plots of the posterior hazards in Figure 7.2. In Figure 7.2, • and ∘ correspond to treatments A and B, respectively. As the antibody titre counts increase, there is a decrease in the posterior hazard estimate for each individual. This phenomenon is observed for both the IgG and IgM antibody measurements, and therefore gives further evidence to the association between antibody titres and relapse-free survival.

The treatment coefficient β_2 has posterior mean and standard deviation of 1.02 and 0.43, respectively, with 95% HPD interval $(0.20, 1.88)$. Thus it is clear that there is also an important treatment difference between A and B. Specifically, we see that treatment A (IFN + GMK) is superior to treatment B (GMK alone). The estimates for β_2 in Table 7.2 confirm this result. Figure 7.1 also confirms this, as we see that, for both the IgG and IgM antibody titres, the posterior hazard estimates are consistently smaller for treatment A then they are for treatment B, indicating a superior treatment A effect.

A sensitivity analysis was conducted with respect to the choice of J. The posterior estimates of β_1, β_2, and the α_j's were also very robust with respect to the choice of J, yielding very similar estimates to those of Table 7.2 for several different values of J. As a result, the posterior hazard and

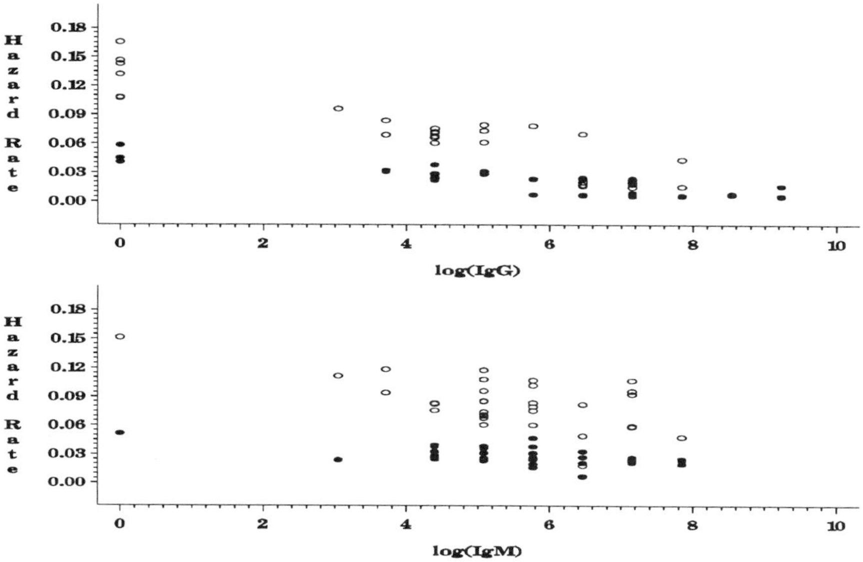

FIGURE 7.2. Estimated hazard rate as a function of log(IgG) and log(IgM) taken at time point of peak measurement.

trajectory function estimates were also robust with respect to the choice of J. In addition to the treatment covariate, other covariates, such as gender, age, and weight, in the model given in (7.3.5) were also considered. Except for the treatment covariate, all of other covariates under consideration are highly insignificant. Fifty thousand Gibbs samples were used to compute all posterior estimates using a burn-in of 1000 samples. The convergence of the Gibbs sampler was checked using several diagnostic procedures as recommended by Cowles and Carlin (1996). The Gibbs sampler converged in less than 1000 iterations.

Wang and Taylor (2000) introduce another Bayesian method for jointly modeling longitudinal and survival data. While their survival model is the same as Ibrahim, Chen, and Sinha (2000) and Faucett and Thomas (1996), they take a more flexible approach to modeling the longitudinal component by incorporating a stochastic process into the trajectory function. The longitudinal model is specified as

$$X_i(t_{ij}) = X_i^*(t_{ij}) + \epsilon_i(t_{ij}) \tag{7.3.10}$$

and

$$X_i^*(t_{ij}) = a_i + bt_{ij} + z_i'(t_{ij})\beta + W_i(t_{ij}), \tag{7.3.11}$$

where $X_i(t_{ij})$ denotes the observed value of the marker for subject i, $X_i^*(t_{ij})$ is the true value of the marker, and $\epsilon_i(t_{ij}) \overset{i.i.d.}{\sim} N(0, \sigma_\epsilon^2)$ at time t_{ij} for $i = 1, 2, \ldots, n$; $j = 1, 2, \ldots, m_i$. In (7.3.11), $\boldsymbol{z}_i$ denotes a $p \times 1$ vector of baseline covariates, $\boldsymbol{\beta}$ is a $p \times 1$ vector of regression coefficients, the random intercepts $a_i \sim N(\mu_a, \sigma_a^2)$, and $W_i(t)$ is an integrated Ornstein-Uhlenbeck (IOU) stochastic process with covariance between times s and t given by

$$\frac{\sigma^2}{2\alpha^3}\left[2\alpha \min(s,t) + \exp(-\alpha t) + \exp(-\alpha s) - 1 - \exp(-\alpha|t-s|)\right]. \quad (7.3.12)$$

The parameters α and σ^2 control the amount of smoothness of the function.

The event time data for subject i is denoted by (y_i, ν_i), where ν_i is the censoring indicator. A proportional hazards model is assumed for the event time:

$$h_i(t) = h(t|X_i^*(t), \boldsymbol{q}_i) = h_0(t)\exp\{\gamma X_i^*(t) + \boldsymbol{\omega}'\boldsymbol{q}_i(t)\}, \quad (7.3.13)$$

where $\boldsymbol{q}_i(t)$ denotes a $q \times 1$ vector of other potential covariates, which may include some or all of the $\boldsymbol{z}_i$ covariates. Similar to Ibrahim, Chen, and Sinha (2000), $h_0(t)$ is taken to be a constant λ_j over the time intervals $I_j = (s_{j-1}, s_j]$, for $j = 1, 2, \ldots, J$, where $s_0 = 0 < s_1 < \ldots < s_J < s_{J+1} = \infty$.

For numerical reasons, σ^2 is reparameterized to $\theta = \sigma^2/\alpha$. The continuous function $W_i(t)$ is approximated by a step function. The parameters are μ_a, σ_a^2, $\boldsymbol{\beta}$, σ_ϵ^2, $\boldsymbol{\omega}$, θ, a_i, γ, λ_j $(j = 1, 2, ..., J)$, and $W_i = (W_{ij}, j = 1, 2, ..., J_i)$, where J_i is the number of grid points for the step function approximation of the IOU process. The likelihood function is given as

$$\prod_{i=1}^{n}\prod_{j_1}^{m_i}[X_i(t_{ij})|a_i, b, W_i, \boldsymbol{\beta}, \sigma_\epsilon^2] \times [y_i|\boldsymbol{\lambda}, \gamma, \boldsymbol{\omega}]. \quad (7.3.14)$$

The conditional posteriors for most of the parameters follow similarly to the previous Bayesian approaches. There are $7 + J + n + \sum_{i=1}^{n} m_i$ parameters, not including covariates in the longitudinal and proportional hazards components of the model. The conditional posteriors for some of these parameters are quite complex and are discussed in detail in Wang and Taylor (2000).

Example 7.3. Multicenter AIDS cohort study. We consider an example from Wang and Taylor (2000) which involves data from the multicenter AIDS cohort study (MACS). The MACS is a cohort study of homosexual men who were recruited between April 1984 and March 1985 from four centers and are followed up at approximately six-month intervals (Kaslow et al., 1987). At each study visit, CD4 counts and other variables are recorded. The date of the occurrence of AIDS is obtained from the participant's medical record. An objective of the MACS is to elucidate the natural history of

HIV and to identify risk factors for clinical expression of AIDS. The data used in this example are from seroconverters in the Los Angeles center. This group consists of 115 people who were observed to change from HIV antibody negative to positive during the follow-up period (up to June 15, 1996). Data from later periods were excluded to avoid the complicating effect of highly effective therapy. The time interval between the last HIV-negative and the first HIV-positive measurement was restricted to be less than 15 months, with 90% being less than nine months. The middle point of the interval was taken as the date of infection. All the CD4 measurements within six months following the infection date were excluded because the CD4 number is known to decrease abruptly on seroconversion (Lang et al., 1989; Margolick et al., 1993). The time to AIDS (y_i) is defined as the time from six months after HIV infection to the date of a clinical AIDS diagnosis for AIDS cases or to last follow-up for non-AIDS cases.

The CD4 data were transformed by a fourth-root power to achieve homogeneity of within-subject variance (Taylor et al., 1991), thus $X_i(t_{ij})$ denotes the observed $\text{CD4}^{1/4}$ for person i at time t_{ij}. The number of CD4 measurements per person ranges from 1 to 32 (median = 11). There are 51 AIDS cases. The time to AIDS (y_i) ranges from 0.5 to 10.5 years (median = 5.5 years). Two covariates ($p = 2$) were used in (7.3.11), which are (i) z_{i1} = average $\text{CD4}^{1/4}$ measurements prior to HIV infection (called pcd4) and (ii) z_{i2} = age at HIV infection. Both have been suggested as factors which influence disease progression (Taylor et al., 1995; Rosenberg, Goedert, and Biggar, 1994).

The joint model is fit after centering the covariates in equations (7.3.11) and (7.3.13) using $\overline{z}_1 = 5.493$ and $\overline{z}_2 = 35$. In (7.3.13), $X_i^*(t)$ is centered by subtracting $\overline{X} = 4.62$, the average of all $\text{CD4}^{1/4}$ data after infection, and $\boldsymbol{q}_i = \boldsymbol{z}_i$. The grid points for W_{ij} are four months apart. Eight steps ($J = 8$) each of length 1.32 years were used for the baseline hazard function.

From the past 15 years of AIDS research there is good qualitative understanding of various aspects of the natural history of CD4 decline and the development of AIDS, particularly regarding the average rate of decline and prognostic importance of CD4, and the incubation period (Lang et al., 1989; Rosenberg, Goedert, and Biggar, 1994; and many others). Based on this collection of knowledge and previous analyses, the following proper priors were used: $N(0.207, 0.011)$ for b, $N(5.233, 0.534)$ for μ_a, $N(0, 100^2)$ for $\beta_1, \beta_2, \omega_1$ and ω_2, $\mathcal{U}(-3.0, -0.1)$ for γ, the inverse gamma $\mathcal{IG}(2, 0.1)$ for σ_a^2, $\mathcal{IG}(2, 0.056)$ for σ_e^2, $\mathcal{IG}(2, 0.13)$ for θ and $\mathcal{IG}(2, 2.5)$ for α. The independent gamma prior $\mathcal{G}(0.04, 1.0)$ was used for $\lambda_1, \lambda_2, \ldots, \lambda_J$. Most of these priors are weakly informative and the inverse gamma priors have finite mean and infinite variance. As a sensitivity analysis (results not presented), Wang and Taylor (2000) repeated the model fitting using a wide range of priors for the regression coefficients and variances, including many "flat" and improper priors as well as alternative mildly informative and proper priors. Other than for the estimation of α, which is quite sensitive to the choice of prior, they found very little dependence on the prior. See Wang (1998) for more details.

For comparison with the joint model, Wang and Taylor (2000) also fit each submodel separately. For the longitudinal model, they obtain the parameter estimates using MCMC sampling. For the hazard model they used two estimation methods. One is the MCMC Bayesian approach and the other is based on partial likelihood. With both methods, the latest observed CD4 count is used as the time-dependent marker. Note that this represents a somewhat different model than equation (7.3.13), because the value used is at a time prior to t, and measurement error is not allowed for. There are other potential methods of incorporating these missing aspects into survival analyses (see Pepe, Self, and Prentice, 1989; Tsiatis, DeGruttola, and Wulfsohn, 1995; Bycott and Taylor, 1998), which may be sensitive to the assumed model or magnitude for the measurement error, but they are not considered here.

For models fitted by MCMC sampling, the chain was run for 400,000 iterations and the first 200,000 iterations were discarded. The remaining sample was used to estimate the posterior densities of the parameters, which is summarized using the sample mean and standard deviation. The computational time was approximately nine hours on a 450 Mhz Pentium III. The Brooks and Gelman (1998) multivariate potential scale reduction factor, based on five parallel chains, was used to assess convergence.

The estimates of the parameters based on the joint and separate modeling methods are presented in the second and third columns of Table 7.3. In Table 7.3, for each parameter, the first row gives the posterior mean, and the second row in parenthesis gives the posterior standard deviation. Based on the joint modeling, the $\text{CD4}^{1/4}$ declines over time and the estimated rate of decline is -0.191 (SD $= 0.015$). This population rate of decline corresponds to a decrease from untransformed CD4 $= 750$ at time zero to untransformed CD4 $= 335$ five years later. The estimate for γ is -1.486 (SD $= 0.201$). It suggests that the relative risk of AIDS development of 1.82 ($\approx \exp\{1.486 * (300^{1/4} - 200^{1/4})\}$) for a subject with a CD4 of 200 compared to a subject with a CD4 of 300. The relative AIDS risk is 2.43 ($\approx \exp\{1.488 * (200^{1/4} - 100^{1/4}\}$) for a CD4 of 100 compared to a CD4 of 200.

The medians and the 95% credible intervals for α are 3.68 and $(1.70, 13.14)$ for joint modeling, and 4.32 and $(1.78, 33.41)$ for separate modeling. The estimate for β_1 of 0.441 (SD $= 0.074$) shows that CD4 levels prior to infection have a significant positive effect on CD4 after infection and higher CD4 before infection leads to higher CD4 after infection. The expected difference in post infection intercept on the untransformed scale between two average people with pre-infection CD4's of 600 and 1200 is 215. The estimate for ω_1 is 0.359 (SD $= 0.348$), indicating that the direct effect of CD4 before infection on the hazard of AIDS is negative but not significant. Hence, it can be concluded that the effect of CD4 prior to infection on AIDS is basically mediated through its effect on the development of CD4 after infection. The estimates for β_2 and ω_2 suggest no effect of age on either CD4 or AIDS. The estimated baseline hazard and its variation increase over time.

The above analysis is based on discretizing time into four-month intervals. Wang and Taylor (2000) repeated the analysis using two-, three- and five-months intervals (results not shown). None of the parameter estimates differed in a substantial way from those in Table 7.3. The only ones which changed in a moderate way were those concerned with the covariance structure of CD4. The joint modeling method and separate modeling method produce very similar results for the parameters in the longitudinal model. For this application, the AIDS event times do not provide much information regarding the process of CD4 change. The joint modeling estimate of the slope is only slightly steeper than the separate modeling estimate. In contrast, there is a noticeable difference in the hazard model parameter estimates. The separate modeling analysis, which ignores measurement error, produces estimates of γ closer to zero.

The IOU process includes Brownian motion (with $\alpha \to \infty$) and a specific random effects model (with $\alpha \to 0$) as special cases. The large estimate of α and its variability shown in Table 7.3 also hints that Brownian motion may be a better choice than a general IOU process for longitudinal CD4 data, agreeing with the conclusion in Taylor et al. (1994). They did, however, find that the posterior distribution for α was dependent on the choice of prior, and in all cases large values are estimated.

The results of the joint modeling analyses with Brownian motion and a general random effects model are summarized in the fourth and fifth columns of Table 7.3. Compared to the results when α is estimated, the joint model with Brownian motion produces very similar estimates for parameters b, β_1, β_2, μ_a, γ, ω_1, ω_2, and λ_j. The estimates for the variance parameters σ_a^2, θ, and σ_e^2 are noticeably smaller than the ones when the IOU process is used. The joint model with random effects yields a slightly larger estimate for the slope, slightly smaller hazard model regression coefficients and slightly larger baseline hazard estimates. The estimates for σ_a^2 and σ_e^2 are also much larger than the ones with the IOU and the Brownian motion.

To check the adequacy of the models, Wang and Taylor (2000) compared the observed CD4 data with their predicted values $(X_i^*(t_{ij}))$, and they found that the random effects model does not fit the marker data as well as the model with the IOU or Brownian motion process. The model with Brownian motion yields the best fit to the data. The predicted values from this model are closest to the observed data. On the basis of a smaller error variance and smaller number of parameters, they recommend using the model with Brownian motion.

TABLE 7.3. Posterior Estimates for MACS Data.

Parameter	IOU Joint	IOU Separate	Brownian Motion	Random Effects
b	−0.191	−0.187	−0.191	−0.212
(slope)	(0.015)	(0.015)	(0.015)	(0.018)
β_1	0.441	0.439	0.433	0.429
(pcd4)	(0.074)	(0.073)	(0.074)	(0.080)
β_2	−0.031	−0.031	−0.021	−0.078
(age)	(0.047)	(0.047)	(0.047)	(0.051)
μ_a	5.099	5.096	5.098	5.141
(intercept)	(0.036)	(0.036)	(0.036)	(0.038)
σ_a^2	0.075	0.073	0.066	0.125
	(0.018)	(0.018)	(0.018)	(0.021)
σ_e^2	0.049	0.048	0.039	0.087
	(0.004)	(0.005)	(0.003)	(0.004)
θ	0.127	0.122	0.116	
	(0.016)	(0.015)	(0.011)	
α	5.28	8.90		
	(4.80)	(11.11)		
Median of α	3.68	4.32		
γ	−1.486	−1.322	−1.485	−1.409
(cd4)	(0.201)	(0.168)	(0.196)	(0.216)
ω_1	0.359	0.249	0.361	0.318
(pcd4)	(0.348)	(0.332)	(0.347)	(0.341)
ω_2	−0.011	−0.016	−0.008	−0.062
(age)	(0.247)	(0.231)	(0.248)	(0.242)
λ_1	0.019	0.018	0.019	0.020
	(0.014)	(0.013)	(0.014)	(0.014)
λ_2	0.025	0.026	0.025	0.026
	(0.012)	(0.013)	(0.012)	(0.013)
λ_3	0.030	0.035	0.030	0.035
	(0.012)	(0.014)	(0.012)	(0.014)
λ_4	0.052	0.063	0.051	0.058
	(0.019)	(0.021)	(0.018)	(0.021)
λ_5	0.046	0.059	0.046	0.057
	(0.019)	(0.023)	(0.019)	(0.022)
λ_6	0.052	0.086	0.053	0.060
	(0.022)	(0.032)	(0.022)	(0.025)
λ_7	0.052	0.083	0.052	0.058
	(0.029)	(0.043)	(0.029)	(0.033)
λ_8	0.034	0.045	0.034	0.040
	(0.036)	(0.044)	(0.035)	(0.041)

Source: Wang and Taylor (2000).

The Bayesian approaches of Faucett and Thomas (1996), Ibrahim, Chen, and Sinha (2000), and Wang and Taylor (2000) discussed here use a proportional hazards model for the survival component of the model. They all differ in their approaches to modeling the longitudinal component. Ibrahim, Chen, and Sinha (2000) use a random effects model with a multivariate longitudinal outcome in contrast to the univariate random effects model of Faucett and Thomas (1996). Wang and Taylor (2000) also use a random effects model but include an IOU process in the trajectory curve, which allows for more flexibility. However, the IOU process greatly increases the number of parameters in the model and the computational complexity.

Exercises

7.1 Show that (7.2.1) is equal to (7.2.3).

7.2 This exercise is related to Bayesian methods based on (7.2.2). We assume that $\varphi(X_i^*(t), \beta) = \exp(\beta X_i^*(t))$.

(a) Show that when $[X_i^*(t), X_i(t)|Y_i \geq t]$ is a jointly Gaussian process, then

$$E\left[\exp(\beta X_i^*(t))|\mathcal{X}_i(t), Y_i \geq t\right] \\ = \exp\left\{\beta\mu(t|\mathcal{X}_i(t)) + \frac{\beta^2}{2}\sigma^2(t|\mathcal{X}_i(t))\right\},$$

where $\mu(t|\mathcal{X}_i(t)) = E(X^*(t)|\mathcal{X}_i(t), Y_i \geq t)$ and $\sigma^2(t|\mathcal{X}_i(t)) = \text{Var}(X^*(t)|\mathcal{X}_i(t), Y_i \geq t)$.

(b) Assume that given $Y_i \geq t$, we have the linear growth curve model

$$X_i^*(t) = \alpha_{0i} + \alpha_{1i}t,$$

where

$$\begin{pmatrix} \alpha_{0i} \\ \alpha_{1i} \end{pmatrix} \sim N_2\left(\begin{pmatrix} \mu_0 \\ \mu_1 \end{pmatrix}, \begin{pmatrix} \sigma_0^2 & \sigma_{01} \\ \sigma_{01} & \sigma_1^2 \end{pmatrix}\right) \equiv N_2(\boldsymbol{\mu}, \Sigma).$$

Moreover, we also assume, given $Y_i \geq t$ and $\mathcal{X}_i(t) = (X_i(t_1), X_i(t_2), \ldots, X_i(t_m))$,

$$X_i(t_j) = X_i^*(t_j) + \epsilon_{ij},$$

where $\epsilon_{ij} \overset{\text{i.i.d.}}{\sim} N(0, \sigma_\epsilon^2)$, and $t_1 < t_2 < \ldots < t_m \leq t$ are the observation times for X_i. Derive the joint distribution of $(X_i^*(t), X_i(t_1), X_i(t_2), \ldots, X_i(t_m)|Y \geq t)$, and show that the mean vector $\boldsymbol{\theta}_i$ has the form

$$\boldsymbol{\theta}_i' = (\mu_0 + \mu_1 t, \mu_0 + \mu_1 t_1, \mu_0 + \mu_1 t_2, \ldots, \mu_0 + \mu_1 t_m)$$

and the covariance matrix Σ_i has the form

$$\Sigma_i = \begin{pmatrix} \Sigma_{11i} & \Sigma_{12i} \\ \Sigma_{21i} & \Sigma_{22i} \end{pmatrix},$$

where

$$\Sigma_{11i} = \sigma_0^2 + 2t\sigma_{01} + t^2\sigma_1^2,$$

$$\Sigma_{12i} = (\sigma_0^2 + (t+t_1)\sigma_{01} + tt_1\sigma_1^2, \sigma_0^2 + (t+t_2)\sigma_{01} + tt_2\sigma_1^2, \ldots, \sigma_0^2 + (t+t_m)\sigma_{01} + tt_m\sigma_0^2),$$

and $\Sigma_{22i} = ((\Sigma_{22i})_{k_1k_2})$ with

$$(\Sigma_{22i})_{k_1k_2} = \sigma_0^2 + \sigma_{01}(t_{k_1} + t_{k_2}) + \sigma_1^2 t_{k_1} t_{k_2}$$

for $k_1 \neq k_2$, $1 \leq k_1, k_2 \leq m$, and

$$(\Sigma_{22i})_{kk} = \sigma_0^2 + 2\sigma_{01}t_k + \sigma_1^2 t_k^2 + \sigma_\epsilon^2$$

for $1 \leq k \leq m$.

(c) Find $\mu(t|\mathcal{X}_i(t))$ and $\sigma^2(t|\mathcal{X}_i(t))$ as functions of $\mathcal{X}_i(t) = (X_i(t_1), X_i(t_2), \ldots, X_i(t_m))$, $\boldsymbol{\mu}$, Σ, and σ_ϵ^2.

(d) Let $\boldsymbol{\psi} = (\boldsymbol{\mu}, \Sigma, \sigma_\epsilon^2)$. Find the marginal distribution of $(\mathcal{X}_i(y_i)|\boldsymbol{\psi})$.

(e) Assume the prior

$$\pi(\boldsymbol{\psi}) = \pi(\boldsymbol{\mu})\pi(\Sigma)\pi(\sigma_\epsilon^2),$$

where $\pi(\boldsymbol{\mu})$ is a $N_2(0, 10^4 I)$, $\pi(\Sigma)$ is Wishart$(3, 10^3 I)$, which denotes the Wishart distribution with 3 degrees of freedom and mean matrix $3(10^3 I)$, and $\pi(\sigma_\epsilon^2)$ is $\mathcal{IG}(0.001, 0.001)$.
Write a BUGS program using the data from the GMK arm only from Example 7.2 (i.e., no fixed covariate) and using the univariate process $\mathcal{X}_i$ of IgM titres.

(f) Define $\pi(\beta|\mathcal{X}, Y, \hat{\Psi}) \propto \text{PL}(\beta|\mathcal{X}, Y, \hat{\boldsymbol{\psi}})\pi(\beta)$, where $\text{PL}(\beta|\mathcal{X}, Y, \hat{\boldsymbol{\psi}})$ is the likelihood of (7.2.2) with $\hat{\boldsymbol{\psi}}$ being the posterior mean of $\boldsymbol{\psi}$ from $\pi(\boldsymbol{\psi}|\mathcal{X})$. After calculating $\hat{\boldsymbol{\psi}}$ from the BUGS output from part (e), write a computer program to simulate β from $\pi(\beta|\mathcal{X}, Y, \hat{\boldsymbol{\psi}})$, where $\pi(\beta)$ is $N(0, 10^6)$. Also construct a 95% credible set for β. Note that $\pi(\beta|\mathcal{X}, Y, \hat{\boldsymbol{\psi}})$ is a univariate density.

(g) Simulate samples from the joint posterior distribution of $(\beta, \boldsymbol{\psi})$, given by

$$\pi(\beta, \boldsymbol{\psi}|D) = \pi(\beta|\mathcal{X}, Y, \boldsymbol{\psi})\pi(\boldsymbol{\psi}|\mathcal{X})$$

using Gibbs samples from $\pi(\boldsymbol{\psi}|\mathcal{X})$, and then in the second step, sampling from $\pi(\beta|\mathcal{X}, Y, \tilde{\boldsymbol{\psi}})$ using the Gibbs sample $\tilde{\boldsymbol{\psi}}$.

(h) Construct a 95% credible set for β from part (g) and compare your answer with that of part (f).

7.3 Using the cancer vaccine data of Example 7.2, fit the partial likelihood model given by (7.2.3) and obtain estimates and standard errors for all parameters.

7.4 Fit the model given by (7.2.4) and obtain estimates and standard errors for all parameters.

7.5 Using the complete log-likelihood in (7.2.4), show that the sufficient statistics for $\boldsymbol{\alpha}$, $\boldsymbol{\zeta}$, $\boldsymbol{\lambda}$, σ^2, ω^2, and Ψ are given by (7.2.5) and (7.2.6).

7.6 Use Louis's formula (Louis, 1982) to derive the covariance matrix of the estimates of the parameters of interest $(\boldsymbol{\alpha}, \boldsymbol{\zeta})$ in (7.2.7) and (7.2.8) at convergence of the EM algorithm.

7.7 Derive the estimating equations in (7.2.7) and (7.2.8).

7.8 Using the cancer vaccine data of Example 7.2, fit the model given by (7.2.11) and obtain estimates and standard errors for all parameters.

7.9 Derive (7.3.1) and (7.3.2).

7.10 For the likelihood function given by (7.3.6) and the corresponding priors specified in Section 7.3, derive the conditional posterior distributions necessary for the Gibbs sampler.

7.11 Derive the expressions given by (7.3.7), (7.3.8), and (7.3.9).

7.12 For the models defined by (7.3.10), (7.3.11), (7.3.12), and (7.3.13), and the priors specified in Example 7.3, derive the conditional posterior distributions for a_i, b, W_i, $\boldsymbol{\beta}$, σ^2_ϵ, $\boldsymbol{\lambda}$, γ, and $\boldsymbol{\omega}$.

7.13 Write a computer program to verify all of the entires in Table 7.2.

7.14 Reproduce Figures 7.1 and 7.2.

7.15 Using the cancer vaccine data of Example 7.2, fit the model given by (7.3.10), (7.3.11), (7.3.12), and (7.3.13), and the priors specified in Example 7.3,

7.16 Derive an explicit form for the likelihood in (7.3.14).

8
Missing Covariate Data

8.1 Introduction

Missing covariate data often arise in various settings, especially in clinical trials, epidemiological studies, and environmental studies. In the frequentist setting, it is well known (see Little and Rubin, 1987) that analyses based on complete cases can and will often result in inaccurate estimates of coefficients and their standard deviations. In the Bayesian context, complete case analyses will often lead to posterior distributions with properties that are quite different than those based on the observed data posterior, i.e., the posterior distribution using all of the cases. Thus, it becomes increasingly important in these situations to develop methods which incorporate the missing data into the analysis. The missing data problem has received much attention under the frequentist paradigm for a wide variety of models because of its common occurrence in many studies. Little (1992) gives an excellent review of developments for a variety of missing data problems. A recent book by Schafer (1997) examines frequentist and Bayesian approaches for missing data problems involving normal and categorical data, but it does not discuss survival models.

In this chapter, we focus on Bayesian methods for missing covariate data in survival analysis, focusing on the cure rate model. The methods we discuss here can be used for other types of models, such as the Cox model, frailty models, semiparametric survival models, and joint models for longitudinal and survival data. Recent advances in computing technology and the development of efficient computational algorithms, such as the Gibbs sampler and other MCMC procedures, have made Bayesian methods more feasible and realistic for solving a wide array of missing data problems in survival analysis. Monte Carlo and Gibbs sampling algorithms such as those of Tanner and Wong (1987), Gelfand and Smith (1990), Wei and Tanner (1990), and Kong, Liu, and Wong (1994) have made likelihood-based methods for complex incomplete data problems much more computationally feasible. However, very little literature exists for Bayesian missing data methods for regression models involving survival data. There have been several articles which examine frequentist procedures using EM-type algorithms for handling missing covariate data for proportional hazards models. These include Schluchter and Jackson (1989), Lin and Ying (1993), Zhou and Pepe (1995), Lipsitz and Ibrahim (1996a, 1998), Paik (1997), Paik and Tsai (1997), Pugh et al. (1993), Chen and Little (1999), Herring and Ibrahim (2001), and Herring, Ibrahim, and Lipsitz (2001).

The literature on Bayesian inference and computation with missing covariate data in survival models is virtually nonexistent. In addition, informative prior elicitation and its implications in the presence of missing covariate data has received relatively little attention. Informative priors can lead to some nice properties in the missing data model and we discuss this in more detail later in this chapter. Specifically, in missing data problems, it often turns out that certain parameters in the likelihood function may not be identifiable, and even if the parameter is identifiable, very little information may be available from the data for making inferences about it due to the missing data patterns and missing data fraction. In these cases, Bayesian approaches to the missing data problem can help overcome these difficulties. First, proper prior distributions can help overcome the problem of non-identifiability, and secondly, informative priors can help in supplying more information about parameters. One way of doing this is to construct informative priors from historical data via the power prior. In this chapter, we follow the approach of Ibrahim, Chen, and Lipsitz (1999) and present a general class of informative priors for semiparametric survival models incorporating a cure fraction with missing covariate data. Several new challenges arise in the construction of the power prior due to the missing covariate data, including model identifiability, model adequacy, and computational techniques. The power priors prove to be especially useful for the missing data problem since they have the potential of providing more information about the parameters in the current study. We demonstrate how these priors can serve these purposes in Section 8.5. We also discuss several theoretical properties of the priors and implied posteriors in Section 8.4. We assume throughout this chapter that the missing data mechanism does not depend on the missing values, but perhaps may depend on the observed values. That is, we assume that the missing data are missing at random (MAR) in the sense of Rubin (1976), and thus the missing data mechanism is ignorable for posterior inferences. In addition, we allow the survival time to be right censored throughout.

8.2 The Cure Rate Model with Missing Covariate Data

Consider the semiparametric cure rate model given by (5.3.2) of Section 5.3. Let $\boldsymbol{x}_i' = (x_{i1}, \ldots, x_{ip})$ denote the $p \times 1$ vector of covariates for the ith subject. Following Ibrahim, Chen, and Lipsitz (1999), we specify a parametric model for the covariates and denote the marginal density of $\boldsymbol{x}_i$ by $f(\boldsymbol{x}_i|\boldsymbol{\alpha})$, where $\boldsymbol{\alpha}$ is a vector of indexing parameters. Let y_i denote the survival time for subject i, which may be right censored, and let ν_i denote the censoring indicator, which equals 1 if y_i is a failure time and 0 if it is right censored. The complete data are $D = (n, \boldsymbol{y}, X, \boldsymbol{\nu})$, where $\boldsymbol{y} = (y_1, y_2, \ldots, y_n)'$, $\boldsymbol{\nu} = (\nu_1, \nu_2, \ldots, \nu_n)'$, and X is an $n \times p$ matrix of covariates with the ith row $\boldsymbol{x}_i'$. Also, let $\boldsymbol{\beta} = (\beta_1, \beta_2, \ldots, \beta_p)'$ denote the vector of regression coefficients and let $\boldsymbol{\lambda} = (\lambda_1, \lambda_2, \ldots, \lambda_J)'$ denote the vector of parameters defining the piecewise baseline hazard function of the promotion time distribution. Then, for model (5.3.2), the complete data likelihood for all subjects is thus given by

$$L(\boldsymbol{\beta}, \boldsymbol{\lambda}, \boldsymbol{\alpha}|D) = L(\boldsymbol{\beta}, \boldsymbol{\lambda}|D) \prod_{i=1}^{n} f(\boldsymbol{x}_i|\boldsymbol{\alpha}). \tag{8.2.1}$$

Our main interest is in posterior inferences about $\boldsymbol{\beta}$ based on the observed data, with $(\boldsymbol{\lambda}, \boldsymbol{\alpha})$ being viewed as *nuisance* parameters, and therefore not as parameters of inferential interest. We write $\boldsymbol{x}_i = (\boldsymbol{x}_{mis,i}', \boldsymbol{x}_{obs,i}')'$, where $\boldsymbol{x}_{mis,i}$ is the $q_i \times 1$ vector of missing components of $\boldsymbol{x}_i$. Further, let $D_{obs} = (n, \boldsymbol{y}, X_{obs}, \boldsymbol{\nu})$ denote the observed data for all subjects in the current study, where X_{obs} denotes the matrix of observed covariates for the current study. We assume that $(\boldsymbol{\beta}, \boldsymbol{\lambda}, \boldsymbol{\alpha})$ are distinct from the parameters of the missing data mechanism, that is, the missing data mechanism is ignorable. To denote the dependence of the complete data likelihood on the covariates, we write

$$L(\boldsymbol{\beta}, \boldsymbol{\lambda}|D) = \prod_{i=1}^{n} L(\boldsymbol{\beta}, \boldsymbol{\lambda}|\boldsymbol{x}_i, y_i, N_i), \tag{8.2.2}$$

where

$$\begin{aligned} L(\boldsymbol{\beta},&\boldsymbol{\lambda}|\boldsymbol{x}_i, y_i, N_i) \\ &= \prod_{j=1}^{J} \exp\left\{-(N_i - \nu_i)\nu_{ij}\left[\lambda_j(y_i - s_{j-1}) + \sum_{g=1}^{j-1} \lambda_g(s_g - s_{g-1})\right]\right\} \\ &\times \prod_{j=1}^{J} (N_i\lambda_j)^{\delta_{ij}\nu_i} \exp\left\{-\nu_i\delta_{ij}\left[\lambda_j(y_i - s_{j-1}) + \sum_{g=1}^{j-1} \lambda_g(s_g - s_{g-1})\right]\right\} \\ &\times \exp\left\{[N_i\boldsymbol{x}_i'\boldsymbol{\beta} - \log(N_i!) - \exp(\boldsymbol{x}_i'\boldsymbol{\beta})]\right\}, \end{aligned} \tag{8.2.3}$$

and N_i and ν_{ij} are defined in (5.3.2). Thus, for model (8.2.1), the joint posterior density of $(\boldsymbol{\beta}, \boldsymbol{\lambda}, \boldsymbol{\alpha})$ based on the observed data is given by

$$\pi(\boldsymbol{\beta}, \boldsymbol{\lambda}, \boldsymbol{\alpha}|D_{obs}) \propto \prod_{i=1}^{n} \left\{ \int_{\boldsymbol{x}_{mis,i}} \sum_{N_i} L(\boldsymbol{\beta}, \boldsymbol{\lambda}|\boldsymbol{x}_i, y_i, N_i) f(\boldsymbol{x}_i|\boldsymbol{\alpha}) \, d\boldsymbol{x}_{mis,i} \right\} \times \pi(\boldsymbol{\beta}, \boldsymbol{\lambda}, \boldsymbol{\alpha}), \tag{8.2.4}$$

where $\pi(\boldsymbol{\beta}, \boldsymbol{\lambda}, \boldsymbol{\alpha})$ denotes the joint prior distribution of $(\boldsymbol{\beta}, \boldsymbol{\lambda}, \boldsymbol{\alpha})$. In (8.2.4), if the ith covariate vector $\boldsymbol{x}_i$ is completely observed, we have

$$\int_{\boldsymbol{x}_{mis,i}} \sum_{N_i} L(\boldsymbol{\beta}, \boldsymbol{\lambda}|\boldsymbol{x}_i, y_i, N_i) f(\boldsymbol{x}_i|\boldsymbol{\alpha}) \, d\boldsymbol{x}_{mis,i} = \sum_{N_i} L(\boldsymbol{\beta}, \boldsymbol{\lambda}|\boldsymbol{x}_i, y_i, N_i) f(\boldsymbol{x}_i|\boldsymbol{\alpha}).$$

A specific form of $\pi(\boldsymbol{\beta}, \boldsymbol{\lambda}, \boldsymbol{\alpha})$ is given in Section 8.4. In general, the multidimensional integrals in (8.2.4) do not have a closed form and are of high dimension if there are several missing covariates for a particular observation. In these cases, the posterior distribution based on the observed data can be quite difficult to compute directly, and analytic approximations to the integrals such as Gaussian quadrature or methods based on splines may not be sufficiently accurate. As an alternative, Markov chain Monte Carlo (MCMC) methods can be used to sample from (8.2.4) and do not require either numerical or analytical evaluation of the integrals.

8.3 A General Class of Covariate Models

We consider a parametric modeling scheme for the covariates that involves writing the joint distribution of the covariates as a sequence of one-dimensional conditional distributions. We allow the covariates to be either categorical or continuous. More generally, we can write the distribution of the p-dimensional covariate vector $\boldsymbol{x}_i = (x_{i1}, \ldots, x_{ip})'$ through a series of one-dimensional conditional distributions, given by

$$\begin{aligned} f(x_{i1}, x_{i2}, \ldots, x_{ip}|\boldsymbol{\alpha}) = & f(x_{ip}|x_{i1}, x_{i2}, \cdots, x_{i,p-1}, \boldsymbol{\alpha}_p) \\ & \times f(x_{i,p-1}|x_{i1}, x_{i2}, \cdots, x_{i,p-2}, \boldsymbol{\alpha}_{p-1}) \\ & \times \cdots \times f(x_{i2}|x_{i1}, \boldsymbol{\alpha}_2) f(x_{i1}|\boldsymbol{\alpha}_1), \end{aligned} \tag{8.3.1}$$

where $\boldsymbol{\alpha}_j$ is a vector of indexing parameters for the jth conditional distribution, $\boldsymbol{\alpha} = (\boldsymbol{\alpha}_1', \boldsymbol{\alpha}_2', \ldots, \boldsymbol{\alpha}_p')'$, and the $\boldsymbol{\alpha}_j$'s are distinct.

Modeling the joint covariate distribution in this fashion will facilitate a more straightforward specification of the covariate distribution, especially in situations where the missing covariates are mixed categorical and continuous. This covariate modeling scheme has been discussed in the frequentist context by Lipsitz and Ibrahim (1996b), Ibrahim, Lipsitz, and

Chen (1999), and Ibrahim, Chen, and Lipsitz (1999). This approach to modeling the covariate distributions is also quite useful in the Bayesian context as discussed by Ibrahim, Chen, and Lipsitz (2000), and Chen, Ibrahim, and Lipsitz (1999) since it greatly reduces the number of nuisance parameters that have to be specified, thus easing the prior elicitation and computational strategies. It facilitates a straightforward elicitation scheme of the prior distributions for all of the nuisance parameters and it also proves to be computationally attractive, as these one-dimensional conditional distributions are typically log-concave. Thus, the complete conditionals of the posterior distributions needed for the Gibbs sampler are log-concave for a wide variety of covariate distributions. This property will allow us to use the adaptive rejection algorithm of Gilks and Wild (1992) within the Gibbs sampler to sample from the observed data posterior, and thus greatly eases the computational implementation. For example, with dichotomous covariates, an obvious choice for $f(\boldsymbol{x}_i|\boldsymbol{\alpha})$ is a joint multinomial log-linear model as described in Agresti (1990). Following Lipsitz and Ibrahim (1996b), one can reduce the number of nuisance parameters by deleting higher order interaction terms from the log-linear model.

We note that (8.3.1) needs to be specified only for those covariates that are missing. Lipsitz and Ibrahim (1996b) call this a conditional-conditional (CC) specification and have successfully implemented it for categorical covariates within the EM algorithm for maximum likelihood estimation. For continuous covariates, (8.3.1) is also quite useful since, in many situations, it is difficult to think of a natural joint continuous distribution. For example, suppose that some missing covariates are strictly positive (as in time to diagnosis of a disease), and some missing covariates take values on the real line. Then in this case, a multivariate normal (or multivariate t) distribution for the covariates is probably not adequate. A more suitable joint distribution may be obtained from (8.3.1) by specifying one-dimensional conditional gamma distributions for the strictly positive covariates and specifying one-dimensional conditional normal distributions for the covariates taking values on the real line. There are many other scenarios, such as situations where some of the missing continuous covariates take values over a finite interval and other missing covariates take values over the entire real line. Again, (8.3.1) would be quite useful in this setting, since there is no natural joint distribution to specify. The modeling of the covariate distributions depends on the order of conditioning. Therefore, to assess the various models, we use the Conditional Predictive Ordinate (CPO) (see Geisser, 1993) for checking the sensitivity in changes in the order of conditioning in the covariate distribution, and for checking the adequacy of the model in general. We show in Example 8.1 that posterior inferences about $\boldsymbol{\beta}$ are generally quite robust with respect changes in the order of conditioning, addition of interaction terms, and changes in the link functions of the categorical covariates.

One of the main motivations of (8.3.1) comes from the problem of trying to specify a joint distribution when categorical and continuous covariates are both missing. The model in (8.3.1) is potentially quite useful since the direct specification of a joint covariate distribution for this case can be quite awkward. The mixed covariate case is perhaps the most common situation in clinical trials and the model in (8.3.1) seems well suited to handle this situation, and will often result in a large reduction in the number of nuisance parameters.

There are many possibilities in modeling the covariate distribution as a sequence of one-dimensional conditional distributions, especially when p is large. Thus, in these cases, some guidance can be given for specifying the joint covariate distribution:

Case 1. For any covariates that are completely observed for all observations, we do not need to specify a covariate distribution for them.

Case 2. If the missing covariates are all categorical, say, dichotomous, then we specify a sequence of logistic regressions for each $f(x_{mis,j}|\boldsymbol{u}_{j-1}, \boldsymbol{\alpha}_j)$, where $\boldsymbol{u}_{j-1} = (x_{mis,j-1}, \ldots, x_{mis,1}, \boldsymbol{x}'_{obs})'$, $j = k, k-1, \ldots, 1$, and k is the dimension of $\boldsymbol{x}_{mis}$. For $j = 1$, we define $\boldsymbol{u}_0 \equiv \boldsymbol{x}_{obs}$. To obtain a reduced model, we would get the best for model for each logistic regression, $f(x_{mis,j}|\boldsymbol{u}_{j-1}, \boldsymbol{\alpha}_j)$, $j = k, k-1, \ldots, 1$, in (8.3.1), by adding interaction terms and comparing nested models via the AIC or BIC criteria, the L measure, or CPO statistic. In Section 8.5, we use the CPO statistic as our model assessment tool. For each $f(x_{mis,j}|\boldsymbol{u}_{j-1}, \boldsymbol{\alpha}_j)$, we use $\boldsymbol{u}_{j-1}$ as the initial set of covariates in the logistic regression models. We note that $f(x_{mis,1}|\boldsymbol{u}_0, \boldsymbol{\alpha}_1)$ only depends on the completely observed covariates, and therefore the main effects and interaction terms for model selection for $f(x_{mis,1}|\boldsymbol{u}_0, \boldsymbol{\alpha}_1)$ only depend on the completely observed covariates $\boldsymbol{x}_{obs}$. We can also use the probit or complementary log-log links to model the categorical covariates.

Case 3. If the missing covariates are categorical with more than two levels, we consider a multinomial logistic regression model (see Agresti, 1990) for those covariates, and carry out the model selection procedure as described in Step 2. If the missing categorical covariates are ordinal with more than two levels, ordinal regression models such as the continuation odds model (Fienberg and Mason, 1979) and the cumulative logit model (McCullagh, 1980) can be used for $f(x_{mis,j}|\boldsymbol{u}_{j-1}, \boldsymbol{\alpha}_j)$. Again, we follow Step 2 and would get the best for model for each logistic regression, $f(x_{mis,j}|\boldsymbol{u}_{j-1}, \boldsymbol{\alpha}_j)$, in (8.3.1), by adding interaction terms and comparing nested models. If the covariates consist of counts, then we can model them via the Poisson distribution.

Case 4. If all of the missing covariates are continuous and take values on the real line, we can specify a joint multivariate normal distribution for them with an arbitrary mean and covariance matrix. Another possibility is to model them via a sequence of univariate normal distributions with $f(x_{mis,j}|\boldsymbol{u}_{j-1},\boldsymbol{\alpha}_j)$ being a univariate normal distribution, $N(\mu_j,\sigma_j^2)$, where $\mu_j = \boldsymbol{\alpha}_j'\boldsymbol{u}_{j-1}$, and σ_j^2 does not depend on any of the covariates. Thus, each μ_j consists of a regression of x_j on $x_{mis,j-1}, x_{mis,j-2}, \ldots, x_{mis,1}, \boldsymbol{x}_{obs}$. This univariate specification is actually equivalent to a joint multivariate normal distribution for the missing covariates. If $x_{mis,j}$ is missing, continuous, and strictly positive, we can transform to logs, $x^*_{mis,j} = \log(x_{mis,j})$, and then specify a normal distribution for $f(x^*_{mis,j}|\boldsymbol{u}_{j-1},\boldsymbol{\alpha}_j)$, or if $\log(x_{mis,j})$ is not approximately normal, we specify an exponential (or gamma) regression with mean $\exp(\boldsymbol{\alpha}_j'\boldsymbol{u}_{j-1})$ for $f(x_{mis,j}|\boldsymbol{u}_{j-1},\boldsymbol{\alpha}_j)$.

Case 5. If we have missing categorical and continuous covariates (i.e., mixed covariates), our usual convention is to specify the joint covariate distribution by specifying the distribution for the categorical covariates by conditioning on the continuous covariates first. For example, suppose that we have three missing covariates $\boldsymbol{x}_{mis} = (x_{mis,1}, x_{mis,2}, x_{mis,3})'$, $x_{mis,1}$ is continuous, and $(x_{mis,2}, x_{mis,3})$ are categorical, say binary. Then our convention for specifying the joint distribution for $\boldsymbol{x}_{mis}$ is

$$f(\boldsymbol{x}_{mis}|\boldsymbol{x}_{obs},\boldsymbol{\alpha}) = f(x_{mis,3}|\boldsymbol{u}_2,\boldsymbol{\alpha}_3)f(x_{mis,2}|\boldsymbol{u}_1,\boldsymbol{\alpha}_2)f(x_{mis,1}|\boldsymbol{u}_0,\boldsymbol{\alpha}_1), \tag{8.3.2}$$

where $\boldsymbol{u}_{j-1} = (x_{mis,j-1}, \ldots, x_{mis,1}, \boldsymbol{x}_{obs}')'$, $j = 1, 2, 3$. We see from (8.3.2) that the distributions for the categorical covariates condition on $x_{mis,1}$ throughout. Following this convention, we can specify logistic regression models for $f(x_{mis,j}|\boldsymbol{u}_{j-1},\boldsymbol{\alpha}_j)$, $j = 3, 2$, and a normal distribution, $N(\mu_1,\sigma_1^2)$ for $f(x_{mis,1}|\boldsymbol{u}_0,\boldsymbol{\alpha}_1)$, where $\mu_1 = \boldsymbol{\alpha}_1'\boldsymbol{u}_0$.

We emphasize here, that even with the guidelines given in Cases 1 – 5, we always recommend that several sensitivity analyses be conducted by

(i) changing the order of conditioning in the covariate distributions;

(ii) fitting main effects models as well as several models with interactions terms between the covariates; and

(iii) examining several link functions, such as the logit, complementary log-log, and probit links for the regressions involving the categorical covariates.

For (i), we recommend that several different conditioning schemes be examined and the posterior summaries of $\boldsymbol{\beta}$ reported along with the CPO statistic for each model. These sensitivity analyses are especially crucial when there are a large number of missing covariates.

The specification in (8.3.1) has other attractive features, such as easing the prior elicitation for $\boldsymbol{\alpha}$, since normal priors are often suitable. We demonstrate this point in more detail in Example 8.1. Also, (8.3.1) eases the computational burden in the Gibbs sampling algorithm required for sampling from (8.2.4). Specifically, if each of the one-dimensional conditional covariate distributions is in the exponential family, and the priors are in the exponential family, then the desired conditional posterior distributions will be log-concave, and this will yield a straightforward implementation of the Gibbs sampler via the adaptive rejection algorithm.

8.4 The Prior and Posterior Distributions

To characterize the properties of the posterior distribution in the presence of missing covariates, we follow Chen, Ibrahim, and Lipsitz (1999) and consider the general mixed covariate situation in which the categorical covariates are Bernoulli and the continuous covariates are normally distributed. To this end, for the ith observation, we partition the covariate vector as $\boldsymbol{x}_i = ((\boldsymbol{x}_i^{(1)})', (\boldsymbol{x}_i^{(2)})', (\boldsymbol{x}_i^{(3)})')'$, where $\boldsymbol{x}_i^{(j)}$ is a p_j-dimensional vector for $j = 1, 2, 3$, and $p = p_1 + p_2 + p_3$. We assume that $\boldsymbol{x}_i^{(1)}$ is completely observed, and $\boldsymbol{x}_i^{(2)}$ and $\boldsymbol{x}_i^{(3)}$ contain missing values. Moreover, we assume that $\boldsymbol{x}_i^{(2)}$ is a vector of continuous covariates, and $\boldsymbol{x}_i^{(3)}$ is a vector of binary covariates. Now let $\boldsymbol{x}_i^{(2)} = (x_{i1}^{(2)}, x_{i2}^{(2)}, \ldots, x_{i,p_2}^{(2)})'$, and take

$$f(\boldsymbol{x}_i^{(2)}|\boldsymbol{x}_i^{(1)}, \boldsymbol{\alpha}_2) = \prod_{j=1}^{p_2} f(x_{ij}^{(2)}|\boldsymbol{x}_i^{(1)}, x_{i1}^{(2)}, \ldots, x_{i,j-1}^{(2)}, \boldsymbol{\alpha}_{2j}), \tag{8.4.1}$$

where

$$f(x_{ij}^{(2)}|\boldsymbol{x}_i^{(1)}, x_{i1}^{(2)}, \ldots, x_{i,j-1}^{(2)}, \boldsymbol{\alpha}_{2j}) \propto \sigma_j^{-1} \exp\left\{-\frac{1}{2\sigma_j^2}(x_{ij}^{(2)} - \left(\boldsymbol{x}_{i(j)}^{(2)}\right)' \boldsymbol{\xi}_{2j})^2\right\}, \tag{8.4.2}$$

$\boldsymbol{x}_{i(j)}^{(2)} = ((\boldsymbol{x}_i^{(1)})', x_{i1}^{(2)}, \ldots, x_{i,j-1}^{(2)})'$, $\boldsymbol{\xi}_{2j} = (\xi_{2j1}, \xi_{2j2}, \ldots, \xi_{2j,p_1+(j-1)})'$, $\sigma_j > 0$, and $\boldsymbol{\alpha}_{2j} = (\boldsymbol{\xi}_{2j}, \sigma_j^2)$. Also let $\boldsymbol{x}_i^{(3)} = (x_{i1}^{(3)}, x_{i2}^{(3)}, \ldots, x_{ip_3}^{(3)})'$, and suppose that

$$f(\boldsymbol{x}_i^{(3)}|\boldsymbol{x}_i^{(1)}, \boldsymbol{x}_i^{(2)}, \boldsymbol{\alpha}_3) = \prod_{j=1}^{p_3} f(x_{ij}^{(3)}|\boldsymbol{x}_i^{(1)}, \boldsymbol{x}_i^{(2)}, x_{i1}^{(3)}, \ldots, x_{i,j-1}^{(3)}, \boldsymbol{\alpha}_{3j}), \tag{8.4.3}$$

where

$$f(x_{ij}^{(3)} = 1|\boldsymbol{x}_i^{(1)}, \boldsymbol{x}_i^{(2)}, x_{i1}^{(3)}, \ldots, x_{i,j-1}^{(3)}, \alpha_{3j}) = \frac{\exp\left\{\left(\boldsymbol{x}_{i(j)}^{(3)}\right)' \boldsymbol{\alpha}_{3j}\right\}}{1 + \exp\left\{\left(\boldsymbol{x}_{i(j)}^{(3)}\right)' \boldsymbol{\alpha}_{3j}\right\}},$$

and $\boldsymbol{x}_{i(j)}^{(3)} = ((\boldsymbol{x}_i^{(1)})', (\boldsymbol{x}_i^{(2)})', x_{i1}^{(3)}, \ldots, x_{i,j-1}^{(3)})'$ is a (p_1+p_2+j-1)-dimensional vector of covariates.

We first consider the noninformative prior

$$\pi(\boldsymbol{\beta}, \boldsymbol{\lambda}, \boldsymbol{\alpha}_2, \boldsymbol{\alpha}_3) \propto \left(\prod_{j=1}^{J} \lambda_j^{\xi_0 - 1} \exp(-\lambda_j \tau_0)\right) \left(\prod_{j=1}^{p_2} \frac{1}{(\sigma_j^2)^{\zeta_0/2}} \exp(-\omega_0/\sigma_j^2)\right), \tag{8.4.4}$$

where ξ_0, τ_0, ζ_0 and ω_0 are prespecified hyperparameters. Let $r_{ij} = 1$ if x_{ij} is missing, and 0 otherwise, $r_i = \sum_{j=1}^{p} r_{ij}$, and write

$$f(\boldsymbol{x}_i|\boldsymbol{\alpha}) = f(\boldsymbol{x}_i^{(2)}|\boldsymbol{x}_i^{(1)}, \boldsymbol{\alpha}_2) f(\boldsymbol{x}_i^{(3)}|\boldsymbol{x}_i^{(1)}, \boldsymbol{x}_i^{(2)}, \boldsymbol{\alpha}_3). \tag{8.4.5}$$

We are now led to the following theorem.

Theorem 8.4.1 *Suppose (i) when $\nu_i = 1$, $y_i > 0$, (ii) there exists $i_1, i_2, \ldots, i_J$ such that $\nu_{i_j} = 1$, and $s_{j-1} < y_{i_j} \le s_j$, $j = 1, 2, \ldots, J$, (iii) the design matrix X^* with i^{th} row equal to $\nu_i^* \boldsymbol{x}_i'$, where $\nu_i^* = 1$ if $\nu_i = 1$ and $r_i = 0$ and $\nu_i^* = 0$ otherwise, is of full rank, (iv) $\xi_0 > 0$, $\tau_0 \ge 0$, $\zeta_0 \ge 0$, and $\omega_0 \ge 0$, (v) X_j^* is an $n \times (p_1 + p_2 + (j-1))$ matrix with i^{th} row $\nu_{ij}^*((\boldsymbol{x}_i^{(1)})', (\boldsymbol{x}_i^{(2)})', x_{i1}^{(3)}, x_{i2}^{(3)}, \ldots, x_{i,j-1}^{(3)})$, where*

$$\nu_{ij}^* = \begin{cases} 1 & \text{if } x_{ij}^{(3)} = 0 \text{ and } r_i = 0 \\ -1 & \text{if } x_{ij}^{(3)} = 1 \text{ and } r_i = 0 \\ 0 & \text{if } r_i = 1, \end{cases}$$

and there exists a vector $\boldsymbol{a}^{(j)} = (a_1^{(j)}, a_2^{(j)}, \ldots, a_n^{(j)})'$, $a_i^{(j)} > 0, i = 1, 2, \ldots, n$, such that $X_j^{*\prime} \boldsymbol{a}^{(j)} = 0$ and X_j^* is of full rank for $j = 1, 2, \ldots, p_3$. Then the posterior distribution of $(\boldsymbol{\beta}, \boldsymbol{\lambda}, \boldsymbol{\alpha})$ is proper, i.e.,

$$\int \left(\prod_{i=1}^{n} \left\{ \int_{\boldsymbol{x}_{mis,i}} \sum_{N_i} L(\boldsymbol{\beta}, \boldsymbol{\lambda}|\boldsymbol{x}_i, y_i, N_i) f(\boldsymbol{x}_i|\boldsymbol{\alpha})\, d\boldsymbol{x}_{mis,i} \right\}\right) \times \pi(\boldsymbol{\beta}, \boldsymbol{\lambda}, \boldsymbol{\alpha})\, d\boldsymbol{\beta}\, d\boldsymbol{\lambda}\, d\boldsymbol{\alpha} < \infty. \tag{8.4.6}$$

A proof of Theorem 8.4.1 is given in the Appendix. Theorem 8.4.1 gives very mild conditions for ensuring propriety of the joint posterior distribution of $(\boldsymbol{\beta}, \boldsymbol{\lambda}, \boldsymbol{\alpha})$ under the semiparametric model with a surviving fraction using the noninformative prior in (8.4.4). This is a useful result when conducting Bayesian analyses with noninformative priors. Conditions (i)–(v) are indeed quite mild and essentially require that all event times are strictly positive, at least one event occurs in each chosen interval $(s_{j-1}, s_j]$, and the covariate matrix corresponding to the completely observed covariates is of full rank. These conditions are easily satisfied in most applications and are quite easy-to-check. Also, condition (iv) allows for general classes of improper priors. We note that condition (v) is essentially equivalent to X_j^* is of full rank and

$$\inf_{\{a_i^{(j)} \geq 1,\ i=1,2,\ldots,n\}} \left\{ (\boldsymbol{a}^{(j)})' X_j^* X_j^{*'} \boldsymbol{a}^{(j)} \right\} = 0, \tag{8.4.7}$$

for $j = 1, 2, \ldots, p_3$. Equation (8.4.7) is a standard quadratic programming problem subject to linear inequality constraints and it can be easily verified by using, for example, the IMSL subroutine QPROG. Theorem 8.4.1 also implies that under these mild conditions, the λ_j's in the semiparametric survival model are identifiable. This is a useful characterization, since it is well known that model identifiability is a crucial and often problematic issue in missing data problems.

Next, we construct the power prior for $(\boldsymbol{\beta}, \boldsymbol{\lambda}, \boldsymbol{\alpha})$. Let $D_0 = (n_0, \boldsymbol{y}_0, X_0, \boldsymbol{\nu}_0, \boldsymbol{N}_0)$ denote the complete historical data, where n_0 is the sample size, $\boldsymbol{y}_0$ is the $n_0 \times 1$ vector of response variables, X_0 is the $n_0 \times p$ complete data covariate matrix, $\boldsymbol{\nu}_0 = (\nu_{01}, \nu_{02}, \ldots, \nu_{0n_0})$ is a vector of censoring indicators, and $\boldsymbol{N}_0 = (N_{01}, N_{02}, \ldots, N_{0n_0})$ is the vector of latent counts for the historical data. As with the current dataset, we allow the historical data to also have missing covariates. Further, denote the ith row of X_0 by $\boldsymbol{x}_{0i}' = (x_{0i1}, x_{0i2}, \ldots, x_{0ip})$, and the ith component of $\boldsymbol{y}_0$ by y_{0i}. Following the notation used earlier in this section, for the ith observation, we partition the covariate vector as $\boldsymbol{x}_{0i} = ((\boldsymbol{x}_{0i}^{(1)})', (\boldsymbol{x}_{0i}^{(2)})', (\boldsymbol{x}_{0i}^{(3)})')'$, where $\boldsymbol{x}_{0i}^{(j)}$ is a p_j-dimensional vector for $j = 1, 2, 3$, and $p = p_1 + p_2 + p_3$. We assume that $\boldsymbol{x}_{0i}^{(1)}$ is completely observed, and $\boldsymbol{x}_{0i}^{(2)}$ and $\boldsymbol{x}_{0i}^{(3)}$ contain missing values. Moreover, we assume that $\boldsymbol{x}_{0i}^{(2)}$ is a vector of continuous covariates, and $\boldsymbol{x}_{0i}^{(3)}$ is a vector of binary covariates. We construct a joint informative prior for $(\boldsymbol{\beta}, \boldsymbol{\lambda}, \boldsymbol{\alpha})$ as follows:

$$\pi(\boldsymbol{\beta}, \boldsymbol{\lambda}, \boldsymbol{\alpha} | a_0, D_{0,obs}) \propto \pi^*(\boldsymbol{\beta}, \boldsymbol{\lambda}, \boldsymbol{\alpha} | a_0, D_{0,obs}) \pi_0(\boldsymbol{\beta}, \boldsymbol{\lambda}, \boldsymbol{\alpha}), \tag{8.4.8}$$

where

$$\pi^*(\boldsymbol{\beta}, \boldsymbol{\lambda}, \boldsymbol{\alpha} | a_0, D_{0,obs}) = \prod_{i=1}^{n_0} \int_{\boldsymbol{x}_{mis,0i}} \left(\left[\sum_{N_{0i}} L(\boldsymbol{\beta}, \boldsymbol{\lambda} | \boldsymbol{x}_{0i}, y_{0i}, N_{0i}) \right]^{a_0} \right.$$
$$\left. \times\, f(x_{0i}^{(2)} | \boldsymbol{x}_{0i}^{(1)}, \boldsymbol{\alpha}_{2j}, a_0) f(x_{0i}^{(3)} | \boldsymbol{x}_{0i}^{(1)}, \boldsymbol{x}_{0i}^{(2)}, \boldsymbol{\alpha}_{3j}, a_0) \right) d\boldsymbol{x}_{mis,0i}. \tag{8.4.9}$$

In (8.4.9), $D_{0,obs} = (n_0, \boldsymbol{y}_0, X_{0,obs}, \boldsymbol{\nu}_0)$ is the observed historical data, $L(\boldsymbol{\beta}, \boldsymbol{\lambda} | \boldsymbol{x}_{0i}, y_{0i}, N_{0i})$ denotes the complete data likelihood for the ith subject defined in (8.2.2) with the current data D replaced by the historical data D_0,

$$f(x_{0i}^{(2)} | \boldsymbol{x}_{0i}^{(1)}, \boldsymbol{\alpha}_{2j}, a_0) = \prod_{j=1}^{p_2} \left[f(x_{0ij}^{(2)} | \boldsymbol{x}_{0i}^{(1)}, x_{0i1}^{(2)}, \ldots, x_{0i,j-1}^{(2)}, \boldsymbol{\alpha}_{2j}) \right]^{a_{0ij}^{(2)}}, \tag{8.4.10}$$

and

$$f(\boldsymbol{x}_{0i}^{(3)}|\boldsymbol{x}_{0i}^{(1)}, \boldsymbol{x}_{0i}^{(2)}, \boldsymbol{\alpha}_{3j}, a_0)$$
$$= \prod_{j=1}^{p_3} \left[f(x_{0ij}^{(3)}|\boldsymbol{x}_{0i}^{(1)}, \boldsymbol{x}_{0i}^{(2)}, x_{0i1}^{(3)}, \ldots, x_{0i,j-1}^{(3)}, \boldsymbol{\alpha}_{3j}) \right]^{a_{0ij}^{(3)}}. \quad (8.4.11)$$

The joint covariate distributions in (8.4.10) and (8.4.11) are given by (8.4.1), (8.4.2), and (8.4.3) with D replaced by D_0. The term $\pi_0(\boldsymbol{\beta}, \boldsymbol{\lambda}, \boldsymbol{\alpha})$ in (8.4.8) is the *initial prior* of $(\boldsymbol{\beta}, \boldsymbol{\lambda}, \boldsymbol{\alpha})$, that is, $\pi_0(\boldsymbol{\beta}, \boldsymbol{\lambda}, \boldsymbol{\alpha})$ is the prior of $(\boldsymbol{\beta}, \boldsymbol{\lambda}, \boldsymbol{\alpha})$ before observing the historical data. The quantity $0 \le a_0 \le 1$ is a scalar prior parameter that weights the historical complete data likelihood relative to the current study. To properly weight the historical complete data likelihood in (8.4.10) and (8.4.11), we let $a_{0ij}^{(l)} = a_0$ if $x_{0ij}^{(l)}$ is observed and $a_{0ij}^{(l)} = 1$ if $x_{0i1}^{(l)}$ is missing, for $i = 1, 2, \ldots, n$, $j = 1, 2, \ldots, p_l$, and $l = 2, 3$.

We specify an improper prior for $\pi_0(\boldsymbol{\beta}, \boldsymbol{\lambda}, \boldsymbol{\alpha})$, and take $\boldsymbol{\beta}$, $\boldsymbol{\lambda}$, and $\boldsymbol{\alpha}$ independent at this stage. To this end, we take

$$\pi_0(\boldsymbol{\beta}, \boldsymbol{\lambda}, \boldsymbol{\alpha}) \propto \left(\prod_{j=1}^{J} \lambda_j^{\xi_0 - 1} \exp(-\lambda_j \tau_0) \right) \left(\prod_{j=1}^{p_2} \frac{1}{(\sigma_j^2)^{\zeta_0/2}} \exp(-\omega_0/\sigma_j^2) \right), \quad (8.4.12)$$

where ξ_0, τ_0, ζ_0 and ω_0 are prespecified hyperparameters. The hyperparameters for $\pi_0(.)$ are chosen so that $\pi_0(.)$ is flat relative to the historical complete data likelihood function.

The prior specification is completed by specifying a prior distribution for a_0. We take a beta prior for a_0, and thus take a joint prior distribution for $(\boldsymbol{\beta}, \boldsymbol{\lambda}, \boldsymbol{\alpha}, a_0)$ to be of the form

$$\pi(\boldsymbol{\beta}, \boldsymbol{\lambda}, \boldsymbol{\alpha}, a_0|D_{0,obs}) \propto \pi^*(\boldsymbol{\beta}, \boldsymbol{\lambda}, \boldsymbol{\alpha}|a_0, D_{0,obs})\pi_0(\boldsymbol{\beta}, \boldsymbol{\lambda}, \boldsymbol{\alpha}) a_0^{\kappa_0 - 1}(1 - a_0)^{\psi_0 - 1}, \quad (8.4.13)$$

where (κ_0, ψ_0) are specified prior parameters. As noted earlier, for the purposes of elicitation, it is often easier to work with $\mu_{a_0} = \kappa_0/(\kappa_0 + \psi_0)$, and $\sigma_{a_0} = (\mu_0(1 - \mu_0))^{1/2} (\kappa_0 + \psi_0 + 1)^{-1/2}$. We emphasize that, in an actual analysis, we recommend that several choices of $(\mu_{a_0}, \sigma_{a_0})$ be used, including ones that give small and large weight to the historical data, and several sensitivity analyses conducted. The prior in (8.4.13) does not have a closed form but it has attractive theoretical and computational properties. We give sufficient conditions for the propriety of (8.4.13) in the following theorem.

Theorem 8.4.2 *Let $r_{0ij} = 1$ if x_{0ij} is missing, and 0 otherwise, $r_{0i} = \sum_{j=1}^{p} r_{0ij}$. Suppose (i) when $\nu_{0i} = 1$, $y_{0i} > 0$, (ii) there exists $i_1, i_2, \ldots, i_J$ such that $\nu_{0i_j} = 1$, and $s_{j-1} < y_{0i_j} \le s_j$, $j = 1, 2, \ldots, J$, (iii) the design matrix X_0^* with i^{th} row equal to $\nu_{0i}^* \boldsymbol{x}_{0i}'$, where $\nu_{0i}^* = 1$ if $\nu_{0i} = 1$ and $r_{0i} = 0$*

and $\nu_{0i}^* = 0$ *otherwise, is of full rank, (iv)* $\xi_0 > 0$, $\tau_0 \geq 0$, $\zeta_0 \geq p_1 + p_2 + 1$, *and* $\omega_0 \geq 0$, *(v)* X_{0j}^* *is an* $n_0 \times (p_1 + p_2 + (j-1))$ *matrix with* i^{th} *row* $\nu_{0ij}^*((\boldsymbol{x}_{0i}^{(1)})', (\boldsymbol{x}_{0i}^{(2)})', x_{0i1}^{(3)}, \ldots, x_{0i,j-1}^{(3)})$, *where*

$$\nu_{0ij}^* = \begin{cases} 1 & \text{if } x_{0ij}^{(3)} = 0 \text{ and } r_{0i} = 0 \\ -1 & \text{if } x_{0ij}^{(3)} = 1 \text{ and } r_{0i} = 0 \\ 0 & \text{if } r_{0i} = 1, \end{cases}$$

and there exists a vector $\boldsymbol{a}^{(j)} = (a_1^{(j)}, a_2^{(j)}, \ldots, a_n^{(j)})'$, $a_i^{(j)} > 0, i = 1, 2, \ldots, n$, *such that* $X_{0j}^{*'}\boldsymbol{a}^{(j)} = 0$ *and* X_{0j}^* *is of full rank for* $j = 1, 2, \ldots, p_3$, *and (vi)* $\kappa_0 > p + J\xi_0^* + \zeta_0^* + \sum_{j=1}^{p_3} dim(\boldsymbol{\alpha}_{3j})$ *and* $\psi_0 > 0$, *where* $\xi_0^* = \xi_0$ *if* $\tau_0 = 0$, *and* 0 *if* $\tau_0 > 0$, $\zeta_0^* = 1$ *if* $\zeta_0 = p_1 + p_2 + 1$, *and* 0 *otherwise, and* $dim(\alpha_{3j})$ *denotes the dimension of* $\boldsymbol{\alpha}_{3j}$. *Then the prior distribution* $\pi(\boldsymbol{\beta}, \boldsymbol{\lambda}, \boldsymbol{\alpha}, a_0 | D_{0,obs})$ *given by* (8.4.13) *is proper, i.e.,*

$$\int \pi^*(\boldsymbol{\beta}, \boldsymbol{\lambda}, \boldsymbol{\alpha} | a_0, D_{0,obs}) \pi_0(\boldsymbol{\beta}, \boldsymbol{\lambda}, \boldsymbol{\alpha}) a_0^{\kappa_0 - 1}(1 - a_0)^{\psi_0 - 1} \, d\boldsymbol{\beta} \, d\boldsymbol{\lambda} \, d\boldsymbol{\alpha} \, da_0 < \infty.$$

Theorem 8.4.2 gives very mild conditions for ensuring propriety of the joint prior distribution of $(\boldsymbol{\beta}, \boldsymbol{\lambda}, \boldsymbol{\alpha})$ under the semiparametric model with a surviving fraction. As in 8.4.1, conditions (i)–(vi) are indeed quite mild and essentially require that all event times are strictly positive, at least one event occur in each chosen interval $(s_{j-1}, s_j]$, and the covariate matrix corresponding to the completely observed covariates be of full rank. These conditions are easily satisfied in most applications and are quite easy to check. Also, condition (iv) allows for general classes of improper priors for $\pi_0(\boldsymbol{\beta}, \boldsymbol{\lambda}, \boldsymbol{\alpha})$.

Finally, we mention that both Theorems 8.4.1 and 8.4.2 are quite general, and can be easily modified for other covariate distributions, including the Poisson and multinomial distributions for the discrete covariates, and gamma and scale mixtures of normal for the continuous covariates. We have presented the binomial-normal case here since this is a common, yet general, situation encountered in practice. The sequence of one-dimensional conditional distributions given in (8.3.1) facilitates many other covariate combinations that could be used and would lead to similar proofs as in Theorems 8.4.1 and 8.4.2.

8.5 Model Checking

In missing data problems, it is crucial to check posterior sensitivity to the choice of the covariate distributions. This is an especially important issue here since the specification of the joint covariate distribution depends on the order of conditioning as seen in (8.3.1). Also, the number of terms included in each of the one-dimensional covariate distributions is an important issue which needs to be addressed. For example, one may want to compare the saturated covariate model against a reduced model. General guidelines for modeling the covariate distributions were given in Section 8.3. In terms of assessing the adequacy of the covariate distributions, we must compare several models. We accomplish this here by reporting posterior summaries of $\boldsymbol{\beta}$ from all the models as well as computing the Conditional Predictive Ordinate (CPO) statistic, which measures the adequacy of a given model. A powerful feature of the CPO statistic is that it does not require proper priors, and thus can be readily used even when we do not have historical data to construct a prior. The CPO statistic is a very useful model assessment tool which has been widely used in the statistical literature under various contexts. For a detailed discussion of the CPO statistic and its applications to model assessment, see Section 6.3. In the complete data setting, for the ith observation, the CPO statistic is defined as

$$\text{CPO}_i^* = f(y_i|D^{(-i)}) = \int f(y_i|\boldsymbol{x}_i, \boldsymbol{\beta}, \boldsymbol{\lambda})\pi(\boldsymbol{\beta}, \boldsymbol{\lambda}|D^{(-i)})\ d\boldsymbol{\beta}\ d\boldsymbol{\lambda}, \tag{8.5.1}$$

where y_i denotes the response variable for case i, $D^{(-i)}$ denotes the data with the ith case deleted, and $\pi(\boldsymbol{\beta}, \boldsymbol{\lambda}|D^{(-i)})$ is the marginal posterior distribution of $(\boldsymbol{\beta}, \boldsymbol{\lambda})$ given the data $D^{(-i)}$. From (8.5.1), we see that CPO_i^* is the marginal posterior predictive density of y_i given $D^{(-i)}$, and can be interpreted as the height of this marginal density at y_i. Thus, large values of CPO_i^* imply a better fit of the model. We mention here that in the complete data setting, a distribution need not specified for the covariates $\boldsymbol{x}_i$, and therefore, (8.5.1) does not depend on $\boldsymbol{\alpha}$. However, in the presence of missing covariate data, (8.5.1) is not well defined. Thus, a new definition of the CPO statistic in the presence of missing covariate data is needed. With missing covariate data, for the ith observation, we define the CPO statistic as

$$\begin{aligned}\text{CPO}_i =& f(y_i|x_{i,obs}, D_{obs}^{(-i)}) \\ =& \int f(y_i|\boldsymbol{x}_{mis,i}, \boldsymbol{x}_{obs,i}, \boldsymbol{\beta}, \boldsymbol{\lambda})\pi(\boldsymbol{\beta}, \boldsymbol{\lambda}, \boldsymbol{\alpha}|D_{obs}^{(-i)}) \\ & \times f(\boldsymbol{x}_{mis,i}|\boldsymbol{x}_{obs,i}, \boldsymbol{\alpha})\ d\boldsymbol{x}_{mis,i}\ d\boldsymbol{\beta}\ d\boldsymbol{\lambda} d\boldsymbol{\alpha},\end{aligned} \tag{8.5.2}$$

where $f(\boldsymbol{x}_{mis,i}|\boldsymbol{x}_{obs,i}, \boldsymbol{\alpha})$ denotes the conditional distribution of the missing covariates given the observed covariates for the ith observation, $D_{obs}^{(-i)}$ denotes the observed data D_{obs} with the ith case deleted, and $D^{(-i)}$ is the complete data D with the ith case deleted. The definition in (8.5.2) makes good intuitive sense. The motivation behind (8.5.2) is that in the presence of missing covariate data, we average (8.5.1) with respect to the conditional distribution of the missing covariates given the observed data. Thus, (8.5.2) can be viewed as a marginal CPO statistic where the marginalization is done with respect to the marginal posterior predictive distribution

$f(\boldsymbol{x}_{mis,i}|\boldsymbol{x}_{obs,i}, D_{obs}^{(-i)})$. If the ith observation does not have any missing covariates, then $\text{CPO}_i = \text{CPO}_i^*$.

As discussed in Section 6.3, a useful summary statistic of the CPO_i's is the logarithm of the pseudomarginal likelihood (LPML), defined as

$$\text{LPML} = \sum_{i=1}^{n} \log(\text{CPO}_i). \tag{8.5.3}$$

In the context of survival data, the LPML statistic was discussed in detail in Section 6.3. A slight modification of (8.5.3) is the average LPML, given by

$$\text{ALPML} = \frac{\text{LPML}}{n}, \tag{8.5.4}$$

where n is the sample size. When comparing two models, the model with the larger value of ALPML is the better fitting model. We see from (8.5.2) that LPML and ALPML are always well defined as long as the posterior predictive density is proper. Thus, LPML and ALPML are well defined under improper priors and in addition, are very computationally stable. Therefore, LPML and ALPML have clear advantages over the Bayes factor and the L measure. Section 6.3 gives a more detailed discussion on this.

Example 8.1. Melanoma data. We consider the ECOG melanoma clinical trials E1684 and E1690, discussed in Example 1.2. Again, E1690 serves as the current study and E1684 serves as the historical data in this example. Models incorporating a surviving fraction have become quite common for cancers such as melanoma, since for these diseases, cure fractions can typically range from 30% to 60%. As shown Figures 1.1, 5.3, and 5.4, Kaplan-Meier RFS plots for E1684 and E1690 both show a pronounced plateau in the survival curve after sufficient follow-up. Thus, the semiparametric model in (8.2.2) is an appropriate model to use for these datasets. We use (8.2.2) with varying numbers of intervals, J, to examine several model fits, as discussed below.

We carry out a Bayesian analysis of E1690 based on the observed data posterior, i.e., the posterior distribution incorporating the missing covariates. We demonstrate the methodology on the E1690 data using the E1684 historical data to construct our prior as in (8.4.13). The response variable is relapse-free survival (RFS) , which may be right censored. The covariates are treatment (x_1: IFN, OBS), age (x_2), sex (x_3), logarithm of Breslow depth (x_4), logarithm of size of primary (x_5), and type of primary (x_6). Covariates x_1, x_3, and x_6 are all binary covariates, whereas x_2, x_4, and x_5 are all continuous. Logarithms of Breslow depth and size were used in all of the analyses to achieve approximate normality in the covariate distributions for the continuous covariates. The regression coefficients corresponding to the covariates $x_1, x_2, \ldots, x_6$, are denoted by $\beta_2, \beta_3, \ldots, \beta_7$, respectively. We also include an intercept term in every model, and thus the regression coefficient vector is denoted by $\boldsymbol{\beta} = (\beta_1, \beta_2, \ldots, \beta_7)'$. Tables 1.1–1.3 summarize the response variable and covariates for the E1690 and E1684 studies, respectively. The covariates x_1, x_2, and x_3 are completely observed, and x_4, x_5, and x_6 have missing values. The total missing data fractions for these three covariates in E1684 and E1690 are 27.4% and 28.6%, respectively. For E1684, 19.6% percent of the cases had exactly one missing covariate, 4.6% had exactly two missing covariates, and 3.2% had exactly three missing covariates. For E1690, 25.3% percent of the cases had exactly one missing covariate, 3.0% had exactly two missing covariates, and

TABLE 8.1. Posterior Estimates of $\boldsymbol{\beta}$ for Complete Case Analysis.

$E(a_0 \mid D_{obs})$	ALPML	$\boldsymbol{\beta}$	Mean	SD	95% HPD Interval
0	−1.2748	β_1	0.154	0.153	(−0.144, 0.454)
		β_2	−0.075	0.154	(−0.377, 0.227)
		β_3	0.023	0.078	(−0.129, 0.177)
		β_4	−0.064	0.164	(−0.389, 0.251)
		β_5	0.028	0.086	(−0.147, 0.189)
		β_6	0.080	0.075	(−0.071, 0.222)
		β_7	−0.196	0.162	(−0.519, 0.119)
0.27	−1.2650	β_1	0.216	0.138	(−0.061, 0.483)
		β_2	−0.106	0.137	(−0.370, 0.171)
		β_3	0.026	0.071	(−0.109, 0.169)
		β_4	−0.067	0.147	(−0.369, 0.207)
		β_5	0.016	0.076	(−0.135, 0.163)
		β_6	0.070	0.068	(−0.060, 0.205)
		β_7	−0.178	0.145	(−0.461, 0.106)
0.43	−1.2624	β_1	0.245	0.132	(−0.013, 0.503)
		β_2	−0.122	0.130	(−0.372, 0.136)
		β_3	0.027	0.067	(−0.105, 0.159)
		β_4	-0.069	0.139	(−0.335, 0.207)
		β_5	0.012	0.072	(−0.134, 0.147)
		β_6	0.066	0.064	(−0.060, 0.191)
		β_7	−0.174	0.138	(−0.446, 0.092)
1	−1.2599	β_1	0.310	0.116	(0.079, 0.531)
		β_2	-0.154	0.112	(−0.364, 0.071)
		β_3	0.033	0.056	(−0.077, 0.144)
		β_4	−0.074	0.118	(−0.309, 0.152)
		β_5	0.004	0.061	(−0.118, 0.122)
		β_6	0.054	0.055	(−0.054, 0.161)
		β_7	−0.160	0.118	(−0.393, 0.069)

Source: Chen, Ibrahim, and Lipsitz (1999).

0.2% had exactly three missing covariates. Tables 1.1 and 1.2 summarize the number of missing values for each covariate.

We first consider a complete case Bayesian analysis using a special case of the priors, in which we delete the cases with missing covariates in the historical and current datasets. For the complete case analysis, we need not specify a covariate distribution, and thus the parameter vector consists of $(\boldsymbol{\beta}, \boldsymbol{\lambda})$. Our main interest is in posterior inferences about $\boldsymbol{\beta}$, with all other parameters being treated as nuisance parameters. We use a special case of the joint prior distribution in (8.4.8) with $a_{ij}^{(2)} = a_{0ij}^{(3)} = 0$ and

$$\pi_0(\boldsymbol{\beta}, \boldsymbol{\lambda}) \propto \prod_{j=1}^{J} \lambda_j^{\xi_0 - 1} \exp(-\lambda_j \tau_0)$$

with $\xi_0 = 1$ and $\tau_0 = 0$. Table 8.1 shows the results for the complete case Bayesian analysis based on several values of posterior means for a_0. From Table 8.1, we first see that for all values of a_0, the 95% HPD interval for the treatment regression coefficient (β_2) gets narrower as more weight is given to the historical data, but it always contains 0. For example, for $a_0 = 1$ with probability 1, the 95% HPD interval for β_2 is $(-0.364, 0.071)$, and for $E(a_0|D_{obs}) = 0.27$, the 95% HPD interval for β_2 is $(-0.370, 0.171)$, which is wider than the one for $a_0 = 1$. In fact, we see from Table 8.1 that all HPD intervals for $(\beta_2, \beta_3, \ldots, \beta_7)$ contain 0 regardless of the posterior mean of a_0. Thus the complete case Bayesian analysis appears to imply that none of the covariates are important predictors of RFS. Table 8.1 also shows that the ALPML statistic is largest when $a_0 = 1$ with probability 1.

This suggests that this model has the best overall fit amongst other models shown in Table 8.1. Thus, the ALPML statistic can be used as a guide here for deciding the appropriate weight to give to the historical data in the analysis.

We next consider an analysis based on the observed data posterior, incorporating the cases with missing covariate values. For this analysis, a covariate distribution needs to be specified, as discussed in Section 8.3. We take

$$\begin{aligned} f(x_4, x_5, x_6|x_1, x_2, x_3, \boldsymbol{\alpha}) =& f(x_6|x_1, x_2, x_3, x_4, x_5, \boldsymbol{\alpha}_6) f(x_5|x_1, x_2, x_3, x_4 \boldsymbol{\alpha}_5) \\ & \times f(x_4|x_1, x_2, x_3, \boldsymbol{\alpha}_4), \end{aligned} \tag{8.5.5}$$

where $\boldsymbol{\alpha} = (\boldsymbol{\alpha}_4', \boldsymbol{\alpha}_5', \boldsymbol{\alpha}_6')'$,

$$\begin{aligned} & f(x_6 = 1|x_4, x_5, x_1, x_2, x_3, \boldsymbol{\alpha}_6) \\ =& \frac{\exp\{\alpha_{60} + \alpha_{61}x_1 + \alpha_{62}x_2 + \alpha_{63}x_3 + \alpha_{64}x_4 + \alpha_{65}x_5\}}{1 + \exp\{\alpha_{60} + \alpha_{61}x_1 + \alpha_{62}x_2 + \alpha_{63}x_3 + \alpha_{64}x_4 + \alpha_{65}x_5\}}, \end{aligned} \tag{8.5.6}$$

$$[x_5|x_1, x_2, x_3, x_4, \alpha_5] \sim N(\xi_{50} + \xi_{51}x_1 + \xi_{52}x_2 + \xi_{53}x_3 + \xi_{54}x_4, \sigma_5^2), \tag{8.5.7}$$

TABLE 8.2. Posterior Estimates of β Based on Observed Data Posterior.

$E(a_0\|D_{obs})$	ALPML	β	Mean	SD	95% HPD Interval
0	−1.2317	β_1	0.129	0.133	(−0.134, 0.384)
		β_2	−0.206	0.132	(−0.469, 0.051)
		β_3	0.097	0.066	(−0.031, 0.229)
		β_4	−0.144	0.138	(−0.411, 0.131)
		β_5	0.025	0.073	(−0.119, 0.167)
		β_6	0.102	0.074	(−0.040, 0.248)
		β_7	−0.034	0.150	(−0.324, 0.266)
0.21	−1.2264	β_1	0.174	0.125	(−0.069, 0.414)
		β_2	−0.231	0.121	(−0.456, 0.017)
		β_3	0.086	0.061	(−0.031, 0.210)
		β_4	−0.128	0.126	(−0.372, 0.121)
		β_5	0.019	0.068	(−0.115, 0.148)
		β_6	0.097	0.067	(−0.034, 0.226)
		β_7	−0.030	0.139	(−0.302, 0.243)
0.34	−1.2247	β_1	0.204	0.120	(−0.026, 0.443)
		β_2	−0.243	0.115	(−0.464, −0.011)
		β_3	0.083	0.059	(−0.030, 0.200)
		β_4	−0.120	0.121	(−0.350, 0.124)
		β_5	0.018	0.064	(−0.104, 0.145)
		β_6	0.094	0.063	(−0.028, 0.221)
		β_7	−0.035	0.129	(−0.292, 0.212)
0.55	−1.2234	β_1	0.229	0.113	(0.013, 0.461)
		β_2	−0.256	0.109	(−0.473, −0.047)
		β_3	0.078	0.055	(−0.029, 0.186)
		β_4	-0.108	0.113	(−0.326, 0.119)
		β_5	0.015	0.060	(−0.103, 0.130)
		β_6	0.090	0.059	(−0.026, 0.204)
		β_7	−0.036	0.122	(−0.270, 0.205)
1	−1.2364	β_1	0.162	0.096	(−0.031, 0.345)
		β_2	−0.270	0.097	(−0.459, −0.082)
		β_3	0.070	0.048	(−0.024, 0.166)
		β_4	−0.081	0.101	(−0.277, 0.122)
		β_5	0.024	0.052	(−0.082, 0.124)
		β_6	0.050	0.051	(−0.051, 0.150)
		β_7	−0.021	0.106	(−0.225, 0.188)

Source: Chen, Ibrahim, and Lipsitz (1999).

$\boldsymbol{\alpha}_5 = (\xi_{50}, \xi_{51}, \ldots, \xi_{54}, \sigma_5^2)'$,

$$[x_4|x_1, x_2, x_3, \alpha_4] \sim N(\xi_{40} + \xi_{41}x_1 + \xi_{42}x_2 + \xi_{43}x_3, \sigma_4^2), \tag{8.5.8}$$

and $\boldsymbol{\alpha}_4 = (\xi_{40}, \xi_{41}, \ldots, \xi_{43}, \sigma_4^2)'$. We will refer to the covariate model specified by (8.5.5) – (8.5.8) as Model 1. For the priors, we use (8.4.13) and (8.4.12) with $J = 10$, $\xi_0 = 1$, $\tau_0 = 0$, $\zeta_0 = 7$, and $\omega_0 = 0$. Again, we emphasize that our main interest is in posterior inferences about $\boldsymbol{\beta}$, with $(\boldsymbol{\lambda}, \boldsymbol{\alpha})$ being treated as nuisance parameters.

Table 8.2 shows results for the Bayesian analysis of Model 1 based on the observed data posterior using several values of a_0. The results in Table 8.2 are in sharp contrast to those of the complete case Bayesian analysis shown in Table 8.1. From Table 8,2, we see that when $E(a_0|D_{obs}) \geq .34$, the 95% HPD interval for the treatment covariate (β_2) does *not* contain 0. Also, the 95% HPD interval for β_2 gets narrower as more weight is given to the historical data. For example, for $E(a_0|D_{obs}) = .55$, the 95% HPD interval for β_2 is $(-0.473, -0.047)$. This indicates that the treatment covariate is an important predictor of RFS, and that IFN may in fact be superior to OBS. This conclusion was not possible based on the complete case Bayesian analysis shown in Table 8.1. Table 8.2 also shows that the ALPML statistic is largest when $E(a_0|D_{obs}) = 0.55$, with a value of ALPML$= -1.2234$. The largest ALPML statistic in Table 8.1 is ALPML$= -1.2599$. Thus, we see here that the ALPML statistics for the analysis based on the observed data posterior are always larger than the ALPML statistics based on the complete case analysis, for all values of a_0. This is also an important finding since it shows that the overall fit of Model 1 is always improved when we incorporate the missing covariates into the analysis compared to a complete case analysis. Table 8.3 shows a Bayesian analysis based on the observed data posterior using Model 1 with $J = 5$ and $E(a_0|D_{obs}) = 0.34$. We see from Table 8.3 that the posterior estimates of $\boldsymbol{\beta}$ are very similar to those of Table 8.2, and thus are quite robust to the choice of J. For example, from Table 8.3 ($J = 5$), we see that the posterior mean and standard deviation of β_2 are -0.241 and 0.116, respectively, and from Table 8.2 ($J = 10$), the posterior mean and standard deviation of β_2 are -0.243 and 0.115, respectively. Also, the last box in Table 8.5, labeled $m = 1$ using $J = 5$, shows the ratios of the posterior means and standard deviations of $(\beta_1, \beta_2, \ldots, \beta_7)$ for Model 1 using $J = 5$ versus $J = 10$ and $E(a_0|D_{obs}) = 0.34$. For example, the ratio of the posterior mean of β_2 using $J = 5$ compared to $J = 10$ is $(-0.241)/(-0.243) = 0.991$, and the ratio of the posterior standard deviations for β_2 is $0.116/0.115 = 1.007$. We see from Table 8.5 that all the ratios for all of the coefficients are very close to 1 indicating robustness for $J = 5$ and $J = 10$. Other values of J yielded similar results to those of Table 8.3, and therefore posterior inferences for $\boldsymbol{\beta}$ are quite robust to the choice of J.

A detailed sensitivity analysis was also conducted on the choice of covariate distribution and the order of conditioning in the covariate distribution.

TABLE 8.3. Posterior Estimates of $\boldsymbol{\beta}$ Based on Observed Data Posterior Using $J = 5$.

$E(a_0 \mid D_{obs})$	$\boldsymbol{\beta}$	Mean	SD	95% HPD Interval
0.34	β_1	0.226	0.119	(−0.022, 0.452)
	β_2	−0.241	0.116	(−0.470, −0.021)
	β_3	0.084	0.058	(−0.033, 0.193)
	β_4	−0.119	0.121	(−0.351, 0.121)
	β_5	0.019	0.064	(−0.103, 0.145)
	β_6	0.093	0.064	(−0.029, 0.219)
	β_7	−0.034	0.128	(−0.289, 0.210)

Source: Chen, Ibrahim, and Lipsitz (1999).

Our main aim in these sensitivity analyses is to show that the posterior estimates of the regression coefficients $\boldsymbol{\beta}$ are robust with respect to changes in the ordering of conditioning as well as the addition of various interaction terms in the covariate distributions. In addition to Model 1 discussed in (8.4.5), five additional models were considered. Thus six different covariate models in all were considered. Covariate Models 1–5 use a logit link for type (x_6), whereas Model 6 uses a complimentary log-log link. Also, $x_i * x_j$ in the notation below denotes the two-way interaction between x_i and x_j.

Model 1: $[x_6|x_1, \ldots, x_5][x_5|x_1, \ldots, x_4][x_4|x_1, x_2, x_3]$ (model (8.4.5))

Model 2: $[x_6|x_1, \ldots, x_5][x_4|x_5, x_1, x_2, x_3][x_5|x_1, x_2, x_3]$

Model 3: $[x_6|x_1, \ldots, x_5][x_5|x_4][x_4]$

Model 4: $[x_6|x_1, \ldots, x_5, x_i * x_j, \text{all } i < j, i \le 3][x_5|x_1, \ldots, x_4, x_1 * x_2, x_1 * x_3, x_2 * x_3][x_4|x_1, x_2, x_3, x_i * x_j, \text{all } i < j]$

Model 5: $[x_6|x_1, x_2, x_3][x_5|x_1, x_2, x_3][x_4|x_1, x_2, x_3]$

Model 6: $[x_6|x_1, \ldots, x_5][x_5|x_1, \ldots, x_4][x_4|x_1, x_2, x_3]$ (log-log link)

For Models 1–6 above, we take normal distributions for all of the continuous covariates and logistic regressions for all of the binary covariates, with the exception of Model 6, in which we use a log-log link for x_6. For example, for Model 3 using the notation above, $[x_6|x_1, x_2, \ldots, x_5]$ is a logistic regression model with covariates $(x_1, x_2, \ldots, x_5)$, $[x_5|x_4]$ is a normal distribution with mean $\xi_{50} + \xi_{51}x_4$ and variance σ_5^2, and $[x_4]$ is a normal distribution with mean ξ_{40} and variance σ_4^2, where ξ_{40} does not depend on other covariates. The other distributions above are defined in a similar fashion. We note that Model 2 consists of the reverse order in conditioning between x_4 and x_5, Model 3 does not depend on the completely observed covariates x_1, x_2, x_3, Model 4 is Model 1 with several two-factor interactions included, Model 5 assumes independence between (x_6, x_5, x_4), and Model 6 is Model 1 with a complimentary log-log link for x_6. Table 8.4

TABLE 8.4. Posterior Summaries Based on Six Covariate Models.

		Model					
		1	2	3	4	5	6
β_1	Mean	0.204	0.199	0.195	0.199	0.194	0.198
	SD	0.120	0.119	0.118	0.122	0.121	0.120
β_2	Mean	−0.243	−0.243	−0.242	−0.241	−0.241	−0.243
	SD	0.115	0.115	0.115	0.116	0.116	0.117
β_3	Mean	0.083	0.083	0.085	0.082	0.083	0.083
	SD	0.059	0.059	0.058	0.058	0.058	0.059
β_4	Mean	−0.120	−0.120	−0.117	−0.121	−0.118	−0.119
	SD	0.121	0.121	0.121	0.120	0.122	0.123
β_5	Mean	0.018	0.018	0.014	0.018	0.017	0.017
	SD	0.064	0.064	0.064	0.064	0.062	0.064
β_6	Mean	0.094	0.092	0.091	0.088	0.093	0.094
	SD	0.063	0.063	0.063	0.062	0.063	0.064
β_7	Mean	−0.035	−0.032	−0.025	−0.031	−0.027	−0.029
	SD	0.129	0.129	0.128	0.130	0.129	0.130

Source: Chen, Ibrahim, and Lipsitz (1999).

summarizes the posterior means and standard deviations of $\boldsymbol{\beta}$, and the posterior mean of a_0. The ALPML statistics are -1.2247, -1.2249, -1.2248, -1.2250, -1.2251, and -1.2249, and $E(a_0|D)$ is 0.34, 0.34, 0.34, 0.34, 0.34, and 0.33 for Models 1–6, respectively. We see from Table 8.4 that the posterior means and standard deviations of the regression coefficients for all six models are remarkably close to one another, even for the complimentary log-log model. Moreover, we see that the ALPML statistics are quite close for all six models, and the posterior means of a_0 are quite similar for the six models. Table 8.5 shows ratios of the posterior means and standard deviations of the regression coefficients based on the six covariate models, namely, Models 1–6. We see that all the ratios are close to 1, indicating that the posterior estimates of $\boldsymbol{\beta}$ are quite robust to the various covariate models. For example, for Model 4, we see that the ratios for β_2 are 0.992 and 1.004. Model 4 is considerably more complex than Model 1, but the ratios are remarkably close to 1. We see the same phenomenon for Model 6, where a completely different link function was used for $[x_6|x_1, x_2, \ldots, x_5]$. We also note in Table 8.5 that the ratios corresponding to covariates with missing values tend to be a bit further away from 1 than the ratios corresponding to completely observed covariates. This makes intuitive sense since one would expect the posterior estimates for those β_j's corresponding to missing covariates to be not quite as good as those β_j's corresponding to completely observed covariates. Thus, we conclude from this investigation that posterior inferences about $\boldsymbol{\beta}$ are very robust to various changes in the

TABLE 8.5. Ratios of Posterior Means and Standard Deviations.

Model m vs Model 1	β	Ratio of Means	Ratio of SD's
$m = 2$	β_1	0.973	0.996
	β_2	1.000	0.999
	β_3	1.000	1.003
	β_4	0.994	0.993
	β_5	1.016	1.003
	β_6	0.980	1.000
	β_7	0.906	1.001
$m = 3$	β_1	0.957	0.986
	β_2	0.995	1.000
	β_3	1.030	0.996
	β_4	0.970	0.994
	β_5	0.775	1.010
	β_6	0.968	0.999
	β_7	0.714	0.996
$m = 4$	β_1	0.976	1.017
	β_2	0.992	1.004
	β_3	0.993	0.998
	β_4	1.010	0.990
	β_5	1.004	1.003
	β_6	0.939	0.982
	β_7	0.885	1.009
$m = 5$	β_1	0.952	1.008
	β_2	0.992	1.007
	β_3	0.999	0.996
	β_4	0.985	1.001
	β_5	0.929	0.981
	β_6	0.992	0.989
	β_7	0.768	1.004
$m = 6$	β_1	0.970	0.998
	β_2	0.998	1.011
	β_3	1.008	1.006
	β_4	0.987	1.012
	β_5	0.950	1.003
	β_6	1.003	1.014
	β_7	0.829	1.010
$m = 1$ using $J = 5$	β_1	1.105	0.996
	β_2	0.991	1.007
	β_3	1.015	0.994
	β_4	0.987	0.994
	β_5	1.084	1.009
	β_6	0.993	1.008
	β_7	0.982	0.991

Source: Chen, Ibrahim, and Lipsitz (1999).

covariate distributions, including changes in the order of conditioning, the addition of interaction terms, and changes in the link function.

Finally, we mention that our intervals $(s_{j-1}, s_j]$, $j = 1, 2, \ldots, J$ were chosen so that with the combined datasets from the historical and current data, at least one failure falls in each interval. This technique for choosing J is quite reasonable and results in a stable Gibbs sampler. We also conducted sensitivity analyses on the construction of the intervals, $(s_{j-1}, s_j]$, $j = 1, 2, \ldots, J$. Three different constructions of $(s_{j-1}, s_j]$ were considered. We chose the subintervals $(s_{j-1}, s_j]$ with

(i) equal numbers of failures or censored observations;

(ii) approximately equal lengths subject to the restriction that at least one failure occurs in each interval; and

(iii) decreasing numbers of failures or censored observations.

The posterior estimates were quite robust with respect to these constructions. Gibbs sampling along with the adaptive rejection algorithm of Gilks and Wild (1992) was used to obtain all posterior estimates. The Gibbs sampling algorithm is similar to that of Chen, Ibrahim, and Yiannoutsos (1999) and Ibrahim, Chen, and MacEachern (1999). All HPD intervals were computed using a Monte Carlo method developed by Chen and Shao (1999). The computing time required to fit all of the models was approximately 2.5 hours on a digital alpha machine, and 20,000 Gibbs iterations were used in all of the computations after a burn-in of 1000 iterations.

Appendix

Proof of Theorem 8.4.1. Let j_i denote an integer so that $\nu_{ij} = 1$ if $j = j_i$ and 0 otherwise. Then

$$f(y_i|\boldsymbol{\lambda}) = \lambda_{j_i} \exp\left\{ -\left[\lambda_{j_i}(y_i - s_{j_i-1}) + \sum_{g=1}^{j_i-1} \lambda_g (s_g - s_{g-1}) \right] \right\}$$

and

$$S(y_i|\boldsymbol{\lambda}) = \exp\left\{ -\left[\lambda_{j_i}(y_i - s_{j_i-1}) + \sum_{g=1}^{j_i-1} \lambda_g (s_g - s_{g-1}) \right] \right\}.$$

In (8.2.2), summing out the unobserved latent variable N_i yields

$$\sum_{N_i} L(\boldsymbol{\beta}, \boldsymbol{\lambda}|x_i, y_i, N_i) = \Big(\theta_i f(y_i|\boldsymbol{\lambda})\Big)^{\nu_i} \exp\{-\theta_i(1 - S(y_i|\boldsymbol{\lambda}))\}, \qquad (8.A.1)$$

where $\theta_i = \exp(\boldsymbol{x}_i'\boldsymbol{\beta})$. Following the proof of Theorem 5.2.1, it can be shown that there exists a constant $M > 1$ such that

$$\left(\theta_i f(y_i|\boldsymbol{\lambda})\right)^{\nu_i} \exp\{-\theta_i(1 - S(y_i|\boldsymbol{\lambda}))\} \leq M, \tag{8.A.2}$$

and

$$\frac{f(y_i|\boldsymbol{\lambda})}{1 - S(y_i|\boldsymbol{\lambda})} \leq M. \tag{8.A.3}$$

Let

$$\pi^*(\boldsymbol{\beta}, \boldsymbol{\lambda}, \boldsymbol{\alpha}|D_{obs}) = \prod_{i=1}^{n} \left\{ \int_{\boldsymbol{x}_{mis,i}} \left(\sum_{N_i} L(\boldsymbol{\beta}, \boldsymbol{\lambda}|\boldsymbol{x}_i, y_i, N_i) \right) f(\boldsymbol{x}_i|\boldsymbol{\alpha})\, d\boldsymbol{x}_{mis,i} \right\} \times \pi^*(\boldsymbol{\beta}, \boldsymbol{\lambda}, \boldsymbol{\alpha}),$$

where

$$\pi^*(\boldsymbol{\beta}, \boldsymbol{\lambda}, \boldsymbol{\alpha}) = \left(\prod_{j=1}^{J} \lambda_j^{\xi_0 - 1} \exp(-\lambda_j \tau_0) \right) \left(\prod_{j=1}^{p_2} \frac{1}{(\sigma_j^2)^{\zeta_0/2}} \exp(-\omega_0/\sigma_j^2) \right),$$

and $I = \{i: \ \nu_i = 1\}$. For each $i \in I$, after some algebra, we get

$$\theta_i f(y_i|\boldsymbol{\lambda}) \exp\{-\theta_i(1 - S(y_i|\boldsymbol{\lambda}))\} \leq K_1 \frac{f(y_i|\boldsymbol{\lambda})}{1 - S(y_i|\boldsymbol{\lambda})}, \tag{8.A.4}$$

for any θ_i, where $K_1 > 0$ is a constant.

Since X^* is of full rank, there exists $i_1^*, i_2^*, \ldots, i_p^*$ such that $\nu_{i_1^*} = \nu_{i_2^*} = \ldots = \nu_{i_p^*} = 1$, $r_{i_1^*} = r_{i_2^*} = \ldots = r_{i_p^*} = 0$, and $X_p^* = (\boldsymbol{x}_{i_1^*}, \boldsymbol{x}_{i_2^*}, \ldots, \boldsymbol{x}_{i_p^*})'$ is of full rank. We make the transformation $\boldsymbol{u} = (u_1, u_2, \ldots, u_p)' = X_p^* \boldsymbol{\beta}$. This linear transformation is one-to-one. Thus, we have

$$\begin{aligned}
&\int_{R^p} \prod_{l=1}^{p} f(y_{i_l^*}|\boldsymbol{\lambda}) \exp(\boldsymbol{x}_{i_l^*}'\boldsymbol{\beta}) \exp\left\{ -\exp(\boldsymbol{x}_{i_l^*}'\boldsymbol{\beta})(1 - S(y_{i_l^*}|\boldsymbol{\lambda})) \right\} d\boldsymbol{\beta} \\
=& |X_p^*|^{-1} \prod_{l=1}^{p} \int_{-\infty}^{\infty} f(y_{i_l^*}|\boldsymbol{\lambda}) \exp(u_l) \exp\left\{ -\exp(u_l)(1 - S(y_{i_l^*}|\boldsymbol{\lambda})) \right\} du_l \\
=& |X_p^*|^{-1} \prod_{l=1}^{p} \frac{f(y_{i_l^*}|\boldsymbol{\lambda})}{1 - S(y_{i_l^*}|\boldsymbol{\lambda})}.
\end{aligned} \tag{8.A.5}$$

Using (8.A.2)–(8.A.5) and after integrating out $\boldsymbol{\beta}$, it can be shown that there exists a constant $K_2 > 0$, such that

$$\begin{aligned}
\int \pi^*(\boldsymbol{\beta}, \boldsymbol{\lambda}, \boldsymbol{\alpha}|D_{obs})\, d\boldsymbol{\beta} \leq& K_2 \left(\prod_{\{i:\ \nu_i = 1\}} \frac{f(y_i|\boldsymbol{\lambda})}{1 - S(y_i|\boldsymbol{\lambda})} \right) \\
&\times \left(\prod_{i=1}^{n} \int_{\boldsymbol{x}_{mis,i}} f(\boldsymbol{x}_i|\boldsymbol{\alpha}) d\boldsymbol{x}_{mis,i} \right) \pi^*(\boldsymbol{\lambda}) \pi^*(\boldsymbol{\alpha}),
\end{aligned} \tag{8.A.6}$$

where

$$\pi^*(\boldsymbol{\lambda}) = \prod_{j=1}^{J} \lambda_j^{\xi_0 - 1} \exp(-\lambda_j \tau_0), \text{ and } \pi^*(\boldsymbol{\alpha}) = \prod_{j=1}^{p_2} \frac{1}{(\sigma_j^2)^{\zeta_0/2}} \exp(-\omega_0/\sigma_j^2).$$

Since $s_{j-1} < y_{i_j} \leq s_j$, we have

$$\begin{aligned}\frac{f(y_{i_j}|\boldsymbol{\lambda})}{1 - S(y_{i_j}|\boldsymbol{\lambda})} =& \frac{\lambda_j \exp\left\{-\left[\lambda_j(y_{i_j} - s_{j-1}) + \sum_{g=1}^{j-1} \lambda_g(s_g - s_{g-1})\right]\right\}}{1 - \exp\left\{-\left[\lambda_j(y_{i_j} - s_{j-1}) + \sum_{g=1}^{j-1} \lambda_g(s_g - s_{g-1})\right]\right\}} \\ \leq& \frac{\lambda_j \exp\left\{-\lambda_j(y_{i_j} - s_{j-1})\right\}}{1 - \exp(-\lambda_j(y_{i_j} - s_{j-1}))} \\ \leq& K_3 \exp\left\{-\frac{\lambda_j}{2}(y_{i_j} - s_{j-1})\right\},\end{aligned} \tag{8.A.7}$$

where $K_3 > 0$ is a constant that is independent of λ_j. We note that (8.A.7) is true even when the $(i_j)^{th}$ observation contains missing covariates. From condition (iv), we have $\xi_0 > 0$ and $\tau_0 \geq 0$, which leads to

$$\prod_{j=1}^{J} \left(\int_0^\infty \exp\left\{-\frac{\lambda_j}{2}(y_{i_j} - s_{j-1})\right\} \lambda_j^{\xi_0 - 1} \exp(-\lambda_j \tau_0)\, d\lambda_j\right) < \infty. \tag{8.A.8}$$

Then, after integrating out $\boldsymbol{\beta}$ and $\boldsymbol{\lambda}$ and combining (8.A.6)–(8.A.8) along with condition (ii) leads to

$$\int \pi^*(\boldsymbol{\beta}, \boldsymbol{\lambda}, \boldsymbol{\alpha}|D_{obs})\, d\boldsymbol{\beta}\, d\boldsymbol{\lambda} \leq K_4 \left[\prod_{i=1}^{n} \int_{\boldsymbol{x}_{mis,i}} f(\boldsymbol{x}_i|\boldsymbol{\alpha})\, d\boldsymbol{x}_{mis,i}\right] \pi^*(\boldsymbol{\alpha}), \tag{8.A.9}$$

where $K_4 > 0$ is a constant. In (8.A.9),

$$f(\boldsymbol{x}_i|\boldsymbol{\alpha}) = f(\boldsymbol{x}_i^{(2)}|\boldsymbol{x}_i^{(1)}, \boldsymbol{\alpha}_2) f(\boldsymbol{x}_i^{(3)}|\boldsymbol{x}_i^{(1)}, \boldsymbol{x}_i^{(2)}, \boldsymbol{\alpha}_3). \tag{8.A.10}$$

We note that $f(\boldsymbol{x}_i^{(2)}|\boldsymbol{x}_i^{(1)}, \boldsymbol{\alpha}_2)$ is a product of one-dimensional normal densities. Thus, it can be shown that conditions (iii) and (iv) guarantee that

$$\int \left(\prod_{i=1}^{n} \int_{\boldsymbol{x}_{mis,i}} f(\boldsymbol{x}_i^{(2)}|\boldsymbol{x}_i^{(1)}, \boldsymbol{\alpha}_2)\, d\boldsymbol{x}_{mis,i}^{(2)}\right) \prod_{j=1}^{p_2} (\sigma_j^2)^{-\zeta_0/2} \exp(-\omega_j/\sigma_j^2)\, d\boldsymbol{\alpha}_2 < \infty. \tag{8.A.11}$$

Since $f(\boldsymbol{x}_i^{(3)}|\boldsymbol{x}_i^{(1)}, \boldsymbol{x}_i^{(2)}, \boldsymbol{\alpha}_3) \leq 1$, (8.A.9) – (8.A.11) yield

$$\int \pi^*(\boldsymbol{\beta}, \boldsymbol{\lambda}, \boldsymbol{\alpha}|D_{obs})\, d\boldsymbol{\beta}\, d\boldsymbol{\lambda}\, d\boldsymbol{\alpha}_2 \leq K_5 \prod_{\{i:\ r_i = 0\}} f(\boldsymbol{x}_i^{(3)}|\boldsymbol{x}_i^{(1)}, \boldsymbol{x}_i^{(2)}, \boldsymbol{\alpha}_3), \tag{8.A.12}$$

where $K_5 > 0$ is a constant. In (8.A.12),

$$\prod_{\{i:r_i=0\}} f(\boldsymbol{x}_i^{(3)}|\boldsymbol{x}_i^{(1)}, \boldsymbol{x}_i^{(2)}, \boldsymbol{\alpha}_3)$$
$$= \prod_{j=1}^{p_3} \prod_{\{i:\ r_i=0\}} f(x_{ij}^{(3)}|\boldsymbol{x}_i^{(1)}, \boldsymbol{x}_i^{(2)}, x_{i1}^{(3)}, \dots, x_{i,j-1}^{(3)}, \boldsymbol{\alpha}_{3j}). \tag{8.A.13}$$

Since $f(x_{ij}^{(3)}|\boldsymbol{x}_i^{(1)}, \boldsymbol{x}_i^{(2)}, x_{i1}^{(3)}, \dots, x_{i,j-1}^{(3)}, \boldsymbol{\alpha}_{3j})$ is the logit probability for a binary response, it directly follows from Chen and Shao (2001) that condition (v) is a necessary and sufficient condition for

$$\int \prod_{\{i:\ r_i=0\}} f(x_{ij}^{(3)}|\boldsymbol{x}_i^{(1)}, \boldsymbol{x}_i^{(2)}, x_{i1}^{(3)}, \dots, x_{i,j-1}^{(3)}, \boldsymbol{\alpha}_{3j})\ d\boldsymbol{\alpha}_{3j} < \infty.$$

This completes the proof. □

Proof of Theorem 8.4.2. From condition (iii), there exists $i_1^*, i_2^*, \dots, i_p^*$ such that $\nu_{0i_1^*} = \nu_{0i_2^*} = \cdots = \nu_{0i_p^*} = 1$, $r_{0i_1^*} = r_{0i_2^*} = \cdots = r_{0i_p^*} = 0$, and $X_{0p}^* = (x_{0i_1^*}, x_{0i_2^*}, \dots, x_{0i_p^*})'$ is of full rank. Let $I_0 = \{i:\ \nu_{0i} = 1\}$ and $I_0^* = \{i_1^*, i_2^*, \dots, i_p^*\}$. Also let $\pi^*(\boldsymbol{\beta}, \boldsymbol{\lambda}, \boldsymbol{\alpha}, a_0|D_{0,obs})$ denote the unnormalized prior density given in (8.4.13). Following the proof of Theorem 8.4.1, it can be shown that there exists a constant $K_{01} > 0$ such that

$$\pi^*(\boldsymbol{\beta},\boldsymbol{\lambda}, \boldsymbol{\alpha}, a_0|D_{0,obs}) \le K_{01} \prod_{\{i:\ i\in I_0 - I_0^*\}} \left[\frac{f(y_{0i}|\boldsymbol{\lambda})}{1 - S(y_{0i}|\boldsymbol{\lambda})}\right]^{a_0}$$
$$\times \prod_{l=1}^{p} \left[f(y_{0i_l^*}|\boldsymbol{\lambda})\right]^{a_0} \exp(a_0\boldsymbol{x}_{0i_l^*}'\boldsymbol{\beta}) \exp\{-a_0 \exp(\boldsymbol{x}_{0i_l^*}'\boldsymbol{\beta})(1 - S(y_{0i}|\boldsymbol{\lambda}))\}$$
$$\times \pi^*(\boldsymbol{\alpha}, a_0|D_{0,obs})\pi_0^*(\boldsymbol{\beta}, \boldsymbol{\lambda}, \boldsymbol{\alpha})a_0^{\kappa_0-1}(1 - a_0)^{\psi_0-1}, \tag{8.A.14}$$

where

$$\pi^*(\boldsymbol{\alpha}, a_0|D_{0,obs})$$
$$= \prod_{i=1}^{n_0} \int_{\boldsymbol{x}_{mis,0i}} \left(\prod_{j=1}^{p_2} \left[f(x_{0ij}^{(2)}|\boldsymbol{x}_{0i}^{(1)}, x_{0i1}^{(2)}, \dots, x_{0i,j-1}^{(2)}, \alpha_{2j})\right]^{a_{0ij}^{(2)}}\right.$$
$$\left.\times \prod_{j=1}^{p_3} \left[f(x_{0ij}^{(3)}|\boldsymbol{x}_{0i}^{(1)}, \boldsymbol{x}_{0i}^{(2)}, x_{0i1}^{(3)}, \dots, x_{0i,j-1}^{(3)}, \alpha_{3j})\right]^{a_{0ij}^{(3)}}\right) d\boldsymbol{x}_{mis,0i}, \tag{8.A.15}$$

and

$$\pi_0^*(\boldsymbol{\beta}, \boldsymbol{\lambda}, \boldsymbol{\alpha}) = \left(\prod_{j=1}^{J} \lambda_j^{\xi_0 - 1} \exp(-\lambda_j \tau_0)\right) \left(\prod_{j=1}^{p_2} \frac{1}{(\sigma_j^2)^{\zeta_0/2}} \exp(-\omega_0/\sigma_j^2)\right). \tag{8.A.16}$$

By taking the transformation $\boldsymbol{u}_0 = (u_{01}, u_{02}, \ldots, u_{0p})' = X_{0p}^* \boldsymbol{\beta}$, we have

$$\begin{aligned}
&\int \prod_{l=1}^{p} \left[f(y_{0i_l^*}|\boldsymbol{\lambda})\right]^{a_0} \exp(a_0 \boldsymbol{x}_{0i_l^*}' \boldsymbol{\beta}) \exp\{-a_0 \exp(\boldsymbol{x}_{0i_l^*}' \boldsymbol{\beta})(1 - S(y_{0i}|\boldsymbol{\lambda}))\}\, d\boldsymbol{\beta} \\
&\propto \prod_{l=1}^{p} \int \left[f(y_{0i_l^*}|\boldsymbol{\lambda})\right]^{a_0} \exp(a_0 u_{0l}) \exp\{-a_0 \exp(u_{0l})(1 - S(y_{0i}|\boldsymbol{\lambda}))\}\, du_{0l} \\
&\propto \prod_{l=1}^{p} \frac{\Gamma(a_0)}{a_0^{a_0}} \left(\frac{f(y_{0i_l^*}|\boldsymbol{\lambda})}{1 - S(y_{0i}|\boldsymbol{\lambda})}\right)^{a_0}.
\end{aligned} \tag{8.A.17}$$

In (8.A.17),

$$a_0^{a_0} = \exp(a_0 \log(a_0)) \le K_{02} \quad \text{and} \quad \Gamma(a_0) = \Gamma(a_0 + 1)/a_0 \le K_{03}/a_0, \tag{8.A.18}$$

for $0 < a_0 < 1$, where $K_{02} > 0$ and $K_{03} > 0$ are constants. After integrating out $\boldsymbol{\beta}$, (8.A.15)–(8.A.18) yield

$$\begin{aligned}
\int \pi^*(\boldsymbol{\beta}, \boldsymbol{\lambda}, \boldsymbol{\alpha}, a_0|D_{0,obs})\, d\boldsymbol{\beta} \le \frac{K_{04}}{a_0^p} &\left(\prod_{\{i:\ i \in I_0\}} \left[\frac{f(y_{0i}|\boldsymbol{\lambda})}{1 - S(y_{0i}|\boldsymbol{\lambda})}\right]^{a_0}\right) \\
&\times \pi^*(\boldsymbol{\alpha}, a_0|D_{0,obs}) \pi_0^*(\boldsymbol{\lambda}, \boldsymbol{\alpha}) a_0^{\kappa_0 - 1}(1 - a_0)^{\psi_0 - 1},
\end{aligned}$$

where $K_{04} > 0$, and $\pi_0^*(\boldsymbol{\lambda}, \boldsymbol{\alpha})$ is given by (8.A.16). Again, following the proof of Theorem 8.4.1, condition (ii) leads to

$$\begin{aligned}
&\int \pi^*(\boldsymbol{\beta}, \boldsymbol{\lambda}, \boldsymbol{\alpha}, a_0|D_{0,obs})\, d\boldsymbol{\beta} \\
&\le \frac{K_{05}}{a_0^p} \prod_{j=1}^{J} \exp\left\{-\frac{a_0 \lambda_j}{2}(y_{0i_j} - s_{j-1})\right\} \lambda_j^{\xi_0 - 1} \exp(-\lambda_j \tau_0) \\
&\quad \times \pi^*(\boldsymbol{\alpha}, a_0|D_{0,obs}) \left(\prod_{j=1}^{p_2} \frac{1}{(\sigma_j^2)^{\zeta_0/2}} \exp(-\omega_0/\sigma_j^2)\right) a_0^{\kappa_0 - 1}(1 - a_0)^{\psi_0 - 1},
\end{aligned}$$

where $K_{05} > 0$. From condition (iv), integrating out $\boldsymbol{\lambda}$ gives

$$\int \pi^*(\boldsymbol{\beta}, \boldsymbol{\lambda}, \boldsymbol{\alpha}, a_0|D_{0,obs})\, d\boldsymbol{\beta}\, d\boldsymbol{\lambda}$$
$$\leq \frac{K_{06}}{a_0^{p+J\xi_0^*}} \pi^*(\boldsymbol{\alpha}, a_0|D_{0,obs}) \left(\prod_{j=1}^{p_2} \frac{1}{(\sigma_j^2)^{\zeta_0/2}} \exp(-\omega_0/\sigma_j^2)\right) a_0^{\kappa_0-1}(1-a_0)^{\psi_0-1}, \tag{8.A.19}$$

where $K_{06} > 0$.

Let $\boldsymbol{x}_{0i(j)}^{(2)} = ((x_{0i}^{(1)})', x_{0i1}^{(2)}, \ldots, x_{0i,j-1}^{(2)})'$, and let $r_{0ij}^{(2)}$ denote the missing indicator corresponding to the covariate $x_{0ij}^{(2)}$. Write $n_{0ij}^{(2)} = n - \sum_{i=1}^{n} r_{0ij}^{(2)}$ for $j = 1, 2, \ldots, p_2$. After some algebra, integrating out the missing covariates yields

$$\pi^*(\boldsymbol{\alpha}, a_0|D_{0,obs})$$
$$\leq K_{07} \prod_{j=1}^{p_2} \left(\frac{1}{(\sigma_j^2)^{a_0 n_{0ij}^{(2)}/2}} \prod_{\{i:\ r_{0i}=0\}} \exp\left\{ -\frac{a_0}{2\sigma_j^2} (x_{0ij}^{(2)} - \left(\boldsymbol{x}_{0i(j)}^{(2)}\right)' \boldsymbol{\xi}_{2j})^2 \right\} \right)$$
$$\times \prod_{j=1}^{p_3} \left(\prod_{\{i:\ r_{0i}=0\}} \left[f(x_{0ij}^{(3)}|\boldsymbol{x}_{0i}^{(1)}, \boldsymbol{x}_{0i}^{(2)}, x_{0i1}^{(3)}, \ldots, x_{0i,j-1}^{(3)}, \boldsymbol{\alpha}_{3j}) \right]^{a_0} \right), \tag{8.A.20}$$

where $K_{07} > 0$. Since $\zeta_0 \geq p_1 + p_2 + 1$, using condition (iii), (8.A.18) and (8.A.21), and integrating out $\boldsymbol{\alpha}_2$ from (8.A.19), we obtain

$$\int \pi^*(\beta, \lambda, \alpha, a_0|D_{0,obs})\, d\beta\, d\lambda\, d\alpha_2$$
$$\leq \frac{K_{08}}{a_0^{p+J\xi_0^*+\zeta_0^*}} \prod_{j=1}^{p_3} \left(\prod_{\{i:\ r_{0i}=0\}} \left[f(x_{0ij}^{(3)}|\boldsymbol{x}_{0i}^{(1)}, \boldsymbol{x}_{0i}^{(2)}, x_{0i1}^{(3)}, \ldots, x_{0i,j-1}^{(3)}, \boldsymbol{\alpha}_{3j}) \right]^{a_0} \right)$$
$$\times a_0^{\kappa_0-1}(1-a_0)^{\psi_0-1}, \tag{8.A.21}$$

where $K_{08} > 0$. Let $c_0^* = \kappa_0 - (p + J\xi_0^* + \zeta_0^*)$. Since $\kappa_0 > p + J\xi_0^* + \zeta_0^*$ (condition (vi)) implies $c_0^* > 0$, it can be shown that integrating out a_0

from (8.A.21) leads to

$$\int \pi^*(\boldsymbol{\beta}, \boldsymbol{\lambda}, \boldsymbol{\alpha}, a_0 | D_{0,obs})\, d\boldsymbol{\beta}\, d\boldsymbol{\lambda}\, d\boldsymbol{\alpha}_2\, da_0$$

$$\leq K_{09}\left(1 - \log\left[\prod_{j=1}^{p_3} \prod_{\{i:\ r_{0i}=0\}} f(x_{0ij}^{(3)} | \boldsymbol{x}_{0i}^{(1)}, \boldsymbol{x}_{0i}^{(2)}, x_{0i1}^{(3)}, \ldots, x_{0i,j-1}^{(3)}, \boldsymbol{\alpha}_{3j})\right]\right)^{-c_0^*}$$

$$\leq K_{09}\left[\left(\prod_{j=1}^{p_3} \prod_{\{i:\ r_{0i}=0\}} f(x_{0ij}^{(3)} | \boldsymbol{x}_{0i}^{(1)}, \boldsymbol{x}_{0i}^{(2)}, x_{0i1}^{(3)}, \ldots, x_{0i,j-1}^{(3)}, \boldsymbol{\alpha}_{3j})\right)\right.$$
$$\left.\times \exp\left\{t_0^*\left(\sum_{j=1}^{p_3} ||\boldsymbol{\alpha}_{3j}|| + 1\right)\right\}\right] + K_{09}(t_0^*)^{-c_0^*}\left[\sum_{j=1}^{p_3} ||\boldsymbol{\alpha}_{3j}|| + 1\right]^{-c_0^*}, \tag{8.A.22}$$

for $t_0^* > 0$, where $K_{09} > 0$, and $||\boldsymbol{\alpha}_{3j}|| = (\boldsymbol{\alpha}_{3j}'\boldsymbol{\alpha}_{3j})^{1/2}$. It directly follows from Chen and Shao (2001) that condition (v) guarantees that the integral of the first term on the right-hand side of (8.A.22) over $\boldsymbol{\alpha}_3$ is finite, while

$$\int \left(\sum_{j=1}^{p_3} ||\boldsymbol{\alpha}_{3j}|| + 1\right)^{-(\kappa_0 - (p + J\xi_0^* + \zeta_0^*))} d\boldsymbol{\alpha}_3 < \infty,$$

provided that $\kappa_0 > p + J\xi_0^* + \zeta_0^* + \sum_{j=1}^{p_3} \dim(\boldsymbol{\alpha}_{3j})$. This completes the proof. □

Exercises

8.1 Show that condition (v) in Theorem 8.4.1 is equivalent to that X_j^* is of full rank and (8.4.7).

8.2 Let $\boldsymbol{x}_{mis}$ denote the vector of all missing covariates. Also let $\pi(\boldsymbol{\beta}, \boldsymbol{\lambda}, \boldsymbol{\alpha} | D_{obs})$ denote the joint posterior distribution of $(\boldsymbol{\beta}, \boldsymbol{\lambda}, \boldsymbol{\alpha})$ defined by (8.4.6).

(a) Write the joint posterior distribution of $(\boldsymbol{\beta}, \boldsymbol{\lambda}, \boldsymbol{\alpha}, \boldsymbol{x}_{mis})$ based on the observed data D_{obs} so that the marginal posterior distribution of $(\boldsymbol{\beta}, \boldsymbol{\lambda}, \boldsymbol{\alpha})$ reduces to $\pi(\boldsymbol{\beta}, \boldsymbol{\lambda}, \boldsymbol{\alpha} | D_{obs})$.

(b) Derive and discuss the properties of the full conditional distributions for $(\boldsymbol{\beta}, \boldsymbol{\lambda}, \boldsymbol{\alpha}, \boldsymbol{x}_{mis})$ required for the Gibbs sampler.

8.3 Consider the power prior given in (8.4.13). Let $\boldsymbol{x}_{mis}$ and $\boldsymbol{x}_{mis,0}$ denote the vectors of missing covariates for the current and historical data, respectively.

(a) Write out the joint posterior distribution of $(\boldsymbol{\beta}, \boldsymbol{\lambda}, \boldsymbol{\alpha}, a_0, \boldsymbol{x}_{mis}, \boldsymbol{x}_{mis,0})$ corresponding to the informative prior based on the historical data.

(b) Derive the full conditional distributions of $(\boldsymbol{\beta}, \boldsymbol{\lambda}, \boldsymbol{\alpha}, a_0, \boldsymbol{x}_{mis}, \boldsymbol{x}_{mis,0})$ necessary for the Gibbs sampler.

8.4 Define the complete-data CPO statistic as

$$\mathrm{CPO}_i^* = \int f(y_i|\boldsymbol{x}_i, \boldsymbol{\beta}, \boldsymbol{\lambda})\pi(\boldsymbol{\beta}, \boldsymbol{\lambda}|\boldsymbol{\alpha}, D^{(-i)})\, d\boldsymbol{\beta}\, d\boldsymbol{\lambda},$$

where $\pi(\boldsymbol{\beta}, \boldsymbol{\lambda}|\boldsymbol{\alpha}, D^{(-i)})$ is the conditional posterior distribution of $(\boldsymbol{\beta}, \boldsymbol{\lambda})$ given $\boldsymbol{\alpha}$ and $D^{(-i)}$. Let $\boldsymbol{x}_{mis}^{(-i)}$ denote the vector of missing covariates with the ith case deleted. Also $f(\boldsymbol{x}_{mis}^{(-i)}|\boldsymbol{\alpha})$ denotes the distribution of $\boldsymbol{x}_{mis}^{(-i)}$, and $\pi(\boldsymbol{\alpha}|D_{obs}^{(-i)})$ denotes the marginal posterior distribution of $\boldsymbol{\alpha}$ based on the observed data $D_{obs}^{(-i)}$.

(a) Write CPO_i given in (8.5.2) in terms of a mixture of CPO_i^*, where the mixture distribution is given by

$$f(\boldsymbol{x}_{mis,i}|\boldsymbol{x}_{obs,i}, \boldsymbol{\alpha}) f(\boldsymbol{x}_{mis}^{(-i)}|\boldsymbol{\alpha})\pi(\boldsymbol{\alpha}|D_{obs}^{(-i)}).$$

(b) Derive an expression of CPO_i in terms of the joint posterior distribution $\pi(\boldsymbol{\beta}, \boldsymbol{\lambda}, \boldsymbol{\alpha}|D_{obs})$ and

$$\int_{\boldsymbol{x}_{mis,i}} f(y_i|\boldsymbol{x}_i, \boldsymbol{\beta}, \boldsymbol{\lambda}) f(\boldsymbol{x}_i|\alpha)\, d\boldsymbol{x}_{mis,i}.$$

(c) Discuss how to compute CPO_i using an MCMC sample of $(\boldsymbol{\beta}, \boldsymbol{\lambda}, \boldsymbol{\alpha})$ from $\pi(\boldsymbol{\beta}, \boldsymbol{\lambda}, \boldsymbol{\alpha}|D_{obs})$.

8.5 Verify (8.A.1).

8.6 Prove (8.A.2) and (8.A.3).

8.7 Show that for any $\lambda > 0$ and $y > 0$,

$$\frac{\lambda \exp(-\lambda y)}{1 - \exp(-\lambda y)} \le K \exp(-\tfrac{1}{2}\lambda y),$$

where $K > 0$ is a constant.

8.8 Verify (8.A.11).

8.9 Let $c_0^* = \kappa_0 - (p + J\xi_0^* + \zeta_0^*)$, and

$$f(\boldsymbol{x}_{0i}^{(3)}|\boldsymbol{x}_{0i}^{(1)}, \boldsymbol{x}_{0i}^{(2)}, \boldsymbol{\alpha}_3) = \prod_{j=1}^{p_3} \prod_{\{i:\ r_{0i}=0\}} f(x_{0ij}^{(3)}|\boldsymbol{x}_{0i}^{(1)}, \boldsymbol{x}_{0i}^{(2)}, x_{0i1}^{(3)}, \ldots, x_{0i,j-1}^{(3)}, \boldsymbol{\alpha}_{3j}).$$

(a) Show that $f(\boldsymbol{x}_{0i}^{(3)}|\boldsymbol{x}_{0i}^{(1)}, \boldsymbol{x}_{0i}^{(2)}, \boldsymbol{\alpha}_3) \le 1$.

(b) Prove that if $c_0^* > 0$ and $\psi_0 > 0$, then

$$\int_0^1 \left[f(\boldsymbol{x}_{0i}^{(3)}|\boldsymbol{x}_{0i}^{(1)}, \boldsymbol{x}_{0i}^{(2)}, \boldsymbol{\alpha}_3)\right]^{a_0} a_0^{c_0^*-1}(1-a_0)^{\psi_0-1}da_0$$
$$\leq K_0\left(1-\log\left[f(\boldsymbol{x}_{0i}^{(3)}|\boldsymbol{x}_{0i}^{(1)}, \boldsymbol{x}_{0i}^{(2)}, \boldsymbol{\alpha}_3)\right]\right)^{-c_0^*},$$

where $K_0 > 0$ is a constant.

8.10 Using the notation defined in Exercise 8.9, show that

$$\int \left(1-\log\left[f(\boldsymbol{x}_{0i}^{(3)}|\boldsymbol{x}_{0i}^{(1)}, \boldsymbol{x}_{0i}^{(2)}, \boldsymbol{\alpha}_3)\right]\right)^{-c_0^*} d\boldsymbol{\alpha}_3$$
$$\leq K_0^* \int f(\boldsymbol{x}_{0i}^{(3)}|\boldsymbol{x}_{0i}^{(1)}, \boldsymbol{x}_{0i}^{(2)}, \boldsymbol{\alpha}_3)\exp\left\{t_0^*(||\boldsymbol{\alpha}_3||+1)\right\} d\boldsymbol{\alpha}_3$$
$$+ K_0^* \int (1+||\boldsymbol{\alpha}_3||)^{-c_0^*} d\boldsymbol{\alpha}_3,$$

for $t_0^* > 0$, where $K_0^* > 0$ is a constant, and $||\boldsymbol{\alpha}_3|| = (\boldsymbol{\alpha}_3'\boldsymbol{\alpha}_3)^{1/2}$.

8.11 Write a computer program to verify all of the entries of Tables 8.1 and 8.2.

8.12 Write a computer program to verify all of the entries of Table 8.4.

8.13 For the E1684 and E1690 data discussed in Example 8.1, consider a piecewise exponential model along with the priors in Example 6.3 using $J = 5$.

(a) Derive the full conditional posterior distributions for all of the parameters.
(b) Write a computer program using the Gibbs sampler and provide tables similar to Tables 8.1, 8.2, and 8.4. From your results, assess which model fits the data better.

8.14 For the E1684 and E1690 data discussed in Example 8.1, consider a Weibull model along with noninformative priors for all of the parameters.

(a) Derive the full conditional posterior distributions for all of the parameters.
(b) Using BUGS, compute posterior summaries for $\boldsymbol{\beta}$, and plot the marginal posterior distribution of $\boldsymbol{\beta}$.
(c) Compare your results to Exercises 8.12 and 8.13.

8.15 Use BUGS to fit the piecewise exponential model in Exercise 8.13 using noninformative priors for all parameters, and compare your results with those of Exercises 8.13 and 8.14.

9
Design and Monitoring of Randomized Clinical Trials

Bayesian approaches to the design and monitoring of randomized trials has received a great deal of recent attention. There are several articles discussing Bayesian design and monitoring of randomized trials, including Berry (1985, 1987, 1993), Freedman and Spiegelhalter (1983, 1989, 1992), Spiegelhalter and Freedman (1986, 1988), Spiegelhalter, Freedman, and Blackburn (1986), Breslow (1990), Jennison and Turnbull (1990), Thall and Simon (1994), Freedman, Spiegelhalter, and Parmar (1994), George et al. (1994), Grossman, Parmar, and Spiegelhalter (1994), Lewis and Berry (1994), Spiegelhalter, Freedman, and Parmar (1993, 1994), Rosner and Berry (1995), Adcock (1997), Joseph and Belisle (1997), Joseph, Wolfson, and DuBerger (1995), Lindley (1997), Pham-Gia (1997), Simon (1999), Simon and Freedman (1997), Weiss (1997), Bernardo and Ibrahim (2000), and Lee and Zelen (2000). In this chapter, we will review some of the basic approaches to the Bayesian design and monitoring of randomized trials as well as compare the Bayesian and frequentist approaches.

9.1 Group Sequential Log-Rank Tests for Survival Data

The majority of phase III clinical trials are designed with a group sequential monitoring plan, which means that accumulating data are analyzed over specified intervals of time. The analyses conducted at the specified intervals are called *interim analyses.* Jennison and Turnbull (2000) is an excellent and comprehensive book on group sequential methods in clinical trials. In most randomized clinical trials, the primary outcome or *endpoint* is time-to-event such as time to relapse, time to treatment failure, or time to death. As discussed in Jennison and Turnbull (2000), in group sequential studies, the effect of right censoring in survival analysis can be seen repeatedly at successive analyses, but with the length of follow-up on each patient increasing as the study progresses.

Suppose that a trial is comparing two treatments, say, A and B, and that the true treatment difference is summarized by a parameter θ, where large values of θ correspond to superiority of the new treatment. Since we focus on survival data in our development here, θ will denote a log hazard ratio. The statistical analysis usually centers on testing the null hypothesis

$$H_0\colon\ \theta = 0. \tag{9.1.1}$$

The log-rank test of Mantel (1966) and Peto and Peto (1972) is commonly used to test (9.1.1).

Following Jennison and Turnbull (2000), let d_k, $k = 1, 2, \ldots, K$ denote the total number of uncensored failures observed when analysis k is conducted in a group sequential study. Here, K denotes the maximum number of interim analyses. Assuming no ties, denote the survival times of these subjects by $y_{1k}, y_{2k}, \ldots, y_{d_k,k}$, where the y_{ik} denote the elapsed times between study entry and failure. Let the numbers known at analysis k to have survived up to time y_{ik} after treatment be $r_{iA,k}$ and $r_{iB,k}$ on treatments A and B, respectively. Then, the log-rank score statistic at analysis k is given by

$$S_k = \sum_{i=1}^{d_k} \left\{ \nu_{iB,k} - \frac{r_{iA,k}}{r_{iA,k} + r_{iB,k}} \right\}, \tag{9.1.2}$$

where $\nu_{iB,k} = 1$ if the failure time at y_{ik} was on treatment B and $\nu_{iB,k} = 0$ otherwise. The variance of S_k when $\theta = 0$ can be approximated by the sum of the conditional variances of the $\nu_{iB,k}$ given the numbers $r_{iA,k}$ and $r_{iB,k}$ at risk on each treatment just before time y_{ik}. Also, $\text{Var}(S_k)$ is the information for θ when $\theta = 0$, and thus we have

$$\mathcal{I}_k \equiv \widehat{\text{Var}}(S_k) = \sum_{i=1}^{d_k} \frac{r_{iA,k} r_{iB,k}}{(r_{iA,k} + r_{iB,k})^2}. \tag{9.1.3}$$

For θ close to 0, we can use the approximation

$$S_k \sim N(\theta \mathcal{I}_k, \mathcal{I}_k) \tag{9.1.4}$$

given a sufficiently large $\mathcal{I}_k$. Furthermore, conditional on the observed information sequence $\{\mathcal{I}_1, \mathcal{I}_2, \ldots, \mathcal{I}_k\}$, the joint distribution of $Z_k = S_k/\sqrt{\mathcal{I}_k}$, $k = 1, 2, \ldots, K$ is multivariate normal, with

$$Z_k \sim N(\theta\sqrt{\mathcal{I}_k}, 1), \tag{9.1.5}$$

and

$$\text{Cov}(Z_{k_1}, Z_{k_2}) = \sqrt{(\mathcal{I}_{k_1}/\mathcal{I}_{k_2})}, \tag{9.1.6}$$

where $k = 1, 2, \ldots, K$, and $1 \leq k_1 \leq k_2 \leq K$. These asymptotic results have been established by Tsiatis (1982) and Harrington, Fleming, and Green (1982). The basic result in (9.1.5) and (9.1.6) implies that the log-rank statistics computed at the interim analyses are asymptotically equivalent to a sequence of normally distributed random variables. Thus, sample size calculations and monitoring for clinical trials with time-to-event endpoints can be based on sequences of normally distributed random variables. That is, sample size calculations and monitoring plans can proceed assuming we have standardized statistics with the asymptotic distribution of (9.1.5) and (9.1.6). This is a key result which makes the design and monitoring of clinical trials with right censored time-to-event data relatively straightforward. We discuss sample size calculations in more detail in the subsequent sections.

9.2 Bayesian Approaches

We now make use of the result (9.1.5) and (9.1.6) to develop the Bayesian paradigm for the design and monitoring of clinical trials. Following Jennison and Turnbull (2000), consider data giving rise to the standardized statistics $Z_1, Z_2, \ldots$ with the asymptotic joint distribution in (9.1.5) and (9.1.6). At any stage k, the likelihood function of θ is a normal density with mean θ and variance $\mathcal{I}_k^{-1}$. Suppose that the prior for θ is

$$\theta \sim N(\mu_0, \sigma_0^2). \tag{9.2.1}$$

Then the posterior distribution of θ at analysis k is

$$N\left(\frac{\hat{\theta}^{(k)}\mathcal{I}_k + \mu_0\sigma_0^{-2}}{\mathcal{I}_k + \sigma_0^{-2}}, \frac{1}{\mathcal{I}_k + \sigma_0^{-2}}\right), \tag{9.2.2}$$

where $\hat{\theta}^{(k)} = \frac{Z_k}{\sqrt{\mathcal{I}_k}}$, and the 95% credible interval for θ is given by

$$\left(\frac{\hat{\theta}^{(k)}\mathcal{I}_k + \mu_0\sigma_0^{-2}}{\mathcal{I}_k + \sigma_0^{-2}} \pm 1.96\sqrt{\frac{1}{\mathcal{I}_k + \sigma_0^{-2}}}\right). \tag{9.2.3}$$

Note that the credible interval is shrunk toward the prior mean μ_0, whereas the usual fixed sample frequentist confidence interval would be centered on the estimate $\hat{\theta}^{(k)}$. Often the prior mean μ_0 is taken be to zero, the value at the null hypothesis. In this case, the credible interval for θ is shrunk toward zero. The limiting case $\sigma_0 \to \infty$ yields a $N(\hat{\theta}^{(k)}, \mathcal{I}_k^{-1})$ posterior for θ. In this case, the credible interval in (9.2.3) is numerically the same as the confidence interval unadjusted for multiple analyses (looks), although its interpretation is different.

Jennison and Turnbull (2000) point out that although Bayesian inference is straightforward, the problems of design and stopping rules are not as clear-cut as in the frequentist paradigm. Under the Bayesian paradigm, one might for example, stop a trial early if, at some intermediate stage k,

$$\text{(i)}\ \ P(\theta < 0|D_k) < \epsilon \ \text{ or }\ \text{(ii)}\ \ P(\theta > 0|D_k) < \epsilon, \tag{9.2.4}$$

where D_k denotes the data accumulated up to stage k. A typical value for ϵ might be 0.025. This is equivalent to stopping when a $1-\epsilon$ credible region excludes zero. Rules similar to this have been proposed by Mehta and Cain (1984), Berry (1985), and Freedman and Spiegelhalter (1989).

The sequential design does not affect the Bayesian inference, so the credible interval in (9.2.3) is still a correct summary under the Bayesian paradigm at the time of the kth analysis, whether or not the trial stops at this point. If, for example, condition (i) in (9.2.4) is satisfied, then the current posterior probability that $\theta < 0$ is less then ϵ, regardless of any stopping rule, formal or informal. This property leads to the statement by Berry (1987, p. 119) that there is "no price to pay for looking at data using a Bayesian approach." Many Bayesians view this as a major advantage of the Bayesian approach over the frequentist approach in monitoring clinical trials. Specifically, the advantage is that the type I error in the Bayesian paradigm is not relevant, and thus one can analyze the data as many times as one wishes, whereas in the frequentist paradigm, the probability of type I error increases as the number of analyses increase. Nevertheless, as Jennison and Turnbull (2000) point out, the frequentist properties of Bayesian monitoring schemes should be examined, therefore making the type I error a relevant issue under the Bayesian paradigm. Specifically, the behavior of the type I error rate should be investigated when examining the frequentist properties of Bayesian procedures.

To illustrate the importance of this issue, we consider an example given in Jennison and Turnbull (2000), which examines the type I error rates under two different priors. In the example in Table 9.1, using a uniform improper prior for θ, we see that the procedure in (9.2.4) with $\epsilon = 0.025$ is equivalent to the repeated significance test rule which stops the first time that $|Z_k| > 1.96$. Table 9.1 shows the type I error rate of this procedure as a function of K, the maximum number of analyses. This is also the probability, assuming $\theta = 0$, that the 95% credible interval on termination, given by $(\hat{\theta}^{(k)} \pm 1.96\mathcal{I}_k^{-1})$, will fail to include the value $\theta = 0$. We note that this credible interval also corresponds to the unadjusted 95% frequentist confidence interval.

Similar behavior of false-positive rates for this procedure also occurs for other choices of priors. For example, if we take a $N(0, 4\mathcal{I}_k^{-1})$, $k = 1, 2, \ldots, K$, then the type I error rates are those given in the last column of Table 9.1. Here the analyses are assumed to be equally spaced with

TABLE 9.1. False Positive Rates for the Stopping Rules in (9.2.4) and (9.2.8) Based on Posterior Probabilities.

Number of Analyses K	Type I Error Probability Non-informative prior*	Prior with 25% Handicap**
1	0.05	0.03
2	0.08	0.04
3	0.11	0.05
4	0.13	0.05
5	0.14	0.05
10	0.19	0.06
20	0.25	0.07
50	0.32	0.09
∞	1.00	1.00

* Boundary: $|Z_k| > 1.96, k = 1, 2, \ldots, K$.
** Boundary given by (9.2.8) with $h = 0.25$ and $\alpha = 0.05$.
Source: Jennison and Turnbull (2000).

$\mathcal{I}_k = (k/K)\mathcal{I}_K, k = 1, 2, \ldots, K$. This prior, sometimes called the "handicap prior," corresponds to the one recommended by Grossman, Parmar, and Spiegelhalter (1994) and Spiegelhalter, Freedman, and Parmar (1994). It is a prior which corresponds to the situation in which a sample equal in size to 25% of the planned study has already been observed, prior to the study, and the mean of this sample is 0. From Table 9.1, we see that for $K \in [3, 5]$, the error rate is approximately 0.05. For $K < 3$, it is less than 0.05 and for $K > 5$, it is greater than 0.05.

Another feature in Table 9.1 is that the false-positive rate approaches unity as $K \to \infty$. This phenomenon is termed "sampling to a foregone conclusion" by Cornfield (1966a,b), who noted that "if one is concerned about the high probability of rejecting H_0, it must be because some probability of its truth is being entertained." Therefore, for the normal prior example, Cornfield (1966b) proposes using a mixed prior with a discrete probability mass q assigned to $\theta = 0$ and the remaining probability $1 - q$ distributed as $N(0, \sigma_0^2)$. With this prior, the posterior odds in favor of the null hypothesis H_0: $\theta = 0$ is given by

$$\lambda = \frac{P(\theta = 0|D_k)}{P(\theta \neq 0|D_k)} = \frac{q}{1-q}B$$

where

$$B = \sqrt{1 + \mathcal{I}_k \sigma_0^2} \exp\left\{\frac{-Z_k^2}{2(1 + (\mathcal{I}_k \sigma_0^2)^{-1})}\right\}$$

is the Bayes factor in favor of H_0. Now suppose that we use the rule to terminate the trial at analysis k if $|Z_k| > 1.96$. Assuming $\mathcal{I}_k$ is sufficiently large, the posterior odds λ, is approximately equal to

$$\left(\frac{q}{1-q}\right)\sigma_0\sqrt{\mathcal{I}_k}\exp(-(1.96)^2/2),$$

which now favors H_0 instead of H_1 if $\mathcal{I}_k$ is sufficiently large. Thus, if decisions are based on the value of the Bayes factor, large numbers of interim analyses do not result in eventual rejection of H_0, and therefore such a procedure is not subject to the notion of "sampling to a foregone conclusion." As a result of this phenomenon, Cornfield (1966b) alternatively proposes a procedure based on the posterior odds, λ. For a procedure based on λ, the stopping boundaries on the standardized Z_k scale diverge with sample size, unlike other commonly used boundaries, such as the Pocock boundary and O'Brien-Fleming boundaries which become narrower with sample size (see Jennison and Turnbull, 1990). Lachin (1981) extends the Cornfield model to a composite null hypothesis by replacing the discrete prior mass at $\theta = 0$ under H_0 by a continuous prior with support in a small interval about 0.

As Spiegelhalter and Freedman (1988) point out, the assignment of a discrete mass to a single point is unrealistic, and as a result, Cornfield designs based on the Bayes factor have not been adopted in practice. As discussed by Freedman, Spiegelhalter, and Parmar (1994) and Fayers, Ashby, and Parmar (1997), current Bayesian recommendations for monitoring have been more along the lines of (9.2.4). We discuss monitoring in more detail in the subsequent subsections. The principal concerns in the application of Bayesian methods discussed in this section are their frequentist properties and the choice of prior. Prior distributions are discussed in greater detail in subsections 9.2.1 and 9.2.2. With regard to the type I error rate and other frequentist properties, some advocates of the Bayesian approach cite the likelihood principle and argue that these issues are largely irrelevant; see, for example, Berry (1987) and Berger and Berry (1988). Other Bayesian advocates, such as Spiegelhalter, Freedman, and Parmar (1994), do admit a concern for controlling the false-positive rate, and propose priors that can overcome the inflation of the type I error rate.

Since Bayesian inference is based on the posterior distribution, it does not depend on monitoring or stopping rules. Thus, under the Bayesian paradigm, it is not necessary to set a maximum sample size in advance. However, this is not practical for most clinical trials, since costs and planning depend on the proposed sample size. Thus, it is essential to determine a target sample size even under the Bayesian paradigm. The sample size can be determined using Bayesian quantities, such as the prior predictive distribution, as discussed in Spiegelhalter and Freedman (1986). Sample size determination is discussed in more detail in Sections 9.3 and 9.4.

9.2.1 Range of Equivalence

Spiegelhalter, Freedman, and Parmar (1994) give a nice discussion of the practical characterizations of different types of priors that can be used in randomized clinical trials. As noted by Spiegelhalter, Freedman, and Parmar (1994), one usually centers on a significance test of the null hypothesis H_0: $\theta = 0$. However, in some circumstances, the treatments may be so unequal in their costs in terms of toxicity and/or monetarily, that it is commonly accepted that the more costly treatment will be required to achieve at least a certain margin of benefit, θ_i, before it can be considered: hence $\theta < \theta_i$ corresponds to clinical inferiority of the new treatment. Another value, θ_s, may be specified where $\theta > \theta_s$ indicates clinical superiority of the new treatment: θ_s is sometimes termed the minimal clinically worthwhile benefit. Spiegelhalter, Freedman, and Parmar (1994) call (θ_i, θ_s) the *range of equivalence*, such that if one is certain that θ lies within this interval, we would be unable to make a definitive choice of treatment. We note that the range of equivalence may change as the trial progresses.

Current statistical practice in clinical trial design involving time-to-event endpoints is to specify a point alternative hypothesis H_1 and to choose a sample size which guarantees, under a value of θ under H_1, denoted θ_a, an adequate statistical power ($\geq 80\%$) of rejecting H_0 by using a specified statistical test, such as the log-rank at a given significance level. Usually H_0 corresponds to $\theta = 0$. Medical statisticians continue to use different prescriptions for specifying the value θ_a which should represent H_1. Following Spiegelhalter, Freedman, and Parmar (1994), θ_a should represent

(a) the smallest clinically worthwhile difference (Lachin, 1981); or

(b) a difference that the investigator thinks is "worth detecting" (Fleiss, 1981); or

(c) a difference that is thought likely to occur (Halperin et al., 1982).

Item (a) above corresponds to taking θ_a as θ_s, whereas the second leaves the choice to the investigator, who may choose θ_a to be unrealistically large to reduce the required sample size. Item (c) differs in stressing the importance of plausibility as a concept for choosing H_1.

Following our discussion in Section 9.1 and the results in (9.1.5) and (9.1.6), suppose we have a set of data and the likelihood after m observations can be summarized by a statistic y_m, where

$$y_m \sim N(\theta, \sigma^2/m). \tag{9.2.5}$$

As stated in Section 9.1, the assumption of a normal likelihood is quite general and covers many situations: For example, in survival analysis with proportional hazards, if m is the total number of events observed and we take $y_m = 4S_m/m$, where S_m is the log-rank test statistic, then y_m has an approximately $N(\theta, 4/m)$ distribution, where θ is the log-hazard ratio (see Tsiatis, 1981b).

Prior elicitation for clinical trials design focuses on the specification of a prior distribution for θ. Following Spiegelhalter, Freedman, and Parmar (1994), the prior for θ is a $N(\theta_0, \sigma^2/n_0)$ distribution, and thus

$$\pi(\theta) \propto \exp\left(-\frac{n_0}{2\sigma^2}(\theta - \theta_0)^2\right). \tag{9.2.6}$$

As Spiegelhalter, Freedman, and Parmar (1994) point out, this prior is equivalent to a normalized likelihood arising from a hypothetical trial of n_0 patients with an observed value θ_0 of the log-hazard ratio. Although any choice of σ^2 is possible in (9.2.6), a common choice of σ^2 for survival data is $\sigma^2 = 4$ since this is consistent with the fact that y_m has an approximately $N(\theta, 4/m)$ distribution. The posterior distribution of θ using the likelihood in (9.2.5) and the prior in (9.2.6) is given by

$$\theta | y_m \sim N\left(\frac{n_0\theta_0 + my_m}{n_0 + m}, \frac{\sigma^2}{n_0 + m}\right). \tag{9.2.7}$$

Example 9.1. Medical research council neutron therapy trial. We consider an example given in Spiegelhalter, Freedman, and Parmar (1994). Errington et al. (1991) report a trial of high-energy neutron therapy for treatment of pelvic cancers. An ad hoc independent monitoring committee was set up by the Medical Research Council (MRC) in January 1990 to review interim results after 151 patients with locally advanced cancers had received either neutron therapy (90 patients) or conventional radiotherapy (61 patients). Interviews were conducted in March 1988 with ten selected clinicians and physicists, knowledgeable about neutron therapy, before the disclosure of any trial results. On average, respondents reported that they would require a change in one-year survival from 50% (assumed standard for photon therapy) to 61.5% for neutron therapy to be recommended as routine treatment. Under a proportional hazards assumption such a change corresponds to a hazard ratio of $\log(0.50)/\log(0.615) = 1.426$ against photon therapy. Using a log-hazard scale, Spiegelhalter, Freedman, and Parmar (1994) take $\theta_s = \log(1.426) = 0.355$ as the upper limit of the range of equivalence and take $\theta_i = 0$ as the lower limit.

Prior information regarding the effect of neutron therapy was available from two sources. First, the consensus belief of those interviewed was 28% that $\theta < 0$ (neutrons worse than photons), with the remaining 46% lying in the range of equivalence. The prior distribution for θ is taken to be normal with a mean of $\theta_0 = 0.169$, and the prior number of events is $n_0 = 48$. We can think of this prior distribution as equivalent to already observing 48 deaths, of which 22 were in the neutron therapy group and 26 in the photon therapy group. The second source of prior information was a statistical analysis of the combined results of previous randomized trials, conducted using different dosage regimens, which gave an estimated odds ratio for one-year mortality of 0.47 (95% interval (0.30, 0.73)) in favor of photon therapy. Relative to the assumed baseline survival of 50% under photons, this can be translated to an approximate normal prior, on the log-hazard ratio scale, with $\theta_0 = -0.569$ and $n_0 = 138$. As Spiegelhalter et al. (1994) point out, this prior distribution is particularly noninformative.

The interim analysis based on 59 observed deaths showed an estimated hazard ratio of conventional over neutron therapy equal to 0.66, with a 95% confidence interval (0.40, 1.10). Thus, the estimated log-hazard ratio is -0.416 with standard error 0.260. Thus, we have $y_m = -0.416$, $\sigma = 2$, and $m = 59$. The posterior probability that the effect of neutron therapy exceeds the minimal clinically worthwhile benefit is small for both priors, and equals 0.004 under the $N(0.169, 4/48)$ prior. In addition, the results of the current trial agree very closely with the combined results from previous trials. The data monitoring committee ratified the decision of the principal investigator to suspend entry of patients into the trial. The Bayesian analysis here supports that decision.

9.2.2 Prior Elicitation

Spiegelhalter, Freedman, and Parmar (1994) present an informative discussion for the different types of priors that can be used in clinical trials. We now summarize these priors.

Reference Priors

Reference priors are supposed to represent minimal prior information, and for the normal prior in (9.2.6), they are obtained by letting $n_0 \to 0$, which leads to a uniform improper prior for θ. These types of priors lead to posterior distributions for θ that are normalized likelihoods. As Spiegelhalter, Freedman, and Parmar (1994) point out, this type of prior in some ways is the most unrealistic of all possible priors. It represents, for example, a belief that it is equally likely that the relative risk associated with neutron therapy is above or below 10. However, it could be argued that such a prior is the least subjective and plays a useful role as a benchmark against which to compare other more plausible priors.

Clinical Priors

A clinical prior is intended to formalize the opinion of well-informed specific individuals, often those taking part in the trial themselves. Deriving such a prior requires asking specific questions concerning a trial, and a variety of sources of evidence may be used as a basis for the opinion. There are many sources of evidence for which to base clinical priors. First, the evidence may come from other randomized trials. If the clinical prior is to represent current knowledge accurately, we should base it, where possible, on objective information. When the results of several similar clinical trials are available, a statistical overview of those results can be used as the basis of a prior distribution. Results from previous randomized trials should generally form the basis for a prior distribution but should not specify the distribution completely. As Kass and Greenhouse (1989) point out, a Bayesian who took this past evidence directly as their prior would be treating the historical and experimental subjects as exchangeable and essentially pooling the results. However, it seems reasonable to combine previous data with some initial skepticism. Belief in heterogeneity of treatment effects across trials or patient subgroups, combined with reasonable skepticism, should suggest either shrinking the apparent treatment effect, expanding the variance, or both. Thus the power prior distribution discussed in earlier chapters can be used to construct such a prior. The hyperparameter a_0 used in the power prior generally achieves the shrinking of the treatment effect and magnifies the variance. It also lowers the effective sample size of the historical data compared to the curent data. In addition to the power prior, another type of prior that could be used here is a prior within a random effects model, in which case the prior distribution would correspond to the predictive distribution for the effect in a new trial (see Carlin, 1992).

Evidence from nonrandomized studies is often available to construct clinical priors when data from randomized trials is not available. The difficulty of constructing a prior distribution from nonrandomized studies is related to the assessment of the possible biases that can exist in such studies (see Byar et al., 1976). Such difficulty may often lead quite reasonably to a prior distribution with a large variance. It may happen that there are no previous randomized or nonrandomized previous studies for which to construct a prior. In this case, clinical priors can be constructed from subjective clinical opinion. One approach to eliciting opinion is to conduct individual interviews with clinicians who will participate in the trial. Individual interviewing is time consuming and it may be preferable to use telephone or postal elicitation. A form of postal elicitation of prior opinion has been designed and used by the Medical Research Council in England.

Skeptical Priors

Spiegelhalter, Freedman, and Parmar (1994) define a class of priors called skeptical priors for use in clinical trials. They note that one step towards incorporation of knowledge into a prior is an attempt to formalize the belief that large treatment differences are unlikely. Such an opinion could be represented with a symmetric prior with mean $\theta_0 = 0$ and suitably spread to represent a wide range of treatment differences. Thus, one example of a skeptical prior is a $N(0, 4/n_0)$ prior for θ, and n_0 is chosen so that $P(\theta > \theta_a) = \gamma$, where θ_a is the specified value of θ under H_1 and γ is small.

Enthusiastic Priors

To counterbalance the skeptical prior, Spiegelhalter, Freedman, and Parmar (1994) introduce what they call an enthusiastic prior. This prior represents individuals who are reluctant to stop when results supporting the null hypothesis are observed. Such a prior may have mean θ_a and the same precision as the skeptical prior, and hence provides a prior belief that $P(\theta < 0) = \gamma$, where γ is small. Thus, a $N(\theta_a, 4/n_0)$ prior, where n_0 is chosen so that $P(\theta < 0) = \gamma$ and γ is small, is an example of an enthusiastic prior.

Handicap Priors

Handicap priors were proposed by Grossman, Parmar, and Spiegelhalter (1994) and Spiegelhalter, Freedman, and Parmar (1994), and discussed in Jennison and Turnbull (2000). Following Jennison and Turnbull (2000), the goal of the prior is to control the frequentist properties of a Bayesian monitoring procedure. The prior is chosen so that the false-positive rate is controlled at a given level α, say, such as $\alpha = 0.05$, assuming a fixed number K of planned interim analyses. Grossman, Parmar, and Spiegelhalter (1994) term this a unified method. For the normal data example presented in Table 9.1, the unified method calls for a normal prior with mean 0 and variance $\sigma_0^2 = (h\mathcal{I}_f)^{-1}$, where $\mathcal{I}_f$ is the information needed for a fixed sample, frequentist test with type I error probability α and power $1-\beta$. The constant h is termed the handicap. To illustrate these priors, consider the Bayesian procedure that stops when the $1-\alpha$ credible region first excludes zero, but with a maximum of K analyses. The handicap h is computed so that this procedure has a type I error rate equal to α. Suppose that the analyses are equally spaced with $\mathcal{I}_k = (k/K)\mathcal{I}_f$, $k = 1, 2, \ldots, K$. Then in terms of standardized test statistics, we have a two-sided test which stops to reject H_0: $\theta = 0$ at analysis k if

$$|Z_k| > \Phi^{-1}(1-\alpha/2)\sqrt{(k+hK)/k}, \tag{9.2.8}$$

where $k = 1, 2, \ldots, K$. For $K = 1$, no handicap is needed, and $h = 0$. This results in an improper uniform prior for θ. Grossman, Parmar, and Spiegelhalter (1994) tabulate the values of h needed to achieve a type I error probability of $\alpha = 0.05$ and $\alpha = 0.01$ with $K = 2, 3, \ldots, 10$ analyses. In the case of $\alpha = 0.05$, the handicap is $h = 0.16$, 0.22, 0.25, 0.27, and 0.33 for $K = 2$, 3, 4, 5, and 10, respectively.

TABLE 9.2. Summary of Results of Questionnaires Completed by Nine Clinicians for CHART Trial.

Clinician	Prior distribution (absolute advantage of CHART) standard therapy in 2-year % disease-free survival							
	(−10,−5)	(−5,0)	0-5	5-10	10-15	15-20	20-25	25-30
1	10	30	50	10				
2	10	10	25	25	20	10		
3			40	40	15	5		
4			20	40	30	10		
5			20	20	60			
6	5	5	10	15	20	25	10	5
7				10	20	40	20	10
8				10	20	40	20	10
9					10	50	30	10
group mean	3	5	18	19	22	20	10	3

Source: Spiegelhalter, Freedman, and Parmar (1994).

Example 9.2. Continuous hyperfractionated accelerated radiotherapy study in head and neck cancer (CHART). Spiegelhalter, Freedman, and Parmar (1994) discuss a trial comparing standard therapy with continuous hyperfractionated accelerated radiotherapy, CHART (Parmar, Spiegelhalter, and Freedman, 1994). In designing the CHART study, baseline two-year disease-free survival under the control therapy was estimated to be 45%. Nine individual prior distributions are summarized in Table 9.2, which presents the proportions of each distribution falling within intervals for treatment difference provided on the questionnaire. The ranges of equivalence provided by clinicians 1, 2, ..., 9 in Table 9.2 are 5–10, 10–10, 5–10, 10–10, 10–15, 15–20, 5–10, 20–20, and 10–15, respectively, and the group mean of the ranges of equivalence for all nine clinicians is 10–13. As expected, there are differences in the positions and shapes of the individual distributions. The opinions of clinicians 1 and 9 do not even intersect, whereas clinician 5 has a very confident (narrow) distribution, in contrast with clinician 6. How should we combine these individual distributions to arrive at a prior distribution for the group? Although many methods are possible, Spiegelhalter, Freedman, and Parmar (1994) discuss arithmetic and logarithmic pooling. These correspond to taking the arithmetic and normalized geometric mean, respectively, within each column of Table 9.2. The former takes the opinions as data and averages, whereas the latter takes the opinions as representing data and pools those implicit data.

Spiegelhalter et al. (1994) give a strong preference for arithmetic pooling, to obtain an estimated opinion of a typical participating clinician. Arithmetic pooling for the CHART study gives a median of 11% improvement, with a prior probability 0.08 that there is deterioration (θ_0). Assuming a 45% two-year disease-free survival rate under standard therapy, and transforming to a log-hazard ratio scale, this corresponds to $\pi(\theta)$ having a median of $\log(\log(0.45)/\log(0.56)) = 0.32$, with $P(\theta < 0) = 0.08$. Assuming a normal prior with these characteristics yields the parameters $\theta_0 = 0.32$ and $n_0 = 77$. This normal prior distribution superimposed on a histogram derived from the group distribution in Table 9.2 shows that the normal prior is a good approximation to the histogram. The mean of the ranges of equivalence that were elicited from the same nine clinicians is (10%, 13%), which corresponds to $(0.29, 0.38)$ on the θ scale.

9.2.3 *Predictions*

A major strength of the Bayesian approach is the ease of making predictions concerning events of interest. Suppose we have m observations and are interested in the possible consequences of continuing the trial for a further n observations. Let y_m denote the statistic computed after m observations. If we denote our future statistic by Y_n, then it has posterior predictive distribution given by

$$\pi(y_n|y_m) = \int f(y_n|\theta)\pi(\theta|y_m)d\theta, \tag{9.2.9}$$

where $f(y_n|\theta)$ is the probability density function of y_n given θ, and $\pi(\theta|y_m)$ is the posterior distribution of θ given y_m. For the normal likelihood in (9.2.5) and the normal prior in (9.2.6) with $\sigma^2 = 4$, a straightforward derivation yields

$$Y_n|y_m \sim N\left(\frac{n_0\theta_0 + my_m}{n_0 + m}, 4\left(\frac{1}{n_0 + m} + \frac{1}{n}\right)\right). \tag{9.2.10}$$

As a special case, the prior predictive distribution is obtained when $m = 0$ (i.e., no observed data), and in this case the prior predictive distribution of Y_n is given by

$$\pi(y_n) = \int f(y_n|\theta)\pi(\theta)d\theta, \tag{9.2.11}$$

where $\pi(\theta)$ is given by (9.2.6). Assuming $\sigma^2 = 4$, this leads to

$$Y_n \sim N\left(\theta_0, 4\left(\frac{1}{n_0} + \frac{1}{n}\right)\right). \tag{9.2.12}$$

Bayesian sample size calculations can be carried out by computing the expected power as discussed by Spiegelhalter, Freedman, and Parmar (1994). This can be computed as follows. Suppose that at the start of a trial ($m = 0$)

we wish to assess the chance that the trial will arrive at a firm positive conclusion, that is, the final interval will exclude a value θ_1 in favor of the new treatment. We first proceed with the standard power formula, which leads us to reject H_0 if

$$Y_n > \theta_1 - \frac{2}{\sqrt{n}} z_\alpha = y^*,$$

where $z_\alpha = \Phi^{-1}(\alpha)$, and $\Phi(.)$ denotes the standard normal cdf. The expected power is thus defined as $P(Y_n > y^*)$, where the probability is computed with respect to the prior predictive distribution of y_n in (9.2.12). Assuming $\sigma^2 = 4$, it can be shown that

$$P(Y_n > y^*) = \Phi\left(z_0(\theta_1)\sqrt{\left(\frac{n}{n_0+n}\right)} + z_\alpha \sqrt{\left(\frac{n_0}{n_0+n}\right)} \right), \qquad (9.2.13)$$

where $z_0(\theta_1) = \sqrt{n_0}(\theta_0 - \theta_1)/2$ is the standardized distance of the prior mean from θ_1. As $n_0 \to \infty$, we obtain the classical power curve

$$\Phi\left((\theta_0 - \theta_1)\sqrt{n}/2 + z_\alpha\right).$$

This use of the full prior distribution for power calculations in (9.2.13) is called predictive power in Spiegelhalter, Freedman, and Blackburn (1986), expected power in Brown et al. (1987), or the strength of the study in Crook and Good (1982). Spiegelhalter and Freedman (1986) provide a detailed example of the predictive power of a study using subjective prior opinions from a group of urological surgeons to form their prior distribution, whereas Moussa (1989) discusses the use of predictive power with group sequential designs based on ranges of equivalence. Brown et al. (1987) have also suggested the use of prior information in power calculations, but conditioning on $\theta > \theta_s$, that is, the predictive probability of concluding the new treatment is superior, given that it truly is superior.

Example 9.3. CHART trial (Example 9.2 continued). Following Spiegelhalter, Freedman, and Parmar (1994), the CHART trial was originally designed with an alternative hypothesis of an absolute 15% improvement of CHART over the 45% two-year disease-free survival estimated for the standard therapy, equivalent to $\theta_a = \log(\log(0.45)/\log(0.60)) = 0.45$ with a two-sided significance level of 0.05, which yielded a power of 90% based on 218 events. Using the prior in (9.2.6), the predictive probability that a final 95% interval will exclude 0 is 83%, 62% that it will be in favor of CHART and 21% that it will be in favor of control. However, as Spiegelhalter et al. (1994) point out, the ranges of equivalence provided by the clinicians indicate that proving "significantly better than no difference" may not be sufficient to change clinical practice. In fact, at the first interim analysis, the data monitoring committee extended accrual to expect 327 events. This decision was made independently of knowledge of interim results: using the same prior opinion the predictive probability of concluding in favor of CHART is now 68%, and in favor of control, 25%.

Spiegelhalter and Freedman (1988) note that the predictive power in (9.2.13) could be expressed as the power against a fixed alternative, θ_u, where

$$\theta_u = \theta_{0.5}\left(1 - \sqrt{\frac{n_0}{n_0 + n}}\right) + \theta_0\sqrt{\frac{n_0}{n_0 + n}},$$

and $\theta_{0.5}$ is the value, which observed, would just lead to rejection of the null hypothesis. Thus θ_u is a weighted average of the prior mean and the point of 50% power. Thus the predictive power will always lie between 0.5 and the power calculated at the prior mean. As Spiegelhalter et al. (1994) point out, the predictive power is a useful addition to simple power calculations for a fixed alternative hypothesis and may be used to compare different sample sizes and designs for a proposed trial, and as an aid to judge competing trials considering different treatments.

9.2.4 Checking Prior-Data Compatibility

Spiegelhalter, Freedman, and Parmar (1994) discuss the importance of checking the compatibility of the historical data with the current data when constructing a prior. This is especially important if one uses a power prior or a clinical prior based on historical data. Spiegelhalter et al. (1994) note that before the trial starts, there should be careful consideration of what the treatment difference is likely to be, and that this should be used as an important component for trial design. As Spiegelhalter et al. (1994) point out, far from invalidating the Bayesian approach, such a conflict between prior and data only emphasizes the importance of pretrial elicitation of belief. Thus, it is useful to have a formal means of assessing the conflict between the prior expectations and observed data. Towards this goal, Box (1980) suggested using the prior predictive distribution as a diagnostic to check prior-data compatibility. Specifically, we first derive the prior predictive distribution of Y_m, and then calculate the chance of a result with lower predictive ordinate than that actually observed. In our context, we would use (9.2.12) but substituting in m for n, to give the prior predictive distribution

$$Y_m \sim N\left(\theta_0, 4\left(\frac{1}{n_0} + \frac{1}{m}\right)\right). \tag{9.2.14}$$

Now given an observed y_m, Box's generalized significance test is given by

$$P(\pi(Y_m) \leq \pi(y_m)) = 2\min\left(\pi(y_m), 1 - \pi(y_m)\right),$$

since the distribution is unimodal.

Example 9.4. Levamisole and 5-fluarouricil in bowel cancer. Moertel et al. (1990) have reported results from a randomized Phase III cancer clinical trial investigating the effect of the drug levamisole (LEV) alone or in combination with 5-fluorouracil (5-FU) for patients with resected cancer of the colon or rectum. Patients entering the trial were allocated to one of three treatment arms: LEV, LEV + 5-FU, or control. The primary endpoint was overall survival measured as time from randomization until death. The study was designed with an alternative $\theta_a = \log(1.35) = 0.30$, and to have 90% power for a one-sided log-rank test with 5% significance, and thus required 380 deaths. The log hazard was based on the comparison between LEV + 5-FU against control. In this context, Fleming and Watelet (1989) suggest a range of equivalence of $(\theta_i, \theta_s) = (0, 0.29)$. This implies that the control treatment would be clinically preferable provided that it had no excess mortality, whereas the new treatment, with its "inconvenience, toxicity, and expense" would only be clinically superior if it reduced the hazard by at least 25%, that is $1 - 1/1.33$.

Spiegelhalter, Freedman, and Parmar (1994) consider both skeptical and enthusiastic priors in this example. The range of equivalence and the interval $(0, \theta_a)$ essentially coincide, as is the case with many other studies. Spiegelhalter et al. (1994) adopt a skeptical prior with $\theta_0 = 0$, $\sigma^2 = 4$, and $n_0 = 120$. They note that this prior is quite reasonable in view of the implausibility of dramatic gains in cancer adjuvant therapy. The enthusiastic prior has the same precision and mean 0.30. From Moertel et al. (1990), $y_m = 0.04$ and $m = 223$ for LEV versus control, and $y_m = 0.40$ with $m = 192$ for LEV + 5-FU versus control. Under a uniform improper prior for θ, the posterior is the normalized likelihood, and $P(\theta > 0|D) = 0.003$, where D denotes the data. There is moderate evidence that the treatment is clinically superior, since $P(\theta > 0.29|D) = 0.78$. The prior predictive distribution shows that there was only a 4% chance of observing such a high result using the skeptical prior. The posterior distributions based on the skeptical and enthusiastic priors were computed for LEV + 5-FU versus control. The skeptical posterior distribution has mean 0.25 with standard deviation 0.11, yielding a posterior mean of 1.28 with 95% HPD interval of $(1.03, 1.60)$ for the hazard ratio. There is a small probability (0.015) of the treatment difference being less than 0, i.e., of the control treatment being superior to LEV + 5-FU, whereas the posterior mean is within the range of equivalence.

Results were also available from an earlier study with historical data D_0 (Laurie et al., 1989) which showed an apparent reduction in the death-rate of patients receiving LEV+ 5-FU, with an estimated log-hazard ratio of 0.14 with standard error 0.17. When including these data in the analysis with a skeptical prior, the posterior mean of the hazard ratio becomes

1.25 with 95% HPD interval $(1.04, 1.49)$. As Spiegelhalter et al. (1994) point out, including these data increases the certainty that $\theta < 0$ since $P(\theta < 0|D) = 0.007$, but decreases the certainty that the treatment is clinically superior, since $P(\theta > 0.29|D, D_0) = 0.22$.

9.3 Bayesian Sample Size Determination

There has been a fair amount of literature on Bayesian sample size determination. Adcock (1997) provides a nice review article on Bayesian methods for sample size. Following Joseph and Belisle (1997), we briefly review three commonly used methods for Bayesian sample size determination. These are the (i) Average Coverage Criterion (ACC), (ii) Average Length Criterion (ALC), and (iii) Worst Outcomes Criterion (WOC).

Average Coverage Criterion (ACC)

For a fixed posterior interval of length l, we can determine the sample size by finding the smallest n such that the equation

$$\int \left(\int_a^{a+l} \pi(\theta|y)\ d\theta \right) \pi(y)dy \geq 1 - \alpha$$

is satisfied. Here,

$$\pi(y) = \int f(y|\theta)\pi(\theta)d\theta$$

is the prior predictive distribution of y. This Average Coverage Criterion (ACC) ensures that the mean coverage of posterior credible intervals of length l, weighted by $\pi(y)$, is at least $1 - \alpha$.

Adcock (1988) first proposed the use of ACC in the context of estimating normal means, where the interval $(a, a + l)$ is chosen to be a symmetric tolerance region around the mean. Joseph, Wolfson, and DuBerger (1995) proposed that the interval $(a, a + l)$ be chosen to be a Highest Posterior Density (HPD) interval for asymmetric posterior distributions.

Average Length Criterion (ALC)

For a fixed posterior credible interval of coverage $1 - \alpha$, we can also determine the sample size by finding the smallest n such that

$$\int l^*(y, n)\pi(y)dy \leq l,$$

where $l^*(y, n)$ is the length of the $100(1-\alpha)\%$ posterior credible interval for data y, determined by solving

$$\int_a^{a+l^*(y,n)} \pi(\theta|y)\, d\theta = 1 - \alpha$$

for $l^*(y, n)$ for each value of $y \in \mathcal{Y}$, where $\mathcal{Y}$ denotes the sample space of y. As before, a can be chosen to give HPD intervals or symmetric intervals. The ALC ensures that the mean length of the $100(1-\alpha)\%$ posterior credible intervals weighted by $\pi(y)$ is at most l. Since most researchers will report intervals of fixed coverage (usually 95%) regardless of their length, it can be argued that the ALC is more conventional than ACC.

Worst Outcome Criterion (WOC)

Cautious investigators may not be satisfied with the "average" assurances provided by the ACC and the ALC criteria. Therefore, a conservative sample size can also be determined by

$$\inf_{y\in\mathcal{Y}^*}\left(\int_a^{a+l(y,n)} \pi(\theta|y)\, d\theta\right) \geq 1 - \alpha,$$

where $\mathcal{Y}^*$ is a suitably chosen subset of the sample space $\mathcal{Y}$. For example, the WOC ensures that if $\mathcal{Y}^*$ consists of the most likely 95% of the possible $y \in \mathcal{Y}$, then there is a 95% assurance that the length of the $100(1-\alpha)\%$ posterior credible interval will be at most l. We let

$$1 - w = \int_{\mathcal{Y}^*} \pi(y) dy \tag{9.3.1}$$

so that $\mathcal{Y}^*$ can be viewed as an $100(1-w)\%$ highest posterior density according to $\pi(y)$.

In the context of sample size determination in clinical trials, it suffices to consider normally distributed observations due to our discussion in Section 9.1. Suppose $y_1, y_2, \ldots, y_n$ are i.i.d. $N(\theta, \sigma^2)$, and let $\tau = 1/\sigma^2$. In the context of survival data in clinical trials, the y_i's can be viewed as log-rank statistics and θ denotes the true log-hazard ratio. Assume the usual conjugate priors for (θ, τ). That is,

$$\theta|\tau \sim N\left(\theta_0, \frac{1}{n_0\tau}\right)$$

and

$$\tau \sim \mathcal{G}(\alpha_0, \lambda_0).$$

We now consider cases for which τ is known and τ is unknown.

Case 1: τ known

In the case that τ is known, we have

$$\theta | \boldsymbol{y} \sim N(\theta_n, \tau_n^{-1}),$$

where $\boldsymbol{y} = (y_1, y_2, \ldots, y_n)'$,

$$\theta_n = \frac{n_0 \theta_0 + n\bar{y}}{n_0 + n},$$

$\bar{y} = \frac{1}{n}\sum_{i=1}^n y_i$, and

$$\tau_n = (n + n_0)\tau.$$

Since the posterior precision depends only on n and does not vary with the observed data vector $\boldsymbol{y}$, all three criteria (ACC, ALC, WOC) lead to the same sample size formula, which is

$$n \geq \frac{4z_{1-\alpha/2}^2}{\tau l^2} - n_0, \tag{9.3.2}$$

where l is the length of the desired posterior interval for θ, and $z_{1-\alpha/2}$ is the $100(1-\alpha/2)$ percentile of a standard normal distribution. If a uniform improper prior for θ is used (i.e., $n_0 = 0$), then inequality (9.3.2) reduces to the usual frequentist formula for the sample size, given by

$$n \geq \frac{4z_{1-\alpha/2}^2}{\tau l^2}. \tag{9.3.3}$$

Case 2: τ unknown

If τ is unknown, then

$$\theta | \boldsymbol{y} \sim t\left(n + n_0, \theta_n, \frac{2\beta_n}{(n + 2\alpha_0)(n + n_0)}\right),$$

where

$$\beta_n = \lambda_0 + \frac{n}{2}s^2 + \frac{nn_0}{2(n + n_0)}(\bar{y} - \theta_0)^2,$$

and $ns^2 = \sum_{i=1}^n (y_i - \bar{y})^2$. Since the posterior precision varies with the data $\boldsymbol{y}$, different criteria will lead to different sample sizes. Here $t(\kappa, b, c)$ denotes the t distribution with κ degrees of freedom, location parameter b, and dispersion parameter c.

For ACC, the formula for the sample size is given by

$$n = \left(\frac{4\lambda_0}{\alpha_0^2 l^2}\right) t_{(2\alpha_0, 1-\alpha/2)}^2 - n_0, \tag{9.3.4}$$

where $t_{(2\alpha_0,1-\alpha/2)}$ is the $100(1-\alpha/2)$ percentile of the standard t distribution with $2\alpha_0$ degrees of freedom. Since α_0/λ_0 is the prior mean for τ, the ACC sample size for unknown τ is similar to that for known τ, in that we only need to substitute the prior mean precision for τ in inequality (9.3.2) and exchange the normal quantile z with a quantile from a $t_{2\alpha_0}$ distribution. Since the degrees of freedom of the t distribution do not increase with the sample size, equation (9.3.4) can lead to sample sizes that are substantially different from those in (9.3.2) or (9.3.3).

For ALC, it can be shown that the required sample size satisfies

$$2t_{(n+2\alpha_0,1-\alpha/2)}\left(\frac{2\lambda_0}{(n+2\alpha_0)(n+n_0)}\right)^{1/2}\frac{\Gamma\left(\frac{n+2\alpha_0}{2}\right)\Gamma\left(\frac{2\alpha_0-1}{2}\right)}{\Gamma\left(\frac{n+2\alpha_0-1}{2}\right)\Gamma(\alpha_0)}\leq l. \tag{9.3.5}$$

For more details on this, see Joseph and Belisle (1997). Although it does not appear feasible to solve the inequality in (9.3.5) explicitly for n, it is straightforward to calculate given α_0, λ_0, n_0, α, and n.

For WOC, it can be shown that n satisfies

$$\frac{l^2(n+2\alpha_0)(n+n_0)}{8\lambda_0\left(1+\frac{n}{2\alpha_0}F_{(n,2\alpha_0,1-w)}\right)}\geq t^2_{(n+2\alpha_0,1-\alpha/2)}, \tag{9.3.6}$$

where $F_{(n,2\alpha_0,1-w)}$ denotes the $100(1-w)$ percentile of the F distribution with $(n,2\alpha_0)$ degrees of freedom and w is defined in (9.3.1). The smallest n satisfying (9.3.5) or (9.3.6) can be found by a search algorithm. See Joseph and Belisle (1997) for more details.

Example 9.5. Comparison of ACC, ALC, and WOC. We compare the three criteria for the case in which τ is unknown, and the results are summarized in Table 9.3.

Cases 1–3 in Table 9.3 show that the Bayesian approach can provide larger sample sizes than the frequentist approach, even though the prior information is incorporated in the final inferences. The same examples also illustrate that the sample size provided by the ALC tends to be smaller than that of the ACC when $1-\alpha$ is near 1 and l is not near 0. This is because coverage probabilities are bounded above by 1, so that maintaining the required average coverage becomes more difficult as $1-\alpha$ becomes larger. Similarly, since l is bounded below by 0, maintaining an average length of l becomes more difficult as l approaches 0, leading to the large sizes for the ALC compared with the ACC in Case 4 in Table 9.3. Case 5 shows that with a large amount of prior information on both (θ,τ), the Bayesian approach leads to smaller sample sizes than the frequentist approach. With an informative prior on τ, but not on θ, similar sample sizes are provided

TABLE 9.3. Comparison of ACC, ALC, and WOC.

	Case					
	1	2	3	4	5	6
α_0	2	2	2	2	100	100
λ_0	2	2	2	2	100	100
n_0	10	10	10	10	100	10
l	0.5	0.2	0.2	0.2	0.2	0.2
$1-\alpha$	0.99	0.95	0.80	0.50	0.95	0.95
frequentist*	107	385	165	46	385	385
ACC	330	761	226	45	289	379
ALC	160	595	248	61	288	378
WOC**	589	2152	914	245	344	436

$*$ The frequentist sample size satisfies (9.3.3).
$**$ $w = 0.05$ for WOC.
Source: Joseph and Belisle (1997).

by all criteria, as suggested by Case 6, with the WOC criterion somewhat higher than the rest.

9.4 Alternative Approaches to Sample Size Determination

Lee and Zelen (2000) present a hybrid Bayes-frequentist procedure for the formulation and calculation of sample sizes for planning clinical trials. As stated in the previous sections, frequentist approaches for sample size calculations are based on choosing a fixed type I error (α) and calculating a sample size consistent with a prespecified power $(1-\beta)$ to detect a prespecified value under the alternative θ_a. Ordinarily, the value of α is taken to be $\alpha = 0.05$ or in rarer cases $\alpha = 0.01$. It is highly unusual to have values of $\alpha > 0.05$, although it is common to have $0.05 < \beta \leq 0.2$. There is no logic in the widespread use of $\alpha = 0.05$ except that there is general agreement that it should not be large. Intuitively it is clear that β should not be large, but there is no general agreement on the widespread use of a fixed β value. Some recommend that the selection of (α, β) should be based in the relative costs of making a wrong decision. However, the consideration of relative costs is rarely done in practice. Lee and Zelen (2000) note that most statistical practitioners of clinical trials acknowledge that the formal procedures for planning sample sizes in clinical trials have many subjective elements. Among these are the choice of the α level and the sensitivity of the trial to have acceptable power for detecting a specified θ_a. Lee and Zelen's approach to the sample size problem leads to specification of (α, β), but requires other subjective elements, which may be easier to estimate or accept.

The basic idea underlying the frequentist theory of planning studies is to control the false-positive and false-negative error rates by specifying them in advance of the study. These are probabilities that are conditional on the true state of the hypotheses under consideration. Lee and Zelen (2000) note that this methodology is inappropriate for the planning of clinical trials. Their views are motivated by the following considerations. If the outcome of the clinical trial is positive (one treatment is declared superior), then the superior therapy is likely to be adopted by other physicians. Similarly if the outcome is neutral (treatments are comparable), then considerations other than outcomes are likely to be important in the adoption of therapy. They note that the consequences of false-negative and positive decisions after the clinical trial has been completed are conditional on the outcome of the clinical trial. Hence the false-positive and -negative error rates should also be conditional on the trial outcome. In their view these latter probabilities should be used in the planning of clinical trials. The contrast is that there are two classes of error rates — the frequentist type I and II error rates which are conditional on the true hypotheses, and the posterior error rates which are conditional on the outcome of the trial. They argue that it is the latter rates that are important in assessing the consequences of wrong decisions. Thus, they note that the two fundamental issues are (a) if the trial is positive, "what is the probability that the therapy is truly beneficial?" and (b) if the trial is negative, "what is the probability that the therapies are comparable?" They note that the frequentist view ignores these fundamental considerations and can result in positive harm because of the use of inappropriate error rates. The positive harm arises because an excessive number of false-positive therapies may be introduced into practice. Many positive trials may be unethical to duplicate and, even if replicated, could require many years to complete. Hence a false-positive trial outcome may generate many years of patients receiving nonbeneficial therapy.

Consider a phase III clinical trial for comparing two treatments. To motivate Lee and Zelen's ideas, suppose the trial is comparing an experimental therapy with the best available therapy. Consider the hypothesis H_0: $\theta = 0$ vs. H_1: $\theta \neq 0$. Lee and Zelen (2000) assign a prior probability of q to the joint event $\theta > 0$ or $\theta < 0$. Hence there will be a prior probability $(1 - q)$ of the null hypothesis being true, and $q/2$ will be the prior probability for each alternative $\theta > 0$ and $\theta < 0$. If the prior probabilities are not equal for $\theta > 0$ vs. $\theta < 0$, then the trial would be unethical as it will be advantageous for the patient to be assigned to the treatment with the larger prior probability of being superior. The quantity q essentially summarizes the prior evidence and/or subjective assessment in favor of differences between treatments. It also reflects the level of clinical innovation that motivated the trial. Numerical values of q are difficult to estimate. In practice it may only be necessary to have a range of values.

As Lee and Zelen (2000) point out, the null hypothesis H_0: $\theta = 0$ can be interpreted as θ is in the neighborhood of $\theta = 0$. That is, there is an "indifference" region in which the treatments are regarded as comparable. Thus they reformulate the hypothesis testing situation by denoting an indifference region $|\theta| < \theta_0$ and a region of importance $|\theta| > \theta_1$. Then the null and alternative hypotheses can be stated as H_0: $|\theta| \leq \theta_0$ and H_1: $|\theta| > \theta_1$, with the region $\theta_0 < |\theta| < \theta_1$ being regarded as an "indecisive" region. A special case in the formulation is to take $\theta_0 = \theta_1$. The usual formulation is to set $\theta_0 = \theta_1 = 0$. Following Lee and Zelen (2000), the outcome of the clinical trial will be idealized as having a positive or negative outcome.

The positive outcome refers to the conclusion that $\theta \neq 0$, whereas a negative outcome refers to the conclusion $\theta = 0$. Define C to be a binary random variable which reflects the outcome of the clinical trial; i.e., $C = +$ or $-$. Also define T to be an indicator random variable which denotes the true state of the hypothesis under evaluation; i.e., "$T = -$" refers to $\theta = 0$ and "$T = +$" signifies $\theta \neq 0$. The assessment of T will be identified with the prior probability associated with $\theta \neq 0$; i.e., $q = P(T = +)$. Also define the usual frequentist probabilities of making false-positive and false-negative conclusions from the data by

$$\begin{aligned} \alpha =& P(C = +|T = -), \\ \beta =& P(C = -|T = +). \end{aligned}$$

In an analogous way define

$$\begin{aligned} \alpha^* =& P(T = +|C = -), \\ \beta^* =& P(T = -|C = +). \end{aligned}$$

These are the posterior probabilities of the true situation being opposite to the outcome of the trial. They are the posterior false-positive and false-negative error probabilities. Also define P_1 and P_2 to be $P_1 = 1 - \alpha^* = P(T = -|C = -)$ and $P_2 = 1 - \beta^* = P(T = +|C = +)$. These quantities are functions of (α, β, θ). Following Lee and Zelen (2000), a direct application of Bayes' theorem results in expressing (P_1, P_2) in terms of (α, β, q); i.e.,

$$\begin{aligned} P_1 =& 1 - \alpha^* = P(T = -|C = -) = (1 - \alpha)(1 - q)/[(1 - \alpha)(1 - q) + \beta q], \\ P_2 =& 1 - \beta^* = P(T = +|C = +) = (1 - \beta)\theta/[(1 - \beta)q + \alpha(1 - q)]. \end{aligned}$$

Note that if $P_1 = P_1(q, \alpha, \beta)$ and $P_2 = P_2(q, \alpha, \beta)$, then $P_2 = P_1(1 - q, \beta, \alpha)$ and $P_1 = P_2(1 - q, \beta, \alpha)$. The quantities P_1 and P_2 also arise in the evaluation of diagnostic tests. They are referred to as the negative and positive predictive values.

Alternatively, (α, β) can be written as a function of (q, P_1, P_2), i.e.,

$$\begin{aligned} \alpha =& (1 - P_2)(q + P_1 - 1)/[(1 - q)(P_1 + P_2 - 1)], \\ \beta =& (1 - P_1)(P_2 - q)/[q(P_1 + P_2 - 1)]. \end{aligned} \tag{9.4.1}$$

The equations in (9.4.1) require $P_1 > 1 - q$ and $P_2 > q$ or $P_1 < 1 - q$ and $P_2 < q$. Otherwise α and β could be negative. Note that $\alpha = \beta = 1 - q$ if $P_1 = P_2 = q$ and $q = 0.5$. The relations $P_1 > 1 - q$ and $P_2 > q$ ensure that the posterior probabilities are larger than the prior probabilities and this hypothesis test is informative. The opposite is true for $P_1 < 1 - q$ and $P_2 < q$. Lee and Zelen (2000) note that this latter condition seems unreasonable and thus only require that $P_1 > 1 - q$ and $P_2 > q$.

The frequentist method of calculating sample size in a clinical trial is to specify (α, β, θ). This specification is sufficient to allow the calculation of the sample size. Lee and Zelen (2000) note that a more relevant way of calculating sample size is to specify (q, P_1, P_2), or equivalently (q, α^*, β^*), yielding the values of (α, β) by use of the equations in (9.4.1).

Given the value of (α, β), Lee and Zelen (2000) obtain the large sample relationship for two-sided tests

$$n(\theta/\sigma)^2 = 2(z_{\alpha/2} + z_\beta)^2, \tag{9.4.2}$$

where

$$Q(z_\gamma) = \int_{z_\gamma}^{\infty} (2\pi)^{-\frac{1}{2}} e^{-t^2/2} dt = \gamma$$

and σ is defined in an appropriate manner based on the test statistic. The relationship given by (9.4.2) is a suitable approximation for the two-sided alternative if the type I error is in the neighborhood of $\alpha \leq 0.05$. Otherwise it may be necessary to use the more accurate normal approximation to the power given by

$$1 - \beta = Q\left(z_{\alpha/2} - \frac{\theta\sqrt{n}}{\sigma\sqrt{2}}\right) + Q\left(z_{\alpha/2} + \frac{\theta\sqrt{n}}{\sigma\sqrt{2}}\right). \tag{9.4.3}$$

Lee and Zelen (2000) note that conditional on the value of (α, β), there is a trade-off between sample size (n) and θ/σ as given by (9.4.2) or its more complicated version (9.4.3). Thus, for fixed (α, β) and given q, one can tabulate values of n and θ/σ which satisfy the error probabilities. In the above it is assumed that σ is approximately known or that θ is expressed as a multiple of σ. In the case of the log-rank statistic, $\sigma = 2$ as noted in earlier sections. Thus, Lee and Zelen (2000) note that instead of choosing (α, β), the investigator chooses (P_1, P_2, q) subject to $P_1 > 1-q$ and $P_2 > q$. This, in turn specifies (α, β) using the equations in (9.4.1), and a table of n vs. θ/σ can be calculated by making use of (9.4.2) or (9.4.2). If there is some uncertainty about q, the entire calculation may be repeated for different values of q. The final sample size is chosen corresponding to a fixed value of θ/σ.

The procedure of Lee and Zelen (2000) requires selecting values of (P_1, P_2) prior to the trial. The trial will be planned so that (P_1, P_2) are the posterior probabilities after the trial is completed. It is clear that these posterior probabilities should be reasonably high. Lee and Zelen (2000) recommend that $P_2 = 0.95$ or higher. They note that values of P_1 can be lower, and could range from 0.85 to values close to unity.

TABLE 9.4. Type I and II Errors for Specified P_1, P_2, and q.

q	P_1	P_2	α	β
0.25	0.85	0.95	0.0083	0.5250
0.25	0.90	0.95	0.0118	0.3294
0.25	0.95	0.95	0.0148	0.1556
0.50	0.85	0.95	0.0438	0.1688
0.50	0.90	0.95	0.0471	0.1059
0.50	0.95	0.95	0.0500	0.0500
0.75	0.85	0.95	0.1500	0.0500
0.75	0.90	0.95	0.1529	0.0314
0.75	0.95	0.95	0.1556	0.0148

Source: Lee and Zelen (2000).

TABLE 9.5. Sample Size vs. (θ/σ) for $P_1 = P_2 = 0.95$.

q	α	β	(θ/σ)	Sample Size (n)
0.5	0.05	0.05	0.1	2599
			0.2	650
			0.3	289
			0.4	162
			0.5	104
0.25	0.015	0.156	0.1	2371
			0.2	593
			0.3	263
			0.4	148
			0.5	95
0.75	0.156	0.015	0.1	2576
			0.2	644
			0.3	286
			0.4	161
			0.5	103

Source: Lee and Zelen (2000).

Table 9.4 summarizes values of (α, β) for $P_1 = 0.85$, 0.90, 0.95, and $P_2 = 0.95$ for values of $q = 0.25$, 0.50, and 0.75, Table 9.5 shows the sample sizes for $P_1 = P_2 = 0.95$ under a range of (θ/σ), and Figure 9.1 shows how these quantities change over a range of prior probabilities. In general, as $q \to 1$, α increases and $\beta \to 0$; similarly as $q \to 0$, β increases and $\alpha \to 0$. When $P_1 = 0.95$ and $P_2 = 0.95$, both α and β remain under 0.156 for q

between 0.25 and 0.75.

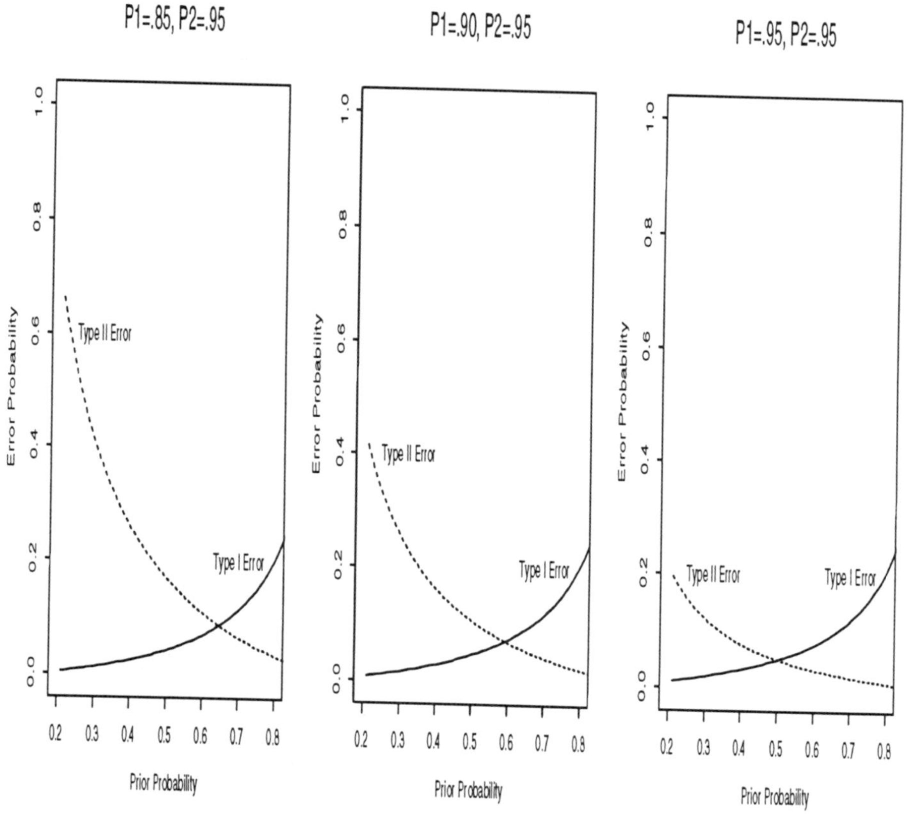

FIGURE 9.1. Type I and type II error probabilities vs. prior probabilities (q) for selected values of P_1 and P_2.

Example 9.6. ECOG studies analysis. Lee and Zelen (2000) illustrate their methods to the planning of cancer clinical trials for the Eastern Cooperative Oncology Group (ECOG). In order to estimate the quantity $q = P(T = +)$, they evaluated outcomes of all phase III clinical trials conducted by ECOG during a recent 15 year period. Between 1980 and 1995, a total of 98 studies were activated and completed. Among these studies, 87 trials had a report on the final outcomes. If any of the major endpoints, such as response rate, overall survival or disease-free survival, was declared significant, the researchers considered that the study had a positive outcome. Most studies used $\alpha = 0.05$ and $0.10 \leq \beta \leq 0.20$. Among the 87 studies, 25 had significant outcomes. Hence an estimate of the probability of having a positive clinical outcome was $P(C = +) = 25/87 = 0.29$. This quantity ranged from 0.25 to 0.43 in major disease sites such as breast (0.38), GI (0.33), GU (0.31), leukemia (0.40), lymphoma (0.43), and melanoma (0.25). In the discussion of the ECOG studies, they use the overall estimate of $P(C = +) = 0.29$. However in planning studies for a particular disease site, Lee and Zelen (2000) recommend that the marginal probability for that site should be utilized.

By using the relationship

$$P(C = +) = P(C = +|T = -)P(T = -) + P(C = +|T = +)P(T = +),$$

they obtain

$$q = \frac{P(C = +) - \alpha}{(1 - \alpha - \beta)}, \tag{9.4.4}$$

thus permitting an estimate of q from knowledge of $P(C = +)$. Under the assumption of $\alpha = 0.05$ and $0.10 \le \beta \le 0.20$, q ranges from 0.28 to 0.32. This leads to the posterior probabilities (P_1, P_2) ranging from $(0.90, 0.88)$ to $(0.96, 0.88)$. Lee and Zelen (2000) thus note that the negative clinical trials have a posterior probability of being true negatives within the range $[0.90, 0.96]$; the positive outcomes have a probability of being true positives equal to 0.88. As a result, among the 25 positive outcomes, 12% (or three trials) are expected to be false-positive trials. Similarly, among the 62 negative (or neutral) outcomes, 4–10% (two to six trials) are expected to be false-negative trials.

Figure 9.2 displays the relationships between the posterior probabilities and β at selected levels of α. Figure 9.3 is a similar plot displaying the relationships between the posterior probabilities and α at selected levels of β. These calculations were made using a value of $q = 0.30$. Note that the posterior negative probability P_1 is not sensitive to the type I error probability α, whereas the posterior positive probability P_2 is not sensitive to the type II error probability β. In order to have the posterior positive probability $P_2 > 0.90$, the type I error should be in the neighborhood of $[0.025, 0.030]$.

Figure 9.4 shows how the posterior probabilities change over a range of β values when $\alpha = 0.025$ and $\alpha = 0.05$. Note that values of β in the neighborhood of 0.20 result in relatively high posterior probabilities. Lee and Zelen (2000) mention that since the ECOG experience is comparable to that of other cancer cooperative clinical control groups, future trials should be planned to have a type I error of less than 0.03 and a power of at least 80%.

Example 9.7. Sample size calculations for the comparison of two survival distributions. Lee and Zelen (2000) illustrate sample size calculations for the comparison of two survival distributions. Following Lee

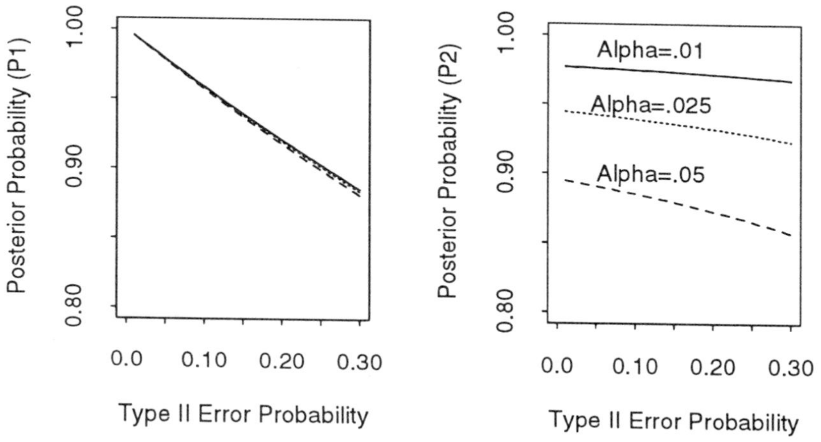

FIGURE 9.2. ECOG posterior probabilities and type II error probability for $\alpha = 0.01, 0.025, 0.05$ (prior probability $q = 0.30$).

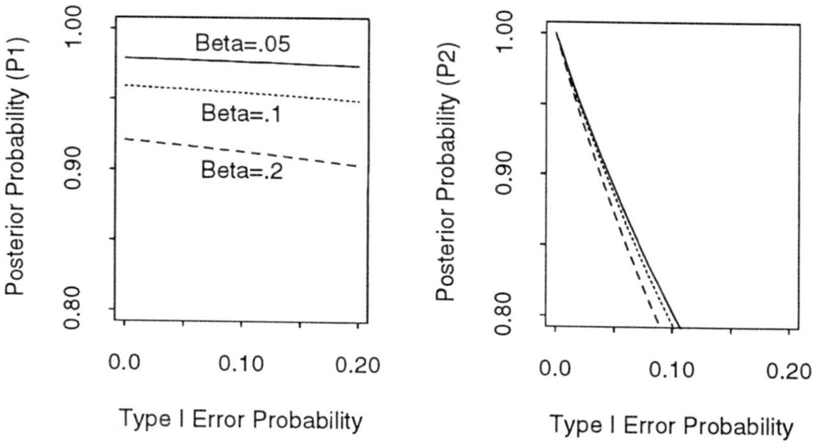

FIGURE 9.3. ECOG posterior probabilities and type I error probability for $\beta = 0.05, 0.1, 0.2$ (prior probability $q = 0.30$).

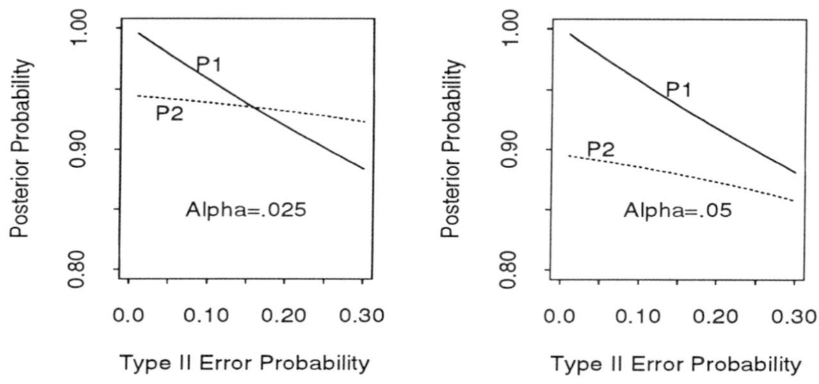

FIGURE 9.4. ECOG posterior probabilities ($q = .30$) for $\alpha = 0.025, 0.05$.

and Zelen (2000), assume that the time-to-event outcome follows the exponential distribution with failure rate λ, and interest is in comparing λ_1 and λ_2 between two treatment groups. In this setting, the total sample size is equivalent to the total number of events from both treatment groups. Denote the number of events by d_1 and d_2 in each treatment group. Using a large sample approximation, they obtain

$$\frac{d_1 d_2}{d_1 + d_2}(\log(\Delta))^2 = (z_{\alpha/2} + z_\beta)^2 \qquad (9.4.5)$$

with $\Delta = \lambda_1/\lambda_2$. In (9.4.5), the type I error is α for two-sided tests. Assuming $d_1 = d_2$, Lee and Zelen (2000) calculate the total number of events (d_1) under a range of Δ values. These are given in Table 9.6 for $P_1 = P_2 = 0.95$.

Lee and Zelen (2000) note that an alternative procedure for sample size calculations for exponentially distributed survival times is to carry out a test using information on the number of failures. If the two groups have equal person years of follow-up, then the total number of observed events to achieve a level α test (two-sided) with a power of $(1-\beta)$ is given by

$$d = \left\{\frac{z_{\alpha/2}}{2} + z_\beta\sqrt{\Delta/(1+\Delta)^2}\right\}^2 \Big/ \left(\frac{\Delta}{1+\Delta} - 0.5\right)^2. \qquad (9.4.6)$$

The above formula assumes the failures follow a Poisson distribution with the same person years of follow-up time for each group. Equation (9.4.6) follows by conditioning on the sum of two Poisson random variables ($d_1 + d_2 = d$), which has a binomial distribution with sample size d and success probability $\Delta/(1+\Delta)$. The value of d in equation (9.4.6) is approximately twice of the value of d_1 in Table 9.6. Although Lee and Zelen (2000) only discuss fixed sample size calculations, their ideas are easily adapted to

TABLE 9.6. Number of Events vs. (Δ) for $P_1 = P_2 = 0.95$.

q	α	β	(Δ)	Number of Events (d_1)
0.5	0.05	0.05	0.4	31
			0.5	54
			0.6	100
			0.7	205
			0.8	522
0.25	0.015	0.156	0.4	28
			0.5	50
			0.6	91
			0.7	187
			0.8	476
0.75	0.156	0.015	0.4	31
			0.5	54
			0.6	99
			0.7	203
			0.8	517

Source: Lee and Zelen (2000).

trials which allow early stopping. One simply uses the calculated (α, β) generated by the prespecified posterior error probabilities to determine the early stopping rules.

Exercises

9.1 Derive (9.1.3).

9.2 Prove (9.1.6).

9.3 Derive (9.2.2).

9.4 Compute and verify all of the entries in Table 9.1.

9.5 Derive (9.2.10) and (9.2.12).

9.6 Derive (9.2.13).

9.7 CONDITIONAL POWER

The frequentist analogue of the predictive power is called the *conditional power*. For the conditional power method, the trial is stopped at the kth interim analysis in favor of the alternative hypothesis whenever the power upon completion of the trial given the value of Z_k, the normalized log-rank statistic at the kth analysis, is greater than some specified value, say γ. Suppose at the kth analysis we have Z_k distributed as in (9.1.5).

(a) Derive the conditional power, that is, compute

$$P(Z_K > c|Z_k, \theta), \tag{9.E.1}$$

where c is a specified constant. Compute (9.E.1) for $\theta = \log(1.5)$, $Z_k = 1.0$, $c = 1.96$, $n_k = 100$, $n_K = 200$, and n_k denotes the number of events at interim analysis k.

(b) Suppose that $\theta \sim N(0, 4/n_0)$. Compute the predictive power using $n = 200$, $n_0 = 20$, and $\theta_1 = \log(1.5)$.

9.8 Compute and verify all of the posterior probability calculations in Example 9.4.

9.9 Derive (9.3.6).

9.10 Verify (9.4.1).

9.11 Derive (9.4.5) and (9.4.6).

9.12 Exponential models are used in designing many of the Eastern Cooperative Oncology Group (ECOG) clinical trials. Consider the following hypothetical study design, which is typical of trials used for studying melanoma. Assume subjects accrue uniformly over four years with an accrual rate of 200 subjects per year. A follow-up period of two years is planned. Let t_a denote the accrual period and t_f be the follow-up period. There are two treatment arms in the trial and subjects are randomized with equal probability to treatment groups 1 and 2. A median survival of 1.5 years is estimated for patients in treatment group 1 and 1.8 years for patients in group 2. At the end of the trial, a total of 413 ($= n_K$) deaths is required to achieve 91% power. Analyses are planned at four information times, i.e., when 25%, 50%, 75% and 100% of the expected deaths are observed. We further assume that we have uniform censoring times with the survivor function

$$G_0(u) = \begin{cases} t_a^{-1}(t_f + t_a - u), & \text{if } t_f < u < t_f + t_a, \\ 1, & \text{if } u \le t_f. \end{cases}$$

Let λ_j denote the hazard rate for group j, $j = 1, 2$. We are interested in testing H_0: $\lambda_1 = \lambda_2$ versus H_1: $\lambda_1 = 0.4621$ and $\lambda_2 = 0.3851$. A two-sided log-rank test with significance level of 0.05 is used to test the null hypothesis.

(a) Compute the stopping rules obtained using the conditional power and predictive power methods at each interim analysis assuming n_k events have been observed at the kth interim analysis. For the predictive power method, assume a $N(0, 4/n_0)$ prior for θ. For the predictive power, compute the stopping rules using $n_0 = 30$, 100, 200. How do the stopping rules for the predictive power change as n_0 increases?

(b) Repeat part (a) assuming that we have a cure rate model for survival, in which for group j, $S_j(t) = \pi_j + (1 - \pi_j)\exp(-\lambda_j t)$, $j = 1, 2$, with H_0: $S_1(t) = S_2(t)$, and H_1: $\pi_1 = 0.30$, $\pi_2 = 0.40$, $\lambda_1 = 0.4621$, and $\lambda_2 = 0.3851$.

(c) Comment on the stopping rules obtained with the cure rate model in part (b) versus those obtained with the exponential model in part (a).

10
Other Topics

In this last chapter, we discuss various other topics in Bayesian survival analysis. Specifically, we discuss semiparametric proportional hazards models built from monotone function, accelerated failure time models, Dirichlet process and Polya tree priors in survival analysis, Bayesian survival analysis using multivariate adaptive regression splines (MARS), change point models, poly-Weibull models, hierarchical survival models using neural networks, Bayesian model diagnostics, and future research topics.

10.1 Proportional Hazards Models Built from Monotone Functions

Gelfand and Mallick (1995) discuss a semiparametric Bayesian approach for proportional hazards models by modeling the cumulative baseline hazard using a dense class of monotone functions. Towards this goal, they write the hazard function for the proportional hazards model as

$$h(t) = h_0(t)\varphi(\boldsymbol{x}) \tag{10.1.1}$$

where φ is positive and $\boldsymbol{x} = (x_1, x_2, \ldots, x_p)'$ is a $p \times 1$ vector of covariates. The cumulative hazard function associated with $h(t)$ is $H(t) = H_0(t)\varphi(\boldsymbol{x})$, where $H_0(t) = \int_0^t h_0(u)du$. The survivor function is $S(t) = \exp\{-H_0(t)\varphi(\boldsymbol{x})\}$ and the density function takes the form

$$f(t) = h_0(t)\varphi(\boldsymbol{x})\exp\{-H_0(t)\varphi(\boldsymbol{x})\}. \tag{10.1.2}$$

Let $T \sim f(t)$. Then, implicit in (10.1.2) is the fact that the random variable $U(T) = H_0(T)\varphi(\boldsymbol{x})$ has an $\mathcal{E}(1)$ distribution. Dating to Cox (1972), the function φ is customarily taken to have the form $\exp(\boldsymbol{x}'\boldsymbol{\beta})$, where $\boldsymbol{\beta}$ is an unknown $p \times 1$ vector of coefficients. Drawing upon terminology for generalized linear models, $\varphi(\boldsymbol{x})$ can be more generally thought of as a covariate link function, which would therefore be strictly increasing to insure that $E(T)$ is strictly increasing in $\eta = \boldsymbol{x}'\boldsymbol{\beta}$. However, since $\boldsymbol{x}'\boldsymbol{\beta}$ can include polynomial forms for any covariate, φ is not assumed to be a monotonic function of any covariate.

Gelfand and Mallick (1995) model H_0 and φ using a dense class of monotone functions from $R^+ = (0, \infty)$ to R^+ and from R^1 to R^+, respectively. These classes arise from mixtures of beta distribution functions. Their approach models the strictly monotone functions H_0 and φ, each transformed by a suitable monotone transformation to have range in $[0, 1]$, as unknown cumulative distribution functions. In the process the resulting domain also becomes $[0, 1]$. Thus each function is characterized as a mixture of beta distribution functions. The motivation behind this formulation is based on the well-known result as in, e.g., Diaconis and Ylvisaker (1985), which says that any continuous density on $[0, 1]$ can be arbitrarily well approximated by a discrete mixture of beta densities. Unlike distributions arising under a Dirichlet process, which could also be used here, we have a continuous, dense class of distributions admitting an explicit form. In practice the number of mixands k, is treated as fixed, and various choices are compared. Alternatively, a discrete prior on k could be attempted. As such Gelfand and Mallick (1995) take an infinite-dimensional model and convert it to a finite-dimensional one. Though arbitrarily large numbers of mixands can be employed, in practice they are usually not needed, since robustness occurs with a small k. In introducing randomness to this finite mixture form, it could be assumed either the mixture weights are random or the parameters of the beta densities are random. Mathematically it is much simpler to work with the former. Hence for a given k they choose a fixed set of beta densities and then assume, a priori, that the mixing weights arise as a random draw on the k-dimensional simplex. Earlier work with this approach (Mallick and Gelfand, 1994) looked at the problem of fitting a generalized linear model when the link function is assumed unknown.

Under the mixture of beta distributions approach, the class of models for H_0 is dense within the family of monotonic functions from $R^+ \rightarrow R^+$. Moreover, each member is smooth, in fact, differentiable, so that $S(t)$ is differentiable and thus $f(t)$ as in (10.1.2) exists.

The nonparametric Bayesian approaches discussed in Chapter 3 proceed in a much different way than the approach of Gelfand and Mallick (1995). The approaches in Chapter 3 basically view the number of events up to time t as a counting process which is specified either by modeling $H(t)$ or $-\log(S(t))$. Then, $H(t)$ or $-\log(S(t))$ is viewed as a random sample path from a Levy process, i.e., a nonnegative nondecreasing process on $[0, \infty)$, which starts at 0 and has independent increments. For example, Kalbfleisch (1978) models $-\log(S(t))$ as a draw from a gamma process, while Hjort (1990) models $H(t)$ as a draw from a beta process. Note that the earliest work, which modeled $1 - S(t)$ by a Dirichlet process (Ferguson, 1974), implies that $-\log(S(t))$ is a Levy process. In all of these fully nonparametric approaches, the resulting $S(t)$ is discrete with probability 1.

As with H_0, the model for φ employs a dense class of functions but from $R^1 \to R^+$. Again, it is parametric and results in a differentiable function. However, fully nonparametric estimation of φ has received attention. For instance, in O'Sullivan (1988) and in Hastie and Tibshirani (1990), it is assumed that $\varphi(\boldsymbol{x}) = \exp\{\theta(\boldsymbol{x})\}$, where $\theta(\boldsymbol{x})$ is an arbitrary additive function in the components of $\boldsymbol{x}$, i.e., $\theta(\boldsymbol{x}) = \sum_{l=1}^{p} \theta_l(x_l)$. Partial likelihoods are used so that φ is estimated independently of H_0. We note that the form $\exp\{\theta(\boldsymbol{x})\}$ does include the assumed form $\varphi(\boldsymbol{x}'\boldsymbol{\beta})$. Moreover, with this approach, an estimate of $\varphi(\eta)$ is easily compared with $\exp(\eta)$, which may be viewed as a "baseline" covariate link. Staniswalis (1989) considers joint estimation of unknown H_0 and φ with only a single dichotomous covariate. She uses kernel estimators resulting in the maximization of a weighted likelihood. Her fitting algorithm is iterative, maximizing over H_0 given φ, then over φ given a new H_0, and so forth.

10.1.1 Likelihood Specification

Suppose we have a total of n subjects. Let t_i and v_i denote the survival time and the censorship time, respectively. For the ith individual let $\nu_i = 1$ if $t_i \leq v_i$, and 0 otherwise. Also, let $y_i = \min(t_i, v_i)$. Then, the joint density of the sample, following from (10.1.2), is

$$\prod_{i=1}^{n} [h_0(y_i)\varphi(\boldsymbol{x}_i'\boldsymbol{\beta})]^{\nu_i} \exp\{-H_0(y_i)\varphi(\boldsymbol{x}_i'\boldsymbol{\beta})\}, \tag{10.1.3}$$

where $\boldsymbol{x}_i = (x_{i1}, x_{i2}, \ldots, x_{ip})'$ is the vector of covariates associated with the ith individual. With H_0, φ and $\boldsymbol{\beta}$ unknown, we may view (10.1.3) as a likelihood function, $L(\boldsymbol{\beta}, H_0, \varphi | D)$, where $D = (\boldsymbol{y}, X, \boldsymbol{\nu})$, $\boldsymbol{y} = (y_1, y_2, \ldots, y_n)'$, X is the $n \times p$ matrix with the ith row $\boldsymbol{x}_i'$, and $\boldsymbol{\nu} = (\nu_1, \nu_2, \ldots, \nu_n)'$. Indeed, L has an infinite-dimensional argument; without further assumptions it need not be identifiable. For example, if $\boldsymbol{x}'\boldsymbol{\beta} = \beta_0 + \beta_1 x$, then H_0, φ, β_0, and β_1 cannot be identified. Let $J_0 = a_0 H_0/(a_0 H_0 + b_0)$, where $a_0, b_0 > 0$ are specified constants. Then J_0 is a differentiable cumulative distribution function (cdf). Modeling J_0 is equivalent to modeling an unknown distribution function. Similarly, let $J_\varphi = a_1\varphi/(a_1\varphi + b_1)$, where $a_1, b_1 > 0$. Then again, J_φ is a differentiable cumulative distribution function.

A rich class of models for J_0 and for J_φ may be created as follows. Associate with H_0 a specific baseline hazard function $\tilde{H}_0$. For instance, $\tilde{H}_0(t)$ might be t^ρ for a given ρ. Similarly, associate with φ a specific baseline covariate link $\tilde{\varphi}$. For instance, $\tilde{\varphi}(\eta)$ might be $\exp(\eta)$. Diaconis and Ylvisaker (1985) argue that discrete mixtures of beta densities provide a continuous dense class of models for densities on $[0,1]$. A member of this class has a density of the form

$$f(u) = \sum_{l=1}^{k} w_l \mathrm{Be}(u|c_l, d_l), \tag{10.1.4}$$

where k denotes the number of mixands, $w_l \geq 0$, $\sum w_l = 1$, and $\mathrm{Be}(u|c_l, d_l)$ denotes the beta density in standard form with parameters, c_l and d_l. If $\mathrm{IB}(u|c_l, d_l)$ denotes the cumulative distribution function associated with $\mathrm{Be}(u|c_l, d_l)$, then let

$$J_0(t; \boldsymbol{w}_1) = \sum_{l=1}^{k_1} w_{1l} \mathrm{IB}(\tilde{J}_0(t)|c_{1l}, d_{1l}) \tag{10.1.5}$$

and

$$J_\varphi(\eta; \boldsymbol{w}_2) = \sum_{l=1}^{k_2} w_{2l} \mathrm{IB}(\tilde{J}_\varphi(\eta)|c_{2l}, d_{2l}), \tag{10.1.6}$$

where $\boldsymbol{w}_1 = (w_{11}, w_{12}, \ldots, w_{1k_1})'$, $\boldsymbol{w}_2 = (w_{21}, w_{22}, \ldots, w_{2k_2})'$, $\tilde{J}_0(t) = a_0\tilde{H}_0(t)/(a_0\tilde{H}_0(t) + b_0)$, and $\tilde{J}_\varphi(\eta) = a_1\tilde{\varphi}(\eta)/(a_1\tilde{\varphi}(\eta) + b_1)$. Clearly J_0 and J_φ are cumulative distribution functions.

It will be of interest to calculate $h_0(t) = \frac{d}{dt}H_0(t)$ and $\varphi'(\eta) = \frac{d\varphi(\eta)}{d\eta}$. Since $H_0 = b_0 J_0/a_0(1 - J_0)$, $\frac{d}{dt}H_0 = b_0\frac{\partial J_0}{\partial t}/a_0(1 - J_0)^2$. Similarly, since $\varphi = b_1 J_\varphi/a_1(1 - J_\varphi)$, $\varphi'(\eta) = b_1\frac{\partial J_\varphi}{\partial \eta}/a_1(1 - J_\varphi)^2$. From (10.1.5) and (10.1.6), $\frac{\partial}{\partial t}J_0(t; \boldsymbol{w}_1) = \frac{d\tilde{J}_0}{dt}\sum_{l=1}^{k_1} w_{1l}\mathrm{Be}(\tilde{J}_0(t)|c_{1l}, d_{1l})$, $\frac{\partial}{\partial \eta}J_\varphi(\eta; \boldsymbol{w}_2) = \frac{d\tilde{J}_\varphi}{d\eta}\sum_{l=1}^{k_2} w_{2l}\mathrm{Be}(\tilde{J}_\varphi(\eta)|c_{2l}, d_{2l})$ with $\frac{d\tilde{J}_0}{dt} = a_0 b_0\frac{d\tilde{H}_0}{dt}/[a_0\tilde{H}_0 + b_0]^2$ and $\frac{d\tilde{J}_\varphi}{d\eta} = a_1 b_1\frac{d\tilde{\varphi}}{d\eta}/[a_1\tilde{\varphi} + b_1]^2$.

Gelfand and Mallick (1995) point out that inference using mixtures with small k is virtually indistinguishable from those with much larger k. In fact, allowing $k \geq n$ does not insure perfect fit since H_0 and φ are restricted to be monotone. Given k, it is natural and certainly mathematically easier to assume that the component beta densities are specified but that the weights are unknown. Here an advantage to employing beta mixtures, i.e., to working on the bounded interval $[0,1]$ is seen; one can choose the finite set of (c_l, d_l) to provide an "equally spaced" collection of beta densities on $[0,1]$. In particular, suppose we set $k_1 = k_2 = k$, $c_{1l} = c_{2l} = c_l$, $d_{1l} = d_{2l} = d_l$ with $c_l = \lambda\, l$, $d_l = \lambda(k+1-l)$ and $\lambda > 0$. As discussed in Subsection 10.1.2, with suitable priors for $\boldsymbol{w}_1$ and $\boldsymbol{w}_2$, an appropriate centering for the prior distributions of H_0 and φ results. In any event, specification of H_0 and φ is equivalent to specification of $\boldsymbol{w}_1$ and $\boldsymbol{w}_2$ and we can denote the likelihood by $L(\boldsymbol{\beta}, \boldsymbol{w}_1, \boldsymbol{w}_2|D)$. Since λ determines how peaked the component beta densities are, experimentation with various choices may prove useful.

Gelfand and Mallick (1995) note that the choice of a_0, b_0 and a_1, b_1 is not a modeling issue but a computational one. For example, if $|\eta|$ is very large, $\tilde{\varphi}(\eta) = \exp(\eta)$ will produce over or underflow. Since, in (10.1.6), only $\tilde{J}_\varphi(\eta)$ is needed, scaling and centering using appropriate a_1 and b_1 can alleviate this problem. In practice, a_1 and b_1 could be obtained by looking at the range of $\boldsymbol{x}'\tilde{\boldsymbol{\beta}}$ over the sample with $\tilde{\boldsymbol{\beta}}$, say, the partial likelihood maximizer. For $\tilde{J}_0(t)$, often $a_0 = b_0$=1 will work. Again, there is no notion of "best" values. Many choices of a_i and b_i will work equally well and, to keep notation simpler, we suppress the a_i and b_i in the sequel.

10.1.2 Prior Specification

Gelfand and Mallick (1995) propose the following specification for the joint prior $\pi(\boldsymbol{\beta}, \boldsymbol{w}_1, \boldsymbol{w}_2)$:

$$\pi(\boldsymbol{\beta}, \boldsymbol{w}_1, \boldsymbol{w}_2) = \pi(\boldsymbol{\beta}|\boldsymbol{w}_1, \boldsymbol{w}_2)\pi(\boldsymbol{w}_1)\pi(\boldsymbol{w}_2), \tag{10.1.7}$$

where $\pi(\boldsymbol{w}_j)$ is a distribution on the k_j dimensional simplex for $j = 1, 2$. Since $\boldsymbol{w}_1$ determines $H_0(t)$, $\pi(\boldsymbol{w}_1)$ is chosen such that the prior density for $H_0(t)$ is *centered* around $\tilde{H}_0(t)$, that is, on the transformed scale, $E[J_0(t; \boldsymbol{w}_1)] = \tilde{J}_0(t)$. From (10.1.4), this holds if

$$\sum_{l=1}^{k_1} E(w_{1l})\text{IB}(u|c_{1l}, d_{1l}) = u. \tag{10.1.8}$$

If $\boldsymbol{w}_1$ has a Dirichlet distribution with all parameters equal to $\varphi > 0$, i.e., $D_{k_1-1}(\varphi, \varphi, \ldots, \varphi)$, then (10.1.8) holds if the average of $\text{IB}(u|c_{1l}, d_{1l})$ equals u. If k_1 is even and c_{1l} and d_{1l} are chosen as in Section 10.1.1, then it can be shown that, to a first-order approximation, (10.1.8) holds. Similar remarks apply to $\pi(\boldsymbol{w}_2)$ with regard to *centering* the prior density for $\varphi(\eta)$ around $\tilde{\varphi}(\eta)$.

The Jeffreys' prior, which is the square root of the determinant of the Fisher information matrix associated with $L(\boldsymbol{\beta}, \boldsymbol{w}_1, \boldsymbol{w}_2|D)$, is used for $\pi(\boldsymbol{\beta}|\boldsymbol{w}_1, \boldsymbol{w}_2)$. Standard calculation yields

$$\pi(\boldsymbol{\beta}|\boldsymbol{w}_1, \boldsymbol{w}_2) \propto |X'MX|^{1/2}, \tag{10.1.9}$$

where X is the $n \times p$ matrix of covariates, M is an $n \times n$ diagonal matrix such that

$$M_{ii} = -[1 - S(v_i|\boldsymbol{x}_i)][\varphi'(\boldsymbol{x}_i'\boldsymbol{\beta})/\varphi(\boldsymbol{x}_i'\boldsymbol{\beta})]^2, \tag{10.1.10}$$

where $S(v_i|\boldsymbol{x}_i) = \exp\{-H_0(v_i)\varphi(\boldsymbol{x}_i'\boldsymbol{\beta})\}$ and φ' was calculated in Subsection 10.1.1. Working with (10.1.3), this calculation is clarified by noting that $E[\nu_i - H_0(y_i)\varphi(\boldsymbol{x}_i'\boldsymbol{\beta})] = 0$. From (10.1.10), it can be seen that Jeffreys's prior depends upon both $\boldsymbol{w}_1$ and $\boldsymbol{w}_2$. In the case of of no censoring, $M_{ii} = -[(\varphi'(\boldsymbol{x}_i'\boldsymbol{\beta})/\varphi(\boldsymbol{x}_i'\boldsymbol{\beta})]^2$ which depends only upon $\boldsymbol{w}_2$.

Given the Bayesian model

$$L(\boldsymbol{\beta}, \boldsymbol{w}_1, \boldsymbol{w}_2|D)\pi(\boldsymbol{\beta}|\boldsymbol{w}_1, \boldsymbol{w}_2)\pi(\boldsymbol{w}_1)\pi(\boldsymbol{w}_2),$$

all inference proceeds from the posterior distribution of $(\boldsymbol{\beta}, \boldsymbol{w}_1, \boldsymbol{w}_2)$, which is proportional to this product. Analytic investigation of this $(p + (k_1 - 1) + (k_2 - 1))$ dimensional nonnormalized joint distribution is infeasible. It would be difficult enough to standardize this form, much less to calculate expectations, marginal distributions, and other posterior quantities of interest. Thus, a sampling based approach using an MCMC algorithm is adopted to obtain random draws essentially from this posterior distribution. Let $\{(\boldsymbol{\beta}_j^*, \boldsymbol{w}_{1j}^*, \boldsymbol{w}_{2j}^*),\ j = 1, 2, \ldots, m\}$ denote an MCMC sample of size m. This sample enables us to carry out any desired posterior inference. The posteriors of H_0 and h_0 can be obtained at any time t using the posterior of $\boldsymbol{w}_1$. The posterior of φ can be obtained at any point η using the posterior of $\boldsymbol{w}_2$.

10.1.3 Time-Dependent Covariates

In the case of time-dependent covariates the hazard in (10.1.1) is replaced by

$$h(t) = h_0(t)\varphi(\boldsymbol{x}(t)) \tag{10.1.11}$$

where $\boldsymbol{x}(t)$ is a known function. As Cox and Oakes (1984, Chapter 8) warn, such covariates must be used with care.

To accommodate such time-dependent covariates, from (10.1.11), $S(t) = \exp\left\{-\int_0^t h_0(u)\varphi(\boldsymbol{x}(u))du\right\}$ and the density becomes

$$f(t) = h_0(t)\varphi(\boldsymbol{x}(t)) \exp\left\{-\int_0^t h_0(u)\varphi(\boldsymbol{x}(u))du\right\}. \tag{10.1.12}$$

From $S(t)$ and (10.1.12), it can be seen that H_0 and φ no longer separate into a product. In fact, H_0 doesn't even appear. However, retaining the same definition as before, h_0 is then well-defined and, thus, so is $A(t) = \int_0^t h_0(u)\varphi(\boldsymbol{x}(u))du$ at a given t. Evaluation of $A(t)$ requires a univariate numerical integration over a bounded domain which presumably can be handled by a trapezoidal or Simpson's integration. Hence, at a given t_i with associated $\boldsymbol{x}_i(t_i)$, $S(t_i)$ and $f(t_i)$ can be calculated so that a likelihood based upon (10.1.12), analogous to (10.1.3), can be calculated. However, each likelihood evaluation requires n numerical integrations and an enormous number of incomplete beta calculations. Thus, sampling from the resulting posterior distribution is very challenging. Since $A(t)$ is strictly increasing, it might appear that modeling A and φ would be preferable to H_0 and φ. Unfortunately, A changes as i, hence $\boldsymbol{x}_i$, changes: one cannot model a single A. Thus, modeling A and φ as well as sampling from the resulting posterior distribution deserves a future research topic.

Example 10.1. Lung cancer survival data. Gelfand and Mallick (1995) considered the lung cancer data given in Lawless (1982, page 6). There are 40 advanced lung cancer patients and three covariates: x_1, performance status at diagnosis, x_2, age of the patient in years, and x_3, months from diagnosis to entry into the study. Note that three of the 40 survival times, are censored. An exponential model, i.e., $h_0(t) = 1$ and $\varphi(\eta) = \exp(-\eta)$, denoted by EE, is used by Lawless (1982). For this four-parameter model (intercept and three coefficients, with intercept needed since $h_0(t)$ is fixed at 1) maximum likelihood estimates and associated standard errors can be obtained from standard statistical packages and are presented in Table 10.1. Accordingly, only x_1 is statistically significant.

TABLE 10.1. Summary of Likelihood Analysis for EE and Posterior Analysis for MM.

	EE		MM	
Parameter	MLE	SD	Mean	SD
β_0	4.742	1.612	3.932	0.167
β_1	0.060	0.009	0.035	0.012
β_2	0.013	0.015	0.001	0.018
β_3	0.003	0.010	0.003	0.009

Source: Gelfand and Mallick (1995).

Gelfand and Mallick (1995) also fit the Bayesian model that incorporates both a monotonic hazard and covariate link each exponentially centered, denoted in Table 10.1 by MM. They took $\lambda = 1$ and $k_i = 3$, and used a flat prior for $\boldsymbol{\beta}$. In addition, they chose a uniform prior on the simplex in three dimensions for $\boldsymbol{w}_j$. Hence, the resulting model has eight parameters. The posterior estimates of $\boldsymbol{\beta} = (\beta_0, \beta_1, \beta_2, \beta_3)'$ are given in Table 10.1, and the posterior correlations amongst the eight parameters under MM are reported in Table 10.2. In terms of inference for $\boldsymbol{\beta}$ alone, there is little qualitative difference between EE and MM. As expected, there is high negative correlation between w_{11} and w_{12} and between w_{21} and w_{22}. Also, not surprisingly, there is strong correlation between β_1 and $\boldsymbol{w}_2$, which determines the link φ. In conclusion, though the present data set does not, perhaps, warrant a more elaborate model than EE, it is attractive to have a broader model formulation along with certain fitting and choice tools to demonstrate this.

TABLE 10.2. Posterior Correlations under Model MM.

	β_0	β_1	β_2	β_3	w_{11}	w_{12}	w_{21}
β_1	0.224						
β_2	0.069	0.012					
β_3	0.094	0.303	0.065				
w_{11}	−0.448	−0.047	−0.025	−0.106			
w_{12}	0.386	0.271	−0.007	0.142	−0.687		
w_{21}	0.777	0.230	0.132	0.057	−0.035	0.264	
w_{21}	−0.739	−0.225	−0.143	−0.067	0.080	−0.275	−0.967

Source: Gelfand and Mallick (1995).

10.2 Accelerated Failure Time models

Kuo and Mallick (1997) and Walker and Mallick (1999) discuss Bayesian semiparametric approaches for accelerated failure time models. Following Walker and Mallick (1999), the accelerated failure time model assumes that failure times, $y_1, y_2, \ldots, y_n$, arise according to the probability model

$$y_i = \exp(-\boldsymbol{x}_i'\boldsymbol{\beta})v_i, \quad i = 1, 2, \ldots, n, \tag{10.2.1}$$

which in log scale becomes the linear model

$$\log y_i = -\boldsymbol{x}_i'\boldsymbol{\beta} + \theta_i, \quad i = 1, 2, \ldots, n, \tag{10.2.2}$$

where $\boldsymbol{x}_i = (x_{i1}, x_{i2}, \ldots, x_{ip})'$ is a vector of known explanatory variables for the ith individual, $\boldsymbol{\beta}$ is a vector of p unknown regression coefficients, and $\theta_i = \log v_i$ is the error term. Usually the distribution of the error term is assumed to be a member of some parametric family. This parametric assumption may be too restrictive and, thus, a Bayesian nonparametric approach is considered here to model the distribution of the error terms. Assume initially that the error terms are independently and identically distributed (i.i.d.) from some unknown distribution F. Later this condition is relaxed by introducing a *partially exchangeable* model.

A classical analysis of an accelerated failure time model is described by Kalbfleisch and Prentice (1980). A semi-Bayesian analysis of this model is given by Christensen and Johnson (1988), where they model v_i, for $i = 1, 2, \ldots, n$, as being i.i.d. from F, with F chosen from a Dirichlet process (Ferguson 1973). The Dirichlet process is popular due to the simple interpretation of its parameters; a base measure with associated precision. However, troubles arise due to the discrete nature of the Dirichlet process, which is discussed in Johnson and Christensen (1989). Even in the uncensored case, a full Bayesian analysis is very difficult, let alone in the presence of censored observations. Christensen and Johnson (1989) present a semi-Bayesian approach in the sense that they first obtain a marginal estimate for $\boldsymbol{\beta}$ after which the analysis is straightforward.

10.2.1 MDP Prior for θ_i

Kuo and Mallick (1997) consider a mixture of Dirichlet processes (MDP) as a prior for θ_i in (10.2.2), and implement the Gibbs sampler for sampling from the posterior distribution. The MDP model provides a more flexible model than a Dirichlet since it models a family of random densities (see Section 3.7.7). It smoothes the Dirichlet process with a continuous known kernel with unknown mixing weights, where prior belief can be incorporated. Moreover, the smoothing on the MDP eliminates the difficulty encountered by Christensen and Johnson (1989) of having to consider the Bayes estimates in many different subcases according to the compatibility of the data with the prior of $\boldsymbol{\beta}$. Specifically, Kuo and Mallick (1997) consider two hierarchical models. The first model, called MDPV, generalizes the Dirichlet process in Christensen and Johnson (1989) to the MDP for the distribution of $v_i = \exp(\theta_i)$. The second model, called MDP-θ, assumes an MDP prior for θ.

In MDPV, assume the random variables $v_i = \exp(\theta_i)$ are independent and identically distributed given G with the following density:

$$f(v_i|G) = \int f(v_i|\psi_i)G(d\psi_i). \tag{10.2.3}$$

Moreover, the unknown G is chosen by a Dirichlet process prior with known parameters M and G_0. Lo (1984) discusses various choices of the kernel density $f(v_i|\psi_i)$. Given G, (10.2.2) and (10.2.3), the density of $y_i = v_i \exp(-\boldsymbol{x}_i'\boldsymbol{\beta})$ is given by

$$f(y_i|\boldsymbol{\beta}, G) = f(v_i|G)\left|\frac{dv_i}{dy_i}\right| = \int \exp(\boldsymbol{x}_i'\boldsymbol{\beta})f(y_i \exp(\boldsymbol{x}_i'\boldsymbol{\beta})|\psi_i)G(d\psi_i). \tag{10.2.4}$$

For the MDP-θ model, the random variables θ_i, $i = 1, 2, \ldots, n$ are independent given G with $f(\theta_i|G)$ given in (10.2.3), where θ_i replaces v_i. In this case, we choose G_0 so that $\int f(\theta|\psi)G_0(d\psi)$ reflects the prior belief about the mean of the random densities of the error term θ, where G_0 can be supported on the whole real line. Since we are more familiar with the distribution of θ_i than v_i, this model seems to be more natural than the previous one. The random variables y_i's are independent given G and $\boldsymbol{\beta}$ with density

$$f(y_i|\boldsymbol{\beta}, G) = f(\theta_i|G)\left|\frac{d\theta_i}{dy_i}\right| = \int \frac{1}{y_i} f(\log(y_i) + \boldsymbol{x}_i'\boldsymbol{\beta}|\psi_i)G(d\psi_i). \tag{10.2.5}$$

For both models, Kuo and Mallick (1997) assume the densities of the error term or its exponential transformation are unknown, and are modeled by the family of MDP.

Next, we discuss how to sample from the posterior distribution. Following Kuo and Mallick (1997), we first consider the case of no censoring. Adapting the algorithm of Escobar (1994) for the MDPV model, we have

$y_1, y_2, \ldots, y_n$ are independent and identically distributed with density (10.2.4). The prior on G is a Dirichlet process with parameters M and G, and the prior density for $\boldsymbol{\beta}$ denoted $\pi(\boldsymbol{\beta})$ can be quite arbitrary, and may be improper. Moreover, G and $\boldsymbol{\beta}$ are independent in the prior specification. Let $\boldsymbol{y} = (y_1, y_2, \ldots, y_n)'$, and let $D = (n, \boldsymbol{y})$ denote the data. The likelihood function of G and $\boldsymbol{\beta}$ given D is

$$L(\boldsymbol{\beta}, G|D) = \prod_{i=1}^{n} \exp(\boldsymbol{x}_i'\boldsymbol{\beta}) \int f(y_i \exp(\boldsymbol{x}_i'\boldsymbol{\beta})|\psi_i) G(d\psi_i).$$

Let δ_{ψ_j} denote a measure with point mass concentrated at ψ_j. It follows from Antoniak (1974) (also see Section 3.7 of Chapter 3) and Lo (1984) that

$$G|\boldsymbol{\beta}, D \sim \int \mathcal{DP}\left(MG_0 + \sum_{j=1}^{n} \delta_{\psi_j}\right) dH(\boldsymbol{\psi}|\boldsymbol{\beta}, \boldsymbol{y}), \tag{10.2.6}$$

where $\boldsymbol{\psi} = (\psi_1, \psi_2, \ldots, \psi_n)'$, and

$$dH(\boldsymbol{\psi}|\boldsymbol{\beta}, \boldsymbol{y}) \propto \prod_{i=1}^{n} f(y_i \exp(\boldsymbol{x}_i'\boldsymbol{\beta})|\psi_i) \left(MG_0 + \sum_{j=1}^{i-1} \delta_{\psi_j}\right)(d\psi_i).$$

Note that y_i is related to v_i and $\boldsymbol{\beta}$ deterministically. Therefore, we can use $\boldsymbol{\beta}$ and $v_1, v_2, \ldots, v_n$ as the conditioning variables in the above expression instead of $\boldsymbol{\beta}$. Let $\boldsymbol{\psi}^{(-i)} = (\psi_1, \psi_2, \ldots, \psi_{i-1}, \psi_{i+1}, \psi_{i+2}, \ldots, \psi_n)'$. The conditional density of ψ_i given $(\boldsymbol{\psi}^{(-i)}, \boldsymbol{\beta}, D)$ is

$$\begin{aligned} &\pi(\psi_i|\boldsymbol{\psi}^{(-i)}, \boldsymbol{\beta}, D) \\ &= \frac{f(y_i \exp(\boldsymbol{x}_i'\boldsymbol{\beta})|\psi_i) MG_0(d\psi_i) + \sum_{j\neq i} f(y_i \exp(\boldsymbol{x}_i'\boldsymbol{\beta})|\psi_j)\delta_{\psi_j}(d\psi_i)}{a_i(\boldsymbol{\beta}) + \sum_{j\neq i} f(y_i \exp(\boldsymbol{x}_i'\boldsymbol{\beta})|\psi_j}, \end{aligned} \tag{10.2.7}$$

where

$$a_i(\boldsymbol{\beta}) = \int f(y_i \exp(\boldsymbol{x}_i'\boldsymbol{\beta})|\psi_i) MG_0(d\psi_i).$$

Moreover, the conditional density of $\boldsymbol{\beta}$ given $(\boldsymbol{\psi}, D)$ is

$$\pi(\boldsymbol{\beta}|\boldsymbol{\psi}, D) \propto \pi(\boldsymbol{\beta}) \prod_{i=1}^{n} \exp(\boldsymbol{x}_i'\boldsymbol{\beta}) f(y_i \exp(\boldsymbol{x}_i'\boldsymbol{\beta})|\psi_i). \tag{10.2.8}$$

The MCMC algorithm can now be summarized as follows.

Step 1. Generate $\boldsymbol{\psi}$ according to $dH(\boldsymbol{\psi}|\boldsymbol{\beta}, \boldsymbol{y})$. This is done by starting with $i = 1$ and selecting $\boldsymbol{\psi}$ from (10.2.7) sequentially for $i = 1, 2, \ldots, n$. Let $b_i(\boldsymbol{\beta}) = \sum_{j\neq i} f(y_i \exp(\boldsymbol{x}_i'\boldsymbol{\beta})|\psi_j)$. Then

(i) with probability $a_i(\boldsymbol{\beta})/(a_i(\boldsymbol{\beta}) + b_i(\boldsymbol{\beta}))$, generate ψ_i from the density that is proportional to $f(y_i \exp(\boldsymbol{x}_i'\boldsymbol{\beta})|\psi_i) MG_0(d\psi_i)$; and

(ii) with probability $b_i(\boldsymbol{\beta})/(a_i(\boldsymbol{\beta})+b_i(\boldsymbol{\beta}))$, generate ψ_i from the set $\{\psi_j : j \neq i\}$ using probabilities proportional to $f(y_i \exp(\boldsymbol{x}_i'\boldsymbol{\beta})|\psi_j)$.

Step 2. Generate $\boldsymbol{\beta}$ given $(\boldsymbol{\psi}, D)$ from (10.2.8) by the Metropolis algorithm. This can be done sequentially for β_l, $l = 1, 2, \ldots, p$, from the density $\pi(\beta_l|\boldsymbol{\beta}^{(-l)}, \boldsymbol{\psi}, D)$, where $\boldsymbol{\beta}^{(-l)} = (\beta_1, \beta_2, \ldots, \beta_{l-1}, \beta_{l+1}, \ldots, \beta_p)'$.

Step 3. Cycle through Steps 1 and 2 until convergence. Moreover, we can also construct multiple chains by replicating the iterations using independent random starting values.

A more efficient algorithm can be obtained by means of the idea of MacEachern (1994), where updating the $\boldsymbol{\psi}$'s is done in clusters. Let $\phi_1^{(-i)}, \phi_2^{(-i)}, \ldots, \phi_{K^{(-i)}}^{(-i)}$ denote the values of the distinct ψ_j's in $\boldsymbol{\psi}^{(-i)}$ with multiplicities $n_1^{(-i)}, n_2^{(-i)}, \ldots, n_{K^{(-i)}}^{(-i)}$. Then, observe

$$\pi(\psi_i|\boldsymbol{\psi}^{(-i)}, \boldsymbol{\beta}, D) = \Bigg[f(y_i \exp(\boldsymbol{x}_i'\boldsymbol{\beta})|\psi_i) M G_0(d\psi_i) + \sum_{j=1}^{K^{(-i)}} n_j^{(-i)} f(y_i \exp(\boldsymbol{x}_i'\boldsymbol{\beta})|\phi_j^{(-i)}) \delta_{\phi_j^{(-i)}}(d\psi_i) \Bigg] \times \left[a_i(\boldsymbol{\beta}) + \sum_{j=1}^{K^{(-i)}} n_j^{(-i)} f(y_i \exp(\boldsymbol{x}_i'\boldsymbol{\beta})|\phi_j^{(-i)}) \right]^{-1}. \tag{10.2.9}$$

This suggests that we can define a configuration index $\boldsymbol{s} = (s_1, s_2, \ldots, s_n)'$, where $s_i = j$ if $\psi_i = \phi_j^{(-i)}$ for $j = 1, 2, \ldots, K^{(-i)}$, and $s_i = 0$ if ψ_i is a new draw from the density proportional to $f(y_i \exp(\boldsymbol{x}_i'\boldsymbol{\beta})|\psi_i) M G_0(d\psi_i)$. From (10.2.9), we can define

$$P(s_i = j|\boldsymbol{\psi}^{(-i)}, \boldsymbol{s}^{(-i)}, K^{(-i)}, D) = q_{ij},$$

where

$$q_{ij} = \begin{cases} c a_i(\boldsymbol{\beta}) & \text{if } j = 0, \\ c n_j^{(-i)} f(y_i \exp(\boldsymbol{x}_i'\boldsymbol{\beta})|\phi_j^{(-i)}) & \text{if } j > 0, \end{cases} \tag{10.2.10}$$

with c being a normalizing constant, and $\boldsymbol{s}^{(-i)}$ is $\boldsymbol{s}$ with s_i deleted. The new $\boldsymbol{s}$ not only determines the structure of the clusters, but also determines a new K that is the number of distinct values on $\boldsymbol{\psi}$. Let $I_j = \{i : s_i = j\}$ denote the set of indices of observations in group j, where $j \geq 1$. Then we have

$$\pi(\boldsymbol{\phi}|\boldsymbol{s}, K, D) \propto \prod_{i \in I_j} f(y_i \exp(\boldsymbol{x}_i'\boldsymbol{\beta})|\psi_i) G_0(d\psi_i). \tag{10.2.11}$$

Therefore, the improved algorithm consists of replacing Step 1 with Steps 1a and 1b:

Step 1a. Generate a new $\boldsymbol{s}$ given an old $\boldsymbol{\psi}$ ($\boldsymbol{\phi}$) using (10.2.10), for each $i = 1, 2, \ldots, n$. The conditioning variables are held fixed for all $i = 1, 2, \ldots, n$. That is, the new $\boldsymbol{s}$ is updated from the old $\boldsymbol{\phi}$. It also determines a new K, the total number of distinct groups.

Step 1b. Given $\boldsymbol{s}$ and K, generate a new set of $\boldsymbol{\phi}$'s as follows. If $s_i = 0$, then draw a new $\boldsymbol{\phi}$ with a density proportional to $f(y_i \exp(\boldsymbol{x}_i'\boldsymbol{\beta})|\psi_i) M G_0(d\psi_i)$, if $s_i = j$, then draw a new ϕ_j from the density that is proportional to (10.2.11). We only need to draw one ϕ_j for all the indices in group I_j. Having exhausted all of the indices, then reconstruct $\boldsymbol{\psi}$ using the updated $\boldsymbol{\phi}$ and $\boldsymbol{s}$.

For the MDPV model, Kuo and Mallick (1997) follow a similar procedure to Steps 1–3, where $f(y_i \exp(\boldsymbol{x}_i'\boldsymbol{\beta})|\psi_i)$ is replaced by $f(\log(y_i) + \boldsymbol{x}_i'\boldsymbol{\beta}|\psi_i)$ and (10.2.8) is replaced by

$$\pi(\boldsymbol{\beta}|\boldsymbol{\psi}, D) \propto \pi(\boldsymbol{\beta}) \prod_{i=1}^{n} f(\log(y_i) + \boldsymbol{x}_i'\boldsymbol{\beta}|\psi_i). \tag{10.2.12}$$

Kuo and Mallick (1997) also discuss how to accommodate right censored data. In addition to the n completely observed survival times, suppose there are m right censored survival times $(y_{n+1}, y_{n+2}, \ldots, y_{n+m})$. Let $\boldsymbol{y} = (y_1, y_2, \ldots, y_n, y_{n+1}, \ldots, y_{n+m})'$. In the MCMC algorithm, they use data augmentation, where they generate the latent variable $\boldsymbol{z} = (z_{n+1}, z_{n+2}, \ldots, z_{n+m})'$ by a rejection algorithm from the truncated density $f(z_{n+1}, z_{n+2}, \ldots, z_{n+m}|\boldsymbol{\beta}, \boldsymbol{\psi}, \boldsymbol{y})$, where $\boldsymbol{\psi}$ is an $n + m$ dimensional vector. Then the complete dataset is $D = (y_1, y_2, \ldots, y_n, z_{n+1}, \ldots, z_{n+m})$. Then, the previous algorithm with size $n + m$ can be applied to generate $\boldsymbol{\psi} = (\psi_1, \psi_2, \ldots, \psi_{n+m})$ given $(\boldsymbol{\beta}, \boldsymbol{y})$ and to generate $\boldsymbol{\beta}$ given $(\boldsymbol{\psi}, \boldsymbol{y})$.

The Gibbs algorithm can be summarized as follows:

Step 0. Generate the complete dataset $D = (y_1, y_2, \ldots, y_n, z_{n+1}, \ldots, z_{n+m})$ given $(\boldsymbol{\psi}, \boldsymbol{\beta}, \boldsymbol{y})$. For each $i = n + 1, \ldots, n + m$, generate z_i given $(\boldsymbol{\psi}, \boldsymbol{\beta}, \boldsymbol{y})$. First generate ψ_i^* from the predictive density of a new ψ_i given the old $\boldsymbol{\psi}$. This is done by generating a new ψ from G_0 with probability $M/(M + n + m)$ and uniformly from the set $\{\psi_1, \psi_2, \ldots, \psi_{n+m}\}$ with probability $(n + m)/(M + n + m)$. Then generate v_i given ψ_i^* according to $f(v_i|\psi_i^*)$ and compute $d_i = v_i \exp(-\boldsymbol{x}_i'\boldsymbol{\beta})$. Let $z_i = d_i$ if $d_i > y_i$; otherwise repeat this process until $d_i > y_i$.

Step 1. Given $D = (y_1, y_2, \ldots, y_n, z_{n+1}, \ldots, z_{n+m})$ generated in Step 0 and $\boldsymbol{\beta}$, apply Step 1 of the previous MCMC algorithm without censoring with size $n + m$ for $i = 1, 2, \ldots n + m$ to generate $\boldsymbol{\psi}$ of size $n + m$.

Step 2. Given ψ and D, apply Step 2 of the previous MCMC algorithm without censoring to generate $\boldsymbol{\beta}$, where the product in (10.2.8) has $n + m$ terms with z_i replacing y_i for $i = n + 1, \ldots, n + m$.

Step 3. Cycle through Steps 0, 1, and 2 until convergence.

For the MDPV model, a similar procedure to Steps 0–3 can be followed, with the kernel changed as discussed earlier; and, in Step 0, v_i is changed to θ_i and $d_i = \exp(-\boldsymbol{x}_i'\boldsymbol{\beta} + \theta_i)$.

10.2.2 Polya Tree Prior for θ_i

Instead of the Dirichlet process, Walker and Mallick (1999) propose a Polya tree distribution (Mauldin, Sudderth, and Williams, 1992; Lavine, 1992) as the prior for the unknown distribution of θ_i. The advantages over the Dirichlet process are:

(i) the conjugate nature of Polya trees make the analysis uncomplicated;

(ii) under some sufficient conditions Polya tree priors assign probability 1 to the set of continuous distributions;

(iii) it is easy to constrain a random Polya tree distribution to have median zero and hence to consider a median regression model given in (10.2.2) (see Ying, Jung, and Wei, 1995, for a frequentist version);

(iv) it is easy to sample a random Polya tree distribution so samples from posterior functionals of the survival curve will be available, permitting a full Bayesian analysis; and

(v) the Dirichlet process and the beta process (Hjort, 1990) are a special case of Polya trees.

Polya tree distributions for random probability measures were recently studied by Lavine (1992, 1994) and Mauldin, Sudderth, and Williams (1992), although an original description was introduced by Ferguson (1974). Walker and Mallick (1996) discuss Polya tree priors for the frailty term in frailty models. Let Ω be a separable measurable space (though only $\Omega = (-\infty, \infty)$ will be considered in this subsection) and let (B_0, B_1) be obtained by splitting Ω into two pieces. Similarly, B_0 splits into (B_{00}, B_{01}) and B_1 splits into (B_{10}, B_{11}). Continue in this fashion *ad infinitum*. Let, for some m, $\epsilon = \epsilon_1\epsilon_2 \ldots \epsilon_m$ with $\epsilon_k \in \{0, 1\}$ for $k = 1, 2, \ldots, m$, so that each ϵ defines a unique set B_ϵ. The number of sets at the mth level is 2^m. Thus, in general, B_ϵ splits into $B_{\epsilon 0}$ and $B_{\epsilon 1}$. Degenerate splits are allowed so that we could have $B_{\epsilon 0} = B_\epsilon$ and $B_{\epsilon 1} = \emptyset$. Imagine a particle cascading through these partitions. It starts in Ω and moves into B_0 with probability C_0, or B_1 with probability $1 - C_0$. In general, on entering B_ϵ, the particle could either move into $B_{\epsilon 0}$ or $B_{\epsilon 1}$. Let it move into the former with probability $C_{\epsilon 0}$ and into the latter with probability $C_{\epsilon 1} = 1 - C_{\epsilon 0}$. For Polya trees these probabilities are random and are beta variables, that is, $C_{\epsilon 0} \sim \mathcal{B}(\alpha_{\epsilon 0}, \alpha_{\epsilon 1})$ with nonnegative $\alpha_{\epsilon 0}$ and $\alpha_{\epsilon 1}$, where $\mathcal{B}(a, b)$ denotes the beta distribution with parameters (a, b). A degenerate beta distribution is allowed, that is, we can take one of $\alpha_{\epsilon 0}$ or $\alpha_{\epsilon 1}$ to be zero. Let $\Pi = (B_0, B_1, B_{00}, \dots)$ and $\mathcal{A} = (\alpha_0, \alpha_1, \alpha_{00}, \dots)$, and so in order to define a particular Polya tree distribution, we need to define Π and $\mathcal{A}$.

According to Lavine (1992), the Polya tree distribution is formally defined as follows. A random probability measure F on Ω is said to have a Polya tree distribution, or a Polya tree prior, with parameter $(\Pi, \mathcal{A})$, written $F \sim \mathcal{PT}(\Pi, \mathcal{A})$, if there exist nonnegative numbers $\mathcal{A} = (\alpha_0, \alpha_1, \alpha_{00}, \dots)$ and random variables $\mathcal{C} = (C_0, C_{00}, C_{10}, \dots)$ such that the following hold:

(i) all the random variables in $\mathcal{C}$ are independent;

(ii) for every ϵ, $C_{\epsilon 0} \sim \mathcal{B}(\alpha_{\epsilon 0}, \alpha_{\epsilon 1})$; and

(iii) for every $m = 1, 2, \dots$ and every $\epsilon = \epsilon_1 \epsilon_2 \dots \epsilon_m$,

$$F(B_{\epsilon_1 \epsilon_2 \dots \epsilon_m}) = \left(\prod_{j=1;\epsilon_j=0}^{m} C_{\epsilon_1 \dots \epsilon_{j-1} 0} \right) \left(\prod_{j=1;\epsilon_j=1}^{m} (1 - C_{\epsilon_1 \dots \epsilon_{j-1} 0}) \right), \tag{10.2.13}$$

where the first terms, i.e., for $j = 1$, are interpreted as C_0 and $1 - C_0$.

A random probability measure $F \sim \mathcal{PT}(\Pi, \mathcal{A})$ is sampled by sampling $\mathcal{C}$ as indicated in the definition of the Polya tree distribution. Since $\mathcal{C}$ is an infinite set, an approximate probability measure from $\mathcal{PT}(\Pi, \mathcal{A})$ is sampled by terminating the process at a finite level M. Lavine (1992) refers to this as a "partially specified Polya tree." Let this finite set be denoted by $\mathcal{C}_M$. From the sampled variates of $\mathcal{C}_M$ we define $F(B_{\epsilon_1 \epsilon_2 \dots \epsilon_M})$ for each $\epsilon = \epsilon_1 \epsilon_2 \dots \epsilon_M$ according to (10.2.13). Walker and Mallick (1999) use the $\kappa = 2^M$ partitions given by $\pi_M = (B_1, B_2, \dots, B_\kappa)$, where π_M is the collection of partitions at level M. If θ_i is an observation from F, then we are only interested in which $B \in \pi_M$ that θ_i is observed. For predictive inference, the κ probabilities $P(\theta_i \in B)$ for each $B \in \pi_M$ are of interest. Assessing the "closeness" of a random Polya tree distribution from a partially specified Polya tree to an exact Polya tree is given in Lavine (1992), where a criterion for the selection of M is given.

The Polya tree prior is centered on a particular probability distribution G by taking the partitions to coincide with percentiles of G and then taking $\alpha_{\epsilon 0} = \alpha_{\epsilon 1}$ for each ϵ. This involves setting $B_0 = (-\infty, G^{-1}(1/2))$, $B_1 = [G^{-1}(1/2), \infty)$, and at level m, setting, for $j = 1, 2, \dots, 2^m$, $B_j = [G^{-1}((j-1)/2^m), G^{-1}(j/2^m))$, with $G^{-1}(0) = -\infty$ and $G^{-1}(1) = +\infty$, where $(B_j; j = 1, 2, \dots, 2^m)$ correspond in order to the 2^m partitions of level m. It is then straightforward to show that $E[F(B_\epsilon)] = G(B_\epsilon)$ for all ϵ. Note here that G defines Π. Therefore, G, $\mathcal{A}$, and M for such a Polya tree need to be specified, and the choices of these parameters are discussed below.

Given an observation θ_1 from F, the posterior Polya tree distribution is easily obtained (Lavine, 1992). Following Walker and Mallick (1999), we write $F|\theta_1 \sim \mathcal{PT}(\Pi, \mathcal{A}|\theta_1)$ and with $\mathcal{A}|\theta_1$ given by

$$\alpha_\epsilon|\theta_1 = \begin{cases} \alpha_\epsilon + 1 & \text{if } \theta_1 \in B_\epsilon, \\ \alpha_\epsilon & \text{otherwise.} \end{cases}$$

For n independent observations given by $\boldsymbol{\theta} = (\theta_1, \theta_2, \ldots, \theta_n)'$, then $\mathcal{A}|\boldsymbol{\theta}$ is given by $\alpha_\epsilon|\boldsymbol{\theta} = \alpha_\epsilon + n_\epsilon$, where n_ϵ is the number of observations from $(\theta_1, \theta_2, \ldots, \theta_n)$ in B_ϵ. It is easy to obtain samples from $P(F(B_\epsilon)|\boldsymbol{\theta})$ by sampling $\mathcal{C}_M$ and using (10.2.13). This permits a full Bayesian analysis.

In many applications, it is the posterior predictive distribution for the next observation that is of interest, i.e., $P(\theta_{n+1} \in B_\epsilon|\boldsymbol{\theta})$ for some ϵ. Let $\epsilon = \epsilon_1\epsilon_2 \ldots \epsilon_M$; then

$$\begin{aligned} P\Big(\theta_{n+1} \in B_\epsilon|\boldsymbol{\theta}\Big) &= E\Big[F(B_\epsilon)|\boldsymbol{\theta}\Big] \\ &= \frac{\alpha_{\epsilon_1} + n_{\epsilon_1}}{\alpha_0 + \alpha_1 + n} \frac{\alpha_{\epsilon_1\epsilon_2} + n_{\epsilon_1\epsilon_2}}{\alpha_{\epsilon_1 0} + \alpha_{\epsilon_1 1} + n_{\epsilon_1}} \cdots \\ &\quad \times \frac{\alpha_{\epsilon_1\ldots\epsilon_m} + n_{\epsilon_1\ldots\epsilon_M}}{\alpha_{\epsilon_1\ldots\epsilon_{M-1}0} + \alpha_{\epsilon_1\ldots\epsilon_{M-1}1} + n_{\epsilon_1\ldots\epsilon_{M-1}}}. \end{aligned} \tag{10.2.14}$$

This result is straightforward to derive from the definition of the Polya tree distribution.

Walker and Mallick (1999) take G to be a normal distribution with mean zero. This is usually the choice for an error distribution in a parametric framework. For the choice of $\mathcal{A}$, as Walker and Mallick (1999) point out, a separate α_ϵ is not assigned for each ϵ in practice. It is convenient therefore to take $\alpha_\epsilon = c_m$ whenever ϵ defines a set at level m. For the higher levels (m small) it is not necessary for $F(B_{\epsilon 0})$ and $F(B_{\epsilon 1})$ to be "close" – on the contrary – it is desirable for a large amount of variability. However, as we move down the levels (m large), we wish $F(B_{\epsilon 0})$ and $F(B_{\epsilon 1})$ to be close, to reflect beliefs in the underlying continuity of F. This can be achieved by allowing c_m to be small for small m and allowing c_m to increase as m increases, for example, $c_m = cm^2$ for some $c > 0$. According to Ferguson (1974), $c_m = m^2$ implies F is absolutely continuous with probability 1 and therefore according to Lavine (1992), this "would often be a sensible canonical choice." Therefore, Walker and Mallick (1999) choose cm^2 for the α's and consider a prior for c. Note the Dirichlet process arises when $c_m = c/2^m$, which means that $c_m \to 0$ as $m \to \infty$ and F is discrete with probability 1 (Blackwell, 1973).

For M, Walker and Mallick (1999) recommend either to fix M or to assign it a prior distribution, for example, a Poisson distribution. This is not difficult to do. However, fixing M such that the partitions around zero were adequately "small" (in the sense that given an observation in one of these partitions it need not to be located any more precisely) is satisfactory. If observations do occur in larger partitions, then M can be increased.

Next, we discuss how to sample from the posterior distributions of $\boldsymbol{\beta}$ and F for both censored and uncensored observations, and derive the predictive survival curve for a patient with a particular set of covariate values. Consider the linear model as described in (10.2.2). For the semiparametric analysis of the model, F is assigned a Polya tree prior and the prior for the parameter $\boldsymbol{\beta}$ is taken to be a p dimensional multivariate normal distribution with mean $\boldsymbol{\mu}$ and covariance matrix Σ. *A priori* F and $\boldsymbol{\beta}$ are assumed to be independent.

If F is allowed to be completely arbitrary, then a source of confounding is the intercept term of $\boldsymbol{\beta}$ with the location of F. To remedy this problem, Walker and Mallick (1999) fix the median of F at $G^{-1}(1/2)$ by defining $F(B_0) = F(B_1) = 1/2$. This does not affect the centering of F. If G is normal with median at 0, then F has its median at 0 almost surely. Here (10.2.2) becomes a median regression model with $\text{med}(\log(y_i)) = -\boldsymbol{x}_i'\boldsymbol{\beta}$ instead of the more usual mean regression model. Newton, Czado, and Chappell (1996) describe a similar procedure for the Dirichlet process. Frequentist semiparametric inference for the median regression model is described by Ying, Jung, and Wei (1995). They do not assume that the error terms are i.i.d. As a consequence their approach only obtains an estimate for $\boldsymbol{\beta}$ and hence only a predictive median survival time. The approach considered by Walker and Mallick (1999) initially assumes exchangeable errors, which allows us to obtain an estimate of $\boldsymbol{\beta}$, an estimate of the predictive survival curve, and the associated uncertainty with this prediction. In the latter part of this subsection, we will allow the restriction on the exchangeable error terms to be relaxed in order to include partially exchangeable error term structures.

There is interest in obtaining the posterior distributions of $\boldsymbol{\beta}$ and F which will then lead to a predictive distribution for the survival curve. This is analytically intractable and hence MCMC sampling is utilized in order to obtain samples from the relevant posterior distributions. Following Walker and Mallick (1999), we first consider the case of no censoring. In this case, the data is represented as $D = (n, \boldsymbol{y}, X)$, where $\boldsymbol{y} = (y_1, y_2, \dots, y_n)'$, all the y_i's are failure times, and X is the $n \times p$ matrix of covariates with i^{th} row $\boldsymbol{x}_i'$. To sample F, as discussed earlier, we consider the distributions for F up to a finite level M for which there will be the partitions $\pi_M = \{B_j : j = 1, 2, \dots, \kappa\}$, where $\kappa = 2^M$. The parameter θ is drawn from F so that the likelihood of θ is given by $F(B_\theta)$, for some $B_\theta \in \pi_M$, where $\theta \in B_\theta$. Likewise as y is an observation, given $\boldsymbol{\beta}$, from an individual with covariate vector $\boldsymbol{x}$, the likelihood of $\boldsymbol{\beta}$ is again given by $F(B_{\boldsymbol{\beta}})$, where $\theta = \log(y) + \boldsymbol{x}'\boldsymbol{\beta}$.

Obtaining the posterior distribution of $f(F|\boldsymbol{\beta}, D)$ for the Polya tree is straightforward, since $\theta_i = \log(y_i) + \boldsymbol{x}_i'\boldsymbol{\beta}$ is an i.i.d. observation from F, for each $i = 1, 2, \ldots, n$. This involves the sampling of a random probability measure F, given the data and the current value of $\boldsymbol{\beta}$ from the Metropolis-Hastings algorithm, which is obtained by sampling $\mathcal{C}_M$ (see the definition of the Polya tree distribution) for which we will obtain the random probabilities $\{u_j = F(B_j) : j = 1, 2, \ldots, \kappa\}$.

To sample $\boldsymbol{\beta}$, we first observe that

$$\pi(\boldsymbol{\beta}|F, D) \propto L(\boldsymbol{\beta}, F|D)\pi(\boldsymbol{\beta}),$$

where $\pi(\boldsymbol{\beta})$ is the prior for $\boldsymbol{\beta}$, and the likelihood $L(\boldsymbol{\beta}, F|D) \propto \prod_{i=1}^n F(B_{\theta_i})$. A Metropolis-Hastings algorithm (see Section 1.6) is used to obtain the required sample.

At the lth iteration of the Markov chain, we obtain the samples $\{u_{jl} : j = 1, 2, \ldots, \kappa\}$ and $\boldsymbol{\beta}_l$. After a suitable burn-in period, for convergence to be obtained, the samples are kept, say, for $l = 1, 2, \ldots, L$, and from which posterior summaries are obtained, for example,

$$E(F(B_j)|D) = L^{-1}\sum_{l=1}^{L} u_{jl} \quad \text{and} \quad E(\boldsymbol{\beta}|D) = L^{-1}\sum_{l=1}^{L} \boldsymbol{\beta}_l.$$

For censored observations the extra step required in the MCMC algorithm is the sampling of the missing data corresponding to those censored observations. In this case, we let $D = (n, \boldsymbol{y}, X, \boldsymbol{\nu})$ denote the observed data, where $\boldsymbol{y} = (y_1, y_2, \ldots, y_n)'$, X is the matrix of covariates as defined as earlier, and $\boldsymbol{\nu} = (\nu_1, \nu_2, \ldots, \nu_n)'$ is the vector of censoring indicators so that $\nu_i = 1$ if y_i is a failure time, and 0 if censored. Assume without loss of generality that $\nu_i = 0$ for $i = 1, 2, \ldots, i'$ and that $\nu_i = 1$ for $i = i' + 1, \ldots, n$ for some i'. Let z_i denote the true failure time for the ith case for $i = 1, 2, \ldots, i'$. The required extra steps are the sampling of random variates from

$$f(z_i|\boldsymbol{x}_i, \boldsymbol{\beta}, F, D^{(-i)}, z_i \geq y_i) = f(z_i|\boldsymbol{x}_i, \boldsymbol{\beta}, F, z_i \geq y_i) \qquad (10.2.15)$$

for $i = 1, 2, \ldots, i'$, where $D^{(-i)}$ is the data D with the ith observation deleted.

Let the partition points of the Polya tree structure Π at level M be $\{a_j : j = 1, 2, \ldots, \kappa - 1\}$. For each $i = 1, 2, \ldots, i'$, select the $a_{j(i)}$ closest to and below $\log(y_i) + \boldsymbol{x}_i'\boldsymbol{\beta}$ and let F_i^* represent the probability measure F restricted to $[a_{j(i)}, \infty)$. If F_i^* is defined by the probabilities u_{ij}^* for $j = j(i), \ldots, \kappa$, then $u_{ij}^* = u_j/u_{(i)}$, where $u_{(i)} = \sum_{j=j(i)}^{\kappa} u_j$. Then θ_i is taken from F_i^* by sampling a set B from $\{B_j : j = j(i), \ldots, \kappa\}$ with corresponding probabilities $\{u_{ij}^* : j = j(i), \ldots, \kappa\}$, and then θ_i is taken uniformly from B. If B is one of the outer partitions, then it is necessary for M to be increased until all of these B's are not outer partitions. Finally, take $\log(z_i) = \theta_i - \boldsymbol{x}_i\boldsymbol{\beta}$ to be the random variate from (10.2.15). Additionally, at the lth iteration of the chain, the samples $\{z_{il} : i = 1, 2, \ldots, i'\}$ are obtained.

To derive the predictive survival curve, i.e., $S(t) = P(y > t|\boldsymbol{x}, D) = 1 - P(y \leq t|\boldsymbol{x}, D)$, for a new patient with covariate vector $\boldsymbol{x}$, Walker and Mallick (1999) first consider the case in which all of the data is uncensored. Since

$$1 - S(t) = P(y \leq t|\boldsymbol{x}, D) = \int P(y \leq t|\boldsymbol{x}, \boldsymbol{\beta}, D)\pi(\boldsymbol{\beta}|D)d\boldsymbol{\beta},$$

$1 - S(t)$ can be approximated, using the samples $\{\boldsymbol{\beta}_l\}$ obtained from the simulated Markov chain, by

$$L^{-1}\sum_{l=1}^{L} P(y \leq t|\boldsymbol{x}, \boldsymbol{\beta}_l, D).$$

Now $P(y \leq t|\boldsymbol{x}, \boldsymbol{\beta}_l, D) = P(\theta \leq \log(t) + \boldsymbol{x}'\boldsymbol{\beta}_l|\boldsymbol{\beta}_l, D)$, which is given by

$$\sum_{j=1}^{\eta} \lambda_j(\boldsymbol{\beta}_l),$$

where $\eta = j(\log(t) + \boldsymbol{x}'\boldsymbol{\beta}_l)$, $\lambda_j(\boldsymbol{\beta}_l) = E(F(B_j)|\theta_{1l}, \theta_{2l}, \ldots, \theta_{nl})$, $\theta_{il} = \log(y_i) + \boldsymbol{x}_i'\boldsymbol{\beta}_l$, and $j(s) \in (1, 2, \cdots, \kappa)$ is such that $s \in B_{j(s)}$. Each λ is evaluated from (10.2.14). The predictive survival curve is then given by

$$\widehat{S}(t) = 1 - \frac{1}{L}\sum_{l=1}^{L}\sum_{j=1}^{\eta} \lambda_j(\boldsymbol{\beta}_l). \tag{10.2.16}$$

For censored data it is straightforward to show that the predictive survival curve is also given by (10.2.16) except for $\lambda_j(\boldsymbol{\beta}_l) = E(F(B_j)|\theta_{1l}, \theta_{2l}, \ldots, \theta_{nl})$, where now $\theta_{il} = \log(z_{il}) + \boldsymbol{x}_i'\boldsymbol{\beta}_l$, for $i = 1, 2, \ldots, i'$, and $\theta_{il} = \log(y_i) + \boldsymbol{x}_i'\boldsymbol{\beta}_l$, for $i = i' + 1, \ldots, n$. It is straightforward to evaluate (10.2.16) with no extra computation since the $\lambda_j(\boldsymbol{\beta}_l)$ can be obtained at each iteration of the chain.

Next, we consider two examples given in Walker and Mallick (1999). For both examples, a noninformative hierarchical normal prior for $\boldsymbol{\beta}$ was used (see Wakefield et al., 1994). For the Polya tree prior, the prior expected probability measure G was taken to be normal with mean 0 and standard deviation of 10, and then $\mathcal{A}$ was taken such that at level m, all of the α's are equal to $0.1 \times m^2$. The MCMC algorithm was first run in order to ascertain at which point suitable convergence was obtained and then a further chain was run in order to obtain 3000 samples for inference. M was taken to be 8 so that the number of partitions considered is 2^8.

Example 10.2. Uncensored leukemia data. Walker and Mallick (1999) considered the dataset presented by Feigl and Zelen (1965), which involves

33 patients suffering from leukemia. Each patient's covariate vector $\boldsymbol{x}_i$ is given by $(x_{i1}, x_{i2}, x_{i3})'$, where $x_{i1} = 1$, x_{i2} is an indicator function which is zero if the patient's AG-factor is positive and one if the patient's AG-factor is negative, and x_{i3} is the natural logarithm of the patient's white blood cell count. The uncensored observations, $\{y_i\}$, are taken to be the patient's survival times in weeks.

The Bayes estimates of $\boldsymbol{\beta}$ and the 95% credible intervals are -5.26 and $(-6.7, -3.69)$ for β_1, 1.46 and $(-0.61, 3.07)$ for β_2, and 0.64 and $(0.31, 0.98)$ for β_3. The estimated failure times for the five patients alongside the true values are given in Table 10.3.

TABLE 10.3. Estimated and True Failure Times.

Patient	Estimated	True
1	60.3	65.0
2	164.0	156.0
3	99.5	100.0
4	134.3	134.0
5	38.9	16.0

Source: Walker and Mallick (1999).

Example 10.3. Small cell lung cancer data. For a censored data example, Walker and Mallick (1999) used the data set given in Ying, Jung, and Wei (1995). This data set involves 121 patients suffering small cell lung cancer and each undertaking one of two treatments, Arm A which had 62 patients, or Arm B which had 59 patients. The survival times are given in days with 98 patients providing exact survival times with the remainder providing right censored survival times. The covariates are the treatment type, 0 or 1, and the natural logarithm of the entry age of the patient.

The posterior means of $\boldsymbol{\beta}$ and 95% credible intervals are -7.57 and $(-8.92, -6.87)$ for β_1 (intercept), 0.30 and $(0.08, 0.91)$ for β_2 (treatment), and 0.31 and $(0.07, 0.78)$ for β_3 (the logarithm of the entry age). The results agree with those of Ying, Jung, and Wei (1995).

Finally, we briefly discuss a partially exchangeable model proposed by Walker and Mallick (1999). In (10.2.2), it is assumed that, given F, all the error terms are i.i.d. F. In some instances, this assumption may be too strong, in particular, when one suspects that the individuals may have varying degrees of similarity or likeness. A more flexible approach, which allows individuals to be thought of as being more alike and unlike, consists of entertaining several *partial exchangeability structures* for the error terms $\{\theta_1, \theta_2, \ldots, \theta_n\}$, and then combining the corresponding inferences by summarizing over the posterior beliefs in these structures.

Briefly the method can be thought of as partitioning the experimental set, $\{i = 1, 2, \ldots, n\}$, to give J subsets denoted by $S_1, S_2, \ldots, S_J$. The basic assumption is to regard as exchangeable only the θ_is associated with individuals belonging to the same partition subset S_j, while the θ_i's associated with individuals in different subsets are taken to be independent. Typically, there will be several partitions g, out of all possible partitions $\mathcal{G}$, whose relative plausibility is described by a prior probability mass function $f(g)$. The final inference on $\{\theta_1, \theta_2, \ldots, \theta_n\}$ will be a mixture over the posterior distribution $\pi(g|D)$ of the inferences on $\boldsymbol{\theta}$ given g. For example, if $n = 4$, two possible partitions are $g_1 = [(1,2,3,4)]$ and $g_2 = [(1,2),(3,4)]$. Clearly $J(g_1) = 1$ with $S_1(g_1) = (1,2,3,4)$. Similarly, $J(g_2) = 2$ with $S_1(g_2) = (1,2)$, $S_2(g_2) = (3,4)$.

The advantage of the analysis lies essentially in its ability to borrow strength from related individuals without imposing a prespecified dependence structure on the error terms.

Survival data typically arises as a result of a study comparing two treatment types which would have the effect of splitting the patient population undergoing therapy into two groups, see, for example, the small cell lung cancer data of Ying, Jung, and Wei (1995). Therefore, an alternative model to (10.2.2) is the *partially exchangeable* model in which it is assumed that the individuals on one of the treatments have error terms which are i.i.d. F_1 and the rest have error terms which are i.i.d. F_2. Therefore, instead of the treatment types causing a median shift in the distribution of the error terms, a possible distribution shift is now being allowed (which is shifting the medians too). The assumption on F_1 and F_2 is that they are i.i.d. from a Polya tree prior distribution. Represent the model given by (10.2.2) as $\mathcal{M}_0$ and the new model by $\mathcal{M}_1$.

These two models can be thought of as competing models. However, instead of selecting one of them as being optimal in some sense, Walker and Mallick (1999) follow Draper (1995) and assign prior weights to the two models to reflect prior opinion concerning their relative plausibility. Inference will then be made by averaging over the posterior weights.

Predictive probabilities are taken as a mixture of the predictive probabilities obtained from each model;

$$\begin{aligned} P(y > t|\boldsymbol{x}, D) =& P(y > t|\boldsymbol{x}, \mathcal{M}_0, D)p(\mathcal{M}_0|D) \\ &+ P(y > t|\boldsymbol{x}, \mathcal{M}_1, D)p(\mathcal{M}_1|D). \end{aligned}$$

Therefore, two analyses are performed under $\mathcal{M}_0$ and $\mathcal{M}_1$. For the final inference, the posterior weights $p(\mathcal{M}_0|D)$ and $p(\mathcal{M}_1|D)$ need to be evaluated. From Bayes theorem, we have

$$p(\mathcal{M}_j|D) = \frac{p(D|\mathcal{M}_j)}{\sum_{i=0}^{1} p(D|\mathcal{M}_i)p(\mathcal{M}_i)} \tag{10.2.17}$$

for $j = 0, 1$, where $p(\mathcal{M}_j)$ is the prior model probability. In (10.2.17), $p(D|\mathcal{M}_j)$ is the marginal distribution of the data D, which is given by

$$p(D|\mathcal{M}_j) = \int L(\boldsymbol{\psi}_j|D, \mathcal{M}_j)\pi(\boldsymbol{\psi}_j|\mathcal{M}_j)\, d\boldsymbol{\psi}_j,$$

where $\boldsymbol{\psi}_j$ is the vector of parameters, $L(\boldsymbol{\psi}_j|D, \mathcal{M}_j)$ is the likelihood, and $\pi(\boldsymbol{\psi}_j|\mathcal{M}_j)$ is the prior under $\mathcal{M}_j$ for $j = 0, 1$. The marginal distribution of the data D, $p(D|\mathcal{M}_j)$, can be easily approximated by using Monte Carlo methods (see, for example, Chen, Shao, Ibrahim, 2000, or Chapter 6).

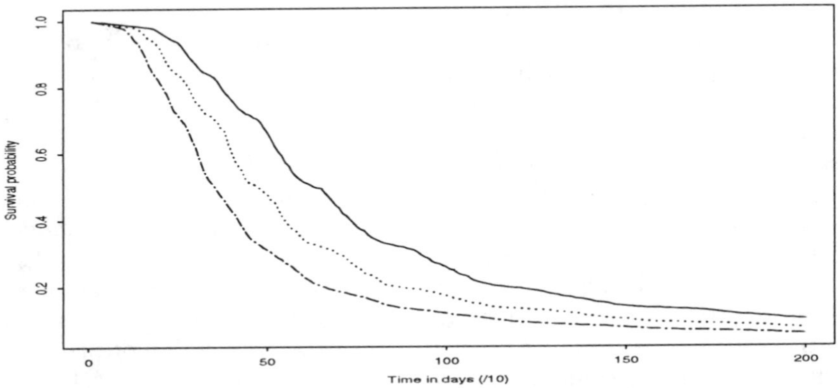

FIGURE 10.1. Predictive survival curves for three new patients with treatment Arm A (exchangeable model).

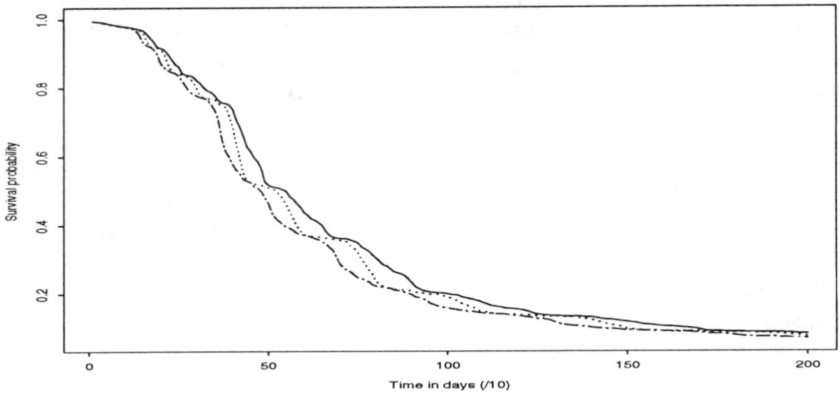

FIGURE 10.2. Predictive survival curves for three new patients with treatment Arm A (partially exchangeable model).

The censored data example (Example 10.3) was re-run with the new mixture model with $\mathcal{M}_1$ being the partially exchangeable model partitioned into the two treatment groups. The prior weights for each of the models was 1/2 and the posterior weights obtained were 0.19 for $\mathcal{M}_0$ and 0.81 for $\mathcal{M}_1$. This suggests that the data "prefers" the *distribution shift* as opposed to the *median shift* for modeling the effect of the treatment types. Figures 10.1 and 10.2 are plots of the predictive survival curves for three new patients with Arm A treatment under the exchangeable model and the partially exchangeable model, respectively. From Figures 10.1 and 10.2, it is easy to see that the three predictive survival curves under the partially exchangeable model are much closer to each other than under the exchangeable model as expected.

10.3 Bayesian Survival Analysis Using MARS

Mallick, Denison, and Smith (1999) discuss Bayesian approaches to survival analysis using a multivariate adaptive regression spline (MARS) model. Various nonparametric Bayesian methods have been devised to address the problems of the classical Cox model. These require some prior random process which generates a cumulative hazard function, $H(t)$, or a hazard function, $h(t)$, and then involves modeling either $H(t)$ or $-\log(S(t))$, where $S(t)$ is the survival function. Note that if $S(t)$ is not differentiable, then these are not the same (Hjort, 1990). Then, $H(t)$, or $-\log(S(t))$, is viewed as a random sample path from a Levy process, i.e., a nonnegative nondecreasing process on $[0, \infty)$ which starts at 0 and has independent increments. For example, Kalbfleisch (1978) models $-\log(S(t))$ as a draw from a gamma process, while Hjort (1990) models $H(t)$ as a draw from a beta process. Note that the earliest work, which models $1 - S(t)$ by a Dirichlet process (Ferguson, 1973), implies that $-\log \mathcal{F}(t)$ is a Levy process. More recently, Gelfand and Mallick (1995) extend these models by introducing a random finite mixture process prior and modifying the usual proportional hazard form as $h(t|\boldsymbol{x}) = h_0(t)\varphi(\boldsymbol{x}'\boldsymbol{\beta})$, where φ is an unknown link function (see Section 10.1 for details). However, all of these methods assume time-independent covariates.

The hazard regression (HARE) model of Kooperberg, Stone, and Truong (1995) is proposed to allow for the considerable possibility of time-varying covariates. This models the conditional log-hazard function, $\alpha(t|\boldsymbol{x}) = \log(h(t|\boldsymbol{x}))$, using truncated linear spline basis functions. These types of basis functions are chosen because they have previously been shown to have good predictive power as well as being easily interpretable (Friedman, 1991), which is an important consideration when performing survival analysis.

The HARE model is nonparametric as the form of the basis functions is determined by the data. This is undertaken in a very similar way to the multivariate adaptive regression spline (MARS) model introduced in Friedman (1991). Both algorithms perform a deterministic search over the model space but Kooperberg, Stone, and Truong (1995) choose to search over a slightly modified space, mainly because of numerical considerations. Thus the aim is to determine basis functions, and their respective coefficients, so that we may model the conditional log-hazard function as

$$\alpha(t|\boldsymbol{x}) = \sum_{j=1}^{k} \beta_j B_j(t|\boldsymbol{x}), \tag{10.3.1}$$

where β_j is the coefficient of the basis function B_j, and k is the number of basis functions that the model chooses to have. Time-varying coefficients are easily accommodated by allowing the basis functions to depend on time, if required.

It has been noted (Denison, Mallick, and Smith 1998; Chipman, George, and McCulloch 1998) that stochastic searches over model spaces can improve on the predictions made from models derived deterministically. This is commonly done by sampling from the posterior distribution of the model given the data. However, this target posterior typically involves an integral which cannot be solved analytically and therefore MCMC methods are employed to generate an approximately independent sample from the required distribution. Denison, Mallick, and Smith (1999) propose a Bayesian MARS model, which is shown to, in general, predict more accurately with fewer basis functions than the classical MARS model (Friedman 1991).

Mallick, Denison, and Smith (1999) extend this Bayesian MARS model to cope with survival data, similar to the way that the HARE model is an extension to the usual MARS methodology. Further, they generalize the methodology to frailty models, and thus their model can be viewed as a generalization of the frailty model of Clayton and Cuzick (1985).

10.3.1 The Bayesian Model

Let $\boldsymbol{x} = (x_1, x_2, \ldots, x_p)'$ denote a $p \times 1$ vector of covariates. Also, let $f(t|\boldsymbol{x})$, $S(t|\boldsymbol{x})$, $h(t|\boldsymbol{x})$, and $\alpha(t|\boldsymbol{x})$ denote the density function, survivor function, hazard function, and log-hazard, respectively. Then, $S(t|\boldsymbol{x}) = \int_t^\infty f(u|\boldsymbol{x})du$, $h(t|\boldsymbol{x}) = \frac{f(t|\boldsymbol{x})}{S(t|\boldsymbol{x})}$, and $\alpha(t|\boldsymbol{x}) = \log h(t|\boldsymbol{x})$. Using (1.4.4), we have

$$S(t|\boldsymbol{x}) = \exp\left[-\int_0^t \exp\{\alpha(u|\boldsymbol{x})\}du\right].$$

Let y_i and ν_i denote the observed survival time and the associated censoring indicator so that $\nu_i = 0$ if y_i is censored and 1 otherwise. In addition, let $\boldsymbol{x}_i$ denote the vector of covariates for the ith observation. Then, the log-likelihood corresponding to individual i, i.e., data point $D_i = (y_i, \boldsymbol{x}_i, \nu_i)$, is given by

$$l(\boldsymbol{\theta}|D_i) = \nu_i \alpha(y_i|\boldsymbol{x}_i) - \int_0^{y_i} \exp\{\alpha(u|\boldsymbol{x}_i)\}\, du, \qquad (10.3.2)$$

for $i = 1, 2, \ldots, n$, where $\boldsymbol{\theta}$ is the vector of the model parameters. The log-likelihood of the parameters for all n observations is given by

$$L_k(\boldsymbol{\theta}|D) = \sum_{i=1}^{n} l(\boldsymbol{\theta}|D_i), \qquad (10.3.3)$$

where $D = (D_i,\ i = 1, 2, \ldots, n)$.

The Bayesian MARS model for survival analysis can be thought of as a stochastic version of the HARE model of Kooperberg, Stone, and Truong (1995) but with some important differences. Both approaches model the conditional log-hazard function using (10.3.1), but they differ in the way the basis functions are determined: the HARE model searches for them in a deterministic manner (stepwise addition followed by stepwise deletion) while the Bayesian MARS model searches for them by an MCMC algorithm. Also the Bayesian model imposes less restrictions on the form of the model, allowing interactions without both main effect terms being present. The HARE model does this to avoid spurious predictors being present and to possibly increase the model's interpretability. For the Bayesian model, we choose to downweight the chance of spurious predictors through the model prior and do not believe that this compromises the interpretability of the model.

The form of the basis functions used in different nonparametric models is important because the relative strengths and weaknesses of each method are directly dependent on it. We now give a brief description of the basis functions used in HARE and Bayesian survival models, which are identical to those proposed by Friedman (1991).

Using the notation of Friedman (1991), the basis functions are of the form

$$B_i(\boldsymbol{x}) = \begin{cases} 1, & i = 1, \\ \prod_{j=1}^{J_i} \left\{ s_{ji} \cdot (x_{v(j,i)} - t_{ji}) \right\}_+, & i = 2, 3, \ldots, \end{cases}$$

where $(\cdot)_+ = \max(0, \cdot)$, J_i is the degree of the interaction of basis B_i, the s_{ji}, which we shall call the sign indicators, equal ± 1, and the $v(j,i)$ give the indices of the predictor variables corresponding to the *knots* t_{ji}. The $v(j,\cdot)$ $(j = 1, 2, \ldots, J_i)$ are constrained to be distinct so that each predictor only appears once in each interaction term to maintain the "linear" nature of the basis functions. Note that in all the work that follows, the maximum number of interactions in a basis function is taken to be 2, so that $J_i \leq 2$ for all i. This is sensible in survival analysis, where interpretation of the model is paramount because interactions between three or more covariates would be difficult to fully understand.

Following Mallick, Denison, and Smith (1999), we perform a stochastic search over the model space $\Theta = \bigcup_{k=0}^{\infty} \Theta_k$, where Θ_k is the subspace of the Euclidean space, $R^{n(k)}$, corresponding to the vector space spanned by all the elements $\boldsymbol{\theta}^{(k)}$ with k basis functions. For this model $\boldsymbol{\theta}^{(k)} = \left[C_i, \beta_i, \{s_{ji}, t_{ji}\}_{j=1}^{J_i}\right]_{i=1}^{k}$, where C_i is the *type* of basis function that B_i is classified as. This is determined from the $v(j,i)$ and just classifies all basis functions involving the same covariates as of the same type no matter what ordering of j is used. With this model structure $n(k) = \sum_{i=1}^{k} 2(1 + J_i)$.

To specify the Bayesian model we must place prior distributions over all the unknown parameters. In this regard, Mallick, Denison, and Smith (1999) assign a Poisson prior with parameter λ over k with discrete uniform priors over the possible values of s_{ji} and t_{ji}, i.e., over $\{-1, 1\}$ and the marginal predictor values of variable $x_{v(j,i)}$, respectively. The prior over C_i is chosen more carefully. The prior is chosen to reflect the fact that among all interaction terms and main effects, each type of basis is equally likely, but also that main effects are favored over interactions. So, for the C_i which represent main effects ($i = 1, 2, \ldots, p$) the prior probability for C_i is ψ/p and for the other C_i, their prior probability of being in the model is $2\psi/\{p(p-1)\}$, where $\psi(> 0.5)$ is the prior proportion of basis functions expected to be main effects. Finally, a zero mean normal prior with variance τ^2 is used for the coefficients.

Since analytic or numerical analyses are intractable, a reversible jump MCMC sampler of the general type discussed by Green (1995) is used to simulate samples from the joint posterior distribution of $\pi(\boldsymbol{\theta}^{(k)}, k|D)$. With multiple parameter subspaces of different dimensionality, it will be necessary to devise different types of moves between the subspaces, Θ_k. These will be combined to form a hybrid sampler which makes a random choice between available moves at each transition in order to traverse freely around the combined parameter space. Mallick, Denison, and Smith (1999) propose the following move types:

(a) a movement in a knot location;

(b) a change in a factor in a basis function;

(c) a change in the basis coefficients;

(d) the addition of a basis function; and

(e) the deletion of a basis function.

The Bayesian MARS model in each of steps (a)–(c) is accepted with a Metropolis-Hastings step and for steps (d)–(e), where dimension is changed, a generalization of this given in Green (1995) is used.

When the MARS structures are changed, as described below, the coefficients of the basis functions β_i $(i = 1, 2, \ldots, k)$ must be drawn simultaneously. This could be done by making a random draw for them around their current values with new basis functions having their coefficient set to zero, but this leads to poor acceptance rates and can hinder the sampler from finding good basis functions. Instead of finding good single models, the MCMC method is used as a search engine to explore the full posterior distribution. Mallick, Denison, and Smith (1999) find that better models are found when a mini-Metropolis-Hastings sampler is run to find the coefficient values every time a new model is proposed. They do this for 100 iterations with the final set of coefficients taken as the new values.

Given the current model, proposing a new model using step (a) is straightforward. First, a basis function, B_i, is selected uniformly at random and one of the factors j is then chosen where the current knot location t_{ji} is altered. A new knot location is chosen uniformly from the marginal predictor values of variable $x_{v(j,i)}$ and this is set to the new t_{ji}. Step (b) is done similarly to step (a) except that when a factor in a basis function to be changed has been picked, a new predictor variable as well as a new knot location for that factor are chosen uniformly. Step (b) contains step (a) as a special case, but its inclusion greatly aids the mixing of the sampler and prevents the sampler from getting "stuck" in local modes by proposing bolder moves than step (a). However, this does not lead to step (a) becoming redundant as these more local moves help us to find local modes in the true probability surface. These local modes correspond to good models so it is important to search intelligently for them. Step (c) is even simpler and just involves resampling all the basis coefficients. The addition of a basis function, step (d), is carried out by choosing a type of basis function to add to the model. This is found by randomly drawing from the prior of the C_i. Then, a knot location and sign indicator for each of the J_i factors in this new basis are chosen uniformly. Step (e), the deletion of a basis function, is constructed in such a way as to make the jump step reversible. This is easily done by choosing a basis function uniformly from those present (except the constant basis function B_1) and removing it.

In the reversible jump algorithm, the five move types described above are used so that we can write the set of moves as $m = \{A, B, C, 1, 2, \ldots\}$. Here, A refers to the movement (alteration) of a knot location, B the changing of a predictor variable and a knot (bigmove), C a change in the basis coefficients and $m = 1, 2, \ldots$ refers to increasing the number of terminal nodes from m to $m+1$ or decreasing it from $m+1$ to m. Independent move types are randomly chosen with probabilities ρ_k, u_k and η_k for $m = A$, B, and C, respectively, b_k for $m = k$ and d_k for $m = k - 1$, which satisfy $\rho_k + u_k + \eta_k + b_k + d_k = 1$ for all k. Specifically, for this Bayesian MARS model, Mallick, Denison, and Smith (1999) take $b_k = c\min\{1, \lambda/(k+1)\}$ and $d_{k+1} = c\min\{1, (k+1)/\lambda\}$ with $\rho_k = u_k = 0.5\eta_k$ for $k = 2, 3, \ldots$ with the constant c, a parameter of the sampler, taken to be 0.25. For $k = 1$, they use $b_1 = 0.8$, $u_k = 0.2$, $d_1 = \rho_1 = \eta_1 = 0$, and when $k > 1$, they determine b_k and d_k and share the rest of the probability between ρ_k, u_k and η_k in the ratio 1:1:2. The acceptance probability for each of the move types in the sampler is the minimum of 1 and the product of the likelihood, prior and proposal ratios. Note that, unlike in Green (1995), a Jacobian term in the acceptance probability is not required, because the new, or changing, components of $\boldsymbol{\theta}^{(k)}$ are drawn independently of their current values. Also, by construction in steps (a)–(b) and (d)–(e), these new components from their prior distributions are proposed so that there is a lot of cancellation of terms when multiplying the prior and posterior ratios. The details of this MCMC algorithm are given as follows:

Step 1. Start with just the constant basis function present.

Step 2. Set k equal to the number of basis functions in the current structure.

Step 3. Generate u uniformly on [0,1].

Step 4. "Go to" move type determined by u (e.g., if $(u \leq b_k)$ then go to step (a) else if $(b_k < u \leq b_k + d_k)$ then go to step (b) etc.).

Step 5. Repeat Step 2 for a suitable number of iterations once there is evidence of convergence.

Example 10.4. Rectal cancer data. Mallick, Denison, and Smith (1999) considered a rectal cancer trial which is reported on, and listed in Harris and Albert (1991). There are only 56 observations, of which a very high number, 31, are censored. The response is survival time in months and there are three covariates: age at entry (in years); gender (0–male, 1–female); and radiation dose (0–$<$ 5000 rad, 1–$\geq$5000 rad).

For this example, they ran the sampler for an initial burn-in period (2000 iterations), after which time the posterior probability of the models had been settled for some time. The generated sample was collected from the next 30,000 iterations, taking every 5th one to be in the sample. They chose the mean of the Poisson prior over the number of basis functions $\lambda = 3$, the proportion of basis functions expected to be main effects $\psi = 0.8$, and the prior variance of the coefficients $\tau^2 = 10$.

The first hyperparameter, λ, was chosen to be 3, since, in this example, one does not expect many basis functions. When there are many predictors, a higher value may be necessary to prevent underfitting. Overfitting seems to be less of a problem in this application, maybe due to the difficulty in finding good coefficient values for large models. The next hyperparameter, $\psi = 0.8$, reflects knowledge that one expects more main effects than interaction terms, in fact, four times as many. The prior variance of the coefficients $\tau^2 = 10$ has less influence and only moderately shrinks the co-

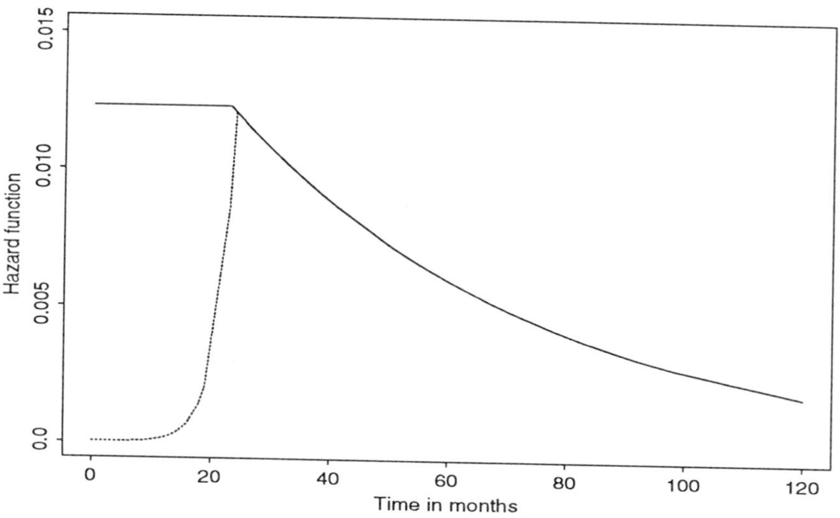

FIGURE 10.3. The Plots of hazard functions for male (dotted line) and female (solid line).

efficient values. Larger values, which are even less informative, could be used and would not affect the results significantly.

For the rectal cancer data, Mallick, Denison, and Smith (1999) found that the maximum *a posteriori* (MAP) estimate of the model, which has the largest posterior model probability, had a lower BIC than that given by HARE (266.5 compared to 276.2) and, interestingly, contains a time-dependent interaction, whereas the HARE model does not find one. The time-sex interaction basis function, $(24 - \text{Time}) \times \text{Male}$ suggests that the hazard function is lower for males just after the radiation treatment is given (see Figure 10.3). In Figure 10.3, for the other covariates, the median values were used, i.e., age $= 62$ and radiation dose $= 1$. From Figure 10.3, it can be seen that after 24 months, the hazard functions for males and females are identical. Mallick, Denison, and Smith (1999) note that one possible problem with this dataset is that many more men were assigned the higher dose of radiation (23 out of 31), whereas the women were much more evenly divided (12 out of 25). This could have led to confounding between sex and dose, possibly leading to the effect of sex being overestimated.

10.3.2 Survival Analysis with Frailties

As discussed in Chapter 4, frailty models introduce a random effect term for each group from which the data is being generated, and these so-called "frailties" (estimates of how likely a group is to fail, i.e., how frail it is) need to be estimated from the data as well as the model. Given the random parameter, the data within the same group are assumed to be independent. Thus, the model for the log-hazard function remains the same as in (10.3.1) but a random effect term is added. Simultaneously, one can estimate the regression function as well as the effect due to frailties.

Mallick, Denison, and Smith (1999) extend the Bayesian MARS model to include terms for the random effects from individuals. The parameter vector $\boldsymbol{\theta}$ is then extended to include the random effects w_i $(i = 1, 2, \ldots, n)$ for each class from which the data is given and the overall variance of the logarithm of the random effects σ^2. The data is of the form $D_{ij} = (y_{ij}, \boldsymbol{x}_{ij}, \nu_{ij})$ $(j = 1, 2, \ldots, m; i = 1, 2, \ldots, n)$, where m is the number of observations in cluster i.

As discussed in Section 10.3.1, we can write the log-likelihood as the sum of the partial log-likelihoods over the observations $(i = 1, \ldots, n; j = 1, 2, \ldots, m)$, but this time $l(\boldsymbol{\theta}|D_{ij})$ is given by

$$l(\boldsymbol{\theta}|D_{ij}) = \nu_{ij}\{\alpha(y_{ij}|\boldsymbol{x}_{ij}) + w_i\} - \exp(w_i)\int_0^{y_{ij}} \exp\{\alpha(u|\boldsymbol{x}_{ij})\}.$$

The variance of the frailty estimates is drawn at each iteration using a Gibbs step because, assuming a $\mathcal{G}(\gamma_1, \gamma_2)$ prior for σ^{-2}, we know the full conditional for the dispersion σ^{-2} is $\mathcal{G}\left(\gamma_1 + \frac{m}{2}, \gamma_2 + \frac{1}{2}\sum(\log w_i)^2\right)$ since it is assumed that $E(\log(w_i)) = 0$. All the random effects can be drawn independently at each iteration using a Metropolis step, making the order in which they are drawn random to prevent any systematic errors in them. Finally, a normal distribution, $N(0, \sigma^2)$, is used to draw the proposed $\log w_i$ for $i = 1, 2, \ldots, m$.

Example 10.5. Kidney infection data. Mallick, Denison, and Smith (1999) reanalyzed the kidney infection data discussed in Example 1.4. The dataset involves infection times for 38 kidney patients. The patients each provide two infection times, some of which are right censored. The other covariates are age in years, sex (0–male, 1–female), and the presence/absence of disease types GN, AN, and PKD (0–absence, 1–presence). As only one disease type can be present in each individual, they do not allow interactions between disease types in the model. Each individual is assumed to have an associated random effect. Using the model with all the covariates, they found the posterior probability of sex being in the model as a main effect to be 0.99, which broadly agrees with the findings of McGilchrist and Aisbett (1991). Also, the model with the highest posterior probability contained only sex and the constant as the only basis functions; this model had a posterior probability of 0.20. However, there is some evidence for a time-sex interaction, as this is present in 18% of the models, including the MAP one.

10.4 Change Point Models

Cancer prevention clinical trials are becoming increasingly common due to recent advances in medical oncology and a better understanding of the disease. Thus it is important to develop and examine statistical models for data arising from such trials. In cancer prevention trials, it is often reasonable to assume that any effect of a treatment is not immediate, but rather the risk of failure is affected only after a time lag. In many cancer prevention trials, the general interest focuses on identifying and evaluating "supplements" which may prevent a specific cancer, and thus these trials are developed for this purpose. For example, Zucker and Lakatos (1990) describe a planned analysis of the Physician's Health Study for testing the effect of beta-carotene on cancer incidence. Zucker and Lakatos (1990), Freedman et al. (1993), and Luo, Turnbull, and Clark (1997) discuss several animal experiments and clinical trials for evaluating the clinical benefits of several compounds for cancer prevention. The Chemoprevention Branch of the National Cancer Institute in the U.S. has developed programs to identify and clinically evaluate compounds which could be useful for cancer prevention. Due to this recent surge in interest for analyzing survival data from clinical trials with preventive compounds, there is a growing need to develop new and useful models, understand their properties, and compare them to existing models.

A popular model for cancer prevention is the lagged regression model of Zucker and Lakatos (1990). We refer to this model as the Z-L model throughout. Zucker and Lakatos (1990) describe several situations, ranging from trials involving cholesterol lowering drugs to cancer prevention trials involving food supplements such as beta-carotene, for which this model is useful. For example, in clinical trials for which beta-carotene is used as a cancer prevention supplement, the preventive treatment would not affect pre-existing tumors and lesions. If the treatment has any effect, then we expect it to reduce the rate of incidence of only new tumors. If we are interested in the effect of treatment on the event time T, which denotes time to cancer from study entry, then we should not expect the treatment to have any effect if the cancer is caused by tumors already existing at study entry. To deal with this problem, investigators often decide not to count any cancer detected during a "waiting period" following entry into the study and use the Z-L model for T. A proportional hazards version of the Z-L method can be defined as

$$h(t|x) = \{(1 - l(t)) + l(t)\varphi(x)\}h_0(t), \tag{10.4.1}$$

where x denotes the treatment covariate, $h_0(t)$ is the baseline hazard of T, and $0 \leq l(t) \leq 1$ is a function of time which captures the lag over time in the treatment effect $\varphi(x)$ on the hazard $h(t|x)$. Further assumptions on the lag-function $l(t)$ are $l(0) = 0$, $l(\infty) = 1$ and $l(t)$ is increasing in t. Therefore, in this model, the treatment effect on the hazard starts at 0 and then increases to $\varphi(x)$ as the lag function increases from 0 to 1. Several papers have discussed different versions of the Z-L model and different forms of the lag-function $l(t)$. For a recent review of articles that examine (10.4.1), see Luo, Turnbull, and Clark (1997) and the references therein.

An alternative method for modeling cancer prevention data is motivated by modeling the stochastic process generating the existing tumors and occurrences of new tumors adjusting for treatment effects. For more on this, see Kokoska (1987), Freedman et al. (1993), and the references therein. The basic idea behind this approach rests on the notion that the cancer is caused by a two-stage process; i) initiation of an undetectable tumor, often called a clonogen, and ii) promotion of a clonogen through an irreversible process to become a detectable cancer. Lesions or tumors are formed at the initiation stage and this initiation process can be modeled by a suitable point process. Then each tumor has some random promotion time to become a cancer. Previous models of this type are only useful for certain types of animal experiments where the treatment is given before the initiation process begins (Freedman et al., 1993). For most clinical trials, we cannot be assured that the treatment is given before the beginning of the tumor initiation process. Thus, models for the initiation stage and promotion stage need to be developed. Sinha, Ibrahim, and Chen (2000) introduce a new model which is based on modeling the initiation and promotion stage. They do not assume that the treatment is allocated before the initiation process. They derive several theoretical properties of the new model and establish its relationship to the Z-L model. In particular, they show how different forms of the lag-function correspond to different modeling assumptions on the initiation and promotion processes. They also establish relationships of the new model and the Z-L model to other models for survival data, such as cure rate models. In addition, they carry out Bayesian inference for the new model and demonstrate several of its advantages over frequentist methods for lagged regression models and the Cox model.

10.4.1 Basic Assumptions and Model

Let T denote the time to cancer from study entry. We assume that every subject is randomized to a treatment, denoted by x, at the time of entry into the trial. A random patient has N_1 undetected tumors (clonogens) at study entry, where N_1 is an unobservable latent quantity. Following Yakovlev et al. (1993) and Chen, Ibrahim, and Sinha (1999), we use the term *clonogen* to represent any undetected lesion or undetected tumor which, if observed, goes through an irreversible promotional process to become a detectable tumor in the future. We let F_1 denote the cumulative distribution function (cdf) of the promotion time from study entry of each existing clonogen.

We make the following assumptions:

(A1) $N_1 \sim \mathcal{P}(\theta)$, i.e., $P(N_1 = k) = \theta^k e^{-\theta}/k!$, for $k = 0, 1, 2, \ldots$.

(A2) The promotion times for N_1 existing clonogens are i.i.d. with common cdf $F_1(t)$.

(A3) New clonogens occur according to a nonhomogeneous Poisson process with intensity $h_2(t|x)$. The process is independent of N_1 and independent of the promotional process of existing clonogens. Note that the intensity does depend on the treatment, x.

(A4) The promotion times for new clonogens are i.i.d. with common cdf $F_2(t)$, and they are independent of the initiation process of other clonogens and the promotional process of existing clonogens. F_2 does not depend on x.

Justifications of the above assumptions can be given as follows. The Poisson distribution is a natural choice for modeling the distribution of the number of clonogens given in (A1). Later, we will relax this assumption by incorporating extra-Poisson variation in the distribution of N_1. The occurrence of new clonogens can be thought of as a superimposition of a large number of point processes, where each one of these point processes may be caused by some environmental and other factors. Each point process generates mutated cells and most of these cells die over time, which corresponds to a "thinning" of the original process. Few of these cells eventually become clonogens, which are still undetectable, but are already in the irreversible paths of becoming detectable tumors in due course. Asymptotic properties of large numbers of superimposed independent point processes, as described in Cox and Isham (1980), justify (A3). Another way of justifying (A3) is based on the cancer epidemiology model of Becker (1989). We can think of the cell mutations as a compound Poisson process. See Karlin and Taylor (1975) for a description of compound Poisson processes. The mutations or "shocks" are assumed to arrive as a Poisson process. At each arrival time, the number of cells mutating (amount of damage to the clonogen host system) follows a distribution G_* and is independent of the mutations in other arrival times. The *cumulative damage* model of Becker (1989) assumes that the host system is able to resist the carcinogenic environmental damage to a certain extent, and the system only fails when the total damage exceeds a certain threshold. Now, given K cells have been mutated at a particular arrival time, following the *cumulative damage* model idea, we can assume that a clonogen is formed with probability $1 - \exp(-e^{\psi K - c})$ where, according to Becker's terminology, ψ is the *load* of the damage caused by a single mutated cell and c is the total resistance of the body to the mutated cells. This process gives rise to a nonhomogeneous Poisson process for the occurrence of new clonogens. The proof follows from the usual properties of compound Poisson processes (see Karlin and Taylor, 1975). A similar justification of (A3) can be also put forward using the DNA damage repair model of carcinogenesis proposed by Kopp-Schneider, Portier, and Rippmann (1991). In most applications involving humans, the resistance parameter $c \equiv c(t)$

is typically a decreasing function of time, t, reflecting a decreasing trend in the body's resistance to cancerous mutations. So, under the additional assumption that the arrival Poisson process of shocks have nearly constant intensity, the resulting intensity $h_2(t|x)$, of the initiation process, should be an increasing function of t.

The literature on multistage stochastic models of carcinogenesis is very rich. The most popular class of models is the so-called Moolgavkar-Venzon-Knudson models of carcinogenesis (Moolgavkar and Venzon, 1979; Moolgavkar and Knudson, 1981; Moolgavkar, Dewanji, and Venzon, 1988). Also, see Chapter 2 of Yakovlev and Tsodikov (1996) for brief descriptions of such models. A Moolgavkar-Venzon-Knudson (MVK) version of the model considered in this section would assume the promotion times from malignant cells to detectable tumors in (A2) and (A4) to be deterministic (nonrandom). In many carcinogenecity experiments and prevention studies, the time of detection of tumor is not the same as the time of first malignant cell generation, and, hence, the assumption of a nonrandom promotion time may not be appropriate. See Yang and Chen (1991), Kopp-Schneider and Portier (1995), and Yakovlev and Tsodikov (1996) for discussions on these issues.

A typical supplement such as selenium may influence the initiation process, or at least the point process for clonogen initiation after study entry. However, it may not influence the promotional stage. Similarly a cholesterol lowering drug may not influence the promotion time of an already formed narrowed passage, due to plack deposition for example, to a full-scale artery blockage. Thus, the independence of F_1 and F_2 from the treatment covariate x are reasonable assumptions in (A2) and (A3). Given that there are N_1 existing clonogens in a patient at study entry, the conditional probability of none of these clonogens becoming a detectable cancer by time t is $\{S_1(t)\}^{N_1}$, where $S_1(t) = 1 - F_1(t)$. Using properties of Poisson processes, the conditional distributions of the exact times of occurrence of new clonogens given N_2 clonogens by time t are i.i.d. with common density

$$f_*(y|x,t) = h_2(y|x)/H_2(t|x) \text{ for } y \in (0,t),$$

where $H_2(t|x) = \int_0^t h_2(y|x)\, dy$ is the cumulative intensity of the initiation process of new clonogens. Now, T, the event time of interest, has a conditional survival function as given by the following theorem.

Theorem 10.4.1 *The conditional survival function of T given N_1 old and N_2 new clonogens by time t is given by*

$$\begin{aligned} P(T > t|x, N_1, N_2) =& S_p(t|x, N_1, N_2)] \\ =& \{S_1(t)\}^{N_1} \left(\int_0^t f_*(y|x,t) S_2(t-y) dy \right)^{N_2}, \end{aligned}$$

where $S_2(t) = 1 - F_2(t)$.

The proof of (10.4.1) is given in the Appendix. Using the fact that $N_1 \sim \mathcal{P}(\theta)$ and $N_2 \sim \mathcal{P}(H_2(t|x))$, where $\mathcal{P}(\mu)$ denotes the Poisson distribution with mean μ, we can get the unconditional survival function as

$$S_p(t|x) = \exp[-\theta F_1(t) - G(t|x)H_2(t|x)], \tag{10.4.2}$$

where $G(t|x) = \int_0^t f_*(y|x,t)F_2(t-y)\,dy$. Alternatively, we can write the above equation as

$$S_p(t|x) = \exp[-\theta F_1(t) - H(t|x)], \tag{10.4.3}$$

where $H(t|x) = G(t|x)H_2(t|x)$. The implicit claim for this result is given by the following theorem.

Theorem 10.4.2 *The function $H(t|x)$ is an increasing nonnegative cumulative intensity function and $\lim_{t\to\infty} H(t|x) < \infty$ if and only if $\lim_{t\to\infty} H_2(t|x) < \infty$.*

The proof of this theorem is given in the Appendix. Thus, $\lim_{t\to\infty} S_p(t|x) > 0$ if and only if $\lim_{t\to\infty} H_2(t|x) < \infty$, which essentially means that there is a positive probability of getting no cancer over a subject's lifetime as long as the expected number of new initiated clonogens is finite.

If we use a proportional intensity model, that is, $h_2(t|x) = \varphi(x)h_{20}(t)$ in (A3), then $f_*(y|x,t)$ and $G(t|x)$ become free of x. The survival function of T in this case is given by

$$S_p(t|x) = \exp[-\theta F_1(t) - \varphi(x)H_0(t)], \tag{10.4.4}$$

where $H_0(t) = \int_0^t \lambda_{20}(y)F_2(t-y)\,dy$ is a cumulative intensity function which is free of x, and $\lambda_{20}(y)$ is a baseline hazard function. We can prove that $H_0(t)$ is a cumulative intensity function by using an argument similar to the proof of (10.4.2). When the densities f_1 and f_2 exist for promotion times of existing and new clonogens, respectively, the hazard function is given by

$$h_p(t|x) = \theta f_1(t) + \varphi(x)h_0(t), \tag{10.4.5}$$

where $\lambda_0(t) = \int_0^t f_2(t-y)h_{20}(y)\,dy$. It is clear that (10.4.5) is not a proportional hazards model, but the hazard function approaches a proportional hazards structure if the tail of $f_1(t)$ decays faster than $h_0(t)$. In Section 10.4.3, we explore the relationship between the model defined by (10.4.5) and the Z-L model given in (10.4.1).

10.4.2 Extra Poisson Variation

We can incorporate extra Poisson variation into (A1) by taking

(A1a) $N_1 \sim \mathcal{P}(\theta W)$, where W is an unobserved patient specific random effect.

(A1b) W varies from subject to subject (i.i.d.) with a common density $g(\cdot|\alpha)$, where α is an unknown parameter of heterogeneity.

One possibility is to assume W has the positive stable distribution. That is, $W \sim S_\alpha(1,1,0)$, where $E[\exp(-sW)] = \exp(-s^\alpha)$, for $0 < \alpha < 1$ (see Section 4.1.4). For the model assuming (A1a) and (A1b), we can show that

$$S_p(t|x) = \exp[-\theta^\alpha [F_1(t)]^\alpha - G(t|x)H_2(t|x)], \tag{10.4.6}$$

where $[F_1(t)]^\alpha$ is a cdf. It is clear that the two models (10.4.6) and (10.4.2) are indistinguishable for all practical purposes. For the sake of parsimony, it is perhaps more desirable to use (10.4.2) for statistical modeling. We note that an interpretation issue arises from (10.4.6) since the infinite mean of the positive stable distribution implies that $E(N_1) = \infty$. To overcome this, we can assume that W follows some finite mean frailty distribution, such as a gamma distribution with unit mean, for example. Let $\mathcal{G}(\alpha, \alpha)$ denote the gamma distribution with shape and scale parameter α. Assuming $W \sim \mathcal{G}(\alpha, \alpha)$, we get

$$S_p(t|x) = \left(\frac{\alpha}{\alpha + \theta F_1(t)}\right)^\alpha \times \exp[-G(t|x)H_2(t|x)]. \tag{10.4.7}$$

Using a proportional intensity in (A3) leads to

$$S_p(t|x) = \left(\frac{\alpha}{\alpha + \theta F_1(t)}\right)^\alpha \times \exp[-\varphi(x)H_0(t)]. \tag{10.4.8}$$

As $t \to \infty$, $S_p(t|x)$ in (10.4.8) approaches a proportional hazards model.

10.4.3 Lag Functions

There is a connection between the model in (10.4.5) and the Z-L model in (10.4.1). This is stated in the following theorem.

Theorem 10.4.3 *Every model in (10.4.5) corresponds to a Z-L model in (10.4.1) provided $h_0(0) = 0$ and $f_1(t)/h_0(t)$ is a decreasing function of t. Also, if $\int_0^\infty [h_0(t)\{1 - l(t)\}]\, dt < \infty$, then every Z-L model can be also written as in (10.4.5).*

To prove the first part of Theorem 10.4.3, note that $h_p(t|x)$ in (10.4.5) can be written as $h_p(t|x) = [1 - l(t)]h_0(t) + l(t)\varphi(x)h_0(t)$, where $h_0(t) = \theta f_1(t) + h_0(t)$ and the lag function is

$$l(t) = \frac{h_0(t)}{\theta f_1(t) + h_0(t)}. \tag{10.4.9}$$

It is clear from (10.4.9) that this lag-function is increasing in t only when $f_1(t)/h_0(t)$ is a decreasing function of t. To insure the condition $l(0) = 0$ in the Z-L model, we need $h_0(0) = 0$. The condition $h_0(0) = 0$ is easily satisfied in practice as long as $h_{20}(t)$ is bounded above near 0. Using similar arguments, we can prove the second part of Theorem 10.4.3.

Luo, Turnbull, and Clark (1997) consider various forms of the lag function in the Z-L model. The main significance of (10.4.9) is that it helps us determine what type of parametric assumptions about F_1, F_2, and h_{20} are needed to obtain different types of lag functions to use in practice. There is a significant practical importance of the lag function $l(t)$. The shape of this function determines how fast $h_p(t|x)$ approaches the proportional hazards structure.

If $f_2(t) \equiv 0$ for $t < \eta$ and $f_1(t) \equiv 0$ for $t > \eta$, then the lag function takes the form $l(t) = 0$ if $t < \eta$, and $l(t) = 1$ if $t > \eta$, where η is a threshold parameter. This is precisely the lag function for the lagged regression model considered by Luo, Turnbull, and Clark (1997), and they present an estimation procedure for the unknown lag threshold parameter η. In general, whenever $f_1(t) \equiv 0$ for $t > \eta$, we get a lag function which is exactly 1 if $t > \eta$, which implies that the hazard function $h_p(t|x)$ has a proportional hazards structure.

Another useful class of lag functions arise when the intensity of new clonogens is a constant, that is, $h_{20}(t) \equiv \lambda$. In this case, the lag function reduces to

$$l(t) = \frac{F_2(t)\lambda}{\theta f_1(t) + F_2(t)\lambda}. \tag{10.4.10}$$

This lag function satisfies the property $l(0) = 0$. It is also increasing as long as $f_1(t)/F_2(t)$ is a decreasing function of t. This last condition is easily satisfied when $f_1(t)$ is a decreasing function of t, as, for example, when $f_1(t)$ is an exponential density. One important class of continuous lag functions is obtained in this situation if and only if $f_1(t)/F_2(t)$ is continuous, and in this case, we get a continuous $l(t)$ which ranges from 0 to 1.

Luo, Turnbull, and Clark (1997) mention that the linear lag-function is an important class of lag-functions. It is given by

$$l(t) = \begin{cases} 0, & \text{if } t < \eta_1, \\ 1, & \text{if } t > \eta_2, \\ \frac{t-\eta_1}{\eta_2-\eta_1}, & \text{if } t \in (\eta_1, \eta_2), \end{cases} \tag{10.4.11}$$

where $0 < \eta_1 < \eta_2$ are range parameters. For the case of constant λ_2, we can obtain the lag function of (10.4.11) as a special case of (10.4.10) by taking $F_2(\cdot)$ to be a uniform cdf in the interval $[\eta_1, \eta_2]$, and taking the density function $f_1(t)$ to be of the form

$$f_1(t) \propto \begin{cases} 0, & \text{if } t \notin (\eta_3, \eta_2), \\ \eta_2 - t, & \text{if } t \in (\eta_3, \eta_2), \end{cases} \tag{10.4.12}$$

where $\eta_3 < \eta_1$. Now these seemingly unrealistic forms of $F_2(t)$ and $f_1(t)$ are in fact useful in practice. The formulation above implies that when we have (i) constant h_2, (ii) a uniformly distributed promotion time over a finite interval for new clonogens, and (iii) a nearly quadratic cdf for the promotion time of old clonogens, then we obtain an approximately linear lag-function, as long as the finite interval support of $f_1(t)$ is placed below the support interval of $F_2(t)$, and the two intervals overlap.

10.4.4 Recurrent Tumors

For many cancers such as melanoma, breast cancer, and gastrointestinal cancer, recurrent occurrences of detectable tumors for a given subject are quite common (see Freedman et al., 1993; Boone, Kelloff, and Malone, 1990). In this subsection, we explore the consequences of modeling such a process of recurrent detectable tumors through the promotion-initiation technique. Instead of looking at time to first detectable tumor, we now allow multiple detectable tumors occurring at different times. Since we are still dealing with cancer prevention models, we use assumptions (A1)–(A4) from Section 10.4.1. However, we now need to consider a point process $N_p(t)$ of recurrent occurrences of detectable tumors. We are led to the following theorem.

Theorem 10.4.4 *The point process $N_p(t)$ for a given subject is a Poisson process with cumulative intensity function*

$$H_p(t|x) = \theta F_1(t) + G(t|x)H_2(t|x), \tag{10.4.13}$$

where F_1, G and H_2 are defined in Sections 10.4.1 and 10.4.2.

The proof of Theorem 10.4.4 is given in the Appendix. Under the proportional intensity assumption in (A3), we get

$$H_p(t|x) = \theta F_1(t) + \varphi(x)H_0(t), \tag{10.4.14}$$

where $H_0(t) = G(t)H_{20}(t)$. It is also clear from (10.4.14) that the intensity function of $N_p(t)$ converges to the proportional intensity function $\varphi(x)\lambda_0(t)$ as $t \to \infty$.

One extension of the model in (10.4.14) is obtained by incorporating extra Poisson variation in the distribution of N_1. Again, this can be done by taking $N_1 \sim \mathcal{P}(\theta W)$, where W has any distribution with finite mean, such as $W \sim \mathcal{G}(\alpha, \alpha)$. Under the $\mathcal{G}(\alpha, \alpha)$ distribution for W, the marginal distribution of $N_p(t)$ is given by the following theorem.

Theorem 10.4.5 *For any finite interval $A = (a, b)$, the distribution of $N_p(A)$ is given by*

$$P(N_p(A) = k|x) = \frac{\exp[-\varphi(x)H_0(A)]\{\varphi(x)H_0(A)\}^k}{k!}$$
$$\times \left[\sum_{j=0}^{k} \binom{k}{j} \frac{\Gamma(\alpha+j)}{\Gamma(\alpha)} \left\{\frac{\alpha\theta F_1(A)}{\varphi(x)H_0(A)}\right\}^j \left\{\frac{1}{\alpha+\theta F_1(A)}\right\}^{j+\alpha}\right], \tag{10.4.15}$$

where $F_1(A)$ and $H_0(A)$ are the increments of F_1 and H_0, respectively, in the interval A.

The proof of Theorem 10.4.5 is given in the Appendix. Clearly, the point process $N_p(t)$ is no longer a Poisson process. However, $N_p(t)$ approaches a Poisson process if the support of F_1 is finite. The survival function for the time to first detectable tumor is the same as the survival function in (10.4.8).

One interesting special case is when there is no new tumor initiation process, that is, the intensity in (A3) is $h_2(t|x) \equiv 0$. This is the situation for most animal experiments with recurrent tumors (see Freedman et al., 1993) when all the animals receive the carcinogenic compound causing the initiation of clonogens at the very beginning of the experiment. For such an experiment, the assumption that $N_1 \sim \mathcal{P}(\theta)$ is often hard to justify. In these cases, one requires the incorporation of extra Poisson variation for the distribution of N_1. In this setting, the conditional point process, $N_p(t)$ given W, has a structure similar to the model for recurrent event-time data presented by Oakes (1992) and Sinha (1993). The difference is that the conditional cumulative intensity $H_p(t|x, W)$ is bounded here. Also, $N_p(t)$ has a Markovian structure in the sense that the conditional hazard of the $(k+1)^{th}$ event given the history of the k previous events depends only on k. Using a calculation similar to the one in Oakes (1992), it can be shown that for $W \sim \mathcal{G}(\alpha, \alpha)$,

$$h_{k+1}(t|\mathcal{H}_p(t-)) = (\alpha + k)\frac{\theta f_1(t)}{\alpha + \theta F_1(t)}, \tag{10.4.16}$$

where $\mathcal{H}_p(t-)$ is the history of $N_p(\cdot)$ at time $t-$ and $\mathcal{H}_p(t-)$ includes information about the exact times of the previous k events.

10.4.5 Bayesian Inference

We now consider Bayesian inference for the model in (10.4.4). Bayesian estimation of the model parameters for the Z-L model is demonstrated in Sinha, Ibrahim, and Chen (2000). Frequentist approaches to hypothesis testing for the Z-L model are given in Zucker and Lakatos (1990) and Luo, Turnbull, and Clark (1997). We introduce covariates through φ as

$\varphi(\boldsymbol{x}) = \exp(\boldsymbol{x}'\boldsymbol{\beta})$, where $\boldsymbol{x}$ is a $p \times 1$ vector of covariates and $\boldsymbol{\beta}$ is a $p \times 1$ vector of regression coefficients. Suppose we have n subjects, and suppose that $F_1(t) = 1 - \exp(-\gamma t)$ and $H(t) = \lambda t^{\alpha}$, so that F_1 corresponds to an exponential distribution and $H(t)$ is the cumulative hazard of a Weibull distribution with parameters α and λ. Let y_i denote the time-to-event for subject i with corresponding censoring indicator ν_i, where $\nu_i = 1$ if the ith subject fails, and 0 otherwise. Also, let $\boldsymbol{x}_i$ denote the $p \times 1$ vector of covariates for the ith subject, X is the $n \times p$ matrix of covariates with i^{th} row $\boldsymbol{x}_i'$, $\boldsymbol{y} = (y_1, y_2, \ldots, y_n)'$, $\boldsymbol{\nu} = (\nu_1, \nu_2, \ldots, \nu_n)'$, and $D = (n, \boldsymbol{y}, X, \boldsymbol{\nu})$ denotes the data. Based on model (10.4.4), the likelihood function for $(\theta, \boldsymbol{\beta}, \gamma, \lambda, \alpha)$ is given by

$$L(\theta, \beta, \gamma, \lambda, \alpha | D)$$
$$= \exp\left\{-\theta \sum_{i=1}^{n} F_1(y_i) - \sum_{i=1}^{n} \exp(\boldsymbol{x}_i'\boldsymbol{\beta}) H(y_i)\right\} \left(\prod_{i=1}^{n} (h_p(y_i))^{\nu_i}\right), \quad (10.4.17)$$

where $F_1(y_i) = 1 - \exp(-\gamma y_i)$, $H(y_i) = \lambda y_i^{\alpha}$, $f_1(y_i) = \gamma \exp(-\gamma y_i)$,

$$h_p(y_i) = \theta f_1(y_i) + \exp(\boldsymbol{x}_i'\boldsymbol{\beta}) h_0(y_i),$$

and $h_0(y_i) = \lambda \alpha y_i^{\alpha - 1}$. We note that (10.4.17) has been written in terms of an arbitrary F_1 and H. We take $\boldsymbol{\beta} \sim N_p(\boldsymbol{\mu}_0, \Sigma_0)$, $\theta \sim \mathcal{G}(\theta_{01}, \theta_{02})$, $\gamma \sim \mathcal{G}(\gamma_{01}, \gamma_{02})$, $\lambda \sim \mathcal{G}(\lambda_{01}, \lambda_{02})$, and $\alpha \sim \mathcal{G}(\alpha_{01}, \alpha_{02})$. We assume that θ, $\boldsymbol{\beta}$, γ, λ, and α are independent *a priori*. We use the Gibbs sampler to carry out Bayesian inference for the model parameters as well as other posterior quantities of interest. The detailed derivations of the full conditional distributions of θ, β, γ, λ, and α for Gibbs sampling are left as an exercise.

Example 10.6. Nutritional prevention of cancer trial data. We consider data from the Nutritional Prevention of Cancer Trial discussed by Clarke et al. (1996) and Luo, Turnbull, and Clark (1997). This trial is a double blind trial in which patients were randomized to placebo or a daily nutritional supplement of selenium. Selenium, an anti-oxidant, was thought to have cancer prevention potential. Full details of the background are given in Clark et al. (1996). The purpose of this example is to (i) demonstrate the usefulness of the model (10.4.4) over the Cox model and the Kaplan-Meier method, and in particular examine the implications of the lag function for these data and (ii) demonstrate the advantages of the Bayesian model based on (10.4.4) over the frequentist approach considered by Luo, Turnbull, and Clark (1997).

The responses (y_i's) are times to first diagnosis of an internal cancer. This category includes mostly lung cancer, 48 cases; prostate cancer, 48 cases; and colorectal cancer, 27 cases; but also some small number of other cancers, e.g., bladder, breast, brain, and leukemia, squamous cell carcinoma, 726; and basal cell carcinomas of the skin, 408. Any subjects who did not have cancer were censored at their last follow-up time. For the analysis here, we consider a subset of the data. The subset consisted of those subjects that were either censored, squamous cell carcinoma (scc) subjects with baseline laboratory values greater than or equal to 130 and not censored, or scc subjects with event times greater than 15 months and baseline laboratory values less than 130. For this subset, we have a total of 1286 observations, of which 1122 are censored. Table 10.4 gives a breakdown of treatment by cancer status.

TABLE 10.4. Data Summary.

Status	Treatment	Frequency
No Cancer	Placebo	542
	Selenium	580
Cancer	Placebo	99
	Selenium	65

Source: Sinha, Ibrahim, and Chen (2000).

TABLE 10.5. Posterior Summaries.

Parameter	Mean	SD	95% HPD interval
β	-0.55	0.20	$(-0.95, -0.17)$
λ	0.007	0.002	$(0.002, 0.011)$
α	1.68	0.16	$(1.39, 2.02)$
γ	0.20	0.48	$(0.00, 0.96)$
θ	0.23	0.44	$(0.00, 1.05)$

Source: Sinha, Ibrahim, and Chen (2000).

We consider model (10.4.4) with likelihood given by (10.4.17). The only covariate in the model is a binary covariate for treatment, which equals 0 for the placebo group, and 1 for the selenium group. Thus β is one-dimensional. We use noninformative priors for all parameters. Specifically, we take $\beta \sim N(0, 100)$, $\alpha \sim \mathcal{G}(1.5, 1)$, $\theta \sim \mathcal{G}(1, 1)$, $\lambda \sim \mathcal{G}(1, 2)$, and $\gamma \sim \mathcal{G}(1, 1)$. Table 10.5 gives posterior estimates of the parameters based on 50,000 Gibbs samples. We see from Table 10.5 that the posterior estimate of β is negative, indicating a positive effect of selenium. Also, we see that the 95% HPD interval for β does not include 0, indicating that the selenium effect is strong. These results are consistent with those of Luo, Turnbull, and Clark (1997). Table 10.6 gives estimates of the lag function and 95% HPD intervals over several time points t. Also Figure 10.4 shows a plot of the posterior lag function for several time points t. We see from Table 10.6 and Figure 10.4 that the lag function estimates are well below 1.0 even at $t = 4$ years, and converge slowly to 1.0 as $t \to \infty$. In fact at $t = 12$ years, the posterior estimate of $l(t)$ is 0.97. These results imply that modeling these data with a lag is more appropriate than using a proportional hazards model. Thus, model (10.4.4) reflects the lagged effect of selenium and thus results in a more accurate assessment of the selenium effect than the Cox model. In fact, the Cox partial likelihood estimate for the treatment covariate is -0.45 (SD $= 0.16$), which yields a P-value of 0.005. In contrast, the posterior mean of β for model (10.4.4) under the noninformative priors above is -0.55 (SD $= 0.20$). The Cox model thus slightly overestimates the treatment effect compared to the lag regression model (10.4.4). Further evidence of the appropriateness of model (10.4.4) can be seen in Figure 10.5, which shows the Kaplan-Meier estimate of survival for both treatments and the posterior estimates of the survival function based on (10.4.17). In this figure we see that the Kaplan-Meier estimates of the survival functions are very similar in shape to the posterior estimates of the survival function based on (10.4.17). Thus, the advantages of the lagged regression model in this case are clear. It provides a more precise estimate of the selenium effect, whereas the Cox model appears to overestimate the treatment effect. Figure 10.6 shows a plot of $\log[-\log(S(t))]$ versus $\log(t)$ based on the Kaplan-Meier estimates and the posterior estimates. First we see from this figure that the posterior estimates of survival produce smoother plots than those based on Kaplan-Meier. Further, we see from this figure that the posterior plot (plot b) is more linear than the Kaplan-Meier plot (plot a). This figure demonstrates that estimates based on the lagged regression model are more appropriate than those based on Kaplan-Meier.

TABLE 10.6. Posterior Summaries for Lag Function.

t (in years)	$E(l(t)\|D)$	SD	95% HPD interval
0.10	0.47	0.25	(0.06, 0.94)
0.30	0.62	0.22	(0.22, 1.00)
0.50	0.69	0.20	(0.33, 1.00)
1.0	0.79	0.16	(0.48, 1.00)
2.0	0.86	0.11	(0.64, 1.00)
3.0	0.90	0.09	(0.73, 1.00)
4.0	0.92	0.07	(0.78, 1.00)
8.0	0.96	0.04	(0.87, 1.00)
12.0	0.97	0.03	(0.91, 1.00)

Source: Sinha, Ibrahim, and Chen (2000).

In addition, one powerful advantage of the Bayesian analysis here over the frequentist approach of Luo, Turnbull, and Clark (1997) is that we are able easily to fit the model for an arbitrary lag function $l(t)$ and obtain posterior estimates of $l(t)$ and $S(t)$ as well as other quantities, such as HPD intervals and standard deviations using the Gibbs sampler, whereas Luo,

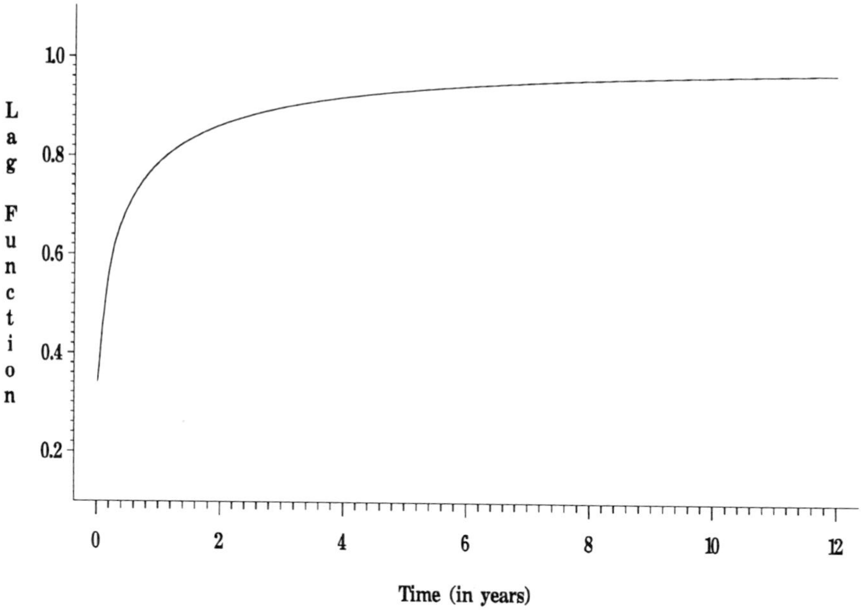

FIGURE 10.4. Plot of posterior estimates of lag function $l(t)$.

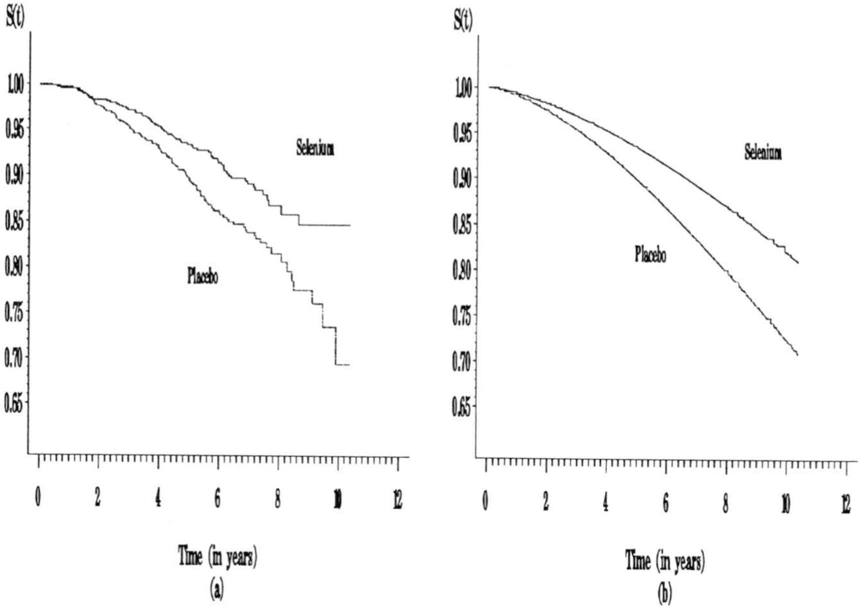

FIGURE 10.5. Plots of survival functions $(S(t))$ versus time (t): (a) Kaplan-Meier and (b) Bayesian.

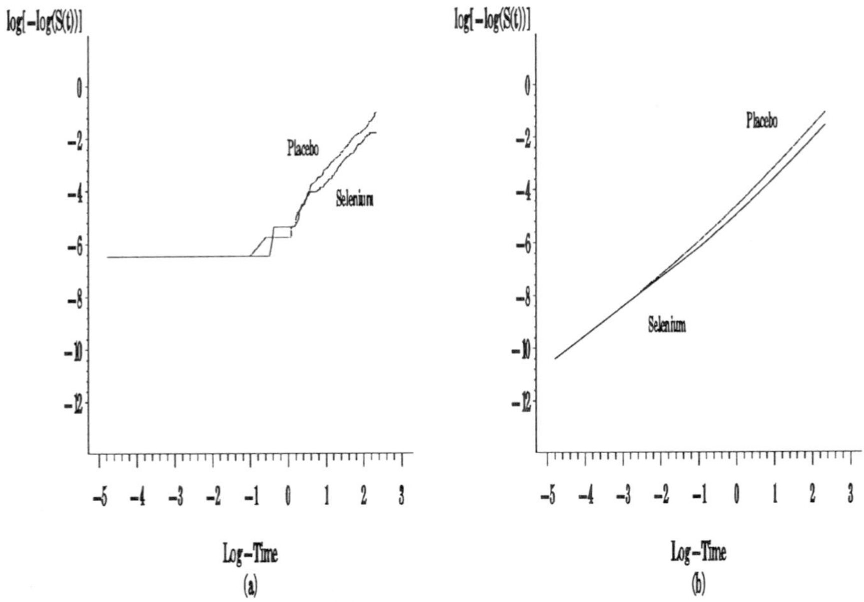

FIGURE 10.6. Plots of log-log survival functions ($\log[-\log(S(t))]$) versus log-time ($\log(t)$): (a) Kaplan-Meier and (b) Bayesian.

Turnbull, and Clark (1997) fit the lagged regression model for piecewise constant values of the lag function with a single unknown change-point. To further emphasize this point, plot (a) in Figure 10.6 clearly indicates that a 0-1 lag function model in (10.4.1) may not be appropriate for our dataset. A rough estimate of η from Figure 10.4 suggests that $\hat{\eta} = 4.8$ for this dataset. In our dataset, 25% of the subjects have failed or were censored before $t = 4.8$. This implies that we would lose information on approximately 25% of the observations for estimating β based on model (10.4.1). That is, 25% of the observations do not contribute to estimating β. In contrast, for the general continuous lag function given in (10.4.5), no information is lost and β is estimated based on the entire dataset. Also, since η is a threshold parameter, precise estimation of η is unlikely in general. Note that even for an optimistic estimate of $\hat{\eta} = 2.8$, approximately 10% of the observations do not contribute to the estimation of β based on model (10.4.1).

Several sensitivity analyses were also conducted by varying the hyperparameters for the prior distributions for $(\beta, \theta, \gamma, \lambda)$. The model was reasonably robust to moderate changes in the hyperparameters. Tables 10.7 and 10.8 show posterior estimates when using $\beta \sim N(0, 100)$, $\alpha \sim \mathcal{G}(1.5, 0.5)$, $\theta \sim \mathcal{G}(1, 0.5)$, $\lambda \sim \mathcal{G}(1, 0.5)$, and $\gamma \sim \mathcal{G}(1, 0.5)$. From Table 10.7, we see that the posterior estimates of (β, λ, α) are quite similar to those of Table 10.5, but posterior estimates of (θ, γ) are somewhat different. Table 10.8 shows posterior estimates of the lag function, which are quite close to the estimates in Table 10.6. It appears that (θ, γ) are the parameters that are affected the most in the sensitivity analyses.

TABLE 10.7. Posterior Summaries: Sensitivity Analysis.

Parameter	Mean	SD	95% HPD interval
β	−0.59	0.60	$(-0.95, -0.17)$
λ	0.006	0.002	$(0.002, 0.011)$
α	1.70	0.17	$(1.39, 2.03)$
γ	0.45	1.23	$(0.00, 2.68)$
θ	0.40	0.85	$(0.00, 1.98)$

Source: Sinha, Ibrahim, and Chen (2000).

TABLE 10.8. Posterior Summaries for Lag Function: Sensitivity Analysis.

t (in years)	$E(l(t)\|D)$	SD	95% HPD interval
0.10	0.45	0.25	(0.02, 0.91)
0.30	0.61	0.23	(0.21, 1.00)
0.50	0.69	0.21	(0.30, 1.00)
1.0	0.78	0.17	(0.46, 1.00)
2.0	0.86	0.13	(0.62, 1.00)
3.0	0.89	0.10	(0.70, 1.00)
4.0	0.92	0.08	(0.76, 1.00)
8.0	0.95	0.05	(0.86, 1.00)
12.0	0.97	0.04	(0.90, 1.00)

Source: Sinha, Ibrahim, and Chen (2000).

10.5 The Poly-Weibull Model

Berger and Sun (1993) discuss an extension of the Weibull distribution, called the poly-Weibull distribution. The poly-Weibull distribution arises in applications involving competing risks in survival analysis. For example, assume that a patient can die either because of a stroke or a heart attack. If the survival times under the two risks are assumed to be independently $\mathcal{W}(\theta_i, \beta_i)$ distributed, then the patient's survival time has a bi-Weibull distribution. Mendenhall and Hadel (1958) and Cox (1959) are among the first authors who addressed competing risks. Other references on competing risks include David and Moeschberger (1978), and Basu and Klein (1982).

The hazard rate corresponding to a bi-Weibull distribution is

$$h(t) = \beta_1 \theta_1^{-\beta_1} t^{\beta_1 - 1} + \beta_2 \theta_2^{-\beta_2} t^{\beta_2 - 1}.$$

The shape of the hazard rate is determined by β_1 and β_2. If $\max(\beta_1, \beta_2) \in (0,1)$ ($\min(\beta_1, \beta_2) > 1$), then the hazard rate is decreasing (increasing). If $\beta_1 < 1$ and $\beta_2 > 1$, then the hazard rate is typically a bathtub curve. It is typically the case that β_1 and β_2 are bounded away from 0.

Now suppose that the survival time depends on m different competing risks, and let the survival time of the jth competing risk be Y_j, which has a $\mathcal{W}(\theta_j, \beta_j)$ distribution, with survival function $S_j(t) = P(Y_j > t) = \exp\left\{-(t/\theta_j)^{\beta_j}\right\}$ for $t > 0$. Now suppose that when the subject has an event, we do not know which competing risk caused the event. This situation is more common in reliability settings in which a system is connected in series. It is less common in biostatistical applications. Thus, failure occurs at $Y = \min(Y_1, Y_2, \ldots, Y_m)$. Assume that the m failure times $Y_1, Y_2, \ldots, Y_m$ are independent, so that the overall survival function is given by

$$S(t) = \prod_{j=1}^{m} S_j(t) = \exp\left\{-\sum_{j=1}^{m}(t/\theta_j)^{\beta_j}\right\}$$

for $t > 0$, and the probability of failure before t is $1 - S(t)$. Then, Y is said to have a poly-Weibull distribution, and its density is given by

$$f(t|\beta_j, \theta_j, j = 1, 2, \ldots, m) = \sum_{j=1}^{m} \frac{\beta_j t^{\beta_j - 1}}{\theta_j^{\beta_j}} \exp\left\{-\sum_{k=1}^{m}(t/\theta_k)^{\beta_k}\right\},$$

for $t > 0$. When the β_j's are all equal, the poly-Weibull distribution is a regular Weibull distribution, and the parameters, $\theta_1, \theta_2, \ldots, \theta_m$ are not identifiable. This is formally not a problem for common Bayesian analyses, because continuous prior densities for the β_j's give probability 0 to parameter equality.

10.5.1 Likelihood and Priors

Let $\boldsymbol{\theta} = (\theta_1, \theta_2, \ldots, \theta_m)'$ and $\boldsymbol{\beta} = (\beta_1, \beta_2, \ldots, \beta_m)'$. Let $y_1, y_2, \ldots, y_d$ be the observed failure times and let $y_{d+1}, y_{d+2}, \ldots, y_n$ be right censored. The likelihood function is then given by

$$L(\boldsymbol{\theta}, \boldsymbol{\beta}|D) = \prod_{i=1}^{d} \sum_{j=1}^{m} \frac{\beta_j y_i^{\beta_j - 1}}{\theta_j^{\beta_j}} \exp\left\{-\sum_{k=1}^{m} \frac{S(\beta_k)}{\theta_k^{\beta_k}}\right\}, \tag{10.5.1}$$

where $D = (n, \boldsymbol{y})$ denotes the data, $\boldsymbol{y} = (y_1, y_2, \ldots, y_n)'$, and

$$S(\beta_k) = \sum_{i=1}^{n} y_i^{\beta_k}. \tag{10.5.2}$$

As Berger and Sun (1993) point out, simple sufficient statistics do not exist and classical approaches to the problem are difficult.

Expanding the product and summation in (10.5.1) results in an expression with m^d terms. For example, if $m = 3$ and $d = 20$, there are 3^{20} terms. Thus, computation via brute force expansion is typically not feasible, and the Gibbs sampler greatly eases the computation.

Assume that $\theta_1, \theta_2, \ldots, \theta_m$, given $\boldsymbol{\beta}$, are independent and that the prior density of θ_j given β_j is

$$\pi_{1j}(\theta_j|\beta_j) \equiv \pi_{1j}(\theta_j|\beta_j, a_j, b_j) = \frac{\beta_j b_j^{a_j}}{\Gamma(a_j)\theta_j^{-(1+\beta_j a_j)}} \exp(-b_j/\theta_j^{\beta_j})$$

where $a_j > 0$ and $b_j > 0$. Thus, $\theta_j^{\beta_j}$ has the inverse gamma $\mathcal{IG}(a_j, b_j)$ distribution. Then the joint density of $\boldsymbol{\theta}$, given $\boldsymbol{\beta}$, is $\pi_1(\boldsymbol{\theta}|\boldsymbol{\beta}) = \prod_{j=1}^m \pi_{1j}(\theta_j|\beta_j)$. Furthermore, assume that $\beta_1, \beta_2, \ldots, \beta_m$ are independent and that β_j has density $\pi_{2j}(\beta_j)$. Thus the prior density of $\boldsymbol{\beta}$ is $\pi_2(\boldsymbol{\beta}) = \prod_{j=1}^m \pi_{2j}(\beta_j)$. Berger and Sun (1993) give guidelines for choosing (a_j, b_j).

10.5.2 Sampling the Posterior Distribution

To facilitate the Gibbs sampler, Berger and Sun (1993) assume that $\pi_{2j}(\beta_j)$ is log-concave in β_j. Furthermore, they introduce the following auxiliary variables. Assume that $y_i = \min(Y_{i1}, Y_{i2}, \ldots, Y_{im})$, where $Y_{i1}, Y_{i2}, \ldots, Y_{im}$ $(1 \le i \le d)$ are independent random variables given $\boldsymbol{\theta}$ and $\boldsymbol{\beta}$ and $Y_{ij} \sim \mathcal{W}(\theta_j, \beta_j)$. Thus, Y_{ij} can be interpreted as the possibly unrealized or unseen failure time for subject i, due to risk j. For $1 \le i \le d$, and $1 \le j \le m$, define $I_{ij} = I(y_i = Y_{ij})$, where $I(.)$ is the indicator function, let $\boldsymbol{I}_j = (I_{i1}, I_{i2}, \ldots, I_{im})'$, $\boldsymbol{I} = (\boldsymbol{I}_1', \boldsymbol{I}_2', \ldots, \boldsymbol{I}_d')'$, $\boldsymbol{Y}_i = (Y_{i1}, Y_{i2}, \ldots, Y_{im})'$, $\boldsymbol{Y} = (\boldsymbol{Y}_1', \boldsymbol{Y}_2', \ldots, \boldsymbol{Y}_d')'$, $\boldsymbol{I}^{(-i)}$ and $\boldsymbol{Y}^{(-i)}$ are the vectors $\boldsymbol{I}$ and $\boldsymbol{Y}$ with $\boldsymbol{I}_i$ and $\boldsymbol{Y}_i'$ deleted, respectively, and $\boldsymbol{\theta}^{(-j)}$ and $\boldsymbol{\beta}^{(-j)}$ are the vectors $\boldsymbol{\theta}$ and $\boldsymbol{\beta}$ with θ_j and β_j deleted, respectively.

There are two possible ways to use these auxiliary variables:

(1) Use the indicators $\boldsymbol{I}$ of cause of failure as auxiliary random variables. Let $\boldsymbol{\xi} = (\boldsymbol{\theta}, \boldsymbol{\beta}, \boldsymbol{I}_1, \boldsymbol{I}_2, \ldots, \boldsymbol{I}_n)$, and sample recursively from the conditional distributions $\pi(\theta_j|\boldsymbol{\theta}^{(-j)}, \boldsymbol{\beta}, \boldsymbol{I}, D)$, $\pi(\beta_j|\boldsymbol{\theta}, \boldsymbol{\beta}^{(-j)}, \boldsymbol{I}, D)$, and $\pi(\boldsymbol{I}_i|\boldsymbol{\theta}, \boldsymbol{\beta}, \boldsymbol{I}^{(-i)}, D)$.

(2) Use the competing risk failure times $\boldsymbol{Y}$ as auxiliary random variables. Let $\boldsymbol{\xi} = (\boldsymbol{\theta}, \boldsymbol{\beta}, \boldsymbol{Y}_1, \boldsymbol{Y}_2, \ldots, \boldsymbol{Y}_d)$, and sample recursively from the conditional distributions $\pi(\theta_j|\boldsymbol{\theta}^{(-j)}, \boldsymbol{\beta}, \boldsymbol{Y}, D)$, $\pi(\beta_j|\boldsymbol{\theta}, \boldsymbol{\beta}^{(-j)}, \boldsymbol{Y}, D)$, and $\pi(\boldsymbol{Y}_i|\boldsymbol{\theta}, \boldsymbol{\beta}, \boldsymbol{Y}^{(-i)}, D)$.

Berger and Sun (1993) note that the first method is considerably more efficient than the second. They show that the conditional density of θ_j given

$(\boldsymbol{\theta}^{(-j)}, \boldsymbol{\beta}, \boldsymbol{I}, D)$ is given by

$$\pi(\theta_j|\boldsymbol{\theta}^{(-j)}, \boldsymbol{\beta}, \boldsymbol{I}, D) = \frac{\beta_j(S(\beta_j)+b_j)^{a_j+N_j}}{\Gamma(a_j+N_j)} \frac{1}{\theta_j^{1+\beta_j(a_j+N_j)}} \times \exp\left(-\frac{S(\beta_j)+b_j}{\theta_j^{\beta_j}}\right), \tag{10.5.3}$$

where $S(\beta_j)$ is given by (10.5.2) and $N_j = \sum_{i=1}^d I_{ij}$. The right side of (10.5.3) does not depend on $\boldsymbol{\theta}^{(-j)}$ and $\boldsymbol{\beta}^{(-j)}$, and the conditional distribution of $\theta_j^{\beta_j}$, given $(\boldsymbol{\theta}^{(-j)}, \boldsymbol{\beta}, \boldsymbol{I}, D)$ is $\mathcal{IG}(a_j+N_j, S(\beta_j)+b_j)$. Hence to simulate an observation from the density (10.5.3), first generate the random variable Y from a $\mathcal{G}(a_j+N_j, 1)$ distribution; then $[S(\beta_j)+b_j)/Y]^{1/\beta_j}$ has the desired density.

Berger and Sun (1993) also show that the conditional density of $\boldsymbol{I}_i$ (with respect to counting measure) given $(\boldsymbol{\theta}, \boldsymbol{\beta}, \boldsymbol{I}^{(-i)}, D)$ is

$$\pi(\boldsymbol{I}_i|\boldsymbol{\theta}, \boldsymbol{\beta}, \boldsymbol{I}^{(-i)}, D) = \beta_j(y_i/\theta_1)^{\beta_1} q^{-1}(y_i|\boldsymbol{\theta}, \boldsymbol{\beta}), \tag{10.5.4}$$

where $\boldsymbol{I}_i = (I_{i1}, \ldots, I_{ij}, \ldots, I_{im})'$ with $I_{ij} = 1$, $I_{ik} = 0$, $k \neq j$, and $q(t|\boldsymbol{\theta}, \boldsymbol{\beta}) = \sum_{k=1}^m \beta_k(t/\theta_k)^{\beta_k}$. Simulating an observation from the discrete distribution (10.5.4) is straightforward. Moreover, Berger and Sun (1993) show that the conditional density of β_j given $(\boldsymbol{\theta}, \boldsymbol{\beta}^{(-j)}, \boldsymbol{I}, D)$ is log-concave.

10.6 Flexible Hierarchical Survival Models

Gustafson (1998) and Carlin and Hodges (2000) present various approaches to developing flexible hierarchical survival models. Following Gustafson (1998), consider modeling the distribution of a survival time T as depending on a vector $\boldsymbol{x} = (x_1, x_2, \ldots, x_p)'$ of explanatory variables. A proportional hazards model specifies the log hazard rate for T as

$$\log h(t|x) = \beta_0(t) + \sum_{j=1}^p x_j\beta_j. \tag{10.6.1}$$

Here $\beta_0(t)$ is the log-baseline hazard function, while $\beta_1, \beta_2, \ldots, \beta_p$ are parameters which modulate the effects of the explanatory variables. The hazard function under any value of $\boldsymbol{x}$ is proportional to the baseline hazard function $\exp\{\beta_0(t)\}$. The partial likelihood method introduced by Cox (1972) permits estimation of $\boldsymbol{\beta} = (\beta_1, \beta_2, \ldots, \beta_p)'$ in the absence of any assumptions about the form of $\beta_0(t)$. The proportional hazards assumption is not appropriate in all situations. Thus the model can be generalized to allow the covariate effects to depend on time, via a log hazard rate of the form

$$\log h(t|x) = \beta_0(t) + \sum_{j=1}^{p} x_j \beta_j(t). \tag{10.6.2}$$

Many authors have suggested particular forms for the time dependence in $\beta_j(t)$, along with corresponding schemes for inference. A smattering of references includes Gamerman (1991), Gray (1992), Kooperberg, Stone, and Truong (1995), and LeBlanc and Crowley (1995).

While (10.6.2) provides more flexibility than (10.6.1), it is still based on the assumption that the covariate effects act additively. That is, the effect on the log hazard associated with the jth explanatory variable is the addition of $x_j\beta_j(t)$, no matter what values the other explanatory variables $(x_1, \ldots, x_{j-1}, x_{j+1}, \ldots, x_p)$ take on. In many situations this is not appropriate, and it is desirable to allow for a log hazard rate that is nonadditive in the covariate effects. Neural networks provide one means to this end. A detailed account of neural networks from a statistical perspective is given by Ripley (1994). A linear predictor of the form $\sum_{j=1}^{p} x_j\beta_j$, with parameters $\{\beta_j\}_{i=1}^{p}$, can be replaced by a predictor of the form $\sum_{j=1}^{k} \varphi(\sum_{i=1}^{p} w_{ij}x_i)$, with parameters $\{w_{ij}\}_{i,j=1}^{p}$. In neural network parlance, this comprises a *multilayer perception network*, with one hidden layer. The nonlinear function φ is the *activation function*, $\sum_{i=1}^{p} w_{ij}x_i$ is the value at the jth of k *hidden units*, and w_{ij} is the *weight* of the connection from the ith input unit (x_i) to the jth hidden unit.

A few authors have applied neural networks to problems of modeling failure time data. Faraggi and Simon (1995) replace the linear term $\sum_j x_j\beta_j$ in (10.6.1) with a neural network. Thus their model permits nonadditive covariate effects, while maintaining proportional hazards. Estimation is carried out by maximizing the partial likelihood function. Liestol, Andersen, and Andersen (1994) consider models for grouped survival data, as well as models based on piecewise constant baseline hazards. They use a neural network structure to relax both the assumption of proportional hazards and the assumption of additive covariate effects.

Because of their flexible nonlinear form, neural networks tend to be valuable for purposes of prediction. However, it can be hard to interpret the unknown parameters or weights. The present goal is to permit nonproportional hazards and nonadditive covariate effects, while retaining some interpretability of the parameters. This is attempted by embedding model (10.6.2) in a modified neural network. Assume that explanatory variables are coded so that a value of $x_j = 0$ represents either an absence of the jth potential risk factor, or an average level of the jth explanatory covariate. This lends specific interpretation to a baseline case corresponding to $\boldsymbol{x} = 0$.

The generalization of (10.6.2) is based on a log hazard rate of the form,

$$\log h(t|x) = \beta_0(t) + \sum_{j=1}^{p} x_j \varphi\left(\sum_{i \neq j} w_{ij} x_i\right) \beta_j(t). \tag{10.6.3}$$

Interpretation is retained in that $\beta_0(t)$ is still the baseline log hazard rate. More importantly, the activation function φ is chosen to satisfy $\varphi(0) = 1$. This makes $\beta_j(t)$ interpretable as the additive effect of the jth explanatory variable, when other explanatory variables are at baseline levels. That is, if $x_i = 0$ for all $i \neq j$, then the log hazard rate is $\beta_0(t) + x_j\beta_j(t)$. Furthermore, the additive model (10.6.2) is recovered in the special case that all the weights are zero.

In fact, model (10.6.3) is used as the first stage of a hierarchical Bayes model. This permits explicit modeling of the degree to which covariate effects are nonadditive and hazards are nonproportional. In particular, the weights w_{ij} and the parameters governing the time-dependence of the covariate effects are assigned priors which favor smaller deviations from additivity and proportionality respectively. This comprises the second stage of the hierarchy. At the third stage, the strength of this favoritism is modeled probabilistically. The net result of stages two and three is a preference for parsimonious models with effects closer to additive and hazards closer to proportional. But the strength of preference is estimated from the data, so that less parsimonious models can receive posterior weight when appropriate. The details of the hierarchical model are presented in the next subsection.

10.6.1 Three Stages of the Hierarchical Model

Gustafson (1998) names the three stages of the hierarchical model as the *data stage*, *smoothing stage*, and *penalty stage*, respectively. Each stage is described in turn.

The Data Stage

The data are modeled as arising from (10.6.3), with specific forms for the covariate effects and activation function. In particular, the log hazard rate for a single observation is modeled as

$$\log h(t|x) = \beta_0(t) + \sum_{j=1}^{p} x_j \varphi\left(\sum_{i \neq j} w_{ij} x_i\right) \beta_j(t),$$

where

$$\beta_j(t) = \begin{cases} \left(1 - \frac{t}{t_0}\right)\beta_{j0} + \frac{t}{t_0}\beta_{j1}, & \text{if } 0 < t < t_0, \\ \beta_{j1}, & \text{if } t \geq t_0, \end{cases} \tag{10.6.4}$$

and

$$\varphi(x) = \frac{2}{1 + \exp\{-x\}}. \tag{10.6.5}$$

In (10.6.4), β_{j0} and β_{j1} are unknown parameters, while t_0 is specified by the user. Since a proportional hazards model obtains when the $\beta_j(t)$'s are constant in time, the linear form in (10.6.4) is an obvious first generalization. The effects are assumed constant to the right of t_0 to ensure that a legitimate log hazard rate obtains. In practice t_0 can be chosen so that most of the data lie in $(0, t_0)$, with the presumption that it is the covariate effects in this interval that are of primary interest. It is unrealistic to expect to make fine distinctions about covariate effects in the right tail of the overall failure time distribution. Consequently, little should be lost by restricting the effects to be constant in this region.

The form of the activation function (10.6.5) follows standard neural network practice of using logistic activation functions. The general form is

$$\varphi(x) = \frac{c}{1 + \exp\{-(x-a)/b\}} + k.$$

As mentioned earlier, we require that $\varphi(0) = 1$ for the sake of interpretability of the covariate effects. The specification $a = 0$, $c = 2$, and $k = 0$ meets this requirement. In addition, it yields limits of 0 and 2 for $\varphi(x)$ as $x \to -\infty$ and $x \to \infty$, respectively. This provides a hard limit on the extent to which the covariate effects can be nonadditive. The specification $b = 1$ is made without loss of generality, since this scale factor can be absorbed into the scale of the weights at the next stage of the hierarchy.

Smoothing Stage

The second stage of the hierarchical model specifies distributions for the neural network weights and the time-effect parameters. As the covariate effects deviate more from additivity and the hazards deviate more from proportionality, the first-stage model is judged to be more complex. The goal is to specify distributions which give less weight to more complex models, as a means of curbing overfit.

The $p(p-1)$ weights w_{ij} are modeled as having independent and identical normal distributions, with mean zero and standard deviation σ. Thus larger-magnitude weights, which correspond to larger deviations from additivity, receive less support from the prior. Models corresponding to strongly nonproportional hazards are penalized similarly. For each i, the conditional distribution of $\beta_{j1}|\beta_{j0}$ is modeled as normal with mean β_{j0} and standard deviation τ. The marginal distribution of β_{j0} is modeled by an improper locally uniform prior density. This is formally equivalent to reversing the roles of the two parameters, in which case $\beta_{j0}|\beta_{j1}$ is normal with mean zero and standard deviation τ, while β_{j1} is locally uniform. The net result is that effects which change more over time, as measured by $|\beta_{j1} - \beta_{j0}|$, receive less support from the prior.

Penalty Stage

The parameters σ and τ can be regarded as penalty parameters which govern the degree to which complexity is penalized. In the extreme case that they are both set equal to zero, the penalty is infinite; a simple model with additive covariate effects and proportional hazards obtains. In the other extreme, as $\sigma \to \infty$ and $\tau \to \infty$, a model which does not penalize complexity obtains. Fixed values of σ and τ can be specified, perhaps based on subjective prior information. But in many instances it is preferable to estimate σ and τ from the current data. In similar models, this is sometimes done using a cross-validation procedure. From the Bayesian viewpoint, however, it is simplest to model the penalty parameters probabilistically as the third stage of the hierarchical model. Then all parameters can be simultaneously estimated via their joint posterior distribution.

Penalty parameter priors are selected to correspond to parsimonious beliefs. Smaller values of σ and τ, which correspond to bigger penalties on complex models, are given more prior weight. Half-Cauchy distributions are used; specifically, the prior densities for σ and τ are taken to be $(2/\sigma_0)[\pi\{1 + (\sigma/\sigma_0)^2\}]^{-1}$ and $(2/\tau_0)[\pi\{1 + (\tau/\tau_0)^2\}]^{-1}$, respectively. The hierarchical model is completely specified once specific values are assigned to the hyperparameters σ_0 and τ_0. There are no obvious default settings for these hyperparameters. It seems necessary to think about their physical meaning in order to arrive at satisfactory specifications. For present purposes, the hyperparameters are set as described below.

Envision a situation where $x_j = 0$ or $x_j = 1$ corresponds to the absence or presence of the jth risk factor thought to be related to the survival time. If only the first risk factor is present, the log hazard rate is $\beta_0(t) + \beta_1(t)$. If only the second risk factor is present, the log hazard is $\beta_0(t)+\beta_2(t)$. Finally, if both risk factors are present, the log hazard is $\beta_0(t) + \varphi(w_{12})\beta_1(t) + \varphi(w_{21})\beta_2(t)$. A parsimonious prior would limit the prior probability of each $\varphi(w_{ij})$ being far from 1, as this limits the deviation of the model from linearity. Therefore, a plausible value of σ is selected to satisfy

$$P\{1 - \epsilon < \varphi(w_{ij}) < 1 + \epsilon\} = \delta, \tag{10.6.6}$$

for particular values of ϵ and δ. We use $\epsilon = 0.05$ and $\delta = 0.5$, which corresponds to a penalty parameter $\sigma = 0.148$. The hyperparameter σ_0 is then determined to make the prior median of σ equal to this plausible value. With the half-Cauchy distributional assumption, this is achieved by setting $\sigma_0 = 0.148$. This sort of argument can be carried over to the case of nonbinary explanatory variables as well, provided the contrast between $x_j = 0$ and $x_j = 1$ is interpretable. For instance, explanatory variables can be coded so that 0 and 1 correspond to minimum and maximum levels, or to particular quantiles of the distribution of levels.

Similar reasoning is applied to set the hyperparameter governing the penalization of nonproportional hazards. It is useful to think temporarily about $h_j(t) = \exp\{\beta_j(t)\}$, the multiplicative effect on the hazard associated with the jth explanatory variable. The degree to which the hazards are nonproportional is then reflected by $h_j(t_0)/h_j(0) = \exp(\beta_{j1} - \beta_{j0})$. So a plausible value of τ can be determined as satisfying

$$P\left\{(1+\epsilon)^{-1} < \frac{h_j(t_0)}{h_j(0)} < (1+\epsilon)\right\} = \delta. \tag{10.6.7}$$

Selecting $\epsilon = 0.05$ and $\delta = 0.5$ gives $\tau = 0.141$. This becomes the prior median for τ, upon choosing $\tau_0 = 0.141$. Again, the rationale for setting the prior is fairly general, since the quantity $h_j(t_0)/h_j(0)$ is invariant under scale changes to the data (from days to months for instance), provided that t_0 scales in the obvious way.

10.6.2 Implementation

Gustafson (1998) uses the hybrid MCMC algorithm, which is discussed in Section 1.6, to sample from the posterior distribution. The primary requirement for implementation of the hybrid MCMC sampler is evaluation of the negative logarithm of the unnormalized posterior density, along with its partial derivatives with respect to parameter components. This breaks down into additive components from the three stages of the hierarchical model.

The first-stage contribution corresponds to the distribution of the response times given the weights $\boldsymbol{w}$ and the effect parameters $\boldsymbol{\beta}$. Let h and H denote hazard and cumulative hazard functions, respectively, and let ν be a censoring indicator. A value of $\nu = 1$ ($\nu = 0$) corresponds to an uncensored (right censored) observation. The contribution of a single case $(t, \boldsymbol{x}, \nu)$ to the negative log posterior density is $H(t|\boldsymbol{x}) - \nu \log h(t|\boldsymbol{x})$. This can be expressed as

$$t_0 e^{d_0}\left\{\frac{e^{d_1 s}-1}{d_1} + \left(\frac{t}{t_0} - s\right)e^{d_1}\right\} - \delta \log(d_0 + s d_1), \tag{10.6.8}$$

where $s = \min\{(t/t_0), 1\}$, and

$$d_0 = \beta_{00} + \sum_{j=1}^{p} x_j \varphi\left(\sum_{i \neq j} w_{ij} x_i\right) \beta_{j0},$$

$$d_1 = (\beta_{01} - \beta_{00}) + \sum_{j=1}^{p} x_j \varphi\left(\sum_{i \neq j} w_{ij} x_i\right) (\beta_{j1} - \beta_{j0}).$$

That is, d_0 and $d_0 + d_1$ correspond to the log hazard rate evaluated at $t = 0$ and $t = t_0$, respectively. Partial derivatives of (10.6.8) with respect to the components of $\boldsymbol{\beta}$ and $\boldsymbol{w}$ can be computed by the chain rule, starting with differentiation of (10.6.8) with respect to d_0 and d_1. The overall first-stage contribution to the negative log posterior density is a sum of terms (10.6.8) over all the cases in the dataset.

A general caution is that the performance of the hybrid algorithm can depend quite strongly on parameterizations chosen for the model. Due to the algorithm's use of tangent approximations, it is desirable to avoid parameterizations under which second partial derivatives of the log posterior density are unbounded. In particular, reciprocal parameterizations are used for σ and τ, which are standard deviations of normal distributions. If we let $\gamma_\sigma = 1/\sigma$, and $\gamma_\tau = 1/\tau$, then the second-stage contribution to the negative log posterior density is

$$\frac{\gamma_\sigma^2}{2}\left(\sum_{j=1}^{p}\sum_{i \neq j} w_{ij}^2\right) - (p)(p-1)\log\gamma_\sigma + \frac{\gamma_\tau^2}{2}\left\{\sum_{i=0}^{p}(\beta_{j0} - \beta_{j1})^2\right\}$$
$$- (p+1)\log\gamma_\tau. \qquad (10.6.9)$$

Finally, upon transforming the half-Cauchy prior distributions for σ and τ to the reciprocal scale, we have a third-stage contribution of

$$\log\{1 + (\sigma_0\gamma_\sigma)^2\} + \log\{1 + (\tau_0\gamma_\tau)^2\}. \qquad (10.6.10)$$

Partial derivatives of both (10.6.9) and (10.6.10) are easily evaluated.

Example 10.7. Simulated data. Gustafson (1998) considers a simulation study of the predictive ability of the hierarchical Bayes model. Regard the p binary components of the explanatory vector $\boldsymbol{x}$ as indicating the absence (0) or presence (1) of p potential risk factors related to the survival time T. The time T itself is simulated as an exponential random variable; for the sake of simplicity, no censoring is introduced. The constant log hazard rate is taken to be of the form $\lambda(s)$, where $s = \sum_{i=1}^{p} x_i$ is the total number of risk factors present. Thus, there is no distinction between the different risk factors. This is not a realistic assumption, but it is a useful test case for the simulation study. Since all hazard rates are constant in time, a proportional hazards model obtains. The model is only additive, however, when λ is specified as a linear function of s.

Three specifications for $\lambda(s)$ are considered:

$$\lambda(s) = (\log 2)s, \tag{10.6.11}$$

$$\lambda(s) = (\log 2)(2 - 2^{1-s}), \tag{10.6.12}$$

$$\lambda(s) = (\log 2)\min\{1, s\}. \tag{10.6.13}$$

The first model (10.6.11) yields a log hazard rate that is additive in the components of the explanatory vector $\boldsymbol{x}$. Thus the effect associated with any particular risk factor does not depend on the presence/absence of the other risk factors. In contrast, under the model (10.6.12) the log hazard is not additive in $\boldsymbol{x}$. Rather, the increase in log hazard associated with $s = j + 1$ risk factors relative to $s = j$ risk factors is half that associated with $s = j$ relative to $s = j - 1$. Thus any particular risk factor has a stronger effect on the log hazard when fewer other risk factors are present. This sort of structure is taken to the extreme in model (10.6.13), where a particular risk factor has an effect on the log hazard only when no other risk factors are present. That is, the hazard rate is 1 when no risk factors are present, and 2 when one or more risk factors are present. All three models yield $\lambda(0) = 0$ and $\lambda(1) = \log 2$, so that the effect of a particular risk factor in the absence of all other risk factors is a doubling of the hazard rate.

Each simulated dataset is split into a *training set* and a *validation set*. Interest focuses on the ability of the hierarchical Bayes model to use the training set to predict the outcomes in the validation set. The quality of the prediction is measured by the log predictive score, $\sum_{(\boldsymbol{x},t)} \log f(t|\boldsymbol{x})$. Here, $f(\cdot|\boldsymbol{x})$ is the density used to predict a survival time associated with the explanatory vector $\boldsymbol{x}$, and the sum is over all $(\boldsymbol{x}, t)$ pairs in the validation set. From the Bayesian point of view, the prediction is based on the predictive density. That is, $f(t|\boldsymbol{x}) = f(t|\boldsymbol{x}, D)$, where D represents the training data. The predictive density is determined as

$$f(t|\boldsymbol{x}, D) = \int f(t|\boldsymbol{x}, \boldsymbol{\theta})\pi(\boldsymbol{\theta}|D)\, d\boldsymbol{\theta},$$

where $\pi(\boldsymbol{\theta}|D)$ is the posterior distribution of the parameter vector $\boldsymbol{\theta}$ given the training sample data D. In comparison, if a model is fitted to the training data using maximum likelihood estimation, then prediction can be based on $f(t|\boldsymbol{x}) = f(t|\boldsymbol{x}, \boldsymbol{\theta})|_{\boldsymbol{\theta}=\hat{\boldsymbol{\theta}}_D}$, where $\hat{\boldsymbol{\theta}}_D$ is the maximum likelihood estimator of $\boldsymbol{\theta}$ based on the training sample data only.

Datasets of 300 cases are simulated for each of the three underlying models, and for each of $p = 3$ and $p = 6$. In every instance, the explanatory $\boldsymbol{x}$ vectors are randomly sampled from the uniform distribution over all 2^p possible binary vectors. Each dataset is split into a training sample of 200 cases, and a validation sample of 100 cases. Predictive scores will depend on the particular partition used to divide the data. Therefore, three splits are considered for each dataset. The splits correspond to the validation sample comprising cases 1 to 100, cases 101 to 200, and cases 201 to 300, respectively.

In all cases, the hierarchical Bayes model is implemented with $t_0 = 1.5$, so that covariate effects are linear on $t \in (0, 1.5)$, and constant thereafter. Under model (10.6.11), which produces the shortest-tailed data of the three models, $t_0 = 1.5$ corresponds approximately to the 95th percentile of the overall data distribution, when $p = 3$. When $p = 6$, $t_0 = 1.5$ corresponds to the 99.1th percentile of the data distribution.

For each training set, three independent Markov chains are simulated to assess the posterior distribution of the parameters. In implementing the hybrid MCMC scheme, a step size of $\epsilon = 0.005$ is utilized in (1.6.3) and (1.6.4). In most instances, this yields a rejection rate in the range of 2% to 10%, though the rejection rate is as high as 16% for a few chains. Each chain is based on 25,000 iterations, the first 5000 of which are discarded. From the remaining 20,000 iterations, parameter values at every fourth iteration are retained for analysis. Thus each posterior sample is of size 5000. For each observation in the validation set, the predictive density is estimated by the model density averaged across parameter values in the posterior sample. The log predictive densities are then summed to arrive at the predictive score.

For the sake of comparison, a parametric Weibull regression model is also fitted to each training sample, using maximum likelihood estimation. In particular, the log survival time is modeled as $\kappa_0 + \sum_{i=1}^{p} x_i \kappa_i + \sigma \epsilon_i$, where ϵ_i follows an extreme value distribution. This is equivalent to a proportional hazards model with a Weibull baseline hazard function. The predictive score for this model is regarded as the baseline predictive score, while predictive scores based on the hierarchical Bayes model are reported as deviations from this baseline score. As a further comparison, the Weibull regression model is expanded to include quadratic interaction terms. That is, terms of the form $x_i x_j$ $(i < j)$ are also included in the model for the log survival time. Again, maximum likelihood is used to fit this model to training samples. The resulting predictive scores are also reported as deviations from the baseline score. Results for all true and fitted models are reported in Tables 10.9 and 10.10, for the cases of $p = 3$ and $p = 6$ explanatory variables, respectively.

In Table 10.9, the rows correspond to different splits of the data, nested within different models generating the data. The first column gives the predictive score of the baseline Weibull regression model, the second column gives the predictive score of the Weibull regression model with interaction terms, relative to the baseline model, and the third through fifth columns are predictive scores of the Bayes model relative to the baseline model, based on three independent Markov chain samples. Table 10.9 indicates that the hierarchical Bayes model always outperforms the baseline model,

TABLE 10.9. Predictive Scores for the Simulation Study, with $p = 3$ Risk Factors.

			Relative Scores			
		Baseline	Interaction		HB Model	
Model	Split	Score	Model	Chain 1	Chain 2	Chain 3
1	1	25.0	−0.2	0.2	0.4	0.1
	2	8.7	0.2	0.8	0.7	0.9
	3	7.2	−0.4	0.4	0.1	0.3
2	1	−25.6	5.0	6.5	6.4	5.8
	2	−10.1	0.4	1.6	1.0	1.5
	3	−29.8	3.5	5.6	6.0	5.8
3	1	−47.5	−3.7	2.1	2.3	1.5
	2	−44.2	2.7	0.5	0.5	1.0
	3	−39.3	0.6	0.8	1.4	0.8

Source: Gustafson (1998).

TABLE 10.10. Predictive Scores for the Simulation Study, with $p = 6$ Risk Factors.

			Relative Scores			
		Baseline	Interaction		HB Model	
Model	Split	Score	Model	Chain 1	Chain 2	Chain 3
1	1	110.3	−5.9	2.1	2.2	0.9
	2	100.3	0.3	2.9	3.4	3.6
	3	17.0	−6.8	−1.9	−1.8	−1.9
2	1	15.0	−2.5	−1.2	−0.5	−0.7
	2	−5.2	-22.3	2.4	1.7	2.6
	3	9.5	0.6	−0.1	0.0	−0.5
3	1	−53.9	−4.8	1.4	3.2	1.6
	2	−38.2	0.3	−0.7	−1.2	0.1
	3	−24.6	−7.8	0.2	0.1	−2.4

Source: Gustafson (1998).

according to the predictive score criterion. Not surprisingly, the gains are slight under model 1, when the baseline model is correctly specified. The gains are more substantial under model 2, at least for two of the three splits of the data. The gains are more modest under model 3, despite the fact that model 3 involves a stronger departure from additive effects than does model 2. With the exception of the second split under model 3, the hierarchical Bayes model outperforms the Weibull model with interaction terms as well. Overall, the performance of the hierarchical Bayes model appears to be quite satisfactory. This is especially true in light of the fact that both Weibull models postulate proportional hazards, which is a correct specification for all the simulated datasets.

The layout in Table 10.10 is the same as in Table 10.9. In Table 10.10, however, the relative performance of the hierarchical model is mixed. There are no clear winners and losers in the comparison between the Bayes model and the two Weibull models. The hierarchical Bayes model, however, does not exhibit the very poor performance of the interaction model for the second split of the model 2 dataset.

Example 10.8. Nursing home usage data. Gustafson (1998) uses the hierarchical Bayes model to fit a dataset on duration of nursing home stays, which was analyzed previously by Morris, Norton, and Zhou (1994). These authors give a more detailed account of the study than the brief description which follows here. The data arose from an experiment, conducted from 1980 to 1982, at 36 for-profit nursing homes in San Diego, California. Data on 1601 patients are available. The response variable is a patient's length of stay in the nursing home, recorded in days. This response is right censored for 20% of the patients in the study. Half of the participating nursing homes were offered financial incentives to accept Medicaid patients, improve their health, and discharge them to their homes in a timely fashion. Patients from these homes compose the treatment group, while patients from the other homes compose the control group. Variables thought to relate to a patient's length of stay are also recorded. These include age, gender, marital status, and health status.

The random allocation to treatment and control groups in this study is made at the nursing home level rather than the patient level. As pointed out by Gustafson (1998), this makes both the present patient-level analysis and the analysis of Morris, Norton, and Zhou (1994) less attractive. Unfortunately, the available data only indicate whether a patient stayed at a treatment or control group nursing home. No information is provided to indicate which patients compose the subsamples corresponding to individual nursing homes. As a result, any nursing-home level analysis is impossible.

As in Morris, Norton, and Zhou (1994), the response variable is slightly transformed for modeling purposes. In particular, the length of stay plus two days is treated as the response time. This avoids problems with the ten cases recorded with stays of zero days. The patient's ages range from 65 years to 104 years. These are transformed to a $[0, 1]$ scale according to $\text{Age}^* = \{\min(\text{Age}, 90) - 65\}/25$. The truncation at 90 years was found to be useful by Morris, Norton, and Zhou (1994). The treatment group indicator is coded as zero for the control group and one for the treatment group. Gender is coded as zero for female and one for male. Marital status is coded as zero for not married and one for married. The original coding of health status is on a scale from one (best) to five (worst), though two is the minimum score amongst patients in the study. Morris, Norton, and Zhou (1994) separately model the effects of the four attained levels using dummy variables. Bearing in mind that the number of weights grows quadratically in the number of variables, we dichotomize health status to be zero (original score two or three) or one (original score four or five).

The data are split into a training sample of 800 cases, and a validation sample of 801 cases. A randomized split is utilized, to avoid any systematic trends with the order in which cases appear in the dataset. The 166 censored cases in the validation sample are discarded, since it is of interest to predict observed duration times rather than censoring times. This leaves 635 cases in the validation set.

In the combined dataset, there are only 129 observed failure times which exceed one year, while there are 448 censoring times which exceed one year. This suggests that estimation of covariate effects will be difficult at times much beyond one year. Therefore, a value of $t_0 = 365$ is specified. That is, covariate effects on the log hazard scale are modeled as linear in time from $t = 0$ to $t = 365$, and constant in time thereafter. Again, the hyperparameters σ_0 and τ_0 are assigned the values determined in Subsection 10.6.1.

To evaluate posterior quantities based on the training set, the hybrid Monte Carlo algorithm is implemented with a step size of $\epsilon = 0.01$. Five independent production chains are implemented, each based on 85,000 iterations, with the first 5000 iterations discarded as a burn-in phase. The negative log unnormalized posterior density (the "potential energy" in the hybrid Monte Carlo algorithm) tends to fall off to a consistent range during this burn-in phase. From the post-burn-in sample paths, every sixteenth iteration is retained for analysis, giving a sample of size 5000 on which to base analysis. Rejection rates in the range of 3% to 11% are observed in the chains.

Gustafson (1998) examined time-series plots of the retained values for σ, τ, and the potential energy for each production chain and found that the posterior marginal distribution of σ appears to be bimodal. Moreover, the Markov chain sampler has considerable difficulty in moving between the small σ and large σ regions of the parameter space. As a result it is not possible reliably to estimate the probabilities associated with the two modes. Moreover, the production chains would fail formal tests for convergence, such as the criterion of Gelman and Rubin (1992). This is not a specific criticism of the hybrid algorithm; most MCMC schemes have difficulty dealing with multimodality. Nevertheless, this sampler behavior weakens posterior inferences. Fortunately, inference about many of the parameters seems to be insensitive to the value of σ. For example, inference about τ is stable across the production chains. To formalize this notion, Gustafson (1998) reported simulation errors, as described below.

Table 10.11 shows estimated posterior means and posterior standard deviations of the parameters. In Table 10.11, the estimated posterior means and standard deviations are averages of estimates from the five independent production chains, and the two rightmost columns give simulation standard errors based on the across–chain averages. In addition, both simulation standard errors are reported as percentages of the estimated posterior standard deviation. For the sake of interpretation, covariate effects are parameterized as $\gamma_j = (\beta_{j0} + \beta_{j1})/2$ and $\nu_j = \beta_{j1} - \beta_{j0}$. Hence, for the jth covariate effect, γ_j is the additive effect on the log hazard at $t = 6$ months, while ν_j is the difference between the effect at $t = 12$ months and the effect at $t = 0$ months. The indices $j = 1$ through $j = 5$ correspond to age, treatment group, gender, marital status, and health status, respectively. For a given parameter, the posterior mean is estimated separately from each chain using a sample average. These five estimates are then pooled to give an overall estimate and a simulation standard error. The same procedure is used to estimate the posterior standard deviation of the parameter. For both the posterior mean and posterior standard deviation, the simulation standard error is reported as a percentage of the estimated posterior standard deviation. This is designed to facilitate quick comparison of the numerical and statistical uncertainty associated with parameter estimates.

Some of the parameter estimates have large simulation standard errors relative to their estimated posterior standard deviations. This is especially true of some of the weights, w_{ij}, and the smoothing parameter σ which governs the weights. For each covariate effect, the estimated posterior mean of ν_j is negative, which suggests that covariate effects decrease with time. In some cases, however, the effect is dwarfed by a relatively large estimated posterior standard deviation. To assess the posterior distributions of the covariate effect parameters more carefully, 80% credible regions are reported for γ_j and ν_j in Table 10.12. In each case, the reported interval is the shortest interval containing 80% of the sampled parameter values. The credible intervals for γ_j suggest that younger patients, patients in the control group, and male patients all tend to have a higher hazard rate at six months. The credible intervals for ν_j suggest that the age effect on the log hazard is decreasing in time.

As with the simulated data, the sum of the log predictive density over the cases in the validation set is considered as a predictive score. The predictive score is estimated separately from each of the five production chains. Somewhat surprisingly, there is very little variation across chains; the across-chain average predictive score is -3854.3, with a simulation stan-

TABLE 10.11. Estimated Posterior Means and Standard Deviations of Parameters for the Nursing Home Usage Data.

	Parameter	Mean	SD	Simulation SE Mean	Simulation SE SD
Intercept	γ_0	−5.61	0.17	12%	3%
Age	γ_1	−0.50	0.24	14%	4%
Treatment	γ_2	−0.33	0.20	11%	9%
Gender	γ_3	0.25	0.17	11%	5%
Married	γ_4	0.13	0.24	11%	13%
Health	γ_5	0.22	0.22	23%	10%
Intercept	ν_0	−1.36	0.32	4%	2%
Age	ν_1	−0.74	0.42	5%	2%
Treatment	ν_2	−0.04	0.39	10%	9%
Gender	ν_3	−0.24	0.35	6%	5%
Married	ν_4	−0.07	0.54	11%	13%
Health	ν_5	−0.56	0.43	17%	9%
	ω_{12}	−0.62	1.21	19%	19%
	ω_{13}	−0.08	1.11	13%	31%
	ω_{14}	−0.47	1.35	22%	30%
	ω_{15}	−0.22	0.87	16%	23%
	ω_{21}	−0.31	0.85	20%	38%
	ω_{23}	−0.65	1.67	30%	33%
	ω_{24}	0.09	1.19	16%	24%
	ω_{25}	−0.72	1.16	27%	31%
	ω_{31}	−0.48	1.17	27%	38%
	ω_{32}	0.38	1.29	23%	37%
	ω_{34}	−0.08	1.22	8%	29%
	ω_{35}	0.45	1.00	16%	25%
	ω_{41}	−0.34	1.19	18%	24%
	ω_{42}	−0.01	1.29	17%	31%
	ω_{43}	−0.33	1.41	21%	29%
	ω_{45}	1.22	1.84	28%	38%
	ω_{51}	−1.09	1.35	23%	21%
	ω_{52}	−0.73	1.26	22%	30%
	ω_{53}	0.31	1.14	18%	29%
	ω_{54}	−0.19	1.38	10%	30%
Weights	σ	1.06	1.03	30%	31%
Effects	τ	0.83	0.31	7%	5%

Source: Gustafson (1998).

TABLE 10.12. Shortest 80% Credible Intervals for Parameters.

	Parameter	80% Credible Interval
Intercept	γ_0	(−5.84, −5.40)
Age	γ_1	(−0.80, −0.16)
Treatment	γ_2	(−0.50, −0.08)
Gender	γ_3	(0.04, 0.45)
Married	γ_4	(−0.16, 0.35)
Health	γ_5	(−0.06, 0.54)
Intercept	ν_0	(−1.77, −0.95)
Age	ν_1	(−1.23, −0.17)
Treatment	ν_2	(−0.47, 0.44)
Gender	ν_3	(−0.64, 0.18)
Married	ν_4	(−0.74, 0.43)
Health	ν_5	(−1.05, 0.06)

Source: Gustafson (1998).

dard error of 0.5. This compares favorably to a predictive score of −3863.9 based on maximum likelihood fitting of a Weibull regression model without interaction terms, and a score of −3865.8 based on an expanded model with interaction terms included. The gain in predictive score of roughly 10 units on the log density scale is somewhat modest, however, in light of the fact that there are over 600 cases in the validation sample.

The hyperparameter specifications described in Subsection 10.6.1. are somewhat arbitrary. Therefore it is of interest to assess the effect of changes in the hyperparameters on posterior quantities. The prospect of redoing the analysis for several sets of hyperparameter values is unappealing in light of the heavy computational burden. Instead, Gustafson (1998) used importance sampling to estimate posterior quantities based on alternate hyperparameters, using the production chains based on the original specifications. The use of importance sampling for this sort of sensitivity analysis is discussed by several authors, including Smith and Gelfand (1992), and Besag, Green, Higdon, and Mengersen (1995). Based on a posterior sample $\{\boldsymbol{\theta}_l\}_{l=1}^{L}$, the posterior mean of $\varphi^*(\boldsymbol{\theta})$, where φ^* is a function of $\boldsymbol{\theta}$, under the original hyperparameters is estimated as the sample average $\sum_{l=1}^{L} \omega_l \varphi^*(\boldsymbol{\theta}_l)$, where $\omega_l = 1/L$. Using the same sample, the posterior mean of $\varphi^*(\boldsymbol{\theta})$ under the alternate hyperparameters is estimated by taking ω_l to be proportional to the ratio of posterior densities evaluated at $\boldsymbol{\theta}_l$. The numerator and denominator posterior densities correspond to the alternate and original hyperparameters, respectively. The normalization $\sum_{l=1}^{L} \omega_l$ is required. Because of the hierarchical structure of the present model, the ratio of posterior densities is proportional to the ratio of prior densities for (σ, τ). This is a substantial computational simplification.

Two alternate values are considered for each of the two hyperparameters. Recall that $\sigma_0 = 0.148$ is based on (10.6.6) with $\epsilon = 0.05$ and $\delta = 0.5$. Alternate values of $\delta = 0.85$ and $\delta = 0.15$ lead to alternate hyperparameter values $\sigma_0 = 0.070$ and $\sigma_0 = 0.529$, respectively. Similarly, replacing $\delta = 0.5$ in (10.6.7) with $\delta = 0.85$ and $\delta = 0.15$ yields $\tau_0 = 0.067$ and $\tau_0 = 0.504$. Table 10.13 shows the estimated posterior means of parameters under the alternate hyperparameter values for four cases: $\sigma_0 = 0.070$ and $\tau_0 = 0.067$ (I); $\sigma_0 = 0.529$ and $\tau_0 = 0.067$ (II); $\sigma_0 = 0.070$ and $\tau_0 = 0.504$ (III); and $\sigma_0 = 0.529$ and $\tau_0 = 0.504$ (IV). The sensitivity of inferences to the prior is a concern when this source of uncertainty is comparable in magnitude to statistical uncertainty due to sampling. Consequently, in Table 10.13, the reported value for each parameter is the deviation between the estimated posterior means under the alternate and original hyperparameters, divided by the corresponding estimated posterior standard deviation under the original hyperparameters. Thus the uncertainty due to the hyperparameter specification is reported relative to the statistical uncertainty about parameter values. In these terms, the most highly sensitive posterior mean is that for σ, which controls the degree of deviation from an additive model. The entries in Table 10.13 should be treated with some caution, in light of the large simulation errors described above. Results of computations not shown indicate that the predictive score for the validation set is very robust to changes in the hyperparameter specification.

10.7 Bayesian Model Diagnostics

Model checking and adequacy play an important role in models for survival data. However, there is very little literature to address this problem. Only recently, Shih and Louis (1995) and Gelfand and Mallick (1995, for univariate survival data), study the issues of goodness of fit in a Bayesian framework. Sahu, Dey, Aslanidou, and Sinha (1997) propose some Bayesian exploratory data analysis (EDA) diagnostic methods for assessing goodness of fit for the proportional hazards frailty model. In this section, we discuss two Bayesian model diagnostic techniques used in survival analysis. These are Bayesian latent residuals (Aslanidou, Dey, and Sinha, 1998) and prequential methods (Arjas and Gasbarra, 1997).

10.7.1 Bayesian Latent Residuals

In the classical framework, an assessment of the hazard function in the presence of censored data is usually done using a hazard plot. More specifically, a parametric specification for the baseline hazard, $h_0(t)$, can be checked using an empirical estimate of $h_0(t)$, say, $\hat{h}_0(t)$. Plots of $\hat{h}_0(t)$ vs. t or $\log(t)$ can be compared with plots assuming the parametric model. An alternative

TABLE 10.13. Sensitivity Analysis for Posterior Means.

		Cases			
	Parameter	I	II	III	IV
Intercept	γ_0	−0.17	0.28	−0.16	0.30
Age	γ_1	0.22	−0.35	0.19	−0.38
Treatment	γ_2	0.12	−0.23	0.11	−0.25
Gender	γ_3	0.00	−0.01	0.00	−0.02
Married	γ_4	−0.04	0.07	−0.03	0.10
Health	γ_5	0.24	−0.46	0.20	−0.50
Intercept	ν_0	−0.03	0.04	−0.06	0.03
Age	ν_1	0.10	−0.13	0.09	−0.15
Treatment	ν_2	0.07	−0.12	0.08	−0.12
Gender	ν_3	−0.02	0.04	−0.02	0.04
Married	ν_4	−0.10	0.16	−0.07	0.20
Health	ν_5	0.19	−0.35	0.14	−0.40
	ω_{12}	0.14	−0.27	0.11	−0.30
	ω_{13}	0.02	−0.05	0.02	−0.06
	ω_{14}	0.13	−0.24	0.11	−0.26
	ω_{15}	0.05	−0.11	0.03	−0.15
	ω_{21}	0.14	−0.22	0.12	−0.23
	ω_{23}	0.07	−0.14	0.05	−0.16
	ω_{24}	−0.04	0.08	−0.05	0.06
	ω_{25}	0.21	−0.39	0.17	−0.42
	ω_{31}	0.15	−0.28	0.12	−0.31
	ω_{32}	−0.12	0.22	−0.11	0.22
	ω_{34}	0.01	0.00	0.00	−0.01
	ω_{35}	−0.16	0.29	−0.13	0.31
	ω_{41}	0.10	−0.21	0.08	−0.23
	ω_{42}	−0.02	0.02	−0.02	0.00
	ω_{43}	0.04	−0.08	0.02	−0.10
	ω_{45}	−0.21	0.40	−0.17	0.43
	ω_{51}	0.24	−0.46	0.20	−0.50
	ω_{52}	0.17	−0.34	0.14	−0.37
	ω_{53}	−0.06	0.11	−0.05	0.10
	ω_{54}	0.03	−0.06	0.01	−0.08
Weights	σ	−0.30	0.55	−0.26	0.58
Effects	τ	−0.10	0.14	0.06	0.30

Source: Gustafson (1998).

approach in the frequentist paradigm for univariate survival data is based on a residual analysis, where a theoretical or empirical Q-Q plot, (see, e.g., Lawless, 1982; Schoenfeld, 1982; Cox and Oakes, 1984; Therneau, Grambsch, and Fleming, 1990) is developed for certain exponential or martingale based residuals. We now examine this framework further under the the Bayesian paradigm.

Under the proportional hazards frailty model (4.1.1), let $H_0(t)$ be the cumulative hazard function. Suppose we divide the time axis into J prespecified intervals $I_k = (s_{k-1}, s_k]$ for $k = 1, 2, \ldots, J$, and assume the baseline hazard to be constant within these intervals. Then it follows that for $t \in I_k = (s_{k-1}, s_k]$, $k = 1, 2, \ldots, J$, $H_0(t) = \sum_{l=1}^{k} \Delta_l \lambda_l$, where $\Delta_l = s_l - s_{l-1}$. Further, we define

$$u_{ij} = u(y_{ij}|w_i, \boldsymbol{x}_{ij}) = H_0(y_{ij})\theta_{ij}w_i, \tag{10.7.1}$$

where $\theta_{ij} = \exp(\boldsymbol{x}_{ij}'\boldsymbol{\beta})$ for $i = 1, 2, \ldots, n$ and $j = 1, 2, \ldots, m_i$. See Section 4.1 for the definition of w_i, $\boldsymbol{x}_{ij}$, y_{ij}, and $\boldsymbol{\beta}$.

Recall $h_0(t) = \lambda_k$ if $t \in I_k = (s_{k-1}, s_k]$. Then the survival function is given by

$$S_{ij}(t|w_i, \boldsymbol{x}_{ij}) = \exp\left\{-\theta_{ij}w_i \int_0^t h_0(u)\, du\right\} = \exp\{-\theta_{ij}w_i H_0(t)\}.$$

Thus, the probability density function of the survival time T_{ij} is given as

$$f_{ij}(t|w_i, \boldsymbol{x}_{ij}) = h_0(t)\theta_{ij}w_i \exp\{-\theta_{ij}w_i H_0(t)\},$$

and

$$\begin{aligned} P(u_{ij} \le u) =& P(\theta_{ij}w_i H_0(t) \le u) = P(T_{ij} \le H_0^{-1}(u/(\theta_{ij}w_i))) \\ =& 1 - \exp\{-\theta_{ij}w_i(u/(\theta_{ij}w_i))\} = 1 - \exp(-u). \end{aligned} \tag{10.7.2}$$

Thus, (10.7.2) implies that given w_i, u_{ij} has a standard exponential distribution. Further, if $v_{ij}(t) = \exp(-u_{ij}(y_{ij}))$, then v_{ij} given w_i has a standard uniform distribution. Also, conditional on w_i, the v_{ij}'s are independent. Then, the v_{ij}'s can be treated as standardized residuals. We call the v_{ij}'s Bayesian latent residuals, since they are functions of the unobserved frailty random variable w_i. Using the v_{ij}'s, Aslanidou, Dey, and Sinha (1998) propose two diagnostic plots, using the output from the MCMC samples. First, if the model is correct, then the v_{ij}'s have a uniform distribution. Thus, a box-plot of the Monte Carlo estimates of the v_{ij}'s can be used to check model adequacy for a given dataset. Alternatively, a Q-Q plot of v_{ij} versus a standard uniform distribution produces similar features. The estimates of the v_{ij}'s are obtained as $\hat{v}_{ij} = L^{-1}\sum_{l=1}^{L} v_{ij}^l$, where the v_{ij}^l's are obtained from (10.7.1) as

$$v_{ij}^l = \exp(-H_0(y_{ij})^l \theta_{ij}^l w_i^l).$$

Here, $H_0(y_{ij})^l$, θ_{ij}^l, and w_i^l are the values of $H_0(y_{ij})$, θ_{ij}, and w_i computed at the lth MCMC iteration for $l = 1, 2, \ldots, L$, and L is the total number of MCMC iterations.

Example 10.9. Kidney infection data. Aslanidou, Dey, and Sinha (1998) reanalyze the kidney infection data given in Example 1.4. They assume that $w_i \sim \mathcal{G}(\kappa^{-1}, \kappa^{-1})$, so that $E(w_i) = 1$ and $\text{Var}(w_i) = \kappa$. The prior for the hyperparameter κ is $\mathcal{G}(6, 1)$ with $E(\kappa) = \text{Var}(\kappa) = 6$ to assure enough heterogeneity among the patients. Since $\eta = \kappa^{-1}$, the prior for η is taken to be an inverse gamma, $\mathcal{IG}(6, 1)$. For the λ_k's, $k = 1, 2, \ldots, J$, they consider the priors $\lambda_1 \sim \mathcal{G}(0.7, 0.7)$ and $\lambda_k|\lambda_{k-1} \sim \mathcal{G}(0.7, 0.7/\lambda_{k-1})$ for $k = 2, 3, \ldots, J$. They consider only one covariate, i.e., sex, in their analysis. For β, the regression coefficient corresponding to sex, they take a $N(-1.2, 100)$ prior. The prior mean is chosen near the estimate found by other analyses of this dataset. The variance of the prior is taken large enough to incorporate sufficient diffuseness. Finally, they divide the survival times of the patients into $J = 20$ equal intervals.

For the model checking analysis, they obtain the following results. The first infection of the 19th, 35th, and 36th patients did not have a nice fit in the model like the rest of the subjects for the first infection. This can be concluded from the box-plots of the quantities $v_{19,1}$, $v_{35,1}$, and $v_{36,1}$, which did not seem to follow a uniform distribution. A similar phenomenon is observed for the second infections of the 14th, 15th, and 22nd patients. For illustration, Figure 10.7 shows the box-plots of the first infection for the 16th patient, who fit nicely to the model, and the first infection of the 36th patient who did not have a nice fit.

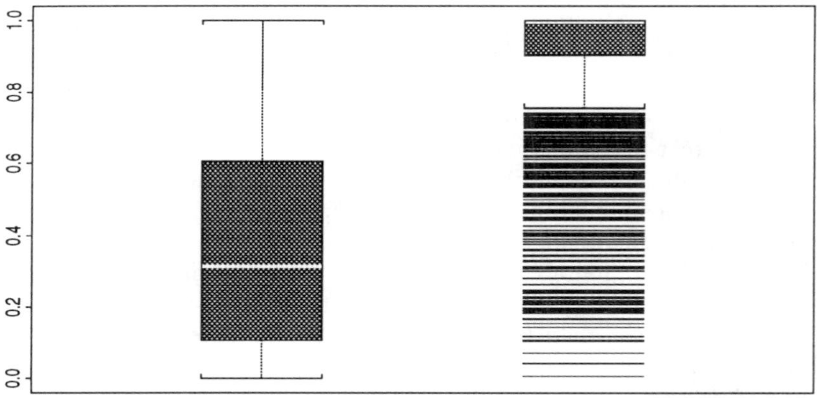

FIGURE 10.7. Box-plots of $v_{16,1}$ and $v_{36,1}$.

10.7.2 Prequential Methods

Arjas and Gasbarra (1997) extend the prequential approach introduced by Dawid (1992) to continuous time marked point processes.

Suppose that the data are of the form $\{(Y_n, X_n)\}$, where $0 \equiv Y_0 < Y_1 < Y_2 < \ldots < Y_N$ are "the observed times of occurrence" and X_n is a description of the event which occurred at Y_n. For simplicity, the marks X_n are assumed to take values in a countable set E. Let $\mathcal{H}_t$ denote the pre-t history

$$\mathcal{H}_t = \{(Y_n, X_n) : \ Y_n \leq t\}, \tag{10.7.3}$$

consisting of the events in the data which occurred before (and including) time t, and by $\mathcal{H}_{t-}$ the corresponding history when the inequality in (10.7.3) is strict. The internal filtration of the process is denoted by $(\mathcal{A}_t)$ with $\mathcal{A}_t = \sigma\{\mathcal{H}_t\}$, the σ-field of $\mathcal{H}_t$.

It is well known that there is a one-to-one correspondence between probabilities on the canonical path space Ω of the marked point process, consisting of the sequences $\omega = \{(y_n, x_n) : n \geq 0\}$ such that $x_n \in E$, where E is some measurable space, $0 \equiv y_0 \leq y_1 \leq y_2 \leq \ldots$, and $y_n < \infty$ implies $y_n < y_{n+1}$, on the one hand, and on the distribution of the "initial mark" X_0 and the $(\mathcal{A}_t)$-compensators of the counting processes

$$N_t(x) = \sum_{Y_n \leq t} 1_{\{Y_n \leq t, X_n = x\}}(t), \quad t \geq 0, x \in E, \tag{10.7.4}$$

on the other. Arjas and Gasbarra (1997) consider absolutely continuous compensators, in which case everything can be expressed in terms of conditional intensities.

In order to tie this framework explicitly with statistical inference, Arjas and Gasbarra (1997) introduce into the notation an unknown parameter θ. It can be viewed as an initial unobserved mark X_0 of the marked point process at time $Y_0 = 0$. The parameter space Θ may be finite-dimensional real or abstract, but its role will always be the same: for any given value $\theta \in \Theta$, we assume that there is a corresponding probability P^θ defined on the path space $\Omega = \{\omega = (X_n)_{n=0}^{\infty} : \ X_i \in E\}$. In practice, as noted above, P^θ is specified most conveniently in terms of an initial distribution for X_0 and the corresponding conditional intensities, $h_t^\theta(x)$, say, $x \in E, t \geq 0$. In classical statistical inference, the family $\mathcal{M} = \{P^\theta : \ \theta \in \Theta\}$ is called a statistical model of the marked point process.

In Bayesian inference, both the parameter θ and the process sample path $\mathcal{H}_\infty$ are viewed as random elements. The joint distribution of $(\theta, \mathcal{H}_\infty)$ is then determined by probabilities of the form

$$P(\theta \in A, \mathcal{H}_\infty \in B) = P_{(\theta, \mathcal{H}_\infty)}(A \times B) = \int_A P^\theta(B)\pi(d\theta), \tag{10.7.5}$$

where $\pi(\theta)$ is the prior distribution for θ.

It is characteristic of Bayesian inference that one is thinking of the dynamical prediction problem in time, where the prediction at time t always concerns the unobserved future sample path, say, $\mathcal{H}_{(t,\infty)}$, and the predictions are updated continuously on the basis of the observed $\mathcal{H}_t$. At time $t = 0$, the predictive distribution is simply the marginal of H_∞, obtained from (10.7.5) by letting $A = \Theta$. Let P denote this probability and also let E denote the corresponding expectation. Then, the updating of P corresponds to viewing $\mathcal{H}_\infty$ as a pair $(\mathcal{H}_t, \mathcal{H}_{(t,\infty)})$ and conditioning the joint distribution (10.7.5) on $\mathcal{H}_t$, again integrating out the parameter θ. This continuous updating of predictions, which in practice is done by applying Bayes' formula on the corresponding posterior distributions for θ, is at the heart of the method proposed by Arjas and Gasbarra (1997) for model assessment.

In more technical terms, this corresponds to focusing interest, not on θ and the $(P^\theta, \mathcal{A}_t)$-intensities $h_t^\theta(x)$, but on the *predictive* $(P, \mathcal{A}_t)$-intensities $\hat{h}_t(x)$. It is well known that these two types of intensity are linked by averaging with respect to the posterior $\pi(d\theta|\mathcal{H}_{t-})$, that is,

$$\hat{h}_t(x) = E(h_t^\theta(x)|\mathcal{H}_{t-}). \tag{10.7.6}$$

However, the real advantage of the predictive intensities is that the existing theory of point processes and martingales can be used, without needing to condition on something unknown. In other words, with this change from $\mathcal{M} = \{P^\theta; \theta \in \Theta\}$ to P, all technical problems arising from an "unknown parameter θ" have been eliminated, and the learning process of a Bayesian statistician working with the postulated model (10.7.5) is actually mimicked. Prequential model assessment means that one is considering these predictions sequentially and comparing them with the actual observed development of the marked point process.

There are several ways of doing this. An obvious one is to use the defining property of the conditional intensities, according to which the differences $N_t(x) - \int_0^t \hat{h}_s(x)\, ds$ are $(P, \mathcal{A}_t)$-martingales. A more intuitive way of writing this is based on the well-known interpretation

$$\hat{h}_t(x)dt = P(dN_t(x) = 1|\mathcal{H}_{t-});$$

the differentials

$$dN_t(x) - P(dN_t(x) = 1|\mathcal{H}_{t-})$$

are then expressing a direct continuous time analogue of the prequential method of Dawid (1984). However, there are also more refined results, which lead to exact reference distributions and statistical tests.

For the sake of simplicity, Arjas and Gasbarra (1997) start by considering a particular mark $x \in E$ and the corresponding sequence of precise times at which x occurs, say

$$\tau_0^x \equiv 0, \quad \tau_{k+1}^x = \inf\{Y_n > \tau_k^x; X_n = x\}.$$

We are led to the following result, given in Arjas and Gasbarra (1997).

Proposition 10.7.1 *The spacings* $\widehat{H}_{\tau_{k+1}^x}(x) - \widehat{H}_{\tau_k^x}(x)$, $k = 0, 1, 2, \ldots$, *of the* $(P, \mathcal{A}_t)$*-compensator* $\widehat{H}_t(x) = \int_0^t \hat{h}_s(x)\, ds$ *form a sequence of independent* $\mathcal{E}(1)$ *random variables.*

Since the result stated in Proposition 10.7.1 is well known, the proof is thus left an exercise. An equivalent alternative formulation of this result is that the time-transformed counting process $\{\hat{N}_u(x),\ u \geq 0\}$, defined by

$$\hat{N}_u(x) = N_{\widehat{H}_u^{-1}(x)}(x), \tag{10.7.7}$$

is a Poisson process with fixed intensity 1. Thus, one can view the statistician as monitoring the occurrences of mark x in time and matching the integrated intensities, obtained from the P-distribution and updated from the previous observations, against the yardstick of spacings of the Poisson (1)-process. Yet another equivalent way is to consider the residuals

$$1 - \exp\{\widehat{H}_{\tau_k^x}(x) - \widehat{H}_{\tau_{k+1}^x}(x)\},$$

which, according to P, are independent and follow the $\mathcal{U}(0,1)$ distribution. These latter random variables have an obvious interpretation in terms of observed residuals of the spacings $\tau_{k+1}^x - \tau_k^x$, obtained sequentially from P and the data. In fact, in this simple case of a univariate process one could just as well denote $\tau_{k+1}^x - \tau_k^x$ by η_{k+1} and then consider the process $\{\eta_k :\ k \geq 1\}$ of a discrete time parameter. This would correspond exactly to the original formulation of the prequential method in Dawid (1984).

However, the $\mathcal{E}(1)$ and independence properties hold in much more generality than that contained in Proposition 10.7.1; see, for example Aalen and Hoem (1978) and Norros (1986). It is somewhat difficult to give a formulation that would both be concise and cover all cases which could arise in an assessment of life history models. The following proposition, which is a direct corollary of Theorem 2.1 in Norros (1986), appears to be sufficiently general for most purposes.

Proposition 10.7.2 *Consider an arbitrary finite collection* $\{\tau_i; 1 \leq i \leq k\}$ *of* $(\mathcal{A}_t)$*-stopping times such that*

(i) $P(0 < \tau_i < \infty) = 1$ *for all* $1 \leq i \leq k$;

(ii) $P(\tau_i = \tau_j) = 0$ *whenever* $i \neq j$; *and*

(iii) each τ_i *is completely unpredictable in the sense that the corresponding* $(P, \mathcal{A}_t)$*-compensator* $(\widehat{H}_t^i)$ *of* τ_i *is continuous.*

Then the random variables $\widehat{H}_{\tau_i}^i$, $1 \leq i \leq k$, *are independent and* $\mathcal{E}(1)$.

Again, as an alternative, one can say that the variables $1 - \exp(-\widehat{H}^i_{\tau_i})$ are independent and $\mathcal{U}(0,1)$. The crucial element of Proposition 10.7.2 is the independence statement, which holds in great generality even if the stopping times τ_i themselves are dependent and, for example, their order of occurrence is not specified in advance (as in Proposition 10.7.1). Note that given a nonnegative and bounded $\{\mathcal{A}_t\}$-predictable process $h(x, t, \mathcal{H}_{t-})$, one can always construct a probability measure P (to be used as a prequential forecasting system) on the canonical space of marked point process histories Ω, such that, under P, $\{N_t(x)\}$ has $(P, \mathcal{A}_t)$-intensity which coincides with $h(x, t, \mathcal{H}_{t-})$ (Brémaud, 1981, IV T4, p. 168).

In particular, Propositions 10.7.1 and 10.7.2 hold for such a constructed P. Therefore, we are not necessarily restricted to Bayesian estimation and prediction. For example, suppose that one has specified a parametric model for the marked point process through a family of $(\mathcal{A}_t)$-intensities $h(x, t, \theta, \mathcal{H}_{t-})$, $\theta \in \Theta$. One could then use a point estimate, such as the maximum likelihood estimate, $\hat{\theta}(t, \mathcal{H}_{t-})$, for $\theta \in \Theta$ at time t, and finally form another probability on $\mathcal{A}$ by means of the plug-in intensity function $\hat{h}(t, x, \mathcal{H}_{t-}) = h(t, x, \hat{\theta}(t, \mathcal{H}_{t-}), \mathcal{H}_{t-})$. A slightly different frequentist idea would be to use the confidence intervals around the maximum likelihood estimate to account for the uncertainty regarding the true value of θ. If the confidence intervals at time t were based on the normal distribution with mean $\hat{\theta}(t, \mathcal{H}_{t-})$ and variance $\sigma^2(t, \mathcal{H}_{t-})$, a prequential forecasting system could then be built on the intensity

$$\tilde{h}(t, x, \mathcal{H}_{t-}) = \frac{1}{\sqrt{2\pi\hat{\sigma}^2(t, \mathcal{H}_{t-})}} \int_R \Bigg[h(t, x, \mathcal{H}_{t-}, \theta) \times \exp\left\{ - \frac{\{\theta - \hat{\theta}(t, \mathcal{H}_{t-})\}^2}{2\hat{\sigma}^2(t, \mathcal{H}_{t-})} \right\} \Bigg] d\theta.$$

Actually, a statistician may use whatever ingredient he or she likes to define h, such as Bayes formula, maximum likelihood, nonparametric methods, and so on. In any case, when a predictable intensity process is given, a nice probability measure on the space of histories is automatically defined.

To assess the predictive performance of the standard Poisson process derived from a possibly highly structured marked point process, Arjas and Gasbarra (1997) check the fit of the corresponding derived process sample path to the standard Poisson process assumption. The following graphical tools are proposed.

Examining the total-time-on-test plot (see, e.g., Andersen et al., 1993) is a good first stage to do a graphical check. The plot process, $n \to S_n = \sum_{i \leq n} \widehat{H}{\tau_i}^i$, under the Bayesian forecasting system, has independent $\mathcal{E}(1)$ increments. Asymptotically, the Kolmogorov law of the iterated logarithm gives a sharp result on the behavior of the random walk process $\{S_n - n\}$:

with probability 1,

$$\limsup \frac{S_n - n}{\sqrt{2n \log\log n}} = +1 \quad \text{and} \qquad \liminf \frac{S_n - n}{\sqrt{2n \log\log n}} = -1. \quad (10.7.8)$$

Note that this result is true for any random walk with independent and identically distributed increments of mean 0 and variance 1. In other words, with probability 1, for any $\varepsilon > 0$,

$$n - (1+\varepsilon)\sqrt{2n \log\log n} \le S_n \le n + (1+\varepsilon)\sqrt{2n \log\log n}$$

will hold eventually for all n, and $\{S_n\}$ will satisfy the inequalities

$$S_n \ge n + (1-\varepsilon)\sqrt{2n \log\log n} \ \text{ and } \ S_n \le n - (1-\varepsilon)\sqrt{2n \log\log n} \qquad (10.7.9)$$

infinitely often. Of course this result does not apply as such in finite samples. Nevertheless, if $\{S_n\}$ infringes the boundaries and does not return, this can be used as evidence against the model.

As an attempt to quantify the evidence for or against the model, Arjas and Gasbarra (1997) consider some functionals of the whole sample path $\{S_n\}$. They observe that S_n with respect to P is gamma distributed with shape parameter n and scale parameter 1. Denoting the corresponding cumulative distribution by $F_\gamma(\cdot\,; n, 1)$ they find that the random variables

$$G_n = F_\gamma(S_n; n, 1)$$

follow the $\mathcal{U}(0,1)$ distribution. Note that the increments of $\{G_n\}$ are not independent, nor is $\{G_n\}$ a martingale. Obvious test statistics are $G_{1n} = \max_{k\le n} G_k$ and $G_{2n} = \min_{k\le n} G_k$. By computing the reference distributions of G_{1N} and G_{2N} we can assign p-values to the whole sample, i.e., $P(G_{1N} \ge g_{1N})$ and $P(G_{2N} \le g_{2N})$ where $g_{1N} = \max_{k\le N} g_k$, $g_{2N} = \min_{k\le N} g_k$ are the observed values of the statistics. Arjas and Gasbarra (1997) call these statistics *prequential P-values*. As usual, P-values close to 0 would be used as evidence against the model P. In principle, the distribution of G_{1n} could be computed recursively. Here approximate prequential P-values were determined by a simple Monte Carlo method, by generating independent identically distributed samples of the process $\{G_n\}$.

Example 10.10. Point process with serially correlated spacings. Arjas and Gasbarra (1997) consider data from a point process, where there is a strong correlation between successive spacings. The purpose of this example is to illustrate how the prequential diagnostic method can be used to reveal the inadequacy of a statistical model in which such correlation has not been accounted for.

Let $0 < Y_1 < Y_2 < \dots$ denote a simple point process and also let $\eta_n = Y_n - Y_{n-1}$ denote the spacings, with $Y_0 \equiv 0$. Arjas and Gasbarra (1997) consider the following two competing Bayesian models:

Model M_0. Suppose that (i) the model parameter θ is a real-valued random variable with prior distribution $F_\gamma(\cdot;\alpha,\beta)$ (α = shape parameter and β = scale parameter), and that (ii) conditionally on θ, the spacings $\{\eta_n; n \geq 1\}$ are independent and distributed as $\mathcal{E}(\theta)$. In other words, $\{Y_n\}$ is a doubly stochastic Poisson process, or Cox process, with conditional intensity given by $h_t^\theta \equiv \theta$.

Model M_1. Suppose that (i) θ is as in model M_0 above, but that (ii), conditionally on $(\theta, \eta_1, \eta_2, \dots, \eta_{n-1})$, the spacings η_n are distributed according to the exponential distribution with parameter θ/η_{n-1}. The conditional intensity is now given by $h_t^\theta = \theta/\eta_{N_{t-}}$. According to model M_1, long (short) spacings are typically followed by long (short) spacings, and therefore the points Y_n tend to be clustered.

Denote by P_i the probabilities induced by the respective models M_i, $i = 0, 1$, on the space $\Theta \times \Omega$. Due to the simplicity of the models we can find analytic expressions for the $(P_0, \mathcal{A}_t)$-compensators. Under M_0, by a well-known conjugacy property of the gamma distribution, the predictive $(P_0, \mathcal{A}_t)$-intensity is given by

$$\tilde{h}_t^0 = E_0(\theta|\mathcal{H}_{t-}) = \frac{\alpha + N_{t-}}{\beta + t},$$

so that the corresponding cumulative intensity from the interval $(Y_{n-1}, Y_n]$ is

$$\int_{Y_{n-1}}^{Y_n} \tilde{h}_s^0 \, ds = (\alpha + n - 1) \log\left(\frac{\beta + Y_n}{\beta + Y_{n-1}}\right).$$

Under M_1, an analogous computation gives the cumulative $(P_1, \mathcal{H}_t)$-intensity

$$\int_{Y_{n-1}}^{Y_n} \tilde{h}_s^1 \, ds = (\alpha + n - 1) \log\left(\frac{\beta + \sum_{i=1}^{n} \frac{\eta_i}{\eta_{i-1}}}{\beta + \sum_{i=1}^{n-1} \frac{\eta_i}{\eta_{i-1}}}\right).$$

A third model (mixture model $M_{0,1}$), which includes both M_0 and M_1 as special cases and which in principle permits consideration of model selection probabilities adapting to the data, would have the following mixture form. Let $q \in [0,1]$ be given and let ξ be a $\{0,1\}$-valued random variable with $P_{0,1}(\xi = 0) = 1 - q$, $P_{0,1}(\xi = 1) = q$. Then define the probability $P_{0,1}$ on $\Theta \times \Omega \times \{0,1\}$ by specifying the conditional probability $P_{0,1}(\cdot|\xi)$ through $P_{0,1}(\cdot|\xi) = \xi P_1(\cdot) + (1-\xi)P_0(\cdot)$.

Define the filtration $\mathcal{G}_t = \mathcal{A}_t \vee \sigma(\xi)$. Then, the $(P_{0,1}, \mathcal{G}_t)$-intensity of the counting process $\{N_t\}$ is $\tilde{h}_t^\xi$. The $(P_{0,1}, \mathcal{A}_t)$-intensity is the mixture

$$\tilde{h}_t^{01} = P_{0,1}(\xi = 1 | \mathcal{H}_{t-})\tilde{h}_t^1 + P_{0,1}(\xi = 0 | \mathcal{H}_{t-})\tilde{h}_t^0,$$

where, from Bayes' formula,

$$P_{0,1}(\xi = 1 | \mathcal{H}_{t-}) = \frac{q \cdot Z_1(t)}{q \cdot Z_1(t) + (1-q) \cdot Z_0(t)},$$

and $Z_\xi(t) = P_\xi(\mathcal{H}_{t-}) = \int_{\Theta_\xi} f_\xi(\mathcal{H}_{t-}|\theta_\xi)\, d\pi(\theta_\xi)$, $\xi = 0, 1$, are the normalizing constants. These quantities can be computed explicitly, leading to the expression

$$P_{0,1}(\xi = 1 | \mathcal{H}_{t-}) = C / \{C + (1-q)(\beta + t)^{-\alpha - N_{t-}}\},$$

where

$$C = q\left(\prod_{i=1}^{N_{t-}-1} \eta_i^{-1}\right)\left(\beta + \sum_{i=1}^{N_{t-}} \frac{\eta_i}{\eta_{i-1}} + \frac{t - Y_{N_{t-}}}{\eta_{N_{t-}}}\right)^{-\alpha - N_{t-}}.$$

The $(P_{0,1}, \mathcal{A}_t)$-compensators can now be evaluated numerically.

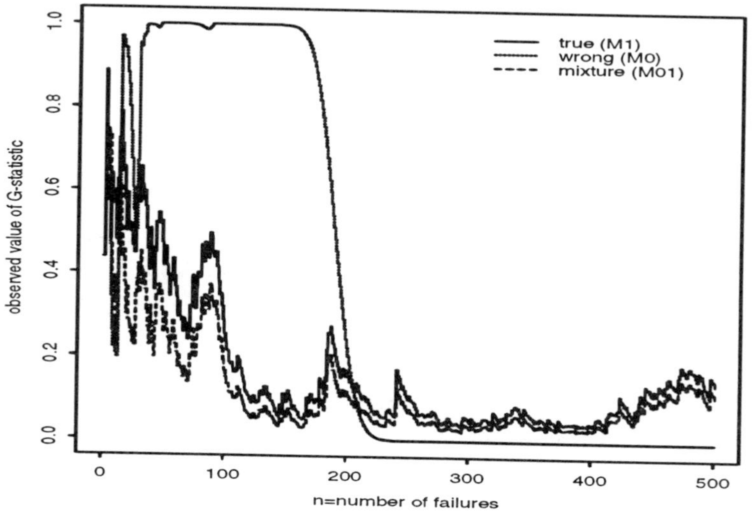

FIGURE 10.8. Observed sample path of the process $\{G_n\}$ under models M_0, $M_{0,1}$, and M_1.

A sample path segment of the process $\{Y_n\}$ consisting of 500 points was generated by a computer from model M_1, with $\theta = 0.6$. The hyperparameters were given the values $\alpha = 0.1$ and $\beta = 0.001$. This prior has a very

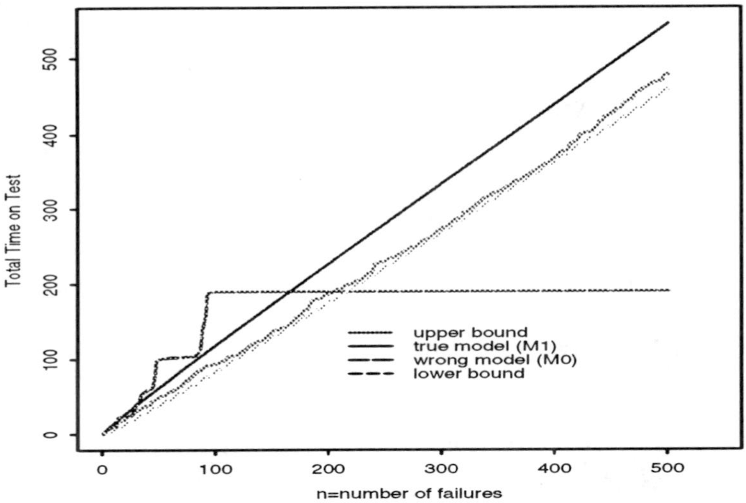

FIGURE 10.9. Total-time-on-test plot for the compensators $\tilde{H}^n_{Y_n}$.

large mean $\alpha/\beta = 100$, compared to the true value of 0.6, but it is also very flat, having variance $\alpha/\beta^2 = 10^5$. Figure 10.8 shows the corresponding sample paths of the process $\{G_n\}$ under the models M_0, M_1, and $M_{0,1}$. It is easy to see that M_0 does not fit the data; $\{G_n\}$ appears to be oscillating between 0 and 1. Under M_1 and $M_{0,1}$, $\{G_n\}$ behaves nicely, and the sample paths are close to each other. This is predictable because, for data from model M_1, $P_{0,1}(\xi = 1|\mathcal{H}_t)$ converges to 1 almost surely according to P_1. The prequential P-values of the generated data under these three different models are shown in Table 10.14. The values of these statistics clearly reinforce the graphical conclusions. The same message is conveyed by the total-time-on-test plots displayed in Figure 10.9, where the compensators $\tilde{H}^n_{Y_n}$ are plotted against n for $n = 1, 2, \ldots, 500$ along with the bounds $n \pm \sqrt{2n \log\log n}$ arising from the law of the iterated logarithm. Again one can see that, while the process $\{S_n - n\}$ has a nice random walk behavior under M_1 and $M_{0,1}$ (graphically coinciding in the figure), staying mostly in the region prescribed by the law of the iterated logarithm, under model M_0 it leaves this region abruptly.

TABLE 10.14. Prequential P-values for Example 10.10.

	M_1	M_0	$M_{0,1}$
$P(G_{1,500} \geq g_{1,500})$	0.594	$\simeq 0$	0.629
$P(G_{2,500} \leq g_{2,500})$	0.337	$\simeq 0$	0.253

Source: Arjas and Gasbarra (1997).

Example 10.11. One-sample versus two-sample model. In this example, Arjas and Gasbarra (1997) consider a sample of right censored survival times from a mixture of two exponential distributions with different parameters, and demonstrate how an attempt to fit a single exponential distribution to such data leads to an obvious rejection of the model.

Consider a right censored sample of survival times $\{(y_i^*, \nu_i); y_i^* \geq 0, \nu_i = 0, 1, i = 1, 2, ..., n\}$. Here y_i^* is the time the ith individual was last seen, and ν_i is the censoring indicator, i.e., $\nu_i = 1$ if the ith individual was dead at y_i^*, and $\nu_i = 0$ if they were censored. Denote the corresponding "complete" lifetimes by Y_i; i.e., $Y_i = y_i^*$ if $\nu_i = 1$, and $Y_i > y_i^*$ if $\nu_i = 0$.

The following two competing parametric models are considered:

Model M_0. (i) The parameter of the model θ is a real-valued random variable with prior distribution $F_\gamma(\cdot\ ; \alpha, \beta)$ with given shape and scale parameters α and β, and (ii) conditionally on θ, the Y_i are independent and identically distributed as $\mathcal{E}(\theta)$. The model assumes also that the censoring mechanism is noninformative with respect to θ.

Model M_1. The individuals are divided into two subsamples, say, "men" and "women". This leads to a marked point process formulation, with two observable marks, say, m and w, and corresponding mark-specific hazard rates $h_t^\theta(m)$ and $h_t^\theta(w)$. These hazard rates are now modeled exactly as in model M_0, assuming independence with respect to the prior: let $\theta = (\theta', \theta'')$, $h_t^\theta(m) = \theta'$ and $h_t^\theta(w) = \theta''$, with θ' and θ'' independent and distributed according $F_\gamma(\cdot\ ; \alpha, \beta)$.

Let

$$N_t = \sum_{i=1}^{n} 1_{\{y_i^* \leq t,\ \nu_i = 1\}}(t) = \#\{\text{ recorded deaths by time } t\},$$

$$R_t = n - \sum_{i=1}^{n} 1_{\{y_i^* < t\}}(t) = \#\{\text{individuals at risk at time } t\},$$

and define analogously N_t', R_t' and N_t'', R_t'' for the subsamples consisting of men and women. Denote by P_0 and P_1 the probabilities on the space $\Theta \times \Omega$ corresponding to the models M_0 and M_1.

Now, under P_0, the intensities corresponding to the "individual counting processes" $N_t(i) = 1_{\{y_i^* \leq t,\ \nu_i = 1\}}$ are given by

$$\tilde{h}_t(i) = E_0(\theta | \mathcal{H}_{t-}) 1_{[0, y_i^*]}(t) = \frac{\alpha + N_{t-}}{\beta + \int_0^t R_s ds} 1_{[0, y_i^*]}(t).$$

The variables $\tilde{H}_{y_i^*}(i) = \int_0^{y_i^*} \tilde{h}_s(i)\, ds$ form a (possibly right censored) simple random sample of $\mathcal{E}(1)$ random variables. Similarly, the counting process $\{N_t\}$ is compensated by $\tilde{H}_t = \sum_{i=1}^n \tilde{H}_t(i)$. Under P_1, similar results will hold for both subgroups. As a result of the assumed prior independence of θ' and θ'', the two groups can be treated separately.

A sample of 500 survival times (250 men and 250 women) was generated from model M_1. The true values of the intensity were $\theta' = 1.2$ for men, and $\theta'' = 0.8$ for women. These survival times were censored from the right by an independent random censoring mechanism, resulting in 46 censored observations for men and 75 for women. The Kaplan-Meier curves for males and females, although not shown here, clearly reflect the poorer survival prospects for males.

Arjas and Gasbarra (1997) computed the prequential forecasts for this two-sample dataset under both models, the true M_1 and the misspecified M_0. In both models the hyperparameters of the priors were given the values $\alpha = \beta = 2$. They considered the stopping times

$$\tau'_k = \inf\{t : N'_t = k\} \quad (k = 1, 2, ..., 250),$$

$$\tau''_l = \inf\{t : N''_t = l\} \quad (l = 1, 2, ..., 250),$$

and

$$\tau_m = \inf\{t : N_t = m\} \quad (m = 1, 2, ..., 500),$$

with the convention that $\inf\{\emptyset\} = +\infty$. Thus, for instance, τ'_k is the time it takes until k men have died. For the particular censored sample considered here, the largest finite values in these sequences were τ'_{204}, τ''_{175}, and τ_{379}.

The compensators of these stopping times admit simple analytic forms. For example, under P_0 the intensity of τ_m is given by

$$R_t(\alpha + N_{t-})\left(\beta + \int_0^t R_s\, ds\right)^{-1} 1_{(\tau_{m-1}, \tau_m]}(t);$$

note that this intensity vanishes outside the interval $(\tau_{l-1}, \tau_l]$. Under P_1, by Proposition 10.7.2, the random variables $H_{\tau'_k}$, $H_{\tau''_l}$ corresponding to the stopping times of τ'_k and τ''_l are independent and $\mathcal{E}(1)$. Also the variables H_{τ_m} corresponding to τ_m are independent and $\mathcal{E}(1)$ under P_1, although dependent on $H_{\tau'_k}$ and $H_{\tau''_l}$.

The advantage of considering these stopping times, rather than simply the individual survival times y_i^*, is that, as the individual observations are censored their cumulative intensities will be censored also, whereas the stopping times τ_m are not censored as long as there are at least m recorded failure times.

Figure 10.10 shows the process-statistics $G'_n = F_\gamma(S'_n; n, 1)$, where $S'_n = \sum_{k=1}^{n} \Lambda_{\tau'_k}$, and the analogously defined $\{G''_n\}$, under both models M_1 and M_0, keeping the same observations previously generated under P_1. The

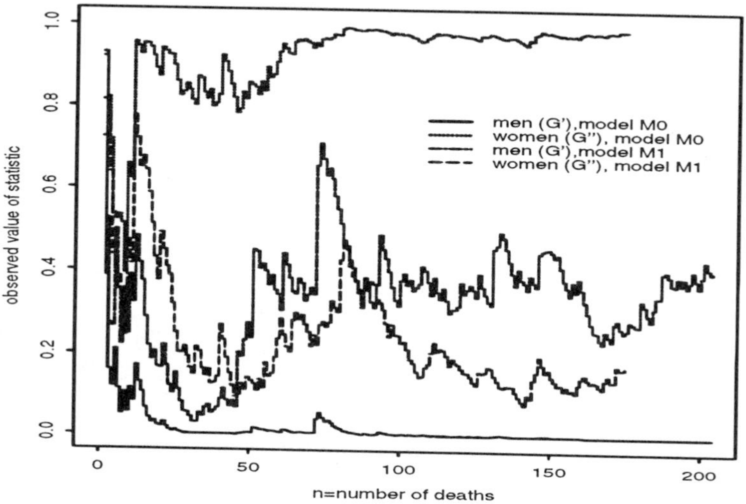

FIGURE 10.10. The G'_n and G''_n statistics for the men's and women's subsamples under models M_0 and M_1.

interpretation is that we are using model M_1 to give separate forecasts for the next failure time within the male and the female subgroups. The prequential P-values under model P_1 are given in Table 10.15. From Figure 10.10 it is clear also that the predictive performance of model M_0 is poor: the sample paths of the processes $\{G'_n\}$ and $\{G''_n\}$ appear to be converging, respectively, to 0 and to 1, whereas, if the model were correct, G'_n and G''_n would be distributed $\mathcal{U}(0, 1)$ for each n. The intuition behind these tests is that in the data, men's survival times tend to be shorter than expected from Model M_0, and women's survival times longer. A statistician who has postulated model M_0 and is monitoring these statistics will soon become puzzled. This "surprise" may be quantified by the prequential P-values under model P_0. The numerical values are in Table 10.15.

In Figure 10.11 the sample paths of the process $\{G_n\}$ are shown, corresponding to the stopping times τ_k, under both models M_1 and M_0. Here we are forecasting the next failure time in the combined sample, i.e., regardless of sex. At the beginning, the two models give almost the same forecasts for τ_m, and both seem to predict successfully. The reason for this is that, as long as the proportions of men and women among the survivors are similar, the opposite effects of the incorrect model specification cancel out. Only later, when the survivors' group is dominated by women, do the two models give different forecasts for the next failure time. The prequential P-values are displayed in Table 10.15, and show that, unless we divide the sample into the two subsamples, our diagnostics do not have enough power to distinguish between the true and the misspecified models. Therefore the

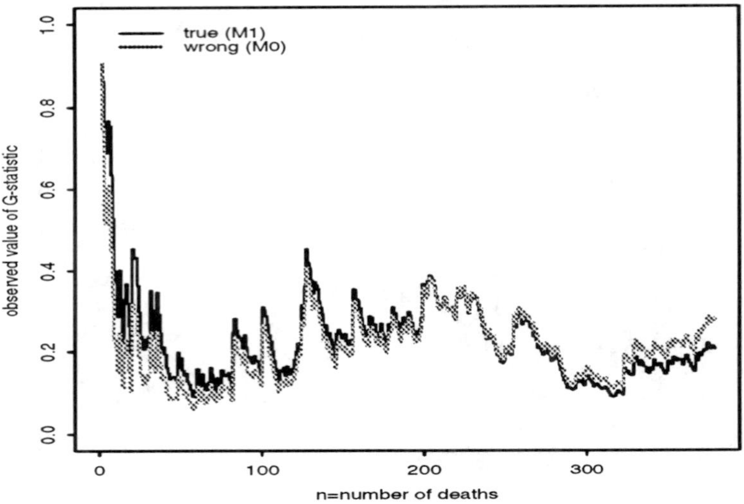

FIGURE 10.11. The G_n statistics for the combined sample under models M_0 and M_1.

choice of the stopping time sequence is crucial for this model checking procedure; a model may give good predictions for some particular sequence of stopping times, but fail completely for another one.

TABLE 10.15. Prequential P-values for Example 10.11.

Model	Men	Women
M_1	$P_1(G'_{1,204} \geq g'_{1,204}) = 0.700$	$P_1(G''_{1,175} \geq g''_{1,175}) = 0.434$
	$P_1(G'_{2,204} \leq g'_{2,204}) = 0.281$	$P_1(G''_{2,175} \geq g''_{2,175}) = 0.484$
M_0	$P_0(G'_{1,204} \geq g'_{1,204}) = 0.821$	$P_0(G''_{1,175} \geq g''_{1,175}) = 0.062$
	$P_0(G'_{2,204} \leq g'_{2,204}) = 0.003$	$P_0(G''_{2,175} \geq g''_{2,175}) = 0.920$
	Combined Sample	
M_1	$P(G_{1,379} \geq g_{1,379}) = 0.279$	$P(G_{2,379} \leq g_{2,379}) = 0.558$
M_0	$P(G_{1,379} \geq g_{1,379}) = 0.460$	$P(G_{2,379} \leq g_{2,379}) = 0.457$

Source: Arjas and Gasbarra (1997).

Figures 10.12 corresponds to the same situations as Figure 10.10, but now the total-time-on-test plots are shown; i.e., (n, S_n), the cumulative intensity process $S_n = \sum H_{\tau_k}$ is plotted against the number of observed deaths n. Figure 10.12 also displays the bounds $n \pm \sqrt{2n \log\log n}$ arising from the law of the iterated logarithm. Under the incorrect model, M_0, the processes $\{S'_n\}$ and $\{S''_n\}$ arising, respectively, from the men's and women's samples exit from the upper and lower boundaries. This corresponds to the

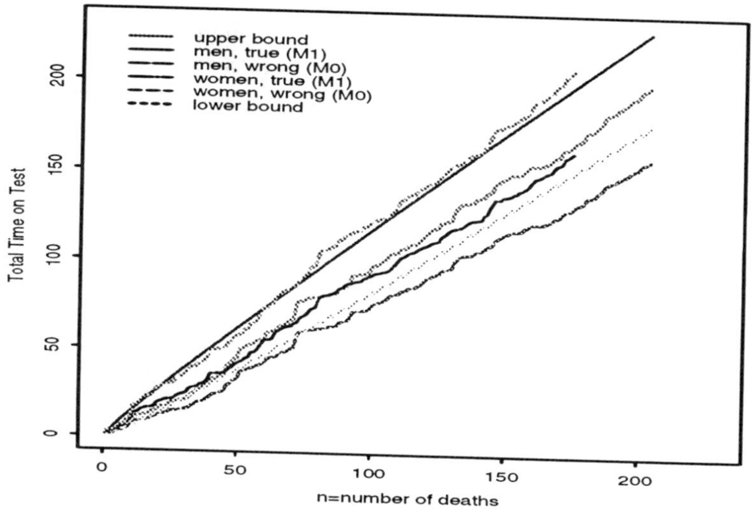

FIGURE 10.12. Total-time-on-test plots of the compensators of the stopping times τ'_k and τ''_l respectively for the men's and women's subsamples under models M_0 and M_1.

convergence of the processes $\{G'_n\}$ and $\{G''_n\}$, respectively, to 0 and 1, as shown in Figure 10.10.

10.8 Future Research Topics

In this section, we discuss some future research problems that may serve as nice thesis topics. The semiparametric Bayesian survival analysis literature has so far put a heavy emphasis on the Cox model (and its different variations) and the accelerated failure time model. Semiparametric versions of models based on residual time, hazard location, and multiple time scales have not been thoroughly explored. Although some work has been done on correlated prior processes as discussed in Section 3.6, more work needs to be done in this area to get more realistic prior processes for semiparametric models. In multivariate survival data, for example, the degree of smoothness could be the only relevant prior information for the baseline hazard. The methods of elicitation of correlated processes in different survival analysis situations have not yet been fully explored. In Section 3.6, we presented one version of a correlated gamma process, but many other versions and developments can be considered. Also, the construction of correlated beta processes as well as correlated Dirichlet processes are of considerable interest in survival analysis. In general, the construction and elicitation of correlated prior processes is an area of considerable importance in which more research is needed.

In Chapter 4, we discussed frailty models and models for multivariate survival data. Bayesian analysis with multivariate survival data using different degrees of association within pairs has not been addressed in the literature. One example is familial studies where association between survival times of spouses is different from association between survival times of a parent and daughter. There are challenges to effective and practical modeling as well as to prior elicitation and computation for such problems. Another area of further research would be the development of multivariate frailty models. Overall, the research for Bayesian multivariate survival analysis has concentrated so far mainly on latent random effects based models, such as the frailty model, but further work needs to be done to explore other types of multivariate survival models.

In Chapter 7, we discussed joint models for longitudinal and survival data, and we assumed that the survival component of the model follows a Cox structure. A useful generalization of this is a joint model for longitudinal and survival data in which the survival component of the model is a cure rate model. Such models have not yet been investigated. Another interesting and useful research problem would be to examine semiparametric joint models for longitudinal and survival data, in which the latent trajectory process follows a generalized linear mixed model and the distribution of the random effects in the trajectory follows some type of prior process, such as the Dirichlet process. Thus a semiparametric model for the trajectory process may be more useful and less restrictive than the fully parametric models for the trajectory that have been currently proposed in the literature.

In Chapter 10, we discussed Bayesian diagnostic methods for survival analysis. There is no literature that we are aware of for Bayesian diagnostic methods in survival analysis in the presence of time-varying covariates, missing covariates, or random covariates. These problems are of great interest since time-varying covariates, missing covariates, and random covariates arise naturally in many applications such as joint models for longitudinal and survival data. Finally, we mention that it is critical to provide flexible as well as sophisticated software to carry out Bayesian inference for complex survival models. Although BUGS is a very useful software package, it is somewhat limited in accommodating semiparametric survival models with the prior processes discussed in Chapter 3, models with time-varying covariates, frailty models, multivariate survival models, joint models for longitudinal and survival data, and general specifications of prior distributions such as the power prior. It will be a great advantage if future versions of BUGS could accommodate the power prior, which arises naturally in many practical problems. A software package to fit various models and priors discussed in this book would be an invaluable asset to the practitioner.

Appendix

Proof of Theorem 10.4.1. Using the property of the Poisson process (see Karlin and Taylor, 1975), new event times are i.i.d. with common density $f_*(y|x,t)$ for $y \in (0,t)$, given the number of events (N_2) by time t. Using the method of convolution, the conditional probability of no cancer from a new clonogen by time t is $\int_0^t f_*(y|x,t)S_2(t-y)dy$. Now, using the fact that there is no cancer by time t if and only if each of N_1 old and N_2 new clonogens do not develop cancer by time t, this proves the theorem. □

Proof of Theorem 10.4.2. It is sufficient to show that for $\delta > 0$,

$$H(t+\delta|x) - H(t|x) \geq 0.$$

Using the definitions of $H(t|x)$ and $f_*(y|x,t)$,

$$H(t|x) = H_2(t|x) \int_0^t f_*(y|x,t)F_2(t-y)\, dy = \int_0^t h_2(y|x)F_2(t-y)\, dy. \tag{10.A.1}$$

Using the above equation,

$$\begin{aligned} H(t+\delta|x) &- H(t|x) \\ &= \int_0^{t+\delta} h_2(y|x)F_2(t+\delta-y)\, dy - \int_0^t h_2(y|x)F_2(t-y)\, dy \\ &= \int_t^{t+\delta} h_2(y|x)F_2(t+\delta-y)dy \\ &\quad + \int_0^t h_2(y|x)[F_2(t+\delta-y) - F_2(t-y)]\, dy. \end{aligned} \tag{10.A.2}$$

Now, $F_2(t+\delta-y) \geq 0$, $F_2(t-y) \geq 0$, and $F_2(t+\delta-y) - F_2(t-y) \geq 0$, because $F_2(\cdot)$ is a cdf. Also, $h_2(y|x) \geq 0$ because it is an intensity function. From (10.A.2), it follows that $H(t+\delta|x) - H(t|x) \geq 0$ for all $\delta > 0$. Because $F_2(t)$ is nondecreasing with $\lim_{t\to\infty} F_2(t) = 1$, from (10.A.1) we can also prove the second part of (10.4.2). □

Proof of Theorem 10.4.4. To prove that $N_p(t)$ is a Poisson process with cumulative intensity $H_p(t)$, it is enough to show that $N_p(0,a)$ and $N_p(a,a+b)$, $a > 0$ and $b > 0$, have independent Poisson distributions with means $H_A = H_p(a)$ and $H_B = H_p(a+b) - H_p(a)$. We use the notation $F_{1B} \equiv F_1(a+b) - F_1(a)$ and $H \equiv H_2(a+b|x)$. We define F_{1A}, F_{2A}, and F_{2B} in a similar fashion. The event $E = \{N_p(0,a) = m, N_p(a,a+b) = n\}$ can be written as a union of disjoint events $E_{k_1k_2N_1N_2}$'s for $k_1 = 0, 2, \ldots, m$, $k_2 = 0, 1, \ldots, n$, $N_1 \geq k_1 + k_2$ and $N_2 \geq m+n-k_1-k_2$. Here $E_{k_1k_2N_1N_2}$ is the event that there are N_1 old clonogens out of which k_1 become tumors in A and k_2 become tumors in B, and there are N_2 new clonogens by time $a+b$ out of which $m-k_1$ become tumors in A and $n-k_2$ become tumors in B. Again, using conditional probability arguments and independence of the two initiation processes, we are led to

$$P(E_{k_1k_2N_1N_2}) = P(E_{1N_1})P(E_{2N_2})P(E_{1k_1k_2}|E_{1N_1})P(E_{2,m-k_1,n-k_2}|E_{2N_2}),$$

since the new clonogens are independent of clonogens existing at the beginning of the trial. Here E_{1N_1} is the event of N_1 old clonogens, E_{2N_2} is the event of N_2 new clonogens by time $a+b$, $E_{1k_1k_2}$ is the event of k_1 tumors in A and k_2 tumors in B only originating from old clonogens, and $E_{2,m-k_1,n-k_2}$ is the event that $m-k_1$ tumors in A and $n-k_2$ tumors in B originating only from new clonogens.

Now, $P(E_{1N_1}) = \exp(-\theta)\theta^{N_1}/N_1!$ and $P(E_{2N_2}) = \exp(-H)H^{N_2}/N_2!$, since the two initiation processes are Poisson. Now we use an argument similar to the one in Theorem 10.4.1 based on the properties of a nonhomogeneous Poisson process of the initiation of new tumors. Thus, the conditional probability, $P(E_{1k_1k_2}|E_{1N_1})$, is the multinomial probability

$$P(E_{1k_1k_2}|E_{1N_1}) = \binom{N_1}{k_1, k_2} F_{1A}^{k_1} F_{1B}^{k_2} (1 - F_{1A} - F_{1B})^{N_1-k_1-k_2},$$

and the conditional probability $P(E_{2,m-k_1,n-k_2}|E_{2N_2})$ is given by

$$\begin{aligned} &P(E_{2,m-k_1,n-k_2}|E_{1k_1k_2}, E_{2N_2}) \\ &= \binom{N_2}{m-k_1, n-k_2} G_A^{m-k_1} G_B^{n-k_2} (1 - G_A - G_B)^{N_1+k_1+k_2-m-n}, \end{aligned}$$

where

$$G_A = P(\text{Promoted in}(0,a)|\text{initiated in}(0,a+b)) \tag{10.A.3}$$

and

$$G_B = P(\text{Promoted in}(a,a+b)|\text{initiated in}(0,a+b)). \tag{10.A.4}$$

Now,

$$\begin{aligned} &P(N_p(0,a) = m, N_p(a,a+b) = n) \\ &= \sum_{k_1=0}^{m} \sum_{k_2=0}^{n} \sum_{N_1=k_1+k_2}^{\infty} \sum_{N_2=m+n-k_1-k_2}^{\infty} P(E_{k_1k_2N_1N_2}). \end{aligned}$$

After some tedious calculations involving summations of multinomial probabilities and exponential series, we are led to

$$\begin{aligned} P\Big(N_p(0,a) = m, N_p(a,a+b) = n\Big) &= \frac{\exp(-\theta F_{1A} - HG_A)(\theta F_{1A} + HG_A)^m}{m!} \\ &\times \frac{\exp(-\theta F_{1B} - HG_B)(\theta F_{1B} + HG_B)^n}{n!}, \end{aligned}$$

where $H = H_2(a+b|x)$. Using (10.A.1) and the definition of G_A from (10.A.3), we can show that

$$\begin{aligned} HG_A =& H_2(a+b|x)\int_0^a f^*(y|x,a+b)F_2(a-y)dy \\ =& \int_0^a h_2(y|x)F_2(a-y)\,dy = H_2(a|x)G(a|x). \end{aligned}$$

Using similar arguments, we can show that

$$HG_B = H_2(a+b|x)G(a+b|x) - H_2(a|x)G(a|x).$$

This completes the proof. □

Proof of Theorem 10.4.5. Using the result of Theorem 10.4.4, we have

$$P\bigg(N_p(A) = k|x,w\bigg) \tag{10.A.5}$$

$$= \frac{\exp[-\{\theta F_1(A)w + H((a,b)|x)\}]\{\theta F_1(A)w + H((a,b)|x)\}^k}{k!}$$

$$= \frac{\exp[-H((a,b)|x)]}{k!}\sum_{j=0}^{k}\Bigg[\binom{k}{j}\{\theta F_1(A)\}^j\{H((a,b)|x)\}^{k-j}$$

$$\times \left(w^j\exp\{-w\theta F_1(A)\}\right)\Bigg], \tag{10.A.6}$$

where $H((a,b)|x) = H(b|x) - H(a|x) = \varphi(x)[H_0(b) - H_0(a)$. The Laplace transform

$$E_*(\exp(-sW)) = \left(\frac{\alpha}{s+\alpha}\right)^{\alpha},$$

where E_* represents the expectation taken with respect to the gamma frailty. Therefore, it follows that

$$E_*[W^j\exp(-sW)] = \frac{\Gamma(\alpha+j)}{\Gamma(\alpha)}\left\{\frac{\alpha^\alpha}{(s+\alpha)^{\alpha+j}}\right\}. \tag{10.A.7}$$

Now we can complete the proof by using (10.A.6), (10.A.7) and noting that $P(N_p(A) = k|x) = E_*[P(N_p(A) = k|x,W)]$. □

Exercises

10.1 Verify (10.1.9).

10.2 Using the E1684 data discussed in Example 5.1 along with the covariates in Table 5.1, fit the accelerated failure time model in (10.2.2) with a Dirichlet process prior on θ_i and a uniform improper prior for the regression coefficients using the MCMC algorithm outlined in Steps 1a and 1b after equation (10.2.11). Compare your results with those in Example 5.1.

10.3 Repeat Exercise 10.2 with a Polya tree prior on θ_i as outlined in Section 10.2.2.

10.4 Verify (10.2.7) and (10.2.9).

10.5 Derive (10.2.14).

10.6 Write a computer program to verify all of the entries in Table 10.3.

10.7 Using the E1684 data discussed in Example 5.1 along with the covariates in Table 5.1, fit the MARS model discussed in Section 10.3. Compare your results with those of Example 5.1.

10.8 Use an argument similar to the proof of (10.4.2) to show $H_0(t)$ given in (10.4.4) is a cumulative intensity function.

10.9 Show (10.4.16).

10.10 Consider the model in (10.4.4) and the likelihood function given by (10.4.17).

(a) Find the joint posterior distribution (up to a normalizing constant) of $(\theta, \boldsymbol{\beta}, \gamma, \lambda, \alpha)$ based on the observed data D using the priors specified in Section 10.4.5.
(b) Derive the full conditional distributions for each of the model parameters θ, $\boldsymbol{\beta}$, γ, λ, and α.
(c) Discuss how to sample from the posterior distribution derived in Part (a).

10.11 Derive (i) (10.4.17) and (ii) the full conditionals for all of the parameters in (10.4.17) using the prior distributions stated in Section 10.4.5.

10.12 Write a computer program to verify all of the entries in Tables 10.5 and 10.6 using the nutritional cancer prevention trial data in Example 10.6.

10.13 Derive (10.5.3) and (10.5.4).

10.14 Using the breast cancer data discussed in Example 1.3 and given in Table 1.4, fit the model given by (10.6.3), using the stages and implementation described in Sections 10.6.1 and 10.6.2.

10.15 Write a computer program to reproduce the results of Table 10.11 using the nursing home usage data.

10.16 Using the kidney infection data in Example 10.9, compute the Bayesian latent residuals for all of the subjects.

10.17 For the kidney infection data given in Example 1.4, consider the proportional hazards frailty model (4.1.1) with the single covariate sex, using the prior distributions and hyperparameters specified in Example 10.9.

(a) Derive the full conditional distributions necessary for the Gibbs sampler.
(b) Draw box-plots for all the Bayesian latent residuals, v_{ij}.
(c) Draw a Q-Q plot for each v_{ij} versus the $\mathcal{U}(0,1)$ distribution for all possible (i,j).

10.18 Prove Proposition 10.7.1.

10.19 Prove Proposition 10.7.2.

10.20 Consider the three models discussed in Example 10.10.

(a) Repeat the simulation experiment described in Example 10.10.
(b) Reproduce Table 10.14 and Figures 10.8 and 10.9.

10.21 Consider the two models given in Example 10.11.

(a) Repeat the simulation experiment described in Example 10.11.
(b) Reproduce Table 10.15 and Figures 10.10–10.12.

List of Distributions

The distributions (name, notation, density) listed below are most frequently cited in the book.

Uniform: $y \sim \mathcal{U}(a, b)$, $a < b$,

$$f(y|a,b) = \begin{cases} \frac{1}{b-a}, & \text{if } a < y < b, \\ 0, & \text{otherwise.} \end{cases}$$

Exponential: $y \sim \mathcal{E}(\lambda)$, $\lambda > 0$,

$$f(y|\lambda) = \begin{cases} \lambda \exp(-\lambda y), & \text{if } y > 0, \\ 0, & \text{otherwise.} \end{cases}$$

Gamma: $y \sim \mathcal{G}(\alpha, \lambda)$, $\alpha > 0$, $\lambda > 0$,

$$f(y|\alpha,\lambda) = \begin{cases} \frac{\lambda^\alpha}{\Gamma(\alpha)} y^{\alpha-1} \exp(-\lambda y), & \text{if } y > 0, \\ 0, & \text{otherwise.} \end{cases}$$

Inverse Gamma: $y \sim \mathcal{IG}(\alpha, \lambda)$, $\alpha > 0$, $\lambda > 0$,

$$f(y|\alpha,\lambda) = \begin{cases} \frac{\lambda^\alpha}{\Gamma(\alpha)} y^{-(\alpha+1)} \exp(-\lambda/y), & \text{if } y > 0, \\ 0, & \text{otherwise.} \end{cases}$$

Beta: $y \sim \mathcal{B}(\alpha, \lambda)$, $\alpha > 0$, $\lambda > 0$,

$$f(y|\alpha, \lambda) = \begin{cases} \frac{\Gamma(\alpha+\lambda)}{\Gamma(\alpha)\Gamma(\lambda)} y^{\alpha-1}(1-y)^{\lambda-1}, & \text{if } 0 < y < 1, \\ 0, & \text{otherwise.} \end{cases}$$

Normal: $y \sim N(\mu, \sigma^2)$, $-\infty < \mu < \infty$, $\sigma > 0$,

$$f(y|\mu, \sigma) = \frac{1}{\sqrt{2\pi}\sigma} \exp\left\{ -\frac{(y-\mu)^2}{2\sigma^2} \right\}.$$

Poisson: $y \sim \mathcal{P}(\lambda)$, $\lambda > 0$,

$$f(k|\lambda) = \frac{\lambda^k}{k!} \exp(-\lambda), \quad k = 0, 1, 2, \ldots$$

Weibull: $y \sim \mathcal{W}(\alpha, \lambda)$, $\alpha > 0$, $-\infty < \lambda < \infty$,

$$f(y|\alpha, \lambda) = \begin{cases} \alpha y^{\alpha-1} \exp\left\{\lambda - \exp(\lambda) y^{\alpha}\right\}, & \text{if } y > 0, \\ 0, & \text{otherwise.} \end{cases}$$

Log-normal: $y \sim \mathcal{LN}(\mu, \sigma^2)$, $-\infty < \mu < \infty$, $\sigma > 0$,

$$f(y|\mu, \sigma) = \begin{cases} \frac{1}{\sqrt{2\pi}\sigma y} \exp\left\{ -\frac{(\log(y)-\mu)^2}{2\sigma^2} \right\}, & \text{if } y > 0, \\ 0, & \text{otherwise.} \end{cases}$$

Extreme Value: $y \sim \mathcal{V}(\alpha, \lambda)$, $\alpha > 0$, $-\infty < \lambda < \infty$,

$$f(y|\alpha, \lambda) = \alpha \exp(\alpha y) \exp\left\{\lambda - \exp(\lambda + \alpha y)\right\}, \quad -\infty < y < \infty.$$

Multivariate Normal: $\boldsymbol{y} = (y_1, y_2, \ldots, y_p)' \sim N_p(\boldsymbol{\mu}, \Sigma)$, $\Sigma > 0$ (positive definite),

$$f(\boldsymbol{y}|\boldsymbol{\mu}, \Sigma) = (2\pi)^{-p/2} |\Sigma|^{-1/2} \exp\left\{ -\frac{1}{2} (\boldsymbol{y} - \boldsymbol{\mu})' \Sigma^{-1} (\boldsymbol{y} - \boldsymbol{\mu}) \right\}.$$

Dirichlet: $\boldsymbol{y} = (y_1, y_2, \ldots, y_{k-1})' \sim D_{k-1}(\alpha_1, \alpha_2, \ldots, \alpha_k)$, $\alpha_j > 0$, $j = 1, 2, \ldots, k$,

$$f(y_1, y_2, \ldots, y_{k-1}|\boldsymbol{\alpha}) = \frac{\Gamma\left(\sum_{j=1}^{k} \alpha_j\right)}{\prod_{j=1}^{k} \Gamma(\alpha_j)} \left(\prod_{j=1}^{k-1} y_j^{\alpha_j - 1} \right) \left(1 - \sum_{j=1}^{k-1} y_j \right)^{\alpha_k - 1},$$

for $0 < y_j < 1$, $\sum_{j=1}^{k-1} y_j \le 1$, where $\boldsymbol{\alpha} = (\alpha_1, \alpha_2, \ldots, \alpha_k)'$.

References

Aalen, O.O. and Hoem, J.M. (1978). Random time changes for multivariate counting processes. *Scandinavian Actuarial Journal*, 81–101.

Achcar, J.A., Bolfarine, H., and Pericchi, L.R. (1987). Transformation of survival data to an extreme value distribution. *The Statistician* **36**, 229–234.

Achcar, J.A., Brookmeyer, R., and Hunter, W.G. (1985). An application of Bayesian analysis to medical follow-up data. *Statistics in Medicine* **4**, 509–520.

Adcock, C.J. (1997). Sample size determination: A review. *The Statistician* **46**, 261–283.

Adcock, C.J. (1988). A Bayesian approach to calculating sample sizes. *The Statistician* **37**, 433–439.

Agresti, A. (1990). *Categorical Data Analysis.* New York: Wiley.

Aitchison, J. and Dunsmore, I.R. (1975). *Statistical Prediction Analysis.* New York: Cambridge University Press.

Akaike, H. (1973). Information theory and an extension of the maximum likelihood principle. In *International Symposium on Information Theory* (Eds. B.N. Petrov and F. Csaki). Budapest: Akademia Kiado, pp. 267–281.

Alder, B.J. and Wainwright, T.E. (1959). Studies in molecular dynamics. I. General method. *Journal of Chemical Physics* **31**, 459–466.

Altman, D.G. and Andersen, P.K. (1989). Bootstrap investigation of the stability of a Cox regression model. *Statistics in Medicine* **8**, 771–783.

Andersen, H.C. (1980). Molecular dynamics simulations at constant pressure and/or temperature. *Journal of Chemical Physics* **72**, 2384–2393.

Anderson, J.A. and Blair, V. (1982). Penalized maximum likelihood estimation in logistic regression and discrimination. *Biometrika* **69**, 123–136.

Anderson, P.K., Borgan, O., Gill, R.D., and Keiding, N. (1993). *Statistical Models Based on Counting Processes.* New York: Springer-Verlag.

Andersen, P.K. and Gill, R.D. (1982). Cox's regression model for counting processes: A large sample study. *The Annals of Statistics* **10**, 1100–1120.

Antoniak, C. (1974). Mixtures of Dirichlet processes with applications to Bayesian nonparametric problems. *The Annals of Statistics* **2**, 1152–1174.

Arjas, E. and Gasbarra, D. (1997). On prequential model assessment in life history analysis. *Biometrika* **84**, 505–522.

Arjas, E. and Gasbarra, D. (1994). Nonparametric Bayesian inference from right censored survival data, using the Gibbs sampler. *Statistica Sinica* **4**, 505–524.

Aslanidou, H., Dey, D.K., and Sinha, D. (1998). Bayesian analysis of multivariate survival data using Monte Carlo methods. *Canadian Journal of Statistics* **26**, 33–48.

Asselain, B., Fourquet, A., Hoang, T., Tsodikov, A.D., and Yakovlev, A.Y. (1996). A parametric regression model of tumor recurrence: An application to the analysis of clinical data on breast cancer. *Statistics and Probability Letters* **29**, 271–278.

Bacchetti, P. (1990). Estimating the incubation period of AIDS by computing population infection and diagnosis patterns. *Journal of the American Statistical Association* **85**, 1002–1008.

Basu, A. and Klein, J.P. (1982). Some recent development in competing risks theory. In *Survival Analysis* (Eds. J. Crowley and R.A. Johnson). Hayward, CA: IMS, pp. 216–229.

Bayarri, M.J. and Berger, J.O. (1999). Quantifying surprise in the data and model verification. In *Bayesian Statistics* **6** (Eds. J.M Bernardo, J.O. Berger, A.P. Dawid, and A.F.M. Smith). Oxford: Oxford University Press, pp. 53–82.

Becker, N. (1989). Cumulative damage models of exposures with limited duration and host factors. *Archives of Environmental Health* **44**, 331–336.

Berger, J.O. (2000). Bayesian analysis: A look at today and thoughts of tomorrow. *Journal of the American Statistical Association* **95**, 1269–1276.

Berger, J.O. and Berry, D.A. (1988). Statistical analysis and the illusion of objectivity. *The American Scientist* **76**, 159–165.

Berger, J.O. and Delampady, M. (1987). Testing precise hypotheses. *Statistical Science* **2**, 317–352.

Berger, J.O. and Mallows, C.L. (1988). Discussion of Bayesian variable selection in linear regression. *Journal of the American Statistical Association* **83**, 1033–1034.

Berger, J.O. and Pericchi, L.R. (1996). The intrinsic Bayes factor for model selection and prediction. *Journal of the American Statistical Association* **91**, 109–122.

Berger, J.O. and Sellke, T. (1987). Testing a point null hypothesis (with Discussion). *Journal of the American Statistical Association* **82**, 112–122.

Berger, J.O. and Sun, D. (1993). Bayesian analysis for the poly-Weibull distribution. *Journal of the American Statistical Association* bf 88, 1412–1418.

Berkson, J. and Gage, R.P. (1952). Survival curve for cancer patients following treatment. *Journal of the American Statistical Association* **47**, 501–515.

Bernardo, P.M.V. and Ibrahim, J.G. (2000). Group sequential designs for cure rate models with early stopping in favour of the null hypothesis. *Statistics in Medicine* **19**, 3023–3035.

Bernhard, J., Cella, D.F. Coates, A.S., Fallowfield, L., Ganz, P.A., Moinpour, C.M., Mosconi, P., Osoba, D., Simes, J., and Hurny, C. (1998). Missing quality of life data in cancer clinical trials: Serious problems and challenges. *Statistics in Medicine* **17**, 517–532.

Berry, D.A. (1993). A case for Bayesianism in clinical trials. *Statistics in Medicine* **12**, 1377–1393.

Berry, D.A. (1987). Interim analysis in clinical trials. *The American Statistician* **41**, 117–122.

Berry, D.A. (1985). Interim analyses in clinical trials: Classical vs. Bayesian approaches. *Statistics in Medicine* **4**, 521–526.

Berzuini, C. and Clayton, D.G. (1994). Bayesian analysis of survival on multiple time scales. *Statistics in Medicine* **13**, 823–838.

Besag, J. and Green, P.J. (1993). Spatial statistics and Bayesian computation. *Journal of the Royal Statistical Society, Series B* **55**, 25–37.

Besag, J., Green, P.J., Higdon, D.M., and Mengersen, K. (1995). Bayesian computation and stochastic systems (with Discussion). *Statistical Science* **10**, 3-36.

Best, N.G., Cowles, M.K., and Vines, K. (1995). CODA: Convergence diagnostics and output analysis software for Gibbs sampling output. Version 30. *Technical Report.* Medical Research Council Biostatistics Unit, Institute of Public Health, Cambridge University.

Blackwell, D. (1973). The discreteness of Ferguson selections. *The Annals of Statistics* **1**, 356–358.

Blackwell, D. and MacQueen, J.B. (1973). Ferguson distributions via Polya urn schemes. *The Annals of Statistics* **1**, 353–355.

Boone, C.W., Kelloff, G.J., and Malone, W. (1990). Identification of candidate cancer chemopreventive agents and their evaluation in animal models and human clinical trials: A review. *Cancer Research* **50**, 2–9.

Box, G.E.P. (1980). Sampling and Bayes' inference in scientific modelling and robustness (with Discussion). *Journal of the Royal Statistical Society, Series A* **143**, 383–430.

Box, G.E.P. and Tiao, G.C. (1992). *Bayesian Inference in Statistical Analysis.* New York: Wiley.

Brémaud, P. (1981). *Point Processes and Queues* . New York: Springer-Verlag.

Breslow, N.E. (1990). Biostatistics and Bayes (with Discussion). *Statistical Science* **5**, 269–298.

Breslow, N.E. (1975). Analysis of survival data under the proportional hazards model. *International Statistical Review* **43**, 45–48.

Breslow, N.E. (1974). Covariance analysis of censored survival data. *Biometrics* **30**, 89–99.

Brooks, S.P. and Gelman, A. (1998). General methods for monitoring convergence of iterative simulations. *Journal of Computational and Graphical Statistics* **7**, 434–455.

Brown, B.W., Herson, J., Atkinson, N, and Rozell, M.E., (1987). Projection from previous studies: A Bayesian and frequentist compromise. *Controlled Clinical Trials* **8**, 29–44.

Brown, E.R., MaWhinney, S., Jones, R.H., Kafadar, K., and Young, B. (2000). Improving the fit of bivariate smoothing splines when estimating longitudinal immunological and virological markers in HIV patients with individual antiretroviral treatment strategies. *Statistics in Medicine.* To appear.

Buckle, D.J. (1995). Bayesian inference for stable distributions. *Journal of the American Statistical Association* **90**, 605–613.

Burridge, J. (1981). Empirical Bayes analysis for survival time data. *Journal of the Royal Statistical Society, Series B* **43**, 65–75.

Bush, C.A. and MacEachern, S.N. (1996). A semiparametric Bayesian model for randomized block designs. *Biometrika* **33**, 275–285.

Byar, D.P., Simoon, R.M., Friedewald, W.T., Schlesselman, J.J., DeMets, D.L., Ellenberg, J.H., Gail, M.H., and Ware, J.H. (1976). Randomized clinical trials: Perspective on some recent ideas. *New England Journal of Medicine* **295**, 74–80.

Bycott, P.W. and Taylor, J.M.G. (1998). A comparison of smoothing techniques for CD4 data measured with error in a time-dependent

Cox proportional hazards model. *Statistics in Medicine* **17**, 2061–2077.

Carlin, B.P. and Hodges, J.S. (1999). Hierarchical proportional hazards regression models for highly stratified data. *Biometrics* **55**, 1162–1170.

Carlin, B.P. and Louis, T.A. (1996). *Bayes and empirical Bayes methods for data analysis.* London: Chapman & Hall.

Carlin, J.B. (1992). Meta-analysis for 2 × 2 tables: A Bayesian approach. *Statistics in Medicine* **11**, 141–158.

Casella, G. and Berger, R.L. (1990). *Statistical Inference.* Belmont, CA: Duxbury Press.

Casella, G. and George, E.I. (1992). Explaining the Gibbs sampler. *The American Statistician* **46**, 167–174.

Chen, H.Y. and Little, R.J.A. (1999). Proportional hazards regression with missing covariates. *Journal of the American Statistical Association* **94**, 896–908.

Chen. M.-H. (1994). Importance-weighted marginal Bayesian posterior density estimation. *Journal of the American Statistical Association* **89**, 818–824.

Chen, M.-H., Dey, D.K., and Sinha, D. (2000). Bayesian analysis of multivariate mortality data with large families. *Applied Statistics* **49**, 129–144.

Chen, M.-H., Harrington, D.P., and Ibrahim, J.G. (1999). Bayesian models for high-risk melanoma: A case study of ECOG trial E1690. *Technical Report MS-06-99-22.* Department of Mathematical Sciences, Worcester Polytechnic Institute.

Chen, M.-H. and Ibrahim, J.G. (2001). Bayesian model comparisons for survival data with a cure fraction. *ISBA 2000, Proceedings*, ISBA and EUROSTAT. To appear.

Chen, M.-H., Ibrahim, J.G., and Lipsitz, S.R. (1999). Bayesian methods for missing covariates in survival data. *Technical Report MS-09-99-28.* Department of Mathematical Sciences, Worcester Polytechnic Institute.

Chen, M.-H., Ibrahim, J.G., and Shao, Q.-M. (2000). Power prior distributions for generalized linear models. *Journal of Statistical Planning and Inference* **41**, 121–137.

Chen, M.-H., Ibrahim, J.G., Shao, Q.-M., and Weiss, R.E. (2001) Prior elicitation for model selection and estimation in generalized linear mixed models. *Journal of Statistical Planning and Inference* **42**. To appear.

Chen, M.-H., Ibrahim, J.G., and Sinha, D. (2001). Bayesian inference for multivariate survival data With a surviving fraction. *Journal of Multivariate Analysis.* To appear.

Chen, M.-H., Ibrahim, J.G., and Sinha, D. (1999). A new Bayesian model For survival data with a surviving fraction. *Journal of the American Statistical Association* **94**, 909–919.

Chen, M.-H., Ibrahim, J.G., and Yiannoutsos, C. (1999). Prior elicitation and Bayesian computation for logistic regression models with applications to variable selection. *Journal of the Royal Statistical Society, Series B* **61**, 223–242.

Chen, M.-H., Manatunga, A.K., and Williams, C.J. (1998). Heritability estimates from human twin data by incorporating historical prior information. *Biometrics* **54**, 1348–1362.

Chen, M.-H. and Shao, Q.-M. (2001). Propriety of posterior distribution for dichotomous quantal response models With general link functions. *Proceedings of the American Mathematical Society* **129**, 293–302.

Chen, M.-H. and Shao, Q.-M. (1999). Monte Carlo estimation of Bayesian credible and HPD intervals. *Journal of Computational and Graphical Statistics* **8**, 69–92.

Chen, M.-H. and Shao, Q.-M. (1997a). On Monte Carlo methods for estimating ratios of normalizing constants. *The Annals of Statistics* **25**, 1563–1594.

Chen, M.-H. and Shao, Q.-M. (1997b). Estimating ratios of normalizing constants for densities with different dimensions. *Statistica Sinica* **7**, 607–630.

Chen, M.-H., Shao, Q.-M., and Ibrahim, J.G. (2000). *Monte Carlo Methods in Bayesian Computation.* New York: Springer-Verlag.

Chen, W.C., Hill, B.M., Greenhouse, J.B., and Fayos, J.V. (1985). Bayesian analysis of survival curves for cancer patients following treatment. In *Bayesian Statistics* **2** (Eds. J.O. Berger, J. Bernardo, and A.F.M. Smith). Amsterdam: North-Holland, pp. 299–328.

Chib, S. (1995). Marginal likelihood from the Gibbs output. *Journal of the American Statistical Association* **90**, 1313–1321.

Chib, S. and Greenberg, E. (1998). Bayesian analysis of multivariate probit models. *Biometrika* **85**, 347–361.

Chib, S. and Greenberg, E. (1995). Understanding the Metropolis–Hastings algorithm. *The American Statistician* **49**, 327–335.

Chipman, H., George, E.I., and McCulloch, R.E. (1998). Bayesian CART model search (with Discussion). *Journal of the American Statistical Association* **93**, 935–960.

Christensen, R. and Johnson, W. (1988). Modelling accelerated failure time with a Dirichlet process. *Biometrika* **75**, 693–704.

Clark, L.C., Combs, G.F., Turnbull, B.W., Slate, E.H., Chlaker, D.K., Chow, J., Curtis, D., Davis, L.S., Glover, R.A., Graham, G.F., Gross, E.G., Krongrad, A., Lesher, J.L., Park, H.K., Sanders, B.B., Smith, C.L., Taylor, J.R., and the Nutritional Prevention of Cancer Study Group (1996). Effects of selenium supplementation for cancer prevention in patients with carcinoma of the skin: A randomized clinical trial. *Journal of the American Medical Association* **276**, 1957–1963.

Clayton, D.G. (1991). A Monte Carlo method for Bayesian inference in frailty models. *Biometrics* **47**, 467–85.

Clayton, D.G. (1978). A model for association in bivariate life tables and its application in epidemiological studies of familiar tendency in chronic disease incidence. *Biometrika* **65**, 141–151.

Clayton, D.G. and Cuzick, J. (1985). Multivariate generalizations of the proportional hazards model (with Discussion). *Journal of the Royal Statistical Society, Series A* **148**, 82–117.

Clyde, M.A. (1999). Bayesian model averaging and model search strategies. In *Bayesian Statistics 6* (Eds. J.M. Bernardo, J.O. Berger, A.P. Dawid, and A.F.M. Smith). Oxford: Oxford University Press, pp. 157–185.

Cornfield, J. (1966a). Sequential trials, sequential analysis and the likelihood principle. *The American Statistician* **20**, 18–23.

Cornfield, J. (1966b). A Bayesian test of some classical hypotheses with applications to sequential clinical trials. *Journal of the American Statistical Association* **61**, 577–594.

Cox, D.R. (1975). Partial likelihood. *Biometrika* **62**, 269–276.

Cox, D.R. (1972). Regression models and life tables. *Journal of the Royal Statistical Society, Series B* **34**, 187–220.

Cox, D.R. (1959). The analysis of exponentially distributed lifetimes with two types of failures. *Journal of the Royal Statistical Society, Series B* **21**, 411–421.

Cox, D.R. and Isham, V. (1980). *Point Processes.* London: Chapman & Hall.

Cox, D.R. and Oakes, D. (1984). *Analysis of Survival Data.* London: Chapman & Hall.

Cowles, M.K. and Carlin, B.P. (1996). Markov chain Monte Carlo convergence diagnostics: A comparative review. *Journal of the American Statistical Association* **91**, 883–904.

Crook, J.F. and Good, I.J. (1982). The powers and strengths of tests for multinomials and contingency tables. *Journal of the American Statistical Association* **77**, 793–802.

Damien, P., Laud, P.W., and Smith, A.F.M. (1996). Implementation of Bayesian non-parametric inference based on beta processes. *Scandinavian Journal of Statistics* **23**, 27–36.

Damien, P., Wakefield, J.C., and Walker, S.G. (1999). Gibbs sampling for Bayesian nonconjugate and hierarchical models using auxiliary variables. *Journal of the Royal Statistical Society, Series B* **61**, 331–344.

David, H.A. and Moeschberger, M.L. (1978). *The Theory of Competing Risks. Griffin's Statistical Monographs & Courses* **39**, London: Charles W. Griffin.

Dawid, A.P. (1992). Prequential analysis, stochastic complexity and Bayesian inference. In *Bayesian Statistics* **4** (Eds. J.M. Bernado, J.O. Berger, A.P. Dawid, and A.F.M. Smith). Oxford: Oxford University Press, pp. 109–125.

Dawid, A.P. (1984). Statistical theory: The prequential approach. *Journal of the Royal Statistical Society, Series A* **147**, 278–292.

DeGruttola, V. and Tu, X.M. (1994). Modeling progression of CD4-lymphocyte count and its relationship to survival time. *Biometrics* **50**, 1003–1014.

DeGruttola, V., Wulfsohn, M.S., Fischl, M.A., and Tsiatis, A.A. (1993). Modeling the relationship between survival and CD4 lymphocytes in patients with AIDS and ARC. *Journal of Acquired Immune Deficiency Syndrome* **6**, 359–365.

Dellaportas, P. and Smith, A.F.M. (1993). Bayesian inference for generalized linear and proportional hazards models via Gibbs sampling. *Applied Statistics* **42**, 443–459.

Dempster, A.P., Laird, N.M., and Rubin, D.B. (1977). Maximum likelihood from incomplete data via the EM algorithm. *Journal of the Royal Statistical Society, Series B* **39**, 1–38.

Denison, D.G.T., Mallick, B.K., and Smith, A.F.M. (1999). Bayesian MARS. *Statistics and Computing* **8**, 337–346.

Denison, D.G.T., Mallick, B.K., and Smith, A.F.M. (1998). A Bayesian CART algorithm. *Biometrika* **85**, 363–377.

Devroye, L. (1986). *Non-Uniform Random Variate Generation.* New York: Springer-Verlag.

Dey, D.K., Chen, M.-H., and Chang, H. (1997). Bayesian approach for nonlinear random effects models. *Biometrics* **53**, 1239–1252.

Diaconis, P. and Freedman, D.A. (1986). On inconsistent Bayes estimates of location. *The Annals of Statistics* **14**, 68–87.

Diaconis, P. and Ylvisaker, D. (1985). Quantifying prior opinion. In *Bayesian Statistics* **2** (Eds. J.O. Berger, J. Bernardo, and A.F.M. Smith). Amsterdam: North-Holland, pp. 133–156.

Dickson, E.R., Fleming, T.R., and Weisner, R.H. (1985). Trial of penicillamine in advanced primary biliary cirrhosis. *New England Journal of Medicine* **312**, 1011–1015.

Doss, H. (1994). Bayesian nonparametric estimation for incomplete data via successive substitution sampling. *The Annals of Statistics* **22**, 1763–1786.

Doss, H. and Huffer, F. (1998). Monte Carlo methods for Bayesian analysis of survival data using mixtures of Dirichlet priors. *Technical Report.* Department of Statistics, Ohio State University.

Doss, H. and Narasimhan, B. (1998). Dynamic display of changing posterior in Bayesian survival analysis. In *Practical Nonparametric and Semiparametric Bayesian Statistics* (Eds. D. Dey, P. Müller, and D. Sinha). New York: Springer-Verlag, pp. 63–84.

Draper, D. (1995). Assessment and propagation of model uncertainty. *Journal of the Royal Statistical Society, Series B* **57**, 45–97.

Duane, S., Kennedy, A.D., Pendleton, B.J., and Roweth, D. (1987). Hybrid Monte Carlo. *Physical Letters B* **195**, 216–222.

Dykstra, R.L. and Laud, P.W. (1981). A Bayesian nonparametric approach to reliability. *The Annals of Statistics* **9**, 356–367.

Edwards, W., Lindman, H., and Savage, L.J. (1963). Bayesian statistical inference for psychological research. *Psychological Review* **70**, 193–242.

Efron, B. (1977). The efficiency of Cox's likelihood function for censored data. *Journal of the American Statistical Association* **72**, 557–565.

Errington, R.D., Ashby, D., Gore, S.M., Abrams, K.R., Myint, S., Bonnett, D.E., Blake, S.W., and Saxton, T.E. (1991). High energy neutron treatment of pelvic cancers: Study stopped because of increased mortality. *British Medical Journal* **302**, 1045–1051.

Escobar, M.D. (1994). Estimating normal means with a Dirichlet process prior. *Journal of the American Statistical Association* **89**, 268–277.

Escobar, M.D. and West, M. (1995). Bayesian density estimation and inference using mixtures. *Journal of the American Statistical Association* **90**, 578–588.

Ettinger, D.S., Finkelstein, D.M., Abeloff, M.D., Ruckdeschel, J.C., Aisner, S.C., and Eggleston, J.C. (1990). A randomized comparison of standard chemotherapy versus alternating chemotherapy and maintenance versus no maintenance therapy for extensive-stage small-cell lung cancer: A phase III study of the Eastern Cooperative Oncology Group. *Journal of Clinical Oncology* **8**, 230–240.

Ewell, M. and Ibrahim, J.G. (1997). The large sample distribution of the weighted log rank statistic under general local alternatives. *Lifetime Data Analysis* **3**, 5–12.

Fahrmeir, L. (1994). Dynamic modelling and penalized likelihood estimation for discrete time survival data. *Biometrika* **81**, 317–330.

Faraggi, D. and Simon, R. (1995). A neural network model for survival data. *Statistics in Medicine* **14**, 73–82.

Farewell, V.T. (1982). The use of mixture models for the analysis of survival data with long-term survivors. *Biometrics* **38**, 1041–1046.

Farewell, V.T. (1986). Mixture models in survival analysis: Are they worth the risk? *Canadian Journal of Statistics* **14**, 257–262.

Faucett, C.J. and Thomas, D.C. (1996). Simultaneously modelling censored survival data and repeatedly measured covariates: A Gibbs sampling approach. *Statistics in Medicine* **15**, 1663–1685.

Fayers, P.M., Ashby, D., and Parmar, M.K.B. (1997). Bayesian data monitoring in clinical trials. *Statistics in Medicine* **16**, 1413–1430.

Feigl, P. and Zelen, M. (1965). Estimation of exponential survival probabilities with concomitant information. *Biometrics* **21**, 826–838.

Feller, W. (1971). *An Introduction to Probability Theory and its Applications.* Volume 2, Second Edition, New York: Wiley.

Ferguson, T.S. (1974). Prior distributions on spaces of probability measures. *The Annals of Statistics* **2**, 615–629.

Ferguson, T.S. (1973). A Bayesian analysis of some non-parametric problems. *The Annals of Statistics* **1**, 209–230

Ferguson, T.S. and Phadia, E.G. (1979). Bayesian nonparametric estimation based on censored data. *The Annals of Statistics* **7**, 163–186.

Fienberg, S.E. and Mason, W.M. (1979). Identification and estimation of age-period-cohort models in the analysis of discrete archival data. *Sociological Methodology*, 1–67.

Finkelstein, D.M. (1986). A proportional hazards model for interval-censored failure time data. *Biometrics* **42**, 845–854.

Finkelstein, D.M. and Wolfe, R.A. (1985). A semiparametric model for regression analysis of interval-censored failure time data. *Biometrics* **41**, 933-945.

Fleiss, J.L. (1981). *Statistical Methods for Rates and Proportions.* Second Edition, New York: Wiley.

Fleming, T.R. and Harrington, D.P. (1991). *Counting Processes and Survival Analysis.* New York: Wiley.

Fleming, T.R. and Watelet, L.F. (1989). Approaches to monitoring clinical trials. *Journal of the National Cancer Institute* **81**, 188–193.

Florens, J.P., Mouchart, M., and Rolin, J.M. (1999). Semi- and non-parametric Bayesian analysis of duration models with Dirichlet priors: A survey. *International Statistical Review* **67**, 187–210.

Freedman, L.S., Midthune, D.N., Brown, C.C., Steele, V., and Kelloff, G.J. (1993). Statistical analysis of animal cancer chemoprevention experiments. *Biometrics* **49**, 259–268.

Freedman, L.S. and Spiegelhalter, D.J. (1992). Application of Bayesian statistics to decision making during a clinical trial. *Statistics in Medicine* **11**, 23–35.

Freedman, L.S. and Spiegelhalter, D.J. (1989). Comparison of Bayesian with group sequential methods for monitoring clinical trials. *Controlled Clinical Trials* **10**, 357–367.

Freedman, L.S. and Spiegelhalter, D.J. (1983). The assessment of of subjective opinion and its use in relation to stopping rules for clinical trials. *The Statistician* **32**, 153–160.

Freedman, L.S., Spiegelhalter D.J., and Parmar, M.K.B. (1994). The what, why and how of Bayesian clinical trials monitoring. *Statistics in Medicine* **13**, 1371–1383.

Fried, L.P., Borhani, N.O., and et al. (1991). The cardiovascular health study: Design and rationale. *Annals of Epidemiology* **1**, 263–276.

Friedman, J.H. (1991). Multivariate adaptive regression splines (with Discussion). *The Annals of Statistics* **19**, 1–141.

Furnival, G.M. and Wilson, R.W. (1974). Regression by leaps and bounds. *Technometrics* **16**, 499–511.

Gail, M.H., Santner, T.J., and Brown, C.C. (1980). An analysis of comparative carcinogenesis experiments with multiple times to tumor. *Biometrics* **36**, 255–266.

Gamerman, D. (1991). Dynamic Bayesian models for survival data. *Applied Statistics* **40**, 63–79.

Geisser, S. (1993). *Predictive Inference: An Introduction.* London: Chapman & Hall.

Geisser, S., and Eddy, W. (1979). A predictive approach to model selection. *Journal of the American Statistical Association* **74**, 153–160.

Gelfand, A.E., Dey, D.K., and Chang, H. (1992). Model determinating using predictive distributions with implementation via sampling-based methods (with Discussion). In *Bayesian Statistics* **4** (Eds. J.M. Bernado, J.O. Berger, A.P. Dawid, and A.F.M. Smith). Oxford: Oxford University Press, pp. 147–167.

Gelfand, A.E. and Ghosh, S.K. (1998). Model choice: A minimum posterior predictive loss approach. *Biometrika* **85**, 1–13.

Gelfand, A.E. and Kuo, L. (1991). Nonparametric Bayesian bioassay including ordered polytomous response. *Biometrika* **78**, 657–666.

Gelfand, A.E. and Mallick, B.K. (1995), Bayesian analysis of proportional hazards models built from monotone functions. *Biometrics* **51**, 843–852.

Gelfand, A.E. and Smith, A.F.M. (1990). Sampling based approaches to calculating marginal densities. *Journal of the American Statistical Association* **85**, 398–409.

Gelfand, A.E., Hills, S.E., Racine-Poon, A., and Smith, A.F.M. (1990). Illustration of Bayesian inference in normal data models using Gibbs sampling. *Journal of the American Statistical Association* **85**, 972–985.

Gelman, A., Carlin, J. B., Stern, H. S., and Rubin, D. B. (1995). *Bayesian Data Analysis.* London: Chapman & Hall.

Gelman, A., Roberts, G.O., and Gilks, W.R. (1995). Efficient Metropolis jumping rules. In *Bayesian Statistics* **5** (Eds. J.O. Berger, J.M. Bernardo, A.P. Dawid, and A.F.M. Smith). Oxford: Oxford University Press, pp. 599–608.

Gelman, A. and Rubin, D.B. (1992). Inference from iterative simulation using multiple sequences. *Statistical Science* **7**, 457–511.

Geman, S. and Geman, D. (1984). Stochastic relaxation, Gibbs distributions and the Bayesian restoration of images. *IEEE Transactions on Pattern Analysis and Machine Intelligence* **6**, 721–741.

George, E.I. and McCulloch, R.E. (1993). Variable selection via Gibbs sampling. *Journal of the American Statistical Association* **88**, 881–889.

George, E.I., and McCulloch, R.E., and Tsay, R.S. (1996). Two approaches to Bayesian model selections with applications. In *Bayesian Analysis in Econometrics and Statistics – Essays in Honor of Arnold Zellner* (Eds. D.A. Berry, K.A. Chaloner, and J.K. Geweke). New York: Wiley, pp. 339–348.

George, S.L., Li, C., Berry, D.A., and Green, M.R. (1994). Stopping a clinical trial early: Frequentist and Bayesian approaches applied to a CALGB trial in non-small-cell lung cancer. *Statistics in Medicine* **13**, 1313–1327.

Geweke, J. (1989). Bayesian inference in econometrics models using Monte Carlo integration. *Econometrica* **57**, 1317–1340.

Geweke, J. (1992). Evaluating the accuracy of sampling-based approaches to the calculation of posterior moments. In *Bayesian Statistics* **4** (Eds. J.M. Bernardo, J.O. Berger, A.P. Dawid, and A.F.M. Smith). Oxford: Oxford University Press, pp. 169–193.

Gilks, W.R. and Wild, P. (1992). Adaptive rejection sampling for Gibbs sampling. *Applied Statistics* **41**, 337–348.

Goel, P.K. (1988). Software for Bayesian analysis: Current status and additional needs. In *Bayesian Statistics 3* (Eds. J.M. Bernardo, M.H. DeGroot, D.V. Lindley, and A.F.M. Smith). Oxford: Oxford University Press, pp. 173–188.

Goldman, A.I. (1984). Survivorship analysis when cure is a possibility: A Monte Carlo study. *Statistics in Medicine* **3**, 153–163.

Good, I.J. (1952). Rational decisions. *Journal of the Royal Statistical Society, Series B* **14**, 107–114.

Good, I.J. and Gaskins, R.A. (1971). Non-parametric roughness penalties for probability densities. *Biometrika* **58**, 255–277.

Gray, R.J. (1994). A Bayesian analysis of institutional effects in a multicenter cancer clinical trial. *Biometrics* **50**, 244–53.

Gray, R.J. (1992). Flexible methods for analyzing survival data using splines, with applications to breast cancer prognosis. *Journal of the American Statistical Association* **87**, 942–951.

Gray, R.J. and Tsiatis, A.A. (1989). A linear rank test for use when the main interest is in differences in cure rates. *Biometrics* **45**, 899–904.

Green, P.J. (1995). Reversible jump Markov chain Monte Carlo computation and Bayesian model determination. *Biometrika* **82**, 711–732.

Greenhouse, J.B. and Wolfe, R.A. (1984). A competing risks derivation of a mixture model for the analysis of survival. *Communications in Statistics – Theory and Methods* **13**, 3133–3154.

Grenander, U. (1983). Tutorial in pattern theorey. *Technical Report.* Providence, R.I.: Division of Applied Mathematics, Brown University.

Grieve, A.P. (1987). Applications of Bayesian software: Two examples. *The Statistician* **36**, 283–288.

Grossman, J., Parmar, M.K.B., and Spiegelhalter, D.J. (1994). A unified method for monitoring and analysing controlled trials. *Statistics in Medicine* 13, 1815–1826.

Gumbel, E.J. (1961). Bivariate logistic distributions. *Journal of the American Statistical Association* **56**, 335–349.

Gustafson, P. (1998). Flexible Bayesian modelling for survival data. *Lifetime Data Analysis* **4**, 281–299.

Gustafson, P. (1997). Large hierarchical Bayesian analysis of multivariate survival data. *Biometrics* **53**, 230–242.

Gustafson, P. (1996). The effect of mixing distribution misspecification in conjugate mixture models. *Canadian Journal of Statistics* **24**, 307–318.

Gustafson, P. (1995). A Bayesian analysis of bivariate survival data from a multicentre cancer clinical trial. *Statistics in Medicine* **14**, 2523–2535.

Halperin, M., Lan, K.K.G., Ware, J.H., Johnson, N.J., and DeMets, D.L. (1982). An aid to data monitoring in long-term clinical trials. *Controlled Clinical Trials* **3**, 311–323.

Halpern, J. and Brown, B.W. Jr. (1987a). Cure rate models: Power of the log rank and generalized Wilcoxon tests. *Statistics in Medicine* **6**, 483–489.

Halpern, J. and Brown, B.W. Jr. (1987b). Designing clinical trials with arbitrary specification of survival functions and for the log rank or generalized Wilcoxon test. *Controlled Clinical Trials* **8**, 177–189.

Hammersley, J.M. and Handscomb, D.C. (1964). *Monte Carlo Methods.* London: Methuen.

Harrington, D.P., Fleming, T.R., and Green, S.J. (1982). Procedures for serial testing in censored survival data. In *Survival Analysis* (Eds. J. Crowley and R.A. Johnson). Hayward, CA: Institute of Mathematical Statistics, pp. 269-286.

Harris, E.K. and Albert, A. (1991). *Survivorship Analysis for Clinical Studies.* New York: Marcel Dekker.

Haseman, J.K., Huff, J., and Boorman, G.A. (1984). Use of historical control data in carcinogenicity studies in rodents. *Toxocologic Pathology* **12**, 126–135.

Hastie, T.J. and Tibshirani, R.J. (1993). Varying-coefficient models (with Discussion). *Journal of the Royal Statistical Society, Series B* **55**, 757–796.

Hastie, T.J. and Tibshirani, R.J. (1990). Exploring the nature of covariate effects in the proportional hazard model. *Biometrics* **46**, 1005–1016.

Hastings, W.K. (1970). Monte Carlo sampling methods using Markov chains and their applications. *Biometrika* **57**, 97–109.

Herring, A.H. and Ibrahim, J.G. (2001). Likelihood-based methods for missing covariates in the Cox proportional hazards model. *Journal of the American Statistical Association.* To appear.

Herring, A.H., Ibrahim, J.G., and Lipsitz, S.R. (2001). Maximum likelihood estimation in random effects models for survival data with missing covariates. *Biometrics.* To appear.

Hickman, J.C. and Miller, R.B. (1981). Bayesian bivariate graduation and forecasting. *Scandinavian Actuarial Journal* **3**, 129–150.

Higdon, D.M. (1998). Auxiliary variable methods for Markov chain Monte Carlo with applications. *Journal of the American Statistical Association* **93**, 585–595.

Hjort, N.L. (1990). Nonparametric Bayes estimators based on beta processes in models of life history data. *The Annals of Statistics* **18**, 1259–1294.

Hodges, J.S. (1998). Some algebra and geometry for hierarchical models, applied to diagnostics (with Discussion). *Journal of the Royal Statistical Society, Series B* **60**, 497–536.

Hogan, J.W. and Laird, N.M. (1997). Mixture models for the joint distribution or repeated measures and event times. *Statistics in Medicine* **16**, 239–257.

Hoover, D.R., Graham, N.M.G., Chen, B., Taylor, J.M.G., Phair, J., Zhou, S.Y.J., Munoz, A. (1992). Effect of CD4+ cell count measurement variability on staging HIV-1 infection. *Journal of Acquired Immune Deficiency Syndrome* **5**, 794–802.

Hougaard, P. (2000). *Analysis of Multivariate Survival Data.* New York: Springer-Verlag.

Hougaard, P. (1995). Frailty models for survival data. *Lifetime Data Analysis* **1**, 255–273.

Hougaard, P. (1986a). Survival models for heterogeneous populations derived from stable distributions. *Biometrika* **73**, 387-396, (Correction, **75** 395).

Hougaard, P. (1986b). A class of multivariate failure time distributions. *Biometrika* **73**, 671–678.

Ibragimov, I.A. and Chernin, K.E. (1959). On the unimodality of stable laws. *Theory of Probability and its Applications* **4**, 417–419.

Ibrahim, J.G. and Chen, M.-H. (2000). Power prior distributions for regression models. *Statistical Science* **15**, 46–60.

Ibrahim, J.G. and Chen, M.-H. (1998). Prior distributions and Bayesian computation for proportional hazards models. *Sankhyā, Series B* **60**, 48–64.

Ibrahim, J.G., Chen, M.-H., Lipsitz, S.R. (2000). Bayesian methods for generalized linear models with missing covariates. *Technical Report.* Department of Mathematical Sciences, Worcester Polytechnic Institute.

Ibrahim, J.G., Chen, M.-H., and Lipsitz, S.R. (1999). Monte Carlo EM for missing covariates in parametric regression models. *Biometrics* **55**, 591–596.

Ibrahim, J.G., Chen, M.-H., and MacEachern, S.N. (1999). Bayesian variable selection for proportional hazards models. *The Canadian Journal of Statistics* **27**, 701–717.

Ibrahim, J.G., Chen, M.-H., and Ryan, L.-M. (2000). Bayesian variable selection for time series count data. *Statistica Sinica* **10**, 971–987.

Ibrahim, J.G., Chen, M.-H., and Sinha, D. (2001a). Bayesian semiparametric models for survival data with a cure fraction. *Biometrics.* To appear.

Ibrahim, J.G., Chen, M.-H., and Sinha, D. (2001b). Criterion based methods for Bayesian model assessment. *Statistica Sinica.* To appear.

Ibrahim, J.G., Chen, M.-H., and Sinha, D. (2000). Bayesian methods for joint modelling of longitudinal and survival data with applications to cancer vaccine studies. *Technical Report MS-01-00-02.* Department of Mathematical Sciences, Worcester Polytechnic Institute.

Ibrahim, J.G. and Laud, P.W. (1994). A predictive approach to the analysis of designed experiments. *Journal of the American Statistical Association* **89**, 309–319.

Ibrahim, J.G., Lipsitz, S.R., and Chen, M.-H. (1999). Missing covariates in generalized linear models when the missing data mechanism is nonignorable. *Journal of the Royal Statistical Society, Series B* **61**, 173–190.

Ibrahim, J.G., Ryan, L.-M., and Chen, M.-H. (1998). Use of historical controls to adjust for covariates in trend tests for binary data. *Journal of the American Statistical Association* **93**, 1282–1293.

Jennison, J. and Turnbull, B.W. (2000). *Group Sequential Clinical Methods with Applications to Clinical Trials.* London: Chapman & Hall.

Jennison, C. and Turnbull, B.W. (1990). Statistical approaches to interim monitoring of medical trials: A review and commentary. *Statistical Science* **5**, 299–317.

Johnson, W. and Christensen, R. (1989). Nonparametric Bayesian analysis of the accelerated failure time model. *Statistics and Probability Letters* **7**, 179–184.

Joseph, L. and Belisle, P. (1997). Bayesian sample size determination for normal means and differences between normal means. *The Statistician* **46**, 209–226.

Joseph, L., Wolfson, D.B., and DuBerger, R. (1995), Sample size calculations for binomial proportions via highest posterior density intervals. *The Statistician* 44, 167–171.

Kalbfleisch, J.D. (1978). Nonparametric Bayesian analysis of survival time data. *Journal of the Royal Statistical Society, Series B* **40**, 214–221.

Kalbfleisch, J.D. and Prentice, R.L. (1980). *The statistical analysis of failure time data.* New York: Wiley.

Kalbfleisch, J.D. and Prentice, R.L. (1973). Marginal likelihood's based on Cox's regression and life model. *Biometrika* **60**, 267–278.

Kalish, L.A. (1992). Phase III multiple myeloma: Evaluation of combination chemotherapy in previously untreated patients. *Technical Report #726E.* Department of Biostatistics, Dana-Farber Cancer Institute.

Kannel, W.B., McGee, D., and Gordon, T. (1976). A general cardiovascular risk profile: The Framingham study. *American Journal of Cardiology* **38**, 46–51.

Kaplan, E.L. and Meier, P. (1958). Nonparametric estimation from incomplete observations. *Journal of the American Statistical Association* **53**, 457–481.

Karlin, S. and Taylor, H.M. (1981). *A Second Course in Stochastic Courses.* London: Academic Press.

Karlin, S. and Taylor, H.M. (1975). *A First Course in Stochastic Processes.* London: Academic Press.

Kaslow, R.A., Ostrow, D.G., Detels, R., Phair, J.P., Polk, B.F., and Rinaldo, C.R. (1987). The multicenter AIDS cohort study: Rationale, organization, and selected characteristics of the participants. *American Journal of Epidemiology* **126**, 310–318.

Kass, R.E. and Greenhouse, J.B. (1989). Comments on "Investigating therapies of potentially great benefit: ECMO" by J.H. Ware. *Statistical Science* **4**, 310–317.

Kass, R.E. and Raftery, A.E. (1995). Bayes factor. *Journal of the American Statistical Association* **90**, 773–795.

Kass, R.E. and Wasserman, L. (1995). A reference Bayesian test for nested hypotheses and its relationship to the Schwarz criterion. *Journal of the American Statistical Association* **90**, 928–934.

Kim, S.W. and Ibrahim, J.G. (2001). On Bayesian inference for proportional hazards models using noninformative priors. *Lifetime Data Analysis.* To appear.

Kirkwood, J.M., Ibrahim, J.G., Sondak, V.K., Richards, J., Flaherty, L.E., Ernstoff, M.S., Smith, T.J., Rao, U., Steele, M., and Blum, R.H. (2000). The role of high- and low-dose interferon Alfa-2b in high-risk melanoma: First analysis of intergroup trial E1690/S9111/C9190. *Journal of Clinical Oncology* **18**, 2444–2458.

Kirkwood, J.M., Strawderman, M.H., Ernstoff, M.S., Smith, T.J., Borden, E.C., Blum, R.H. (1996). Interferon alfa-2b adjuvant therapy of high-risk resected cutaneous melanoma: The Eastern Cooperative

Oncology Group trial EST 1684. *Journal of Clinical Oncology* **14**, 7–17.

Klein, J.P. and Moeschberger, M.L. (1997). *Survival Analysis.* New York: Springer-Verlag.

Kleinman, K.P. and Ibrahim, J.G. (1998a). A semiparametric Bayesian approach to the random effects model. *Biometrics* **54**, 921–938.

Kleinman, K.P. and Ibrahim, J.G. (1998b). A semi-parametric Bayesian approach to generalized linear mixed models. *Statistics in Medicine* **17**, 2579–2596.

Knuth, D.E. (1992). *Literate Programming.* Center for Study of Language and Information, Stanford University.

Kokoska, S.M. (1987). The analysis of cancer chemoprevention experiments. *Biometrics* **43**, 525–534.

Kong, A., Liu, J.S., and Wong, W.H. (1994). Sequential imputations and Bayesian missing data problems. *Journal of the American Statistical Association* **89**, 278–288.

Kooperberg, C., Stone, C.J., and Truong, Y.K. (1995). Hazard regression. *Journal of the American Statistical Association* **90**, 78–94.

Kopp-Schneider, A. and Portier, C.J. (1995). Carcinoma formation in NMRI mouse skin painting studies is a process suggesting greater than two stages. *Carcinogenesis* **16**, 53–59.

Kopp-Schneider, A., Portier, C.J., and Rippmann, F. (1991). The application of a multistage model that incorporates DNA damage and repair to the analysis of initiation/promotion experiments. *Mathematical Biosciences* **105**, 139–166.

Krall, J.M., Uthoff, V.A., and Harley, J.B. (1975). A step-up procedure for selecting variables associated with survival. *Biometrics* **31**, 49–57.

Kuk, A.Y.C. and Chen, C.-H. (1992). A mixture model combining logistic regression with proportional hazards regression. *Biometrika* **79**, 531–541.

Kuo, L. and Mallick, B.K. (1997). Bayesian semiparametric inference for the accelerated failure-time model. *The Canadian Journal of Statistics* **25**, 457–472.

Kuo, L. and Smith, A.F.M. (1992). Bayesian computations in survival models via the Gibbs sampler. In *Survival Analysis: State of the Art* (Eds. J.P. Klein and P.K. Goel). Boston: Kluwer Academic, pp. 11–24.

Lachin, J.M. (1981). Introduction to sample size determination and power analysis for clinical trials. *Controlled Clinical Trials* **1**, 13–28.

Lam, K.F. and Kuk, A.Y.C. (1997). A marginal likelihood approach to estimation in frailty models. *Journal of the American Statistical Association* **92**, 985–990.

Lang, W., Perkins, H., Anderson, R.E., Royce, R., Jewell, N.P., and Winkelstein, W. (1989). Patterns of T lymphocyte changes with human immunodeficiency virus infection: From seroconversion to the development of AIDS. *Journal of Acquired Immune Deficiency Syndrome* **2**, 63–69.

Lange, N., Carlin, B.P., and Gelfand, A.E. (1992). Hierarchical Bayes models for the progression of HIV infection using longitudinal CD4 T-cell numbers. *Journal of the American Statistical Association* **87**, 615–633.

Laska, E.M. and Meisner, M.J. (1992). Nonparametric estimation and testing in a cure rate model. *Biometrics* **48**, 1223–1234.

Laud, P.W. and Ibrahim, J.G. (1995). Predictive model selection. *Journal of the Royal Statistical Society, Series B* **57**, 247–262.

Laud, P.W., Smith, A.F.M., and Damien, P. (1996). Monte Carlo methods for approximating a posterior hazard rate process. *Statistics and Computing* **6**, 77–83.

Laurie, J.A., Moertel, C.G., Fleming, T.R., Wieand, H.S., Leigh, J.E., Rubin, J., McCormack, G., Gerstner, J.B., Krook, J.E., Mailliard, J., Twllo, D.I., Merton, R.F., Tschelter, L.K., and Barlow, J.F. (1989). Surgical adjuvant therapy of large-bowel carcinoma: An evaluation of levamisole and the combination of levamisole and fluorouracil. *Journal of Clinical Oncology* **7**, 1447–1456.

LaValley, M.P. and DeGruttola, V. (1996). Models for empirical Bayes estimators of longitudinal CD4 counts. *Statistics in Medicine* **15**, 2289–2305.

Lavine, M. (1994). More aspects of Polya tree distributions for statistical modelling. *The Annals of Statistics* **22**, 1161–1176.

Lavine, M. (1992). Some aspects of polya tree distributions for statistical modeling. *The Annals of Statistics* **20**, 1222–1235.

Lawless, J.F. (1982). *Statistical Models and Methods for Life Time Data.* New York: Wiley.

Lawless, J.F., and Singhal, K. (1978). Efficient screening of nonnormal regression models. *Biometrics* **34**, 318–327.

LeBlanc, M. and Crowley, J. (1995). Step-function covariate effects in the proportional hazards model. *Canadian Journal of Statistics* **23**, 109–129,

Lee, E.T. (1992). *Statistical Methods for Survival Data Analysis.* New York: Wiley.

Lee, S.J. and Zelen, M. (2000). Clinical trials and sample size considerations: Another perspective (with Discussion). *Statistical Science* **15**, 95–110.

Leonard, T. (1978). Density estimation, stochastic processes and prior information. *Journal of the Royal Statistical Society, Series B* **40**, 113–146.

Leonard, T. (1973). A Bayesian method for histograms. *Biometrika* **60**, 297–308.

Lewis, R.J. and Berry, D.A. (1994). Group sequential clinical trials: A classical evaluation of Bayesian decision-theoretic designs. *Journal of the American Statistical Association* **89**, 1528–1534.

Liestol, K., Andersen, P.K., and Andersen, U. (1994). Survival analysis and neural nets. *Statistics in Medicine* **13**, 1189–1200.

Lin, D.Y. and Ying, Z. (1993). Cox regression with incomplete covariate measurements. *Journal of the American Statistical Association* **88**, 1341–1349.

Lindley, D.V. (1997). The choice of sample size. *The Statistician* **46**, 129–138.

Lipsitz, S.R. and Ibrahim, J.G. (1998). Estimating equations with incomplete categorical covariates in the Cox model. *Biometrics* **54**, 1002–1013.

Lipsitz, S.R. and Ibrahim, J.G. (1996a). Using the EM algorithm for survival data With incomplete categorical covariates. *Lifetime Data Analysis* **2**, 5–14.

Lipsitz, S.R. and Ibrahim, J.G. (1996b). A conditional model for incomplete covariates in parametric regression models. *Biometrika* **83**, 916–922.

Little, R.J.A. (1992). Regression with missing X's: A review. *Journal of the American Statistical Association* **87**, 1227–1237.

Little, R.J.A. and Rubin, D.B. (1987). *Statistical Analysis With Missing Data.* New York: Wiley.

Liu, J.S. (1994). The collapsed Gibbs sampler in Bayesian computations with applications to a gene regulation problem. *Journal of the American Statistical Association* **89**, 958–966.

Lo, A. (1984). On a class of Bayesian nonparametric estimates: I. density estimates. *The Annals of Statistics* **12**, 351–357.

Louis, T.A. (1982). Finding the observed information matrix when using the EM algorithm. *Journal of the Royal Statistical Society, Series B* **44**, 226–233.

Luo, X., Turnbull, B.W., and Clark, L.C. (1997). Likelihood ratio test for a changepoint with survival data. *Biometrika* **84**, 555–565.

MacEachern, S.N. (1994). Estimating normal means with a conjugate style Dirichlet process prior. *Communications in Statistics – Theory and Methods* **23**, 727–741.

MacEachern, S.N. and Müller, P. (1998). Estimating mixture of Dirichlet process models. *Journal of Computational and Graphical Statistics* **7**, 223–238.

Madigan, D., Andersson, S.A., Perlman, M., and Volinsky, C.T. (1996). Bayesian model averaging and model selection for Markov equivalence classes of acyclic digraphs. *Communications in Statistics – Theory and Methods* **25**, 2493–2520.

Madigan, D. and Raftery, A.E. (1994). Model selection and accounting for model uncertainty in graphical models using Occam's window. *Journal of the American Statistical Association* **89**, 1535–1546.

Madigan, D. and York, J. (1995), Bayesian graphical models for discrete data. *International Statistical Review* **63**, 215–232.

Maller, R. and Zhou, X. (1996). *Survival Analysis with Long-Term Survivors.* New York: Wiley.

Mallick, B.K., Denison, D.G.T., and Smith, A.F.M. (1999). Bayesian survival analysis using a MARS model. *Biometrics* **55**, 1071–1077.

Mallick, B.K. and Gelfand, A.E. (1994). Generalised linear models with unknown link functions. *Biometrika* **81**, 237–245.

Manatunga, A.K. (1989). Inference for multivariate survival distributions generated by stable frailties. *Unpublished Ph.D. Dissertation.* Department of Biostatistics, University of Rochester, Rochester, New York.

Manolio, T.A., Kronmal, R.A., Burke, G.L., and et al. (1996). Short-term predictors of incident stroke in older adults. *Stroke* **27**, 1479–1486.

Mantel, N. (1966). Evaluation of survival data and two new rank order statistics arising in its consideration. *Cancer Chemotherapy Reports* **50**, 163-170.

Margolick, J.B., Donnenberg, A.D., Muñoz, A., Park, L.P., Bauer, K.D., Giorgi, J.V., Ferbas, J., and Saah, A.J. (1993). Changes in T and non-T lymphocytes subsets following seroconversion to HIV-1: Stable $CD3^+$ and declining $CD3^-$ populations suggest regulatory responses linked to loss of CD4 lymphocytes. *Journal of Acquired Immune Deficiency Syndrome* **6**, 153–161.

Mauldin, R.D., Sudderth, W.D., and Williams, S.C. (1992). Polya trees and random distributions. *The Annals of Statistics* **20**, 1203–1221.

Mehta, C.R. and Cain, K.C. (1984). Charts for early stopping of pilot studies. *Journal of Clinical Oncology* **2**, 676–682.

McCullagh, P. (1980). Regression models for ordinal data. *Journal of the Royal Statistical Society, Series B* **42**, 109–142.

McGilchrist, C.A. (1993). REML estimation for survival models with frailty. *Biometrics* **49**, 221–225.

McGilchrist, C.A. and Aisbett, C.W. (1991) Regression with frailty in survival analysis. *Biometrics* **47**, 461–466.

Meinhold, R.J. and Singpurwalla, N.D. (1983). Understanding the Kalman filter. *The American Statistician* **37**, 123–127.

Mendenhall, W. and Hadel, R.J. (1958). Estimation of parameters of mixed exponentially distributed failure time distributions from censored life test data. *Biometrika* **45**, 504–520.

Metropolis, N., Rosenbluth, A.W., Rosenbluth, M.N., Teller, A.H., and Teller, E. (1953). Equations of state calculations by fast computing machines. *Journal of Chemical Physics* **21**, 1087–1092.

Mezzetti, M. and Ibrahim, J.G. (2000). Bayesian inference for the Cox model using correlated gamma process priors. *Technical Report.* Department of Biostatistics, Harvard School of Public Health.

Mitchell, T.J. and Beauchamp, J.J. (1988). Bayesian variable selection in linear regression (with Discussion). *Journal of the American Statistical Association* **83**, 1023–1036.

Moertel, C.G., Fleming, T.R., MacDonald, J.S., Haller, D.G., Laurie, J.A., Goodman, P.J., Ungerleider, J.S., Emerson, W.A., Tormey, D.C., Glick, J.H., Veeder, M.H., and Mailliard, J.A. (1990). Levamisole and flurouracil for adjuvant therapy of resected colon carcinoma. *New England Journal of Medicine* **322**, 352–358.

Moolgavkar, S.H., Dewanji, A., and Venzon, D.J. (1988). A stochastic two-stage model for cancer risk assessment: The hazard function and the probability of tumor. *Risk Analysis* **8**, 383–392.

Moolgavkar, S.H. and Knudson, A.G. (1981). Mutation and cancer: A model for human carcinogenesis. *Journal of National Cancer Institute* **66**, 1037–1052.

Moolgavkar, S.H. and Venzon, D.J. (1979). Two events model for carcinogenesis: Incidence curves for childhood and adult tumors. *Mathematical Biosciences* **47**, 55–77.

Morris, C.N., Norton, E.C., and Zhou, X.H. (1994). Parametric duration analysis of nursing home usage. In *Case Studies in Biometry* (Eds. N. Lange, L. Ryan, L. Billard, D. Brillinger, L. Conquest, and J. Greenhouse). New York: Wiley, pp. 231–248.

Moussa, M.A.A. (1989). Exact, conditional, and predictive power in planning clinical trials. *Controlled Clinical Trials* **10**, 378–385.

Neal, R.M. (1993a). Bayesian learning via stochastic dynamics. In *Advances in Neural Information Processing Systems 5* (Eds. C.L. Giles, S.J. Hanson, and J.D. Cowan). San Mateo, California: Morgan Kaufmann, pp. 475–482.

Neal, R.M. (1993b). Probabilistic inference using Markov chain Monte Carlo methods. *Technical Report CRG-TR-93-1*, Department of Computer Science, University of Toronto.

Neal, R.M. (1994a). An improved acceptance procedure for the hybrid Monte Carlo algorithm. *Journal of Computational Physics* **111**, 194–204.

Neal, R.M. (1994b). Bayesian learning for neural networks. *Unpublished Ph.D. Dissertation.* Department of Computer Science, University of Toronto.

Neuhaus, J.M., Hauck, W.W., and Kalbfleisch, J.D. (1992). The effects of mixture distribution misspecification when fitting mixed–effects logistic models. *Biometrika* **79**, 755–762.

Neuhaus, J.M., Kalbfleisch, J.D., and Hauck, W.W. (1994). Conditions for consistent estimation in mixed–effects models for binary matched pairs data. *Canadian Journal of Statistics* **22**, 139–148.

Newton, M.A., Czado, C., and Chappell, R. (1996). Bayesian inference for semiparametric binary regression. *Journal of the American Statistical Association* **91**, 142–153.

Nieto-Barajas, L.E. and Walker, S.G. (2000). Markov beta and gamma processes for modeling hazard rates. *Technical Report.* Imperial College, London.

Norros, I. (1986). A compensator representation of multivariate life lengths distributions, with applications. *Scandinavian Journal of Statistics* **13**, 99–112.

Oakes, D. (1994). Use of frailty models for multivariate survival data. *Proceedings of the XVII th International Biometrics Conference*, Hamilton, Ontario, Canada, 275–286.

Oakes, D. (1992). Frailty models for multiple event times. In *Survival Analysis: State of the Art* (Eds. J.P. Klein and P.K. Goel). Netherlands: Kluwer Academic, pp. 371–379.

Oakes, D. (1989). Bivariate survival models induced by frailties. *Journal of the American Statistical Association* **84**, 487–493.

Oakes, D. (1986). Semiparametric inference in a model for association in bivariate survival data. *Biometrika* **73**, 353–361.

O'Neill, R. (1971). Algorithm AS47-function minimization using a simplex procedure. *Applied Statistics* **20**, 338–345.

O'Sullivan, F. (1988). Nonparametric estimation of relative risk using splines and crossvalidation. *SIAM Journal on Scientific and Statistical Computing* **9**, 531–542.

Padgett, W.J. and Wei, L.J. (1980). Maximum likelihood estimation of a distribution function with increasing failure rate based on censored observations. *Biometrika* **67**, 470–474.

Paik, M.C. (1997). Multiple imputation for the Cox proportional hazards model with missing covariates. *Lifetime Data Analysis* **3**, 289–298.

Paik, M.C. and Tsai, W.-Y. (1997). On using the Cox proportional hazards model with missing covariates. *Biometrika* **84**, 579–593.

Parmar, M.K.B., Spiegelhalter, D.J., and Feedman, L.S. (1994). The CHART trials: Bayesian design and monitoring in practice. *Statistics in Medicine* **13**, 1297–1312.

Pepe, M.S., Self, S., and Prentice, R.L. (1989). Further results on covariate measurement errors in cohort studies with time to response data. *Statistics in Medicine* **8**, 1167–1178.

Peto, R. and Peto, J. (1972). Asymptotically efficient rank invariant procedures. *Journal of the Royal Statistical Society, Series A* **135**, 185–206.

Pham-Gia, T. (1997). On Bayesian analysis, Bayesian decision theory and the sample size problem. *The Statistician* **46**, 139–144.

Pickles, A. and Crouchley, R. (1995). A comparison of frailty models for multivariate survival data. *Statistics in Medicine* **14**, 1447–1461.

Poon, M.A., O'Connell, M.J., Moertel, C.G., Wieand, H.S., Cullinan, S.A., Everson, L.K., Krook, J.E., Mailliard, J.A., Laurie, J.A., Tschetter, L.K., and Wiesenfeld, M. (1989). Biochemical modulation of fluorouracil: Evidence of significant improvement of survival and quality of life in patients with advanced colorectal carcinoma. *Journal of Clinical Oncology* **7**, 1407–1418.

Prentice, R. L. (1989). Surrogate endpoints in clinical trials: Definition and operational criteria. *Statistics in Medicine*, **8**, 431–440.

Prentice, R.L. (1982). Covariate measurement errors and parameter estimation in a failure time regression model. *Biometrika* **69**, 331–342.

Prentice, R.L. (1973). Exponential survivals with censoring and explanatory variables. *Biometrika* **60**, 279–288.

Pugh, M., Robins, J., Lipsitz, S.R., and Harrington, D.P. (1993). Inference in the Cox proportional hazards model with missing covariate data. *Technical Report.* Department of Biostatistical Science, Dana-Farber Cancer Institute, Boston, Massachusetts.

Qiou, Z. (1996). Bayesian inference for stable processes. *Unpublished Ph.D. Dissertation.* Department of Statistics, University of Connecticut.

Qiou, Z., Ravishanker, N., and Dey, D.K. (1999). Multivariate survival analysis with positive frailties. *Biometrics* **55**, 637–644.

Raftery, A.E. (1996). Approximate Bayes factors and accounting for model uncertainty in generalised linear models. *Biometrika* **83**, 251–266.

Raftery, A.E. (1995). Bayesian model selection in social research (with Discussion). In *Sociological Methodology* (Ed. P.V. Marsden). Cambridge, Massachusetts: Blackwells, pp. 111-195.

Raftery, A.E. and Lewis, S. (1992). How many iterations in the Gibbs sampler? In *Bayesian Statistics* **4** (Eds. J.M. Bernardo, J.O. Berger, A.P. Dawid, and A.F.M. Smith). Oxford: Oxford University Press, pp. 763–773.

Raftery, A.E., Madigan, D., and Hoeting, J.A. (1997). Bayesian model averaging for linear regression models. *Journal of the American Statistical Association* **92**, 179–191.

Raftery, A.E., Madigan, D., and Volinsky, C.T. (1995). Accounting for model uncertainty in survival analysis improves predictive performance. In *Bayesian Statistics 5* (Eds. J.M. Bernardo, J.O. Berger, A.P. Dawid, and A.F.M. Smith). Oxford: Oxford University Press, pp. 323–350.

Ramgopal, P., Laud, P.W., and Smith, A.F.M. (1993). Nonparametric Bayesian bioassay with prior constraints on the shape of the potency curve. *Biometrika* **80**, 489–498.

Ravishanker, N. and Dey, D.K. (2000). Multivariate survival models with a mixture of positive stable frailties. *Methodology and Computing in Applied probability.* To appear.

Ripley, B.D. (1994). Neural networks and related methods for classification. *Journal of the Royal Statistical Society, Series B* **56**, 409–456.

Ripley, B.D. (1987). *Stochastic Simulation.* New York: Wiley.

Roberts, G.O., Gelman, A., and Gilks, W.R. (1997). Weak convergence and optimal scaling of random walk Metropolis algorithms. *Annals of Applied Probability* **7**, 110–120.

Roberts, G.O. and Polson, N.G. (1994). On the geometric convergence of the Gibbs sampler. *Journal of the Royal Statistical Society, Series B* **56**, 377–384.

Rosenberg, P.S., Goedert, J.J., and Biggar, R.J. (1994). Effect of age at seroconversion on the natural AIDS incubation distribution. *AIDS* **8**, 803–810.

Rosner, G.L. and Berry, D.A. (1995). A Bayesian group sequential design for a multiple arm randomized Clinical trial. *Statistics in Medicine* **14**, 381–394.

Rubin, D.B. (1976). Inference and missing data. *Biometrika* **63**, 581–592.

Sahu, S.K., Dey, D.K., Aslanidou, H., and Sinha, D. (1997). A Weibull regression model with gamma frailties for multivariate survival data. *Lifetime Data Analysis* **3**, 123–137.

Samorodnitsky, G. and Taqqu, M.S. (1994). *Stable Non-Gaussian Random Processes: Stochastic Models with Infinite Variance.* London: Chapman & Hall.

Sargent, D.J. (1998). A general framework for random effects survival analysis in the Cox proportional hazards setting. *Biometrics* **54**, 1486–1497.

Schafer, J. (1997). *Analysis of Incomplete Multivariate Data by Simulation.* London: Chapman & Hall.

Schervish, M.J. and Carlin, B.P. (1992). On the convergence of successive substitution sampling. *Journal of Computational and Graphical Statistics* **1**, 111–127.

Schluchter, M.D. (1992). Methods for the analysis of informatively censored longitudinal data. *Statistics in Medicine* **11**, 1861–1870.

Schluchter, M.D., Greene, T., and Beck, G.J. (2000). Analysis of change in the presence of informative censoring — Application to a longitudinal clinical trial of progressive renal disease. *Statistics in Medicine.* To appear.

Schluchter, M.D. and Jackson, K. (1989). Log-linear analysis of censored survival data with partially observed covariates. *Journal of the American Statistical Association* **84**, 42–52.

Schoenfeld, D. (1982). Partial residuals for proportional hazards regression model. *Biometrika* **69**, 239–241.

Schwarz, G. (1978). Estimating the dimension of a model. *The Annals of Statistics* **6**, 461–464.

Scott, D.W., Tapia, R.A., and Thompson, J.R. (1980). Nonparametric probability density estimation by discrete maximum penalized-likelihood criteria. *The Annals of Statistics* **4**, 820–832.

Shih, J.A. and Louis, T.A. (1995). Assessing gamma frailty models for clustered failure time data. *Lifetime Data Analysis* **1**, 205–220.

Silverman, B.W. (1985). Some aspects of the spline smoothing approach to nonparametric regression curve fitting (with Discussion). *Journal of the Royal Statistical Society, Series B* **47**, 1–52.

Simon, R. (1999). Bayesian design and analysis of active control trials. *Biometrics* **55**, 484–487.

Simon, R. and Freedman, L.S. (1997). Bayesian design and analysis of two × two factorial clinical trials. *Biometrics* **53**, 456–464.

Sinha, D. (1998). Posterior likelihood methods for multivariate survival data. *Biometrics* **54**, 1463–1474.

Sinha, D. (1997). Time-discrete beta process model for interval-censored survival data. *Canadian Journal of Statistics* **25**, 445–456.

Sinha, D. (1993). Semiparametric Bayesian analysis of multiple event time data. *Journal of the American Statistical Association* **88**, 979–983.

Sinha, D., Chen, M.-H., and Ghosh, S.K. (1999). Bayesian analysis and model selection for interval-censored survival data. *Biometrics* **55**, 585–590.

Sinha, D. and Dey, D.K. (1997). Semiparametric Bayesian analysis of survival data. *Journal of the American Statistical Association* **92**, 1195–1212.

Sinha, D., Ibrahim, J.G., and Chen, M.-H. (2000). Bayesian models for survival data from cancer prevention studies. *Technical Report MS-11-00-14*. Department of Mathematical Sciences, Worcester Polytechnic Institute.

Skene, A.M. and Wakefield, J.C. (1990). Hierarchical models for multicentre binary response studies. *Statistics in Medicine* **9**, 919–929.

Smith, A.F.M. and Gelfand, A.E. (1992). Bayesian statistics without tears: A sampling–resampling perspective. *The American Statistician* **46**, 84–88.

Smith, A.F.M. and Spiegelhalter, D.J. (1980). Bayes factors and choice criteria for linear models. *Journal of the Royal Statistical Society, Series B* **43**, 213–220.

Spiegelhalter, D.J. and Freedman, L.S. (1988). Bayesian approaches to clinical trials. In *Bayesian Statistics 3* (Eds. J.M. Bernardo, M.H. DeGroot, D.V. Lindley, and A.F.M. Smith). Oxford: Oxford University Press, pp. 453–477.

Spiegelhalter, D.J. and Freedman, L.S. (1986). A predictive approach to selecting the size of a clinical trial, based upon subjective clinical opinion. *Statistics in Medicine* **5**, 1–13.

Spiegelhalter, D.J., Freedman, L.S., and Blackburn, P.R. (1986). Monitoring clinical trials: Conditional or predictive power. *Controlled Clinical Trials* **7**, 8–17.

Spiegelhalter, D.J., Freedman, L.S., and Parmar, M.K.B. (1994). Bayesian approaches to randomized trials (with Discussion). *Journal of the Royal Statistical Society, Series A* **157**, 357–416.

Spiegelhalter, D.J., Freedman, L.S., and Parmar, M.K.B. (1993). Applying Bayesian thinking in drug development in clinical trials. *Statistics in Medicine* **12**, 1501–1511.

Spiegelhalter, D.J., Thomas, A., Best, N.G., and Gilks, W.R. (1995). *BUGS: Bayesian Inference Using Gibbs Sampling, Version 0.50.* MRC Biostatistics Unit, Cambridge, England.

Sposto, R., Sather, H.N., and Baker, S.A. (1992). A comparison of tests of the difference in the proportion of patients who are cured. *Biometrics* **48**, 87–99.

Stangl, D.K. (1995). Prediction and decision making using Bayesian hierarchical models. *Statistics in Medicine* **14**, 2173–2190.

Stangl, D.K. and Greenhouse, J.B. (1998). Assessing placebo response Using Bayesian hierarchical survival models. *Lifetime Data Analysis* **4**, 5–28.

Stangl, D.K. and Greenhouse, J.B. (1995). Assessing placebo response using Bayesian hierarchical survival models. *Discussion paper 95-01.* Institute of Statistics and Decision Sciences, Duke University.

Staniswalis, J. (1989). The kernel estimate of a regression function in likelihood-based models. *Journal of the American Statistical Association* **84**, 276–283.

Sun, J. and Wei, L.J. (2000). Regression analysis of panel count data with covariate-dependent observation and censoring times. *Journal of the Royal Statistical Society, Series B* **62**, 293–302.

Susarla, V. and Van Ryzin, J. (1976). Nonparametric Bayesian estimation of survival curves from incomplete observations. *Journal of the American Statistical Association* **71**, 897–902.

Sy, J.P. and Taylor, J.M.G. (2000). Estimation in a proportional hazards cure model. *Biometrics* **56**, 227–336.

Tanner, M.A. and Wong, W.H. (1987). The calculation of posterior distributions by data augmentation. *Journal of the American Statistical Association* **82**, 528–549.

Taplin, R.H. (1993). Robust likelihood calculation for time series. *Journal of the Royal Statistical Society, Series B* **55**, 829–836.

Taplin, R.H. and Raftery, A.E. (1994). Analysis of agricultural field trials in the presence of outliers and fertility jumps. *Biometrics* **50**, 764–781.

Taylor, J.M.G. (1995). Semi-parametric estimation in failure time mixture models. *Biometrics* **51**, 899–907.

Taylor, J.M.G., Cumberland, W.G., and Sy, J.P. (1994). A stochastic model for analysis of longitudinal AIDS data. *Journal of the American Statistical Association* **89**, 727–736.

Taylor, J.M.G., Fahey, J., Detels, R., and Giorgi, J. (1989). CD4 percentage, CD4 number, and CD4:CD8 ratio in HIV infection: Which to choose and how to use. *Journal of Acquired Immune Deficiency Syndrome* **2**, 114–124.

Taylor, J.M.G., Sy, J.P., Visscher, B., and Giorgi, J.V. (1995). CD4 T-cell number at the time of AIDS. *American Journal of Epidemiology* **141**, 645–651.

Taylor, J.M.G., Tan, S.J., Detels, R., and Giorgi, J.V. (1991). Applications of computer simulation model of the natural history of CD4 T-cell number in HIV-infected individuals. *AIDS* **5**, 159–167.

Thall, P.F. and Simon, R. (1994). A Bayesian approach to establishing sample size and monitoring criteria for phase II clinical trials. *Controlled Clinical Trials* **15**, 463–481.

Therneau, T.M. (1994). A package of survival functions for S. *Technical Report No. 53.* Section of Biostatistics, Mayo Clinic.

Therneau, T.M., Grambsch, P.M., and Fleming, T.R. (1990). Martingale based residuals for survival models. *Biometrika* **77**, 147–160.

Tierney, L. (1994). Markov chains for exploring posterior distributions (with Discussions). *The Annals of Statistics* **22**, 1701–1762.

Troxel, A.B., Fairclough, D.L., Curran, D., and Hahn, E.A. (1998). Statistical analysis of quality of life with missing data in cancer clinical trials. *Statistics in Medicine* **17**, 653–666.

Tsiatis, A.A. (1982). Group sequential methods for survival analysis with staggered entry. In *Survival Analysis* (Eds. J. Crowley and R.A. Johnson). Hayward, CA: Institute of Mathematical Statistics, pp. 257-268.

Tsiatis, A.A. (1981a). A large sample study of Cox's regression model. *The Annals of Statistics* **9**, 93–108.

Tsiatis, A.A. (1981b). The asymptotic joint distribution of the efficient scores test for proportional hazards models calculated over time. *Biometrika* **68**, 311–315.

Tsiatis, A.A., DeGruttola, V., and Wulfsohn, M.S. (1995). Modeling the relationship of survival to longitudinal data measured with error. Applications to survival and CD4 counts in patients with AIDS. *Journal of the American Statistical Association* **90**, 27–37.

Tsiatis, A.A., Dafni, U., DeGruttola, V., Propert, K.J., Strawderman, R.L., and Wulfsohn, M.S. (1992). The relationship of CD4 counts over time to survival in patients with AIDS: Is CD4 a good surrogate marker? In *AIDS epidemiology: Methodological issues* (Eds.

N.P. Jewell, K. Kietz, and V.T. Farewell). Boston: Birkhäuser, pp. 256–274.

Turnbull, B.W. (1976). The empirical distribution function with arbitrary grouped, censored and truncated data. *Journal of the Royal Statistical Society, Series B* **38**, 290–295.

Vaupel, J.M., Manton, K.G., and Stallard, E. (1979). The impact of heterogeneity in individual frailty on the dynamics of mortality. *Demography* **16**, 439–454.

Volinsky, C.T., Madigan, D., Raftery, A.E., and Kronmal, R.A. (1997). Bayesian model averaging in proportional hazards models: Assessing the risk of a stroke. *Applied Statistics* **46**, 433–448.

Volinsky, C.T. and Raftery, A.E. (2000). Bayesian information criterion for censored survival models. *Biometrics* **56**, 256–262.

Wahba, G. (1993). Discussion of "Varying-coefficient models" by T.J. Hastie and R.J. Tibshirani. *Journal of the Royal Statistical Society, Series B* **55**, 757–796.

Wahba, G. (1983). Bayesian 'confidence intervals' for the cross-validated smoothing spline. *Journal of the Royal Statistical Society, Series B* **45**, 133–150.

Wakefield, J.C., Gelfand, A.E., and Smith, A.F.M. (1991). Efficient generation of random variates via the ratio-of-uniforms method. *Statistics and Computing* **1**, 129–133.

Wakefield, J.C., Smith, A.F.M., Racine-Poon, A., and Gelfand, A.E. (1994). Bayesian analysis of linear and non-linear population models using the Gibbs sampler. *Applied Statistics* **43**, 201–221.

Walker, S.G., Damien, P., Laud, P.W., and Smith, A.F.M. (1999). Bayesian nonparametric inference for random distributions and related functions (with Discussion). *Journal of the Royal Statistical Society, Series B* **61**, 485–528.

Walker, S.G. and Mallick, B.K. (1999). A Bayesian semiparametric accelerated failure time model. *Biometrics* **55**, 477–483.

Walker, S.G. and Mallick, B.K. (1996). Hierarchical generalized linear models and frailty models with Bayesian nonparametric mixing. *Journal of the Royal Statistical Society, Series B* **59**, 845–860.

Wang, Y. (1998). A comprehensive joint model for longitudinal data and survival time data: Application in AIDS studies. *Unpublished Ph.D. Dissertation.* Department of Biostatistics, School of Public Health, University of California at Los Angeles.

Wang, Y. and Taylor, J.M.G. (2000). Jointly modelling longitudinal and event time data with application to AIDS studies. *Technical Report.* Department of Biostatistics, School of Public Health, University of California at Los Angeles.

Wei, G.C. and Tanner, M.A. (1990). A Monte Carlo implementation of the EM algorithm and the poor man's data augmentation algorithms. *Journal of the American Statistical Association* **85**, 699–704.

Weiss, R.E. (1997). Bayesian sample size calculations for hypothesis testing. *The Statistician* **46**, 185–191.

Wolpert, R.L. (1991). Monte Carlo importance sampling in Bayesian statistics. In *Statistical Multiple Integration* (Eds. N. Flournoy and R. Tsutakawa). *Contemporary Mathematics* **116**, 101–115.

Wulfsohn, M.S. and Tsiatis, A.A. (1997). A joint model for survival and longitudinal data measured with error. *Biometrics* **53**, 330–339.

Yakovlev, A.Y. (1994). Letter to the Editor. *Statistics in Medicine* **13**, 983–986.

Yakovlev, A.Y., Asselain, B., Bardou, V.J., Fourquet, A., Hoang, T., Rochefediere, A., and Tsodikov, A.D. (1993). A simple stochastic model of tumor recurrence and its applications to data on premenopausal breast cancer. In *Biometrie et Analyse de Dormees Spatio-Temporelles* **12** (Eds. B. Asselain, M. Boniface, C. Duby, C. Lopez, J.P. Masson, and J. Tranchefort). Société Francaise de Biométrie, ENSA Renned, France, pp. 66–82.

Yakovlev, A.Y. and Tsodikov, A.D. (1996). *Stochastic Models of Tumor Latency and Their Biostatistical Applications.* New Jersey: World Scientific.

Yamaguchi, K. (1992). Accelerated failure-time regression models with a regression model of surviving fraction: An application to the analysis of "permanent employment" in Japan. *Journal of the American Statistical Association* **87**, 284–292.

Yang, G.L. and Chen, C.W. (1991). A stochastic two-stage carcinogenesis model: A new approach to computing the probability of observing tumor in animal bioassays. *Mathematical Biosciences* **104**, 247–258.

Yau, K.K.W. and McGilchrist, C.A. (1994). ML and REML estimation in survival analysis with time-dependent correlated frailty. *Technical Report.* NCEPH, Australian National University, Canberra ACT 0200, Australia.

Ying, Z., Jung, S.H., and Wei, L.J. (1995). Survival analysis with median regression models. *Journal of the American Statistical Association* **90**, 178–184.

Zeger, S.L. and Karim, M.R. (1991). Generalized linear models with random effects: A Gibbs sampling approach. *Journal of the American Statistical Association* **86**, 79–86.

Zhou, H. and Pepe, M.S. (1995). Auxiliary covariate data in failure time regression. *Biometrika* **82**, 139–149.

Zucker, D.M. and Lakatos, E. (1990). Weighted log rank type statistics for comparing survival curves when there is a time lag in the effectiveness of a treatment. *Biometrika* **77**, 853–864.

Author Index

Subject Index

Springer Series in Statistics

(continued from p. ii)

Küchler/Sørensen: Exponential Families of Stochastic Processes.
Le Cam: Asymptotic Methods in Statistical Decision Theory.
Le Cam/Yang: Asymptotics in Statistics: Some Basic Concepts, 2nd edition.
Liu: Monte Carlo Strategies in Scientific Computing.
Longford: Models for Uncertainty in Educational Testing.
Mielke/Berry: Permutation Methods: A Distance Function Approach.
Miller, Jr.: Simultaneous Statistical Inference, 2nd edition.
Mosteller/Wallace: Applied Bayesian and Classical Inference: The Case of the Federalist Papers.
Parzen/Tanabe/Kitagawa: Selected Papers of Hirotugu Akaike.
Politis/Romano/Wolf: Subsampling.
Ramsay/Silverman: Functional Data Analysis.
Rao/Toutenburg: Linear Models: Least Squares and Alternatives.
Read/Cressie: Goodness-of-Fit Statistics for Discrete Multivariate Data.
Reinsel: Elements of Multivariate Time Series Analysis, 2nd edition.
Reiss: A Course on Point Processes.
Reiss: Approximate Distributions of Order Statistics: With Applications to Non-parametric Statistics.
Rieder: Robust Asymptotic Statistics.
Rosenbaum: Observational Studies.
Rosenblatt: Gaussian and Non-Gaussian Linear Time Series and Random Fields.
Särndal/Swensson/Wretman: Model Assisted Survey Sampling.
Schervish: Theory of Statistics.
Shao/Tu: The Jackknife and Bootstrap.
Siegmund: Sequential Analysis: Tests and Confidence Intervals.
Simonoff: Smoothing Methods in Statistics.
Singpurwalla and Wilson: Statistical Methods in Software Engineering: Reliability and Risk.
Small: The Statistical Theory of Shape.
Sprott: Statistical Inference in Science.
Stein: Interpolation of Spatial Data: Some Theory for Kriging.
Taniguchi/Kakizawa: Asymptotic Theory of Statistical Inference for Time Series.
Tanner: Tools for Statistical Inference: Methods for the Exploration of Posterior Distributions and Likelihood Functions, 3rd edition.
Tong: The Multivariate Normal Distribution.
van der Vaart/Wellner: Weak Convergence and Empirical Processes: With Applications to Statistics.
Verbeke/Molenberghs: Linear Mixed Models for Longitudinal Data.
Weerahandi: Exact Statistical Methods for Data Analysis.
West/Harrison: Bayesian Forecasting and Dynamic Models, 2nd edition.

ALSO AVAILABLE FROM SPRINGER!

CHRISTIAN P. ROBERT

THE BAYESIAN CHOICE

From Decision-Theoretic Foundations to Computational Implementation
Second Edition

This book introduces Bayesian statistics and decision theory. Its scope covers both the basic ideas of statistical theory, and also some of the more modern and advanced topics of Bayesian statistics such as complete class theorems, the Stein effect, Bayesian model choice, hierarchical and empirical Bayes modeling, Monte Carlo integration including Gibbs sampling, and other MCMC techniques.

2001/624 PAGES/HARDCOVER
ISBN 0-387-95231-4
SPRINGER TEXTS IN STATISTICS

TERRY M. THERNEAU and **PATRICIA GRAMBSCH**

MODELING SURVIVAL DATA:

Extending the Cox Model

This is a book for statistical practitioners, particularly those who design and analyze studies for survival and event history data. Its goal is to extend the toolkit beyond the basic triad provided by most statistical packages: the Kaplan-Meier estimator, log-rank test, and Cox regression model. Building on recent developments motivated by counting process and martingale theory, it shows the reader how to extend the Cox model to analyze multiple/correlated event data using marginal and random effects (frailty) models. The methods are now readily available in SAS and S-Plus and this book gives a hands-on introduction, with worked examples for many data sets.

2000/376 PP./HARDCOVER
ISBN 0-387-98784-3
STATISTICS FOR BIOLOGY AND HEALTH

PHILIP HOUGAARD

ANALYSIS OF MULTIVARIATE SURVIVAL DATA

Survival data or more general time-to-event data occurs in many areas, including medicine, biology, engineering, economics, and demography, but previously standard methods have required that all time variables are univariate and independent. This book extends the field by allowing for multivariate times. Applications where such data appear are survival of twins, survival of married couples and families, time to failure of right and left kidney for diabetic patients, life history data with time to outbreak of disease, complications and death, recurrent episodes of diseases and cross-over studies with time responses.

2000/480 PP./HARDCOVER
ISBN 0-387-98873-4
STATISTSICS FOR BIOLOGY AND HEALTH

To Order or for Information:

In North America: **CALL:** 1-800-SPRINGER or **FAX:** (201) 348-4505 • **WRITE:** Springer-Verlag New York, Inc., Dept. S2568, PO Box 2485, Secaucus, NJ 07096-2485 • **VISIT:** Your local technical bookstore • **E-MAIL:** orders@springer-ny.com

Outside North America: **CALL:** +49/30/8/27 87-3 73 • +49/30/8 27 87-0 • **FAX:** +49/30 8 27 87 301 • **WRITE:** Springer-Verlag, P.O. Box 140201, D-14302 Berlin, Germany • **E-MAIL:** orders@springer.de

PROMOTION: S2568

Springer
www.springer-ny.com